Chevrolet S-10 & Blazer
GMC Sonoma, Jimmy & Envoy
Oldsmobile Bravada
Isuzu Hombre
Automotive Repair Manual

by Robert Maddox
and John H Haynes
Member of the Guild of Motoring Writers

Models covered:
1994 through 2004 Chevrolet S-10 & GMC Sonoma pick-ups
1995 through 2005 Chevrolet Blazer and GMC Jimmy
1998 through 2001 GMC Envoy
1996 through 2001 Oldsmobile Bravada
1996 through 2000 Isuzu Hombre

(24071-3W35)

Haynes Group Limited
Haynes North America, Inc.
www.haynes.com

Disclaimer

There are risks associated with automotive repairs. The ability to make repairs depends on the individual's skill, experience and proper tools. Individuals should act with due care and acknowledge and assume the risk of performing automotive repairs.

The purpose of this manual is to provide comprehensive, useful and accessible automotive repair information, to help you get the best value from your vehicle. However, this manual is not a substitute for a professional certified technician or mechanic.

This repair manual is produced by a third party and is not associated with an individual vehicle manufacturer. If there is any doubt or discrepancy between this manual and the owner's manual or the factory service manual, please refer to the factory service manual or seek assistance from a professional certified technician or mechanic.

Even though we have prepared this manual with extreme care and every attempt is made to ensure that the information in this manual is correct, neither the publisher nor the author can accept responsibility for loss, damage or injury caused by any errors in, or omissions from, the information given.

The manufacturer's authorised representative in the EU for product safety is:

HaynesPro BV
Stationsstraat 79 F, 3811MH Amersfoort
The Netherlands
gpsr@haynes.co.uk

About this manual

Its purpose

The purpose of this manual is to help you get the best value from your vehicle. It can do so in several ways. It can help you decide what work must be done, even if you choose to have it done by a dealer service department or a repair shop; it provides information and procedures for routine maintenance and servicing; and it offers diagnostic and repair procedures to follow when trouble occurs.

We hope you use the manual to tackle the work yourself. For many simpler jobs, doing it yourself may be quicker than arranging an appointment to get the vehicle into a shop and making the trips to leave it and pick it up. More importantly, a lot of money can be saved by avoiding the expense the shop must pass on to you to cover its labor and overhead costs. An added benefit is the sense of satisfaction and accomplishment that you feel after doing the job yourself.

Using the manual

The manual is divided into Chapters. Each Chapter is divided into numbered Sections, which are headed in bold type between horizontal lines. Each Section consists of consecutively numbered paragraphs.

At the beginning of each numbered Section you will be referred to any illustrations which apply to the procedures in that Section. The reference numbers used in illustration captions pinpoint the pertinent Section and the Step within that Section. That is, illustration 3.2 means the illustration refers to Section 3 and Step (or paragraph) 2 within that Section.

Procedures, once described in the text, are not normally repeated. When it's necessary to refer to another Chapter, the reference will be given as Chapter and Section number. Cross references given without use of the word "Chapter" apply to Sections and/or paragraphs in the same Chapter. For example, "see Section 8" means in the same Chapter.

References to the left or right side of the vehicle assume you are sitting in the driver's seat, facing forward.

Even though we have prepared this manual with extreme care, neither the publisher nor the author can accept responsibility for any errors in, or omissions from, the information given.

NOTE

A **Note** provides information necessary to properly complete a procedure or information which will make the procedure easier to understand.

CAUTION

A **Caution** provides a special procedure or special steps which must be taken while completing the procedure where the Caution is found. Not heeding a Caution can result in damage to the assembly being worked on.

WARNING

A **Warning** provides a special procedure or special steps which must be taken while completing the procedure where the Warning is found. Not heeding a Warning can result in personal injury.

Acknowledgements

Technical writers who contributed to this project include Jeff Killingsworth, Jay Storer, Mike Stubblefield and Larry Warren. Wiring diagrams originated exclusively for Haynes North America, Inc. by CAD Counsel.

© Haynes North America, Inc. 1996, 1999, 2001, 2005, 2008, 2018

With permission from Haynes Group Limited

A book in the Haynes Automotive Repair Manual Series

All rights reserved. No part of this book may be reproduced or transmitted in any form or by any means, electronic or mechanical, including photocopying, recording or by any information storage or retrieval system, without permission in writing from the copyright holder.

ISBN-13: 978-1-62092-327-6
ISBN-10: 1-62092-327-0

Library of Congress Control Number: 2018939120

While every attempt is made to ensure that the information in this manual is correct, no liability can be accepted by the authors or publishers for loss, damage or injury caused by any errors in, or omissions from, the information given.

Contents

Introductory pages

About this manual	0-2
Introduction	0-4
Vehicle identification numbers	0-5
Buying parts	0-6
Maintenance techniques, tools and working facilities	0-6
Jacking and towing	0-12
Anti-theft audio system	0-13
Booster battery (jump) starting	0-13
Automotive chemicals and lubricants	0-14
Conversion factors	0-15
Safety first!	0-16
Troubleshooting	0-17

Chapter 1
Tune-up and routine maintenance — 1-1

Chapter 2 Part A
2.2L four-cylinder engine — 2A-1

Chapter 2 Part B
4.3L V6 engine — 2B-1

Chapter 2 Part C
General engine overhaul procedures — 2C-1

Chapter 3
Cooling, heating and air conditioning systems — 3-1

Chapter 4
Fuel and exhaust systems — 4-1

Chapter 5
Engine electrical systems — 5-1

Chapter 6
Emissions and engine control systems — 6-1

Chapter 7 Part A
Manual transmission — 7A-1

Chapter 7 Part B
Automatic transmission — 7B-1

Chapter 7 Part C
Transfer case — 7C-1

Chapter 8
Clutch and driveline — 8-1

Chapter 9
Brakes — 9-1

Chapter 10
Suspension and steering systems — 10-1

Chapter 11
Body — 11-1

Chapter 12
Chassis electrical system — 12-1

Wiring diagrams — 12-16

Index — IND-1

Haynes photographer, author and mechanic with 1995 GMC Jimmy

Introduction to the Chevrolet S-10 and Blazer, GMC Sonoma, Jimmy and Envoy, Oldsmobile Bravada and Isuzu Hombre

These models are available in pickup and 2-and 4-door sport utility body styles.

Engines are either fuel-injected MPFI 2.2L four-cylinder or 4.3L V6. Later models are equipped with the On Board Diagnostic Second-Generation (OBD-II) computerized engine management system that controls virtually every aspect of engine operation. In Canada, this system is called Enhanced Diagnostics. OBD-II is designed to keep the emissions system operating at the federally specified level for the life of the vehicle. OBD-II monitors emissions system components for signs of degradation and engine operation for any malfunction that could affect emissions, turning on the Service Engine Soon light if any faults are detected.

Chassis layout is conventional, with the engine mounted at the front and the power being transmitted through either a 5-speed manual or 4-speed automatic transmission and a driveshaft to the solid rear axle. On 4WD models, a transfer case transmits power to a front differential by way of a driveshaft and then to the front wheels through independent driveaxles.

These models feature independent front suspension with torsion bars (4WD) or coil springs (2WD) and shock absorber at the front and solid axle with leaf springs and shock absorbers at the rear.

The brakes are disc on the front and drum on the rear wheels, with an Anti-Lock Brake System (ABS) standard on most models.

The power-assisted recirculating ball-type steering is mounted on the chassis frame rail to the left of the engine.

Vehicle identification numbers

Modifications are a continuing and unpublicized part of vehicle manufacturing. Since spare parts manuals and lists are compiled on a numerical basis, the individual vehicle numbers are essential to correctly identify the component required.

Vehicle Identification Number (VIN)

This very important identification number is stamped on a plate attached to the left side of the dashboard and is visible through the driver's side of the windshield (see illustration). The VIN also appears on the Vehicle Certificate of Title and Registration. It contains valuable information such as where and when the vehicle was manufactured, the model year and the body style.

VIN engine and model year codes

Two particularly important pieces of information found in the VIN are the engine code and the model year code. Counting from the left, the engine code letter designation is the 8th digit and the model year code letter designation is the 10th digit.

On the models covered by this manual the engine codes are:

W .... 4.3L V6 with Central Port Fuel Injection (CPI)
X .... 4.3L V6 with Sequential Electronic Fuel Injection (SEFI)
Z .... 4.3L V6 with Throttle Body Injection (TBI)
4 or 5 ... 2.2L four-cylinder engine with Multi-Port Injection (MPI)

On the models covered by this manual the model year codes are:

R 1994
S 1995
T 1996
V 1997
W 1998
X 1999
Y 2000
1 2001
2 2002
3 2003
4 2004
5 2005

Service parts identification label

This label is located in the glove box (see illustration). It lists the VIN, paint number, options and other information specific to the vehicle. You may sometimes need to refer to this label when you order parts (see illustration).

The Vehicle Identification Number (VIN) is at the front of the driver's side of the dashboard, visible from outside the vehicle, looking through the windshield

Vehicle certification label

This label is located at the rear edge of the driver's side door. It lists the VIN number, wheelbase, paint number, options and other information specific to the vehicle it's attached to. Refer to this label when ordering parts.

Engine identification numbers

On 2.2L four-cylinder engines this number is found on a pad at the rear of the block, just behind the oil dipstick opening. On 4.3L V6 engines the numbers are found on the lower left side of the block, just above the oil pan rail and on the right rear side of the block on a casting adjacent to the cylinder head.
On 1999 and later model engines built in the Tonawanda or Romulus plants, the 4.3L V6 engine number is at the top/right front of the engine.

Automatic transmission number

The automatic transmission ID number is stamped into the passenger's side of the case, just above the fluid pan, toward the rear.

The service parts identification label is located on the glove box (arrow)

On later-model 4L60-E transmissions, the ID number is stamped into the rear of the case, near the rear pan rail.

Manual transmission number

The ID number can be found on a pad on the right side of the case on the Borg-Warner T5 transmission and on the left side of the New Venture Gear NV3500 transmission.

Transfer case number

The ID number on the Borg-Warner transfer case can be found on a pad on the right side of the case while on the New Process transfer case it is on the left side.

Rear axle number

The ID number is stamped on front side of the right (passenger's side) axle tube.

Vehicle Emissions Control Information label

This label is found in the engine compartment. See Chapter 6 for more information on this label.

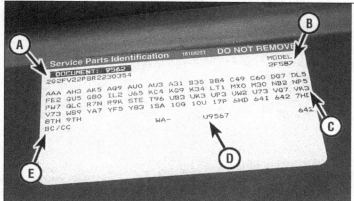

Much important information is on the service parts identification label

A Vehicle identification number
B Body type and style
C Options
D Paint codes
E Paint type (some models)

Buying parts

Replacement parts are available from many sources, which generally fall into one of two categories - authorized dealer parts departments and independent retail auto parts stores. Our advice concerning these parts is as follows:

Retail auto parts stores: Good auto parts stores will stock frequently needed components which wear out relatively fast, such as clutch components, exhaust systems, brake parts, tune-up parts, etc. These stores often supply new or reconditioned parts on an exchange basis, which can save a considerable amount of money. Discount auto parts stores are often very good places to buy materials and parts needed for general vehicle maintenance such as oil, grease, filters, spark plugs, belts, touch-up paint, bulbs, etc. They also usually sell tools and general accessories, have convenient hours, charge lower prices and can often be found not far from home.

Authorized dealer parts department: This is the best source for parts which are unique to the vehicle and not generally available elsewhere (such as major engine parts, transmission parts, trim pieces, etc.).

Warranty information: If the vehicle is still covered under warranty, be sure that any replacement parts purchased - regardless of the source - do not invalidate the warranty!

To be sure of obtaining the correct parts, have engine and chassis numbers available and, if possible, take the old parts along for positive identification.

Maintenance techniques, tools and working facilities

Maintenance techniques

There are a number of techniques involved in maintenance and repair that will be referred to throughout this manual. Application of these techniques will enable the home mechanic to be more efficient, better organized and capable of performing the various tasks properly, which will ensure that the repair job is thorough and complete.

Fasteners

Fasteners are nuts, bolts, studs and screws used to hold two or more parts together. There are a few things to keep in mind when working with fasteners. Almost all of them use a locking device of some type, either a lockwasher, locknut, locking tab or thread adhesive. All threaded fasteners should be clean and straight, with undamaged threads and undamaged corners on the hex head where the wrench fits. Develop the habit of replacing all damaged nuts and bolts with new ones. Special locknuts with nylon or fiber inserts can only be used once. If they are removed, they lose their locking ability and must be replaced with new ones.

Rusted nuts and bolts should be treated with a penetrating fluid to ease removal and prevent breakage. Some mechanics use turpentine in a spout-type oil can, which works quite well. After applying the rust penetrant, let it work for a few minutes before trying to loosen the nut or bolt. Badly rusted fasteners may have to be chiseled or sawed off or removed with a special nut breaker, available at tool stores.

If a bolt or stud breaks off in an assembly, it can be drilled and removed with a special tool commonly available for this purpose. Most automotive machine shops can perform this task, as well as other repair procedures, such as the repair of threaded holes that have been stripped out.

Flat washers and lockwashers, when removed from an assembly, should always be replaced exactly as removed. Replace any damaged washers with new ones. Never use a lockwasher on any soft metal surface (such as aluminum), thin sheet metal or plastic.

Bolt strength marking (standard/SAE/USS; bottom - metric)

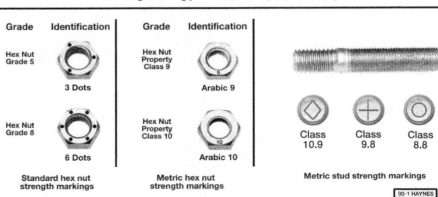

Standard hex nut strength markings

Metric hex nut strength markings

Metric stud strength markings

Fastener sizes

For a number of reasons, automobile manufacturers are making wider and wider use of metric fasteners. Therefore, it is important to be able to tell the difference between standard (sometimes called U.S. or SAE) and metric hardware, since they cannot be interchanged.

All bolts, whether standard or metric, are sized according to diameter, thread pitch and length. For example, a standard 1/2 - 13 x 1 bolt is 1/2 inch in diameter, has 13 threads per inch and is 1 inch long. An M12 - 1.75 x 25 metric bolt is 12 mm in diameter, has a thread pitch of 1.75 mm (the distance between threads) and is 25 mm long. The two bolts are nearly identical, and easily confused, but they are not interchangeable.

In addition to the differences in diameter, thread pitch and length, metric and standard bolts can also be distinguished by examining the bolt heads. To begin with, the distance across the flats on a standard bolt head is measured in inches, while the same dimension on a metric bolt is sized in millimeters (the same is true for nuts). As a result, a standard wrench should not be used on a metric bolt and a metric wrench should not be used on a standard bolt. Also, most standard bolts have slashes radiating out from the center of the head to denote the grade or strength of the bolt, which is an indication of the amount of torque that can be applied to it. The greater the number of slashes, the greater the strength of the bolt. Grades 0 through 5 are commonly used on automobiles. Metric bolts have a property class (grade) number, rather than a slash, molded into their heads to indicate bolt strength. In this case, the higher the number, the stronger the bolt. Property class numbers 8.8, 9.8 and 10.9 are commonly used on automobiles.

Strength markings can also be used to distinguish standard hex nuts from metric hex nuts. Many standard nuts have dots stamped into one side, while metric nuts are marked with a number. The greater the number of dots, or the higher the number, the greater the strength of the nut.

Metric studs are also marked on their ends according to property class (grade). Larger studs are numbered (the same as metric bolts), while smaller studs carry a geometric code to denote grade.

It should be noted that many fasteners, especially Grades 0 through 2, have no distinguishing marks on them. When such is the case, the only way to determine whether it is standard or metric is to measure the thread pitch or compare it to a known fastener of the same size.

Standard fasteners are often referred to as SAE, as opposed to metric. However, it should be noted that SAE technically refers to a non-metric fine thread fastener only. Coarse thread non-metric fasteners are referred to as USS sizes.

Since fasteners of the same size (both standard and metric) may have different

	Ft-lbs	Nm
Metric thread sizes		
M-6	6 to 9	9 to 12
M-8	14 to 21	19 to 28
M-10	28 to 40	38 to 54
M-12	50 to 71	68 to 96
M-14	80 to 140	109 to 154
Pipe thread sizes		
1/8	5 to 8	7 to 10
1/4	12 to 18	17 to 24
3/8	22 to 33	30 to 44
1/2	25 to 35	34 to 47
U.S. thread sizes		
1/4 - 20	6 to 9	9 to 12
5/16 - 18	12 to 18	17 to 24
5/16 - 24	14 to 20	19 to 27
3/8 - 16	22 to 32	30 to 43
3/8 - 24	27 to 38	37 to 51
7/16 - 14	40 to 55	55 to 74
7/16 - 20	40 to 60	55 to 81
1/2 - 13	55 to 80	75 to 108

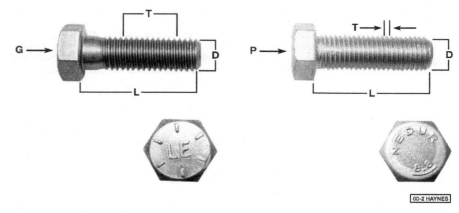

Standard (SAE and USS) bolt dimensions/grade marks

- G Grade marks (bolt strength)
- L Length (in inches)
- T Thread pitch (number of threads per inch)
- D Nominal diameter (in inches)

Metric bolt dimensions/grade marks

- P Property class (bolt strength)
- L Length (in millimeters)
- T Thread pitch (distance between threads in millimeters)
- D Diameter

strength ratings, be sure to reinstall any bolts, studs or nuts removed from your vehicle in their original locations. Also, when replacing a fastener with a new one, make sure that the new one has a strength rating equal to or greater than the original.

Tightening sequences and procedures

Most threaded fasteners should be tightened to a specific torque value (torque is the twisting force applied to a threaded component such as a nut or bolt). Overtightening the fastener can weaken it and cause it to break, while undertightening can cause it to eventually come loose. Bolts, screws and studs, depending on the material they are made of and their thread diameters, have specific torque values, many of which are noted in the Specifications at the beginning of each Chapter. Be sure to follow the torque recommendations closely. For fasteners not assigned a specific torque, a general torque value chart is presented here as a guide. These torque values are for dry (unlubricated) fasteners threaded into steel or cast iron (not aluminum). As was previously mentioned, the size and grade of a fastener determine the amount of torque that can safely be applied to it. The figures listed here are approximate for Grade 2 and Grade 3 fasteners. Higher grades can tolerate higher torque values.

Fasteners laid out in a pattern, such as cylinder head bolts, oil pan bolts, differential cover bolts, etc., must be loosened or tightened in sequence to avoid warping the com-

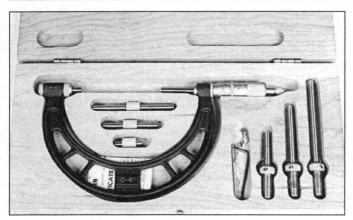
Micrometer set

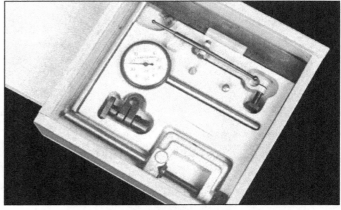
Dial indicator set

ponent. This sequence will normally be shown in the appropriate Chapter. If a specific pattern is not given, the following procedures can be used to prevent warping.

Initially, the bolts or nuts should be assembled finger-tight only. Next, they should be tightened one full turn each, in a criss-cross or diagonal pattern. After each one has been tightened one full turn, return to the first one and tighten them all one-half turn, following the same pattern. Finally, tighten each of them one-quarter turn at a time until each fastener has been tightened to the proper torque. To loosen and remove the fasteners, the procedure would be reversed.

Component disassembly

Component disassembly should be done with care and purpose to help ensure that the parts go back together properly. Always keep track of the sequence in which parts are removed. Make note of special characteristics or marks on parts that can be installed more than one way, such as a grooved thrust washer on a shaft. It is a good idea to lay the disassembled parts out on a clean surface in the order that they were removed. It may also be helpful to make sketches or take instant photos of components before removal.

When removing fasteners from a component, keep track of their locations. Sometimes threading a bolt back in a part, or putting the washers and nut back on a stud, can prevent mix-ups later. If nuts and bolts cannot be returned to their original locations, they should be kept in a compartmented box or a series of small boxes. A cupcake or muffin tin is ideal for this purpose, since each cavity can hold the bolts and nuts from a particular area (i.e. oil pan bolts, valve cover bolts, engine mount bolts, etc.). A pan of this type is especially helpful when working on assemblies with very small parts, such as the carburetor, alternator, valve train or interior dash and trim pieces. The cavities can be marked with paint or tape to identify the contents.

Whenever wiring looms, harnesses or connectors are separated, it is a good idea to identify the two halves with numbered pieces of masking tape so they can be easily reconnected.

Gasket sealing surfaces

Throughout any vehicle, gaskets are used to seal the mating surfaces between two parts and keep lubricants, fluids, vacuum or pressure contained in an assembly.

Many times these gaskets are coated with a liquid or paste-type gasket sealing compound before assembly. Age, heat and pressure can sometimes cause the two parts to stick together so tightly that they are very difficult to separate. Often, the assembly can be loosened by striking it with a soft-face hammer near the mating surfaces. A regular hammer can be used if a block of wood is placed between the hammer and the part. Do not hammer on cast parts or parts that could be easily damaged. With any particularly stubborn part, always recheck to make sure that every fastener has been removed.

Avoid using a screwdriver or bar to pry apart an assembly, as they can easily mar the gasket sealing surfaces of the parts, which must remain smooth. If prying is absolutely necessary, use an old broom handle, but keep in mind that extra clean up will be necessary if the wood splinters.

After the parts are separated, the old gasket must be carefully scraped off and the gasket surfaces cleaned. Stubborn gasket material can be soaked with rust penetrant or treated with a special chemical to soften it so it can be easily scraped off. A scraper can be fashioned from a piece of copper tubing by flattening and sharpening one end. Copper is recommended because it is usually softer than the surfaces to be scraped, which reduces the chance of gouging the part. Some gaskets can be removed with a wire brush, but regardless of the method used, the mating surfaces must be left clean and smooth. If for some reason the gasket surface is gouged, then a gasket sealer thick enough to fill scratches will have to be used during reassembly of the components. For most applications, a non-drying (or semi-drying) gasket sealer should be used.

Hose removal tips

Warning: *If the vehicle is equipped with air conditioning, do not disconnect any of the A/C hoses without first having the system depressurized by a dealer service department or a service station.*

Hose removal precautions closely parallel gasket removal precautions. Avoid scratching or gouging the surface that the hose mates against or the connection may leak. This is especially true for radiator hoses. Because of various chemical reactions, the rubber in hoses can bond itself to the metal spigot that the hose fits over. To remove a hose, first loosen the hose clamps that secure it to the spigot. Then, with slip-joint pliers, grab the hose at the clamp and rotate it around the spigot. Work it back and forth until it is completely free, then pull it off. Silicone or other lubricants will ease removal if they can be applied between the hose and the outside of the spigot. Apply the same lubricant to the inside of the hose and the outside of the spigot to simplify installation.

As a last resort (and if the hose is to be replaced with a new one anyway), the rubber can be slit with a knife and the hose peeled from the spigot. If this must be done, be careful that the metal connection is not damaged.

If a hose clamp is broken or damaged, do not reuse it. Wire-type clamps usually weaken with age, so it is a good idea to replace them with screw-type clamps whenever a hose is removed.

Tools

A selection of good tools is a basic requirement for anyone who plans to maintain and repair his or her own vehicle. For the owner who has few tools, the initial investment might seem high, but when compared to the spiraling costs of professional auto maintenance and repair, it is a wise one.

To help the owner decide which tools are needed to perform the tasks detailed in this manual, the following tool lists are offered: *Maintenance and minor repair, Repair/overhaul* and *Special.*

The newcomer to practical mechanics

Maintenance techniques, tools and working facilities

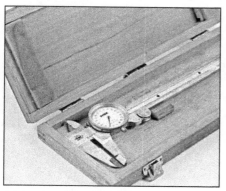
Dial caliper

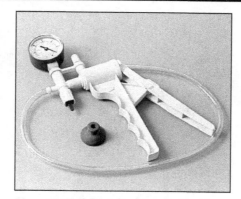
Hand-operated vacuum pump

Timing light

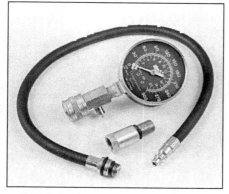
Compression gauge with spark plug hole adapter

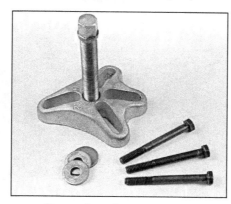
Damper/steering wheel puller

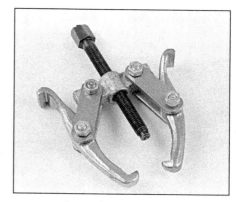
General purpose puller

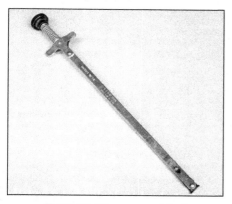
Hydraulic lifter removal tool

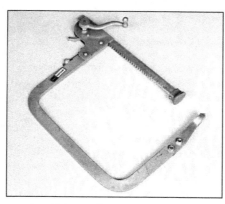
Valve spring compressor

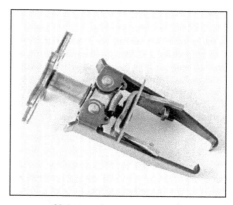
Valve spring compressor

Ridge reamer

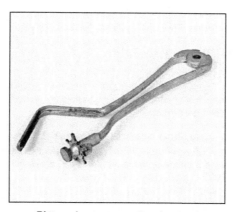
Piston ring groove cleaning tool

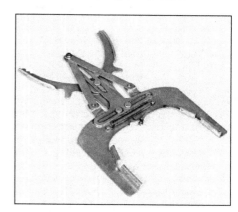
Ring removal/installation tool

Ring compressor

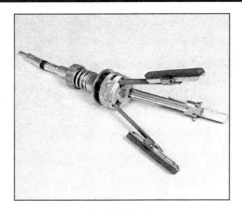
Cylinder hone

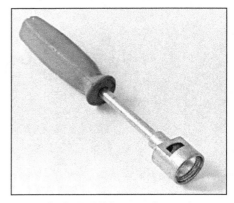
Brake hold-down spring tool

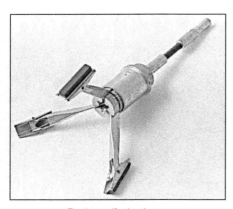
Brake cylinder hone

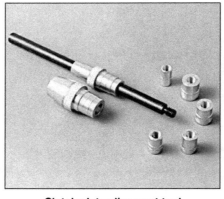
Clutch plate alignment tool

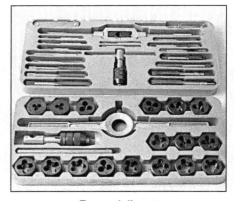
Tap and die set

should start off with the *maintenance and minor repair* tool kit, which is adequate for the simpler jobs performed on a vehicle. Then, as confidence and experience grow, the owner can tackle more difficult tasks, buying additional tools as they are needed. Eventually the basic kit will be expanded into the *repair and overhaul* tool set. Over a period of time, the experienced do-it-yourselfer will assemble a tool set complete enough for most repair and overhaul procedures and will add tools from the special category when it is felt that the expense is justified by the frequency of use.

Maintenance and minor repair tool kit

The tools in this list should be considered the minimum required for performance of routine maintenance, servicing and minor repair work. We recommend the purchase of combination wrenches (box-end and open-end combined in one wrench). While more expensive than open end wrenches, they offer the advantages of both types of wrench.

Combination wrench set (1/4-inch to 1 inch or 6 mm to 19 mm)
Adjustable wrench, 8 inch
Spark plug wrench with rubber insert
Spark plug gap adjusting tool
Feeler gauge set
Brake bleeder wrench
Standard screwdriver (5/16-inch x 6 inch)
Phillips screwdriver (No. 2 x 6 inch)
Combination pliers - 6 inch
Hacksaw and assortment of blades
Tire pressure gauge
Grease gun
Oil can
Fine emery cloth
Wire brush
Battery post and cable cleaning tool
Oil filter wrench
Funnel (medium size)
Safety goggles
Jackstands (2)
Drain pan

Note: *If basic tune-ups are going to be part of routine maintenance, it will be necessary to purchase a good quality stroboscopic timing light and combination tachometer/dwell meter. Although they are included in the list of special tools, it is mentioned here because they are absolutely necessary for tuning most vehicles properly.*

Repair and overhaul tool set

These tools are essential for anyone who plans to perform major repairs and are in addition to those in the maintenance and minor repair tool kit. Included is a comprehensive set of sockets which, though expensive, are invaluable because of their versatility, especially when various extensions and drives are available. We recommend the 1/2-inch drive over the 3/8-inch drive. Although the larger drive is bulky and more expensive, it has the capacity of accepting a very wide range of large sockets. Ideally, however, the mechanic should have a 3/8-inch drive set and a 1/2-inch drive set.

Socket set(s)
Reversible ratchet
Extension - 10 inch
Universal joint
Torque wrench (same size drive as sockets)
Ball peen hammer - 8 ounce
Soft-face hammer (plastic/rubber)
Standard screwdriver (1/4-inch x 6 inch)
Standard screwdriver (stubby - 5/16-inch)
Phillips screwdriver (No. 3 x 8 inch)
Phillips screwdriver (stubby - No. 2)
Pliers - vise grip
Pliers - lineman's
Pliers - needle nose
Pliers - snap-ring (internal and external)
Cold chisel - 1/2-inch
Scribe
Scraper (made from flattened copper tubing)
Centerpunch
Pin punches (1/16, 1/8, 3/16-inch)
Steel rule/straightedge - 12 inch

Maintenance techniques, tools and working facilities

Allen wrench set (1/8 to 3/8-inch or
   4 mm to 10 mm)
A selection of files
Wire brush (large)
Jackstands (second set)
Jack (scissor or hydraulic type)

Note: *Another tool which is often useful is an electric drill with a chuck capacity of 3/8-inch and a set of good quality drill bits.*

Special tools

The tools in this list include those which are not used regularly, are expensive to buy, or which need to be used in accordance with their manufacturer's instructions. Unless these tools will be used frequently, it is not very economical to purchase many of them. A consideration would be to split the cost and use between yourself and a friend or friends. In addition, most of these tools can be obtained from a tool rental shop on a temporary basis.

This list primarily contains only those tools and instruments widely available to the public, and not those special tools produced by the vehicle manufacturer for distribution to dealer service departments. Occasionally, references to the manufacturer's special tools are included in the text of this manual. Generally, an alternative method of doing the job without the special tool is offered. However, sometimes there is no alternative to their use. Where this is the case, and the tool cannot be purchased or borrowed, the work should be turned over to the dealer service department or an automotive repair shop.

Valve spring compressor
Piston ring groove cleaning tool
Piston ring compressor
Piston ring installation tool
Cylinder compression gauge
Cylinder ridge reamer
Cylinder surfacing hone
Cylinder bore gauge
Micrometers and/or dial calipers
Hydraulic lifter removal tool
Balljoint separator
Universal-type puller
Impact screwdriver
Dial indicator set
Stroboscopic timing light (inductive
   pick-up)
Hand operated vacuum/pressure pump
Tachometer/dwell meter
Universal electrical multimeter
Cable hoist
Brake spring removal and installation
   tools
Floor jack

Buying tools

For the do-it-yourselfer who is just starting to get involved in vehicle maintenance and repair, there are a number of options available when purchasing tools. If maintenance and minor repair is the extent of the work to be done, the purchase of individual tools is satisfactory. If, on the other hand, extensive work is planned, it would be a good idea to purchase a modest tool set from one of the large retail chain stores. A set can usually be bought at a substantial savings over the individual tool prices, and they often come with a tool box. As additional tools are needed, add-on sets, individual tools and a larger tool box can be purchased to expand the tool selection. Building a tool set gradually allows the cost of the tools to be spread over a longer period of time and gives the mechanic the freedom to choose only those tools that will actually be used.

Tool stores will often be the only source of some of the special tools that are needed, but regardless of where tools are bought, try to avoid cheap ones, especially when buying screwdrivers and sockets, because they won't last very long. The expense involved in replacing cheap tools will eventually be greater than the initial cost of quality tools.

Care and maintenance of tools

Good tools are expensive, so it makes sense to treat them with respect. Keep them clean and in usable condition and store them properly when not in use. Always wipe off any dirt, grease or metal chips before putting them away. Never leave tools lying around in the work area. Upon completion of a job, always check closely under the hood for tools that may have been left there so they won't get lost during a test drive.

Some tools, such as screwdrivers, pliers, wrenches and sockets, can be hung on a panel mounted on the garage or workshop wall, while others should be kept in a tool box or tray. Measuring instruments, gauges, meters, etc. must be carefully stored where they cannot be damaged by weather or impact from other tools.

When tools are used with care and stored properly, they will last a very long time. Even with the best of care, though, tools will wear out if used frequently. When a tool is damaged or worn out, replace it. Subsequent jobs will be safer and more enjoyable if you do.

How to repair damaged threads

Sometimes, the internal threads of a nut or bolt hole can become stripped, usually from overtightening. Stripping threads is an all-too-common occurrence, especially when working with aluminum parts, because aluminum is so soft that it easily strips out.

Usually, external or internal threads are only partially stripped. After they've been cleaned up with a tap or die, they'll still work. Sometimes, however, threads are badly damaged. When this happens, you've got three choices:

1) *Drill and tap the hole to the next suitable oversize and install a larger diameter bolt, screw or stud.*
2) *Drill and tap the hole to accept a threaded plug, then drill and tap the plug to the original screw size. You can also buy a plug already threaded to the original size. Then you simply drill a hole to the specified size, then run the threaded plug into the hole with a bolt and jam nut. Once the plug is fully seated, remove the jam nut and bolt.*
3) *The third method uses a patented thread repair kit like Heli-Coil or Slimsert. These easy-to-use kits are designed to repair damaged threads in straight-through holes and blind holes. Both are available as kits which can handle a variety of sizes and thread patterns. Drill the hole, then tap it with the special included tap. Install the Heli-Coil and the hole is back to its original diameter and thread pitch.*

Regardless of which method you use, be sure to proceed calmly and carefully. A little impatience or carelessness during one of these relatively simple procedures can ruin your whole day's work and cost you a bundle if you wreck an expensive part.

Working facilities

Not to be overlooked when discussing tools is the workshop. If anything more than routine maintenance is to be carried out, some sort of suitable work area is essential.

It is understood, and appreciated, that many home mechanics do not have a good workshop or garage available, and end up removing an engine or doing major repairs outside. It is recommended, however, that the overhaul or repair be completed under the cover of a roof.

A clean, flat workbench or table of comfortable working height is an absolute necessity. The workbench should be equipped with a vise that has a jaw opening of at least four inches.

As mentioned previously, some clean, dry storage space is also required for tools, as well as the lubricants, fluids, cleaning solvents, etc. which soon become necessary.

Sometimes waste oil and fluids, drained from the engine or cooling system during normal maintenance or repairs, present a disposal problem. To avoid pouring them on the ground or into a sewage system, pour the used fluids into large containers, seal them with caps and take them to an authorized disposal site or recycling center. Plastic jugs, such as old antifreeze containers, are ideal for this purpose.

Always keep a supply of old newspapers and clean rags available. Old towels are excellent for mopping up spills. Many mechanics use rolls of paper towels for most work because they are readily available and disposable. To help keep the area under the vehicle clean, a large cardboard box can be cut open and flattened to protect the garage or shop floor.

Whenever working over a painted surface, such as when leaning over a fender to service something under the hood, always cover it with an old blanket or bedspread to protect the finish. Vinyl covered pads, made especially for this purpose, are available at auto parts stores.

Jacking and towing

Jacking

Warning: *The jack supplied with the vehicle should only be used for raising the vehicle when changing a tire or placing jackstands under the frame. Never work under the vehicle or start the engine while the jack is being used as the only means of support.*

The vehicle must be on a level surface with the wheels blocked and the transmission in Park. Apply the parking brake if the front of the vehicle must be raised. Make sure no one is in the vehicle as it's being raised with the jack.

Remove the jack and lug nut wrench and spare tire from the compartment in the right rear corner of the rear passenger compartment.

To replace the tire, use the tapered end of the lug wrench to pry loose the wheel cover (if equipped). **Note:** *If the vehicle is equipped with aluminum wheels, it may be necessary to pry out the special lug nut covers. Also, aluminum wheels normally have anti-theft lug nuts (one per wheel) which require using a special "key" between the lug wrench and lug nut. The key is usually in the glove compartment.* Loosen the lug nuts one-half turn, but leave them in place until the tire is raised off the ground.

Position the jack under the side of the vehicle at the jacking points (refer to your vehicle's owner's manual or the jacking instructions with the jack). There's a front and rear jacking point on each side of the vehicle consisting of hole in the chassis or spring hanger for the pin in the lifting surface of the jack.

Operate the jack until the tire clears the ground. Remove the lug nuts and pull the tire off. Clean the mating surfaces of the hub and wheel, then install the spare. Replace the lug nuts with the beveled edges facing in and tighten them snugly. Don't attempt to tighten them completely until the vehicle is lowered or it could slip off the jack.

Lower the vehicle with the jack. Remove the jack and tighten the lug nuts in a criss-cross pattern. If possible, tighten the nuts with a torque wrench (see Chapter 1 for the torque values). If you don't have access to a torque wrench, have the nuts checked by a service station or repair shop as soon as possible.

Caution: *The compact spare included with some models is intended for temporary use only. Have the tire repaired and reinstall it on the vehicle at the earliest opportunity and don't exceed 50 mph with the spare tire on the car.*

Install the wheel cover, then stow the tire, jack and wrench and unblock the wheels.

Towing

We recommend these vehicles (except all-wheel drive models) be towed from the rear, with the rear wheels off the ground. If it's absolutely necessary, these vehicles can be towed from the front with the front wheels off the ground, provided that speeds don't exceed 35 mph and the distance is less than 50 miles; the transmission can be damaged if these mileage/speed limitations are exceeded. Vehicles with all-wheel drive must not be towed with all four wheels on the ground. They must only be towed with all four wheels off the ground.

Equipment specifically designed for towing should be used. It must be attached to the main structural members of the vehicle, not the bumpers or brackets.

Safety is a major consideration when towing and all applicable state and local laws must be obeyed. A safety chain must be used at all times.

The parking brake must be released and the transmission must be in Neutral. The steering must be unlocked (ignition switch in the Off position). Remember that power steering and power brakes won't work with the engine off.

Anti-theft audio system

General information

1 Some models are equipped with the anti-theft audio system which includes an anti-theft feature that will render the stereo inoperative if stolen. If the power source to the stereo is cut with the anti-theft feature activated, the stereo will be inoperative. Even if the power source is immediately re-connected, the stereo will not function. If your vehicle is equipped with this anti-theft system, do not disconnect the battery, remove the stereo or disconnect related components unless you have either turned off the feature or have the individual ID (code) number for the stereo.
2 Refer to your vehicle's owner's manual for more complete information on this audio system and its anti-theft feature.

Disabling the anti-theft feature

3 Press the stereo's 1 and 4 buttons at the same time for five seconds with the ignition on and the radio power off. The display will show SEC, indicating the unit is in the secure mode (anti-theft feature enabled).
4 Press the SET button. The display will show "000."
5 Press the SCAN button to make the first number appear.
6 Press the SEEK right or SEEK left arrow until the second and third digit of your code appear. The numbers will be displayed as entered.
7 Press the BAND knob. "000" will be displayed.
8 Enter the second three digits of the code are displayed.
9 Press the lower BAND knob. If the display shows "_ _," you have successfully disabled the anti-theft feature. If SEC is displayed, the code you entered was incorrect and the anti-theft feature is still enabled.

Unlocking the stereo after a power loss

10 When power is restored to the stereo, the stereo won't turn on and LOC will appear on the display. Enter your ID code as follows; pause no more than 15 seconds between Steps.
11 Turn the ignition switch to ON, but leave the stereo off.
12 Press the SET button. "000" should display.
13 Press the SCAN button to make the first number appear, then release it.
14 Press the SEEK right or SEEK left arrows and make sure the second and third numbers agree with your code.
15 Repeat Steps 13 and 14 for the last three digits of your code.
16 Press the BAND knob. If SEC appears, the numbers you entered were correct and the stereo will work. If LOC appears, the numbers you entered were not correct and the stereo is still inoperative

Booster battery (jump) starting

Observe these precautions when using a booster battery to start a vehicle:

a) Before connecting the booster battery, make sure the ignition switch is in the Off position.
b) Turn off the lights, heater and other electrical loads.
c) Your eyes should be shielded. Safety goggles are a good idea.
d) Make sure the booster battery is the same voltage as the dead one in the vehicle.
e) The two vehicles MUST NOT TOUCH each other!
f) Make sure the transaxle is in Neutral (manual) or Park (automatic).
g) If the booster battery is not a maintenance-free type, remove the vent caps and lay a cloth over the vent holes.

Connect the red jumper cable to the positive (+) terminals of each battery (see illustration).

Connect one end of the black jumper cable to the negative (-) terminal of the booster battery. The other end of this cable should be connected to a good ground on the vehicle to be started, such as a bolt or bracket on the body.

Start the engine using the booster battery, then, with the engine running at idle speed, disconnect the jumper cables in the reverse order of connection.

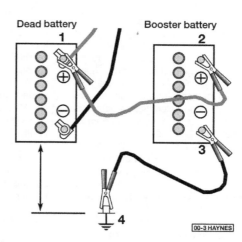

Make the booster battery cable connections in the numerical order shown (note that the negative cable of the booster battery is NOT attached to the negative terminal of the dead battery)

Automotive chemicals and lubricants

A number of automotive chemicals and lubricants are available for use during vehicle maintenance and repair. They include a wide variety of products ranging from cleaning solvents and degreasers to lubricants and protective sprays for rubber, plastic and vinyl.

Cleaners

Carburetor cleaner and choke cleaner is a strong solvent for gum, varnish and carbon. Most carburetor cleaners leave a dry-type lubricant film which will not harden or gum up. Because of this film it is not recommended for use on electrical components.

Brake system cleaner is used to remove brake dust, grease and brake fluid from the brake system, where clean surfaces are absolutely necessary. It leaves no residue and often eliminates brake squeal caused by contaminants.

Electrical cleaner removes oxidation, corrosion and carbon deposits from electrical contacts, restoring full current flow. It can also be used to clean spark plugs, carburetor jets, voltage regulators and other parts where an oil-free surface is desired.

Demoisturants remove water and moisture from electrical components such as alternators, voltage regulators, electrical connectors and fuse blocks. They are non-conductive and non-corrosive.

Degreasers are heavy-duty solvents used to remove grease from the outside of the engine and from chassis components. They can be sprayed or brushed on and, depending on the type, are rinsed off either with water or solvent.

Lubricants

Motor oil is the lubricant formulated for use in engines. It normally contains a wide variety of additives to prevent corrosion and reduce foaming and wear. Motor oil comes in various weights (viscosity ratings) from 0 to 50. The recommended weight of the oil depends on the season, temperature and the demands on the engine. Light oil is used in cold climates and under light load conditions. Heavy oil is used in hot climates and where high loads are encountered. Multi-viscosity oils are designed to have characteristics of both light and heavy oils and are available in a number of weights from 0W-20 to 20W-50.

Gear oil is designed to be used in differentials, manual transmissions and other areas where high-temperature lubrication is required.

Chassis and wheel bearing grease is a heavy grease used where increased loads and friction are encountered, such as for wheel bearings, balljoints, tie-rod ends and universal joints.

High-temperature wheel bearing grease is designed to withstand the extreme temperatures encountered by wheel bearings in disc brake equipped vehicles. It usually contains molybdenum disulfide (moly), which is a dry-type lubricant.

White grease is a heavy grease for metal-to-metal applications where water is a problem. White grease stays soft under both low and high temperatures (usually from -100 to +190-degrees F), and will not wash off or dilute in the presence of water.

Assembly lube is a special extreme pressure lubricant, usually containing moly, used to lubricate high-load parts (such as main and rod bearings and cam lobes) for initial start-up of a new engine. The assembly lube lubricates the parts without being squeezed out or washed away until the engine oiling system begins to function.

Silicone lubricants are used to protect rubber, plastic, vinyl and nylon parts.

Graphite lubricants are used where oils cannot be used due to contamination problems, such as in locks. The dry graphite will lubricate metal parts while remaining uncontaminated by dirt, water, oil or acids. It is electrically conductive and will not foul electrical contacts in locks such as the ignition switch.

Moly penetrants loosen and lubricate frozen, rusted and corroded fasteners and prevent future rusting or freezing.

Heat-sink grease is a special electrically non-conductive grease that is used for mounting electronic ignition modules where it is essential that heat is transferred away from the module.

Sealants

RTV sealant is one of the most widely used gasket compounds. Made from silicone, RTV is air curing, it seals, bonds, waterproofs, fills surface irregularities, remains flexible, doesn't shrink, is relatively easy to remove, and is used as a supplementary sealer with almost all low and medium temperature gaskets.

Anaerobic sealant is much like RTV in that it can be used either to seal gaskets or to form gaskets by itself. It remains flexible, is solvent resistant and fills surface imperfections. The difference between an anaerobic sealant and an RTV-type sealant is in the curing. RTV cures when exposed to air, while an anaerobic sealant cures only in the absence of air. This means that an anaerobic sealant cures only after the assembly of parts, sealing them together.

Thread and pipe sealant is used for sealing hydraulic and pneumatic fittings and vacuum lines. It is usually made from a Teflon compound, and comes in a spray, a paint-on liquid and as a wrap-around tape.

Chemicals

Anti-seize compound prevents seizing, galling, cold welding, rust and corrosion in fasteners. High-temperature anti-seize, usually made with copper and graphite lubricants, is used for exhaust system and exhaust manifold bolts.

Anaerobic locking compounds are used to keep fasteners from vibrating or working loose and cure only after installation, in the absence of air. Medium strength locking compound is used for small nuts, bolts and screws that may be removed later. High-strength locking compound is for large nuts, bolts and studs which aren't removed on a regular basis.

Oil additives range from viscosity index improvers to chemical treatments that claim to reduce internal engine friction. It should be noted that most oil manufacturers caution against using additives with their oils.

Gas additives perform several functions, depending on their chemical makeup. They usually contain solvents that help dissolve gum and varnish that build up on carburetor, fuel injection and intake parts. They also serve to break down carbon deposits that form on the inside surfaces of the combustion chambers. Some additives contain upper cylinder lubricants for valves and piston rings, and others contain chemicals to remove condensation from the gas tank.

Miscellaneous

Brake fluid is specially formulated hydraulic fluid that can withstand the heat and pressure encountered in brake systems. Care must be taken so this fluid does not come in contact with painted surfaces or plastics. An opened container should always be resealed to prevent contamination by water or dirt.

Weatherstrip adhesive is used to bond weatherstripping around doors, windows and trunk lids. It is sometimes used to attach trim pieces.

Undercoating is a petroleum-based, tar-like substance that is designed to protect metal surfaces on the underside of the vehicle from corrosion. It also acts as a sound-deadening agent by insulating the bottom of the vehicle.

Waxes and polishes are used to help protect painted and plated surfaces from the weather. Different types of paint may require the use of different types of wax and polish. Some polishes utilize a chemical or abrasive cleaner to help remove the top layer of oxidized (dull) paint on older vehicles. In recent years many non-wax polishes that contain a wide variety of chemicals such as polymers and silicones have been introduced. These non-wax polishes are usually easier to apply and last longer than conventional waxes and polishes.

Conversion factors

Length (distance)
Inches (in)	X	25.4	= Millimeters (mm)	X 0.0394	= Inches (in)
Feet (ft)	X	0.305	= Meters (m)	X 3.281	= Feet (ft)
Miles	X	1.609	= Kilometers (km)	X 0.621	= Miles

Volume (capacity)
Cubic inches (cu in; in³)	X	16.387	= Cubic centimeters (cc; cm³)	X 0.061	= Cubic inches (cu in; in³)
Imperial pints (Imp pt)	X	0.568	= Liters (l)	X 1.76	= Imperial pints (Imp pt)
Imperial quarts (Imp qt)	X	1.137	= Liters (l)	X 0.88	= Imperial quarts (Imp qt)
Imperial quarts (Imp qt)	X	1.201	= US quarts (US qt)	X 0.833	= Imperial quarts (Imp qt)
US quarts (US qt)	X	0.946	= Liters (l)	X 1.057	= US quarts (US qt)
Imperial gallons (Imp gal)	X	4.546	= Liters (l)	X 0.22	= Imperial gallons (Imp gal)
Imperial gallons (Imp gal)	X	1.201	= US gallons (US gal)	X 0.833	= Imperial gallons (Imp gal)
US gallons (US gal)	X	3.785	= Liters (l)	X 0.264	= US gallons (US gal)

Mass (weight)
Ounces (oz)	X	28.35	= Grams (g)	X 0.035	= Ounces (oz)
Pounds (lb)	X	0.454	= Kilograms (kg)	X 2.205	= Pounds (lb)

Force
Ounces-force (ozf; oz)	X	0.278	= Newtons (N)	X 3.6	= Ounces-force (ozf; oz)
Pounds-force (lbf; lb)	X	4.448	= Newtons (N)	X 0.225	= Pounds-force (lbf; lb)
Newtons (N)	X	0.1	= Kilograms-force (kgf; kg)	X 9.81	= Newtons (N)

Pressure
Pounds-force per square inch (psi; lbf/in²; lb/in²)	X	0.070	= Kilograms-force per square centimeter (kgf/cm²; kg/cm²)	X 14.223	= Pounds-force per square inch (psi; lbf/in²; lb/in²)
Pounds-force per square inch (psi; lbf/in²; lb/in²)	X	0.068	= Atmospheres (atm)	X 14.696	= Pounds-force per square inch (psi; lbf/in²; lb/in²)
Pounds-force per square inch (psi; lbf/in²; lb/in²)	X	0.069	= Bars	X 14.5	= Pounds-force per square inch (psi; lbf/in²; lb/in²)
Pounds-force per square inch (psi; lbf/in²; lb/in²)	X	6.895	= Kilopascals (kPa)	X 0.145	= Pounds-force per square inch (psi; lbf/in²; lb/in²)
Kilopascals (kPa)	X	0.01	= Kilograms-force per square centimeter (kgf/cm²; kg/cm²)	X 98.1	= Kilopascals (kPa)

Torque (moment of force)
Pounds-force inches (lbf in; lb in)	X	1.152	= Kilograms-force centimeter (kgf cm; kg cm)	X 0.868	= Pounds-force inches (lbf in; lb in)
Pounds-force inches (lbf in; lb in)	X	0.113	= Newton meters (Nm)	X 8.85	= Pounds-force inches (lbf in; lb in)
Pounds-force inches (lbf in; lb in)	X	0.083	= Pounds-force feet (lbf ft; lb ft)	X 12	= Pounds-force inches (lbf in; lb in)
Pounds-force feet (lbf ft; lb ft)	X	0.138	= Kilograms-force meters (kgf m; kg m)	X 7.233	= Pounds-force feet (lbf ft; lb ft)
Pounds-force feet (lbf ft; lb ft)	X	1.356	= Newton meters (Nm)	X 0.738	= Pounds-force feet (lbf ft; lb ft)
Newton meters (Nm)	X	0.102	= Kilograms-force meters (kgf m; kg m)	X 9.804	= Newton meters (Nm)

Vacuum
Inches mercury (in. Hg)	X	3.377	= Kilopascals (kPa)	X 0.2961	= Inches mercury
Inches mercury (in. Hg)	X	25.4	= Millimeters mercury (mm Hg)	X 0.0394	= Inches mercury

Power
Horsepower (hp)	X	745.7	= Watts (W)	X 0.0013	= Horsepower (hp)

Velocity (speed)
Miles per hour (miles/hr; mph)	X	1.609	= Kilometers per hour (km/hr; kph)	X 0.621	= Miles per hour (miles/hr; mph)

*Fuel consumption**
Miles per gallon, Imperial (mpg)	X	0.354	= Kilometers per liter (km/l)	X 2.825	= Miles per gallon, Imperial (mpg)
Miles per gallon, US (mpg)	X	0.425	= Kilometers per liter (km/l)	X 2.352	= Miles per gallon, US (mpg)

Temperature
Degrees Fahrenheit = (°C x 1.8) + 32   Degrees Celsius (Degrees Centigrade; °C) = (°F - 32) x 0.56

*It is common practice to convert from miles per gallon (mpg) to liters/100 kilometers (l/100km), where mpg (Imperial) x l/100 km = 282 and mpg (US) x l/100 km = 235

Safety first!

Regardless of how enthusiastic you may be about getting on with the job at hand, take the time to ensure that your safety is not jeopardized. A moment's lack of attention can result in an accident, as can failure to observe certain simple safety precautions. The possibility of an accident will always exist, and the following points should not be considered a comprehensive list of all dangers. Rather, they are intended to make you aware of the risks and to encourage a safety conscious approach to all work you carry out on your vehicle.

Essential DOs and DON'Ts

DON'T rely on a jack when working under the vehicle. Always use approved jackstands to support the weight of the vehicle and place them under the recommended lift or support points.

DON'T attempt to loosen extremely tight fasteners (i.e. wheel lug nuts) while the vehicle is on a jack - it may fall.

DON'T start the engine without first making sure that the transmission is in Neutral (or Park where applicable) and the parking brake is set.

DON'T remove the radiator cap from a hot cooling system - let it cool or cover it with a cloth and release the pressure gradually.

DON'T attempt to drain the engine oil until you are sure it has cooled to the point that it will not burn you.

DON'T touch any part of the engine or exhaust system until it has cooled sufficiently to avoid burns.

DON'T siphon toxic liquids such as gasoline, antifreeze and brake fluid by mouth, or allow them to remain on your skin.

DON'T inhale brake lining dust - it is potentially hazardous (see *Asbestos* below).

DON'T allow spilled oil or grease to remain on the floor - wipe it up before someone slips on it.

DON'T use loose fitting wrenches or other tools which may slip and cause injury.

DON'T push on wrenches when loosening or tightening nuts or bolts. Always try to pull the wrench toward you. If the situation calls for pushing the wrench away, push with an open hand to avoid scraped knuckles if the wrench should slip.

DON'T attempt to lift a heavy component alone - get someone to help you.

DON'T rush or take unsafe shortcuts to finish a job.

DON'T allow children or animals in or around the vehicle while you are working on it.

DO wear eye protection when using power tools such as a drill, sander, bench grinder, etc. and when working under a vehicle.

DO keep loose clothing and long hair well out of the way of moving parts.

DO make sure that any hoist used has a safe working load rating adequate for the job.

DO get someone to check on you periodically when working alone on a vehicle.

DO carry out work in a logical sequence and make sure that everything is correctly assembled and tightened.

DO keep chemicals and fluids tightly capped and out of the reach of children and pets.

DO remember that your vehicle's safety affects that of yourself and others. If in doubt on any point, get professional advice.

Asbestos

Certain friction, insulating, sealing, and other products - such as brake linings, brake bands, clutch linings, torque converters, gaskets, etc. - may contain asbestos. Extreme care must be taken to avoid inhalation of dust from such products, since it is hazardous to health. If in doubt, assume that they do contain asbestos.

Fire

Remember at all times that gasoline is highly flammable. Never smoke or have any kind of open flame around when working on a vehicle. But the risk does not end there. A spark caused by an electrical short circuit, by two metal surfaces contacting each other, or even by static electricity built up in your body under certain conditions, can ignite gasoline vapors, which in a confined space are highly explosive. Do not, under any circumstances, use gasoline for cleaning parts. Use an approved safety solvent.

Always disconnect the battery ground (-) cable at the battery before working on any part of the fuel system or electrical system. Never risk spilling fuel on a hot engine or exhaust component. It is strongly recommended that a fire extinguisher suitable for use on fuel and electrical fires be kept handy in the garage or workshop at all times. Never try to extinguish a fuel or electrical fire with water.

Fumes

Certain fumes are highly toxic and can quickly cause unconsciousness and even death if inhaled to any extent. Gasoline vapor falls into this category, as do the vapors from some cleaning solvents. Any draining or pouring of such volatile fluids should be done in a well ventilated area.

When using cleaning fluids and solvents, read the instructions on the container carefully. Never use materials from unmarked containers.

Never run the engine in an enclosed space, such as a garage. Exhaust fumes contain carbon monoxide, which is extremely poisonous. If you need to run the engine, always do so in the open air, or at least have the rear of the vehicle outside the work area.

If you are fortunate enough to have the use of an inspection pit, never drain or pour gasoline and never run the engine while the vehicle is over the pit. The fumes, being heavier than air, will concentrate in the pit with possibly lethal results.

The battery

Never create a spark or allow a bare light bulb near a battery. They normally give off a certain amount of hydrogen gas, which is highly explosive.

Always disconnect the battery ground (-) cable at the battery before working on the fuel or electrical systems.

If possible, loosen the filler caps or cover when charging the battery from an external source (this does not apply to sealed or maintenance-free batteries). Do not charge at an excessive rate or the battery may burst.

Take care when adding water to a non maintenance-free battery and when carrying a battery. The electrolyte, even when diluted, is very corrosive and should not be allowed to contact clothing or skin.

Always wear eye protection when cleaning the battery to prevent the caustic deposits from entering your eyes.

Household current

When using an electric power tool, inspection light, etc., which operates on household current, always make sure that the tool is correctly connected to its plug and that, where necessary, it is properly grounded. Do not use such items in damp conditions and, again, do not create a spark or apply excessive heat in the vicinity of fuel or fuel vapor.

Secondary ignition system voltage

A severe electric shock can result from touching certain parts of the ignition system (such as the spark plug wires) when the engine is running or being cranked, particularly if components are damp or the insulation is defective. In the case of an electronic ignition system, the secondary system voltage is much higher and could prove fatal.

Troubleshooting

Contents

Symptom	Section
Engine	
Engine backfires	13
Engine diesels (continues to run) after switching off	15
Engine hard to start when cold	4
Engine hard to start when hot	5
Engine lacks power	12
Engine lopes while idling or idles erratically	8
Engine misses at idle speed	9
Engine misses throughout driving speed range	10
Engine rotates but will not start	2
Engine stalls	11
Engine starts but stops immediately	7
Engine will not rotate when attempting to start	1
Pinging or knocking engine sounds during acceleration or uphill	14
Starter motor noisy or excessively rough in engagement	6
Starter motor operates without rotating engine	3
Engine electrical system	
Battery light fails to come on when key is turned on	18
Battery will not hold a charge	16
Ignition light fails to go out	17
Fuel system	
Excessive fuel consumption	19
Fuel leakage and/or fuel odor	20
Cooling system	
Coolant loss	25
External coolant leakage	23
Internal coolant leakage	24
Overcooling	22
Overheating	21
Poor coolant circulation	26
Clutch	
Clutch pedal stays on floor when disengaged	32
Clutch slips (engine speed increases with no increase in vehicle speed)	28
Fails to release (pedal pressed to the floor - shift lever does not move freely in and out of Reverse)	27
Grabbing (chattering) as clutch is engaged	29
Squeal or rumble with clutch fully disengaged (pedal depressed)	31
Squeal or rumble with clutch fully engaged (pedal released)	30
Manual transmission	
Difficulty in engaging gears	37
Noisy in all gears	34
Noisy in Neutral with engine running	33

Symptom	Section
Noisy in one particular gear	35
Oil leakage	38
Slips out of high gear	36
Automatic transmission	
Fluid leakage	42
General shift mechanism problems	39
Transmission slips, shifts rough, is noisy or has no drive in forward or reverse gears	41
Transmission will not downshift with accelerator pedal pressed to the floor	40
Transfer case	
Lubricant leaks from the vent or output shaft seals	46
Noisy or jumps out of four-wheel drive Low range	45
Transfer case is difficult to shift into the desired range	43
Transfer case noisy in all gears	44
Driveshaft	
Knock or clunk when the transmission is under initial load (just after transmission is put into gear)	48
Metallic grinding sound consistent with vehicle speed	49
Oil leak at front of driveshaft	47
Vibration	50
Axles	
Noise	51
Oil leakage	53
Vibration	52
Brakes	
Brake pedal feels spongy when depressed	57
Brake pedal pulsates during brake application	60
Excessive brake pedal travel	56
Excessive effort required to stop vehicle	58
Noise (high-pitched squeal with the brakes applied)	55
Pedal travels to the floor with little resistance	59
Vehicle pulls to one side during braking	54
Suspension and steering systems	
Excessive pitching and/or rolling around corners or during braking	63
Excessive play in steering	65
Excessive tire wear (not specific to one area)	67
Excessive tire wear on inside edge	69
Excessive tire wear on outside edge	68
Excessively stiff steering	64
Lack of power assistance	66
Shimmy, shake or vibration	62
Tire tread worn in one place	70
Vehicle pulls to one side	61

This section provides an easy reference guide to the more common problems which may occur during the operation of your vehicle. These problems and possible causes are grouped under various components or systems; i.e. Engine, Cooling System, etc., and also refer to the Chapter and/or Section which deals with the problem.

Remember that successful troubleshooting is not a mysterious black art practiced only by professional mechanics. It's simply the result of a bit of knowledge combined with an intelligent, systematic approach to the problem. Always work by a process of elimination, starting with the simplest solution and working through to the most complex - and never overlook the obvious. Anyone can forget to fill the gas tank or leave the lights on overnight, so don't assume that you are above such oversights.

Finally, always get clear in your mind why a problem has occurred and take steps to ensure that it doesn't happen again. If the electrical system fails because of a poor connection, check all other connections in the system to make sure that they don't fail as well. If a particular fuse continues to blow, find out why - don't just go on replacing fuses. Remember, failure of a small component can often be indicative of potential failure or incorrect functioning of a more important component or system.

0-18　Troubleshooting

Engine

1　Engine will not rotate when attempting to start

1   Battery terminal connections loose or corroded. Check the cable terminals at the battery. Tighten the cable or remove corrosion as necessary.
2   Battery discharged or faulty. If the cable connections are clean and tight on the battery posts, turn the key to the On position and switch on the headlights and/or windshield wipers. If they fail to function, the battery is discharged.
3   Automatic transmission not completely engaged in Park or Neutral or clutch pedal not completely depressed.
4   Broken, loose or disconnected wiring in the starting circuit. Inspect all wiring and connectors at the battery, starter solenoid and ignition switch.
5   Starter motor pinion jammed in flywheel ring gear. If manual transmission, place transmission in gear and rock the vehicle to manually turn the engine. Remove starter and inspect pinion and flywheel at earliest convenience (Chapter 5).
6   Starter solenoid faulty (Chapter 5).
7   Starter motor faulty (Chapter 5).
8   Ignition switch faulty (Chapter 12).

2　Engine rotates but will not start

1   Fuel tank empty.
2   Fault in the fuel injection system (Chapter 4).
3   Battery discharged (engine rotates slowly). Check the operation of electrical components as described in the previous Section.
4   Battery terminal connections loose or corroded (see previous Section).
5   Fuel pump faulty (Chapter 4).
6   Excessive moisture on, or damage to, ignition components (see Chapter 5).
7   Worn, faulty or incorrectly gapped spark plugs (Chapter 1).
8   Broken, loose or disconnected wiring in the starting circuit (see previous Section).
9   Broken, loose or disconnected wires at the distributor (V6) or ignition coil pack (four-cylinder) (Chapters 1 and 5).
10 Distributor cap wet, cracked or carbon tracked (V6) (Chapter 1)
11 Ignition coil or coil pack faulty (Chapter 5).
12 Engine mechanical failure.

3　Starter motor operates without rotating engine

1   Starter pinion sticking. Remove the starter (Chapter 5) and inspect.
2   Starter pinion or flywheel teeth worn or broken. Remove the flywheel/driveplate access cover and inspect.

4　Engine hard to start when cold

1   Battery discharged or low. Check as described in Section 1.
2   Fault in the fuel or electrical systems (Chapters 4 and 5).

5　Engine hard to start when hot

1   Air filter clogged (Chapter 1).
2   Fault in the fuel or electrical systems (Chapters 4 and 5).
3   Fuel not reaching the injectors (see Chapter 4).

6　Starter motor noisy or excessively rough in engagement

1   Pinion or flywheel gear teeth worn or broken. Remove the cover at the rear of the engine (if equipped) and inspect.
2   Starter motor mounting bolts loose or missing.

7　Engine starts but stops immediately

1   Loose or faulty electrical connections at distributor, coil or alternator.
2   Fault in the fuel or electrical systems (Chapters 4 and 5).
3   Vacuum leak at the gasket surfaces of the intake manifold or throttle body. Make sure all mounting bolts/nuts are tightened securely and all vacuum hoses connected to the manifold are positioned properly and in good condition.

8　Engine lopes while idling or idles erratically

1   Vacuum leakage. Check the mounting bolts/nuts at the throttle body and intake manifold for tightness. Make sure all vacuum hoses are connected and in good condition. Use a stethoscope or a length of fuel hose held against your ear to listen for vacuum leaks while the engine is running. A hissing sound will be heard. A soapy water solution will also detect leaks.
2   Fault in the fuel or electrical systems (Chapters 4 and 5).
3   Plugged PCV valve or hose (see Chapters 1 and 6).
4   Air filter clogged (Chapter 1).
5   Fuel pump not delivering sufficient fuel to the fuel injectors (see Chapter 4).
6   Leaking head gasket. Perform a compression check (Chapter 2).
7   Camshaft lobes worn (Chapter 2).

9　Engine misses at idle speed

1   Spark plugs worn, fouled or not gapped properly (Chapter 1).
2   Fault in the fuel or electrical systems (Chapters 4 and 5).
3   Faulty spark plug wires (Chapter 1).
4   Vacuum leaks at intake or hose connections. Check as described in Section 8.
5   Uneven or low cylinder compression. Check compression as described in Chapter 1.

10　Engine misses throughout driving speed range

1   Fuel filter clogged and/or impurities in the fuel system (Chapter 1).
2   Faulty or incorrectly gapped spark plugs (Chapter 1).
3   Fault in the fuel or electrical systems (Chapters 4 and 5).
4   Defective spark plug wires (Chapter 1).
5   Faulty emissions system components (Chapter 6).
6   Low or uneven cylinder compression pressures. Remove the spark plugs and test the compression with a gauge (Chapter 2).
7   Weak or faulty ignition system (Chapter 5).
8   Vacuum leaks at the throttle body, intake manifold or vacuum hoses (see Section 8).

11　Engine stalls

1   Idle speed incorrect. Refer to the VECI label.
2   Fuel filter clogged and/or water and impurities in the fuel system (Chapter 1).
3   Fault in the fuel system or sensors (Chapters 4 and 6).
4   Faulty emissions system components (Chapter 6).
5   Faulty or incorrectly gapped spark plugs (Chapter 1). Also check the spark plug wires (Chapter 1).
6   Vacuum leak at the throttle body, intake manifold or vacuum hoses. Check as described in Section 8.

12　Engine lacks power

1   Fault in the fuel or electrical systems (Chapters 4 and 5).
2   Faulty or incorrectly gapped spark plugs (Chapter 1).
3   Faulty coil (Chapter 5).
4   Brakes binding (Chapter 1).
5   Automatic transmission fluid level incorrect (Chapter 1).
6   Clutch slipping (Chapter 8).
7   Fuel filter clogged and/or impurities in the fuel system (Chapter 1).
8   Emissions control system not functioning properly (Chapter 6).
9   Use of substandard fuel. Fill the tank with the proper octane fuel.
10 Low or uneven cylinder compression pressures. Test with a compression tester, which will detect leaking valves and/or a blown head gasket (Chapter 2).

Troubleshooting 0-19

13 Engine backfires

1 Emissions system not functioning properly (Chapter 6).
2 Fault in the fuel or electrical systems (Chapters 4 and 5).
3 Faulty secondary ignition system (cracked spark plug insulator or faulty plug wires) (Chapters 1 and 5).
4 Vacuum leak at the throttle body, intake manifold or vacuum hoses. Check as described in Section 8.
5 Valves sticking (Chapter 2).
6 Crossed plug wires (Chapter 1).

14 Pinging or knocking engine sounds during acceleration or uphill

1 Incorrect grade of fuel. Fill the tank with fuel of the proper octane rating.
2 Fault in the fuel or electrical systems (Chapters 4 and 5).
3 Improper spark plugs. Check the plug type against the VECI label located in the engine compartment. Also check the plugs and wires for damage (Chapter 1).
4 Faulty emissions system (Chapter 6).
5 Vacuum leak. Check as described in Section 9.

15 Engine diesels (continues to run) after switching off

Fault in the fuel or electrical systems (Chapters 4 and 5).

Engine electrical system

16 Battery will not hold a charge

1 Alternator drivebelt defective or not adjusted properly (Chapter 1).
2 Electrolyte level low or battery discharged (Chapter 1).
3 Battery terminals loose or corroded (Chapter 1).
4 Alternator not charging properly (Chapter 5).
5 Loose, broken or faulty wiring in the charging circuit (Chapter 5).
6 Short in the vehicle wiring causing a continuous drain on the battery (refer to Chapter 12 and the Wiring Diagrams).
7 Battery defective internally.

17 Ignition light fails to go out

1 Fault in the alternator or charging circuit (Chapter 5).
2 Alternator drivebelt defective or not properly adjusted (Chapter 1).

18 Battery light fails to come on when key is turned on

1 Instrument cluster warning light bulb defective (Chapter 12).
2 Alternator faulty (Chapter 5).
3 Fault in the instrument cluster printed circuit, dashboard wiring or bulb holder (Chapter 12).

Fuel system

19 Excessive fuel consumption

1 Dirty or clogged air filter element (Chapter 1).
2 Emissions system not functioning properly (Chapter 6).
3 Fault in the fuel or electrical systems (Chapters 4 and 5).
4 Low tire pressure or incorrect tire size (Chapter 1).

20 Fuel leakage and/or fuel odor

1 Leak in a fuel feed or vent line (Chapter 4).
2 Tank overfilled. Fill only to automatic shut-off.
3 Evaporative emissions system canister clogged (Chapter 6).
4 Vapor leaks from system lines (Chapter 4).
5 Fault in the fuel system (Chapter 4).

Cooling system

21 Overheating

1 Insufficient coolant in the system (Chapter 1).
2 Water pump drivebelt defective (Chapter 1).
3 Radiator core blocked or radiator grille dirty and restricted (see Chapter 3).
4 Thermostat faulty (Chapter 3).
5 Fan blades broken or cracked (Chapter 3).
6 Radiator cap not maintaining proper pressure. Have the cap pressure tested by a service station or repair shop.
7 Also see causes for "poor coolant circulation" in Section 26.

22 Overcooling

1 Thermostat faulty (Chapter 3).
2 Inaccurate temperature gauge (Chapter 12).

23 External coolant leakage

1 Deteriorated or damaged hoses or loose clamps. Replace hoses and/or tighten the clamps at the hose connections (Chapter 1).
2 Water pump seals defective. If this is the case, water will drip from the weep hole in the water pump body (Chapter 3).
3 Leakage from the radiator core or side tank(s). This will require the radiator to be professionally repaired (see Chapter 3 for removal procedures).
4 Engine drain plug leaking (Chapter 1) or water jacket core plugs leaking (see Chapter 2).

24 Internal coolant leakage

Note: *Internal coolant leaks can usually be detected by examining the oil. Check the dipstick and inside of the valve cover for water deposits and an oil consistency like that of a milkshake.*

1 Leaking cylinder head gasket. Have the cooling system pressure tested.
2 Cracked cylinder bore or cylinder head. Dismantle the engine and inspect (Chapter 2).
3 Leaking intake manifold gasket (V6) (Chapter 2).

25 Coolant loss

1 Too much coolant in the system (Chapter 1).
2 Coolant boiling away due to overheating (see Section 15).
3 External or internal leakage (see Sections 23 and 24).
4 Faulty radiator cap. Have the cap pressure tested.

26 Poor coolant circulation

1 Thermostat sticking (Chapter 3).
2 Restriction in the cooling system. Drain, flush and refill the system (Chapter 1). If necessary, remove the radiator (Chapter 3) and have it reverse flushed.
3 Water pump drivebelt defective or not adjusted properly (Chapter 1).
4 Inoperative water pump (Chapter 3). A good water pump should circulate water well enough to maintain a continuous flow of water to the heater core - with the heater on, and the engine at normal operating temperature, feel the inlet and outlet hoses to the heater core, which both should be hot to the touch.

Clutch

27 Fails to release (pedal pressed to the floor - shift lever does not move freely in and out of Reverse)

1 Leak in the clutch hydraulic system. Check the master cylinder, slave cylinder and lines (Chapter 8).
2 Clutch plate warped or damaged (Chapter 8).

Troubleshooting

28 Clutch slips (engine speed increases with no increase in vehicle speed)

1 Clutch plate oil soaked or lining worn. Remove clutch (Chapter 8) and inspect.
2 Clutch plate not seated. It may take 30 or 40 normal starts for a new one to seat.
3 Pressure plate worn (Chapter 8).

29 Grabbing (chattering) as clutch is engaged

1 Oil on clutch plate lining. Remove (Chapter 8) and inspect. Correct any leakage source.
2 Worn or loose engine or transmission mounts. These units move slightly when the clutch is released. Inspect the mounts and bolts (Chapter 2).
3 Worn splines on clutch plate hub. Remove the clutch components (Chapter 8) and inspect.
4 Warped pressure plate or flywheel. Remove the clutch components and inspect.

30 Squeal or rumble with clutch fully engaged (pedal released)

1 Release bearing binding on transmission bearing retainer. Remove clutch components (Chapter 8) and check bearing. Remove any burrs or nicks; clean and relubricate bearing retainer before installing.

31 Squeal or rumble with clutch fully disengaged (pedal depressed)

1 Worn, defective or broken release bearing (Chapter 8).
2 Worn or broken pressure plate springs (or diaphragm fingers) (Chapter 8).

32 Clutch pedal stays on floor when disengaged

1 Linkage or release bearing binding. Inspect the linkage or remove the clutch components as necessary.
2 Make sure proper pedal stop (bumper) is installed.

Manual transmission

Note: *All the following references are in Chapter 7, unless noted.*

33 Noisy in Neutral with engine running

1 Input shaft bearing worn.
2 Damaged main drive gear bearing.
3 Worn countershaft bearings.
4 Worn or damaged countershaft endplay shims.

34 Noisy in all gears

1 Any of the above causes, and/or:
2 Insufficient lubricant (see the checking procedures in Chapter 1).

35 Noisy in one particular gear

1 Worn, damaged or chipped gear teeth for that particular gear.
2 Worn or damaged synchronizer for that particular gear.

36 Slips out of high gear

1 Transmission loose on clutch housing.
2 Dirt between the transmission case and engine or misalignment of the transmission (Chapter 7).

37 Difficulty in engaging gears

1 Clutch not releasing completely (see clutch adjustment in Chapter 1).
2 Loose, damaged or out-of-adjustment shift linkage. Make a thorough inspection, replacing parts as necessary (Chapter 7).

38 Oil leakage

1 Excessive amount of lubricant in the transmission (see Chapter 1 for correct checking procedures). Drain lubricant as required.
2 Transmission oil seal or speedometer oil seal in need of replacement (Chapter 7).

Automatic transmission

Note: *Due to the complexity of the automatic transmission, it's difficult for the home mechanic to properly diagnose and service this component. For problems other than the following, the vehicle should be taken to a dealer service department or a transmission shop.*

39 General shift mechanism problems

1 Chapter 7 deals with checking and adjusting the shift linkage on automatic transmissions. Common problems which may be attributed to poorly adjusted linkage are:
 a) *Engine starting in gears other than Park or Neutral.*
 b) *Indicator on shifter pointing to a gear other than the one actually being selected.*
 c) *Vehicle moves when in Park.*
2 Refer to Chapter 7 to adjust the linkage.

40 Transmission will not downshift with accelerator pedal pressed to the floor

Since these transmissions are electronically controlled, your dealer or a professional shop with the proper equipment will have to diagnose the probable cause.

41 Transmission slips, shifts rough, is noisy or has no drive in forward or reverse gears

1 There are many probable causes for the above problems, but the home mechanic should be concerned with only one possibility - fluid level.
2 Before taking the vehicle to a repair shop, check the level and condition of the fluid as described in Chapter 1. Correct fluid level as necessary or change the fluid and filter if needed. If the problem persists, have a professional diagnose the probable cause.
3 If the transmission shifts late and the shifts are harsh, suspect a faulty transmission pressure control solenoid valve. Check for Diagnostic Trouble Codes (Chapter 6).

42 Fluid leakage

1 Automatic transmission fluid is a deep red color. Fluid leaks should not be confused with engine oil, which can easily be blown by air flow to the transmission.
2 To pinpoint a leak, first remove all built-up dirt and grime from around the transmission. Degreasing agents and/or steam cleaning will achieve this. With the underside clean, drive the vehicle at low speeds so air flow will not blow the leak far from its source. Raise the vehicle and determine where the leak is coming from. Common areas of leakage are:
 a) **Pan:** *Tighten the mounting bolts and/or replace the pan gasket as necessary (see Chapter 7).*
 b) **Filler pipe:** *Replace the rubber seal where the pipe enters the transmission case.*
 c) **Transmission oil lines:** *Tighten the connectors where the lines enter the transmission case and/or replace the lines.*
 d) **Vent pipe:** *Transmission overfilled and/or water in fluid (see checking procedures, Chapter 1).*
 e) **Speedometer connector:** *Replace the O-ring where the speedometer sensor enters the transmission case (Chapter 7).*

Transfer case

43 Transfer case is difficult to shift into the desired range

1 Speed may be too great to permit

Troubleshooting 0-21

engagement. Stop the vehicle and shift into the desired range.
2   Shift linkage loose, bent or binding. Check the linkage for damage or wear and replace or lubricate as necessary (Chapter 7).
3   If the vehicle has been driven on a paved surface for some time, the driveline torque can make shifting difficult. Stop and shift into two-wheel drive on paved or hard surfaces.
4   Insufficient or incorrect grade of lubricant. Drain and refill the transfer case with the specified lubricant. (Chapter 1).
5   Worn or damaged internal components. Disassembly and overhaul of the transfer case may be necessary (Chapter 7).

44   Transfer case noisy in all gears

Insufficient or incorrect grade of lubricant. Drain and refill (Chapter 1).

45   Noisy or jumps out of four-wheel drive Low range

1   Transfer case not fully engaged. Stop the vehicle, shift into Neutral and then engage 4L.
2   Shift linkage loose, worn or binding. Tighten, repair or lubricate linkage as necessary.
3   Shift fork cracked, inserts worn or fork binding on the rail. Disassemble and repair as necessary (Chapter 7).

46   Lubricant leaks from the vent or output shaft seals

1   Transfer case is overfilled. Drain to the proper level (Chapter 1).
2   Vent is clogged or jammed closed. Clear or replace the vent.
3   Output shaft seal incorrectly installed or damaged. Replace the seal and check contact surfaces for nicks and scoring.

Driveshaft

47   Oil leak at seal end of driveshaft

Defective transmission or transfer case oil seal. See Chapter 7 for replacement procedures. While this is done, check the splined yoke for burrs or a rough condition which may be damaging the seal. Burrs can be removed with crocus cloth or a fine whetstone.

48   Knock or clunk when the transmission is under initial load (just after transmission is put into gear)

1   Loose or disconnected rear suspension components. Check all mounting bolts, nuts and bushings (see Chapter 10).
2   Loose driveshaft bolts. Inspect all bolts and nuts and tighten them to the specified torque.
3   Worn or damaged universal joint bearings. Check for wear (see Chapter 8).

49   Metallic grinding sound consistent with vehicle speed.

Pronounced wear in the universal joint bearings. Check as described in Chapter 8.

50   Vibration

Note: *Before assuming that the driveshaft is at fault, make sure the tires are perfectly balanced and perform the following test.*
1   Install a tachometer inside the vehicle to monitor engine speed as the vehicle is driven. Drive the vehicle and note the engine speed at which the vibration (roughness) is most pronounced. Now shift the transmission to a different gear and bring the engine speed to the same point.
2   If the vibration occurs at the same engine speed (rpm) regardless of which gear the transmission is in, the driveshaft is NOT at fault since the driveshaft speed varies.
3   If the vibration decreases or is eliminated when the transmission is in a different gear at the same engine speed, refer to the following probable causes.
4   Bent or dented driveshaft. Inspect and replace as necessary (see Chapter 8).
5   Undercoating or built-up dirt, etc. on the driveshaft. Clean the shaft thoroughly and recheck.
6   Worn universal joint bearings. Remove and inspect (see Chapter 8).
7   Driveshaft and/or companion flange out of balance. Check for missing weights on the shaft. Remove the driveshaft (see Chapter 8) and reinstall 180-degrees from original position, then retest. Have the driveshaft professionally balanced if the problem persists.

Axles

51   Noise

1   Road noise. No corrective procedures available.
2   Tire noise. Inspect tires and check tire pressures (Chapter 1).
3   Rear wheel bearings loose, worn or damaged (Chapter 8).

52   Vibration

See probable causes under Driveshaft. Proceed under the guidelines listed for the driveshaft. If the problem persists, check the rear wheel bearings by raising the rear of the vehicle and spinning the rear wheels by hand. Listen for evidence of rough (noisy) bearings. Remove and inspect (see Chapter 8).

53   Oil leakage

1   Pinion seal damaged (see Chapter 8).
2   Axleshaft oil seals damaged (see Chapter 8).
3   Differential inspection cover leaking. Tighten the bolts or replace the gasket as required (see Chapters 1 and 8).

Brakes

Note: *Before assuming that a brake problem exists, make sure that the tires are in good condition and inflated properly (see Chapter 1), that the front end alignment is correct and that the vehicle is not loaded with weight in an unequal manner.*

54   Vehicle pulls to one side during braking

1   Defective, damaged or oil contaminated disc brake pads or shoes on one side. Inspect as described in Chapter 9.
2   Excessive wear of brake shoe or pad material or drum/disc on one side. Inspect and correct as necessary.
3   Loose or disconnected front suspension components. Inspect and tighten all bolts to the specified torque (Chapter 10).
4   Defective drum brake or caliper assembly. Remove the drum or caliper and inspect for a stuck piston or other damage (Chapter 9).
5   Inadequate lubrication of front brake caliper slide rails. Remove caliper and lubricate slide rails (Chapter 9).

55   Noise (high-pitched squeal with the brakes applied)

1   Disc brake pads worn out. The noise comes from the wear sensor rubbing against the disc (does not apply to all vehicles) or the actual pad backing plate itself if the material is completely worn away. Replace the pads with new ones immediately (Chapter 9). If the pad material has worn completely away, the brake discs should be inspected for damage as described in Chapter 9.
2   Missing or damaged brake pad insulators (disc brakes). Replace pad insulators (see Chapter 9).
3   Linings contaminated with dirt or grease. Replace pads or shoes.
4   Incorrect linings. Replace with correct linings.

56   Excessive brake pedal travel

1   Partial brake system failure. Inspect the

0-22 Troubleshooting

entire system (Chapter 9) and correct as required.
2 Insufficient fluid in the master cylinder. Check (Chapter 1), add fluid and bleed the system if necessary (Chapter 9).
3 Rear brakes not adjusting properly. Make a series of starts and stops while the vehicle is in Reverse. If this does not correct the situation, remove the drums and inspect the self-adjusters (Chapter 9).
4 Problem with the anti-lock brake system (Chapter 9).

57 Brake pedal feels spongy when depressed

1 Air in the hydraulic lines. Bleed the brake system (Chapter 9).
2 Faulty flexible hoses. Inspect all system hoses and lines. Replace parts as necessary.
3 Master cylinder mounting bolts/nuts loose.
4 Master cylinder defective (Chapter 9).
5 Problem with the anti-lock brake system (Chapter 9).

58 Excessive effort required to stop vehicle

1 Power brake booster not operating properly (Chapter 9).
2 Excessively worn linings or pads. Inspect and replace if necessary (Chapter 9).
3 One or more caliper pistons or wheel cylinders seized or sticking. Inspect and rebuild as required (Chapter 9).
4 Brake linings or pads contaminated with oil or grease. Inspect and replace as required (Chapter 9).
5 New pads or shoes installed and not yet seated. It will take a while for the new material to seat against the drum (or rotor).
6 Problem with the anti-lock brake system (Chapter 9).

59 Pedal travels to the floor with little resistance

1 Little or no fluid in the master cylinder reservoir caused by leaking wheel cylinder(s), leaking caliper piston(s), loose, damaged or disconnected brake lines. Inspect the entire system and correct as necessary.
2 Worn master cylinder seals (Chapter 9).
3 Problem with the anti-lock brake system (Chapter 9).

60 Brake pedal pulsates during brake application

1 Caliper improperly installed. Remove and inspect (Chapter 9).
2 Disc or drum defective. Remove (Chapter 9) and check for excessive lateral runout and parallelism. Have the disc or drum resurfaced or replace it with a new one.

Suspension and steering systems

61 Vehicle pulls to one side

1 Tire pressures uneven (Chapter 1).
2 Defective tire (Chapter 1).
3 Excessive wear in suspension or steering components (Chapter 10).
4 Front end in need of alignment.
5 Front brakes dragging. Inspect the brakes as described in Chapter 9.

62 Shimmy, shake or vibration

1 Tire or wheel out-of-balance or out-of-round. Have professionally balanced.
2 Loose, worn or out-of-adjustment front wheel bearings (Chapter 1).
3 Shock absorbers and/or suspension components worn or damaged (Chapter 10).

63 Excessive pitching and/or rolling around corners or during braking

1 Defective shock absorbers. Replace as a set (Chapter 10).
2 Broken or weak springs and/or suspension components. Inspect as described in Chapter 10.

64 Excessively stiff steering

1 Lack of fluid in power steering fluid reservoir (Chapter 1).
2 Incorrect tire pressures (Chapter 1).
3 Lack of lubrication at steering joints (see Chapter 1).
4 Front end out of alignment.
5 Lack of power assistance (see Section 62).

65 Excessive play in steering

1 Loose front wheel bearings (Chapters 1 and 10).
2 Excessive wear in suspension or steering components (Chapter 10).
3 Steering gearbox damaged or out of adjustment (Chapter 10).

66 Lack of power assistance

1 Steering pump drivebelt faulty or not adjusted properly (Chapter 1).
2 Fluid level low (Chapter 1).
3 Hoses or lines restricted. Inspect and replace parts as necessary.
4 Air in power steering system. Bleed the system (Chapter 10).

67 Excessive tire wear (not specific to one area)

1 Incorrect tire pressures (Chapter 1).
2 Tires out-of-balance. Have professionally balanced.
3 Wheels damaged. Inspect and replace as necessary.
4 Suspension or steering components excessively worn (Chapter 10).

68 Excessive tire wear on outside edge

1 Inflation pressures incorrect (Chapter 1).
2 Excessive speed in turns.
3 Front end alignment incorrect (excessive toe-in). Have professionally aligned.
4 Suspension arm bent or twisted (Chapter 10).

69 Excessive tire wear on inside edge

1 Inflation pressures incorrect (Chapter 1).
2 Front end alignment incorrect (toe-out). Have professionally aligned.
3 Loose or damaged steering components (Chapter 10).

70 Tire tread worn in one place

1 Tires out-of-balance.
2 Damaged or buckled wheel. Inspect and replace if necessary.
3 Defective tire (Chapter 1).

Chapter 1 Tune-up and routine maintenance

Contents

	Section
Air filter replacement	26
Automatic transmission fluid and filter change	30
Automatic transmission fluid level check	6
Battery check, maintenance and charging	9
Brake and Transmission Shift Interlock (BTSI) system check (1996 and later models with automatic transmission)	18
Brake check	23
Chassis lubrication	13
Cooling system check	10
Cooling system servicing (draining, flushing and refilling)	34
Cylinder compression check	See Chapter 2A
Differential lubricant change	32
Differential lubricant level check	21
Driveaxle boot check (4WD models)	24
Drivebelt check and replacement	28
Engine oil and filter change	8
Evaporative emissions control system check	36
Exhaust system check	15
Fluid level checks	4
Front wheel bearing check, repack and adjustment (2WD models)	29
Fuel filter replacement	27
Fuel system check	25

	Section
Ignition timing check and adjustment (1994 and 1995 VIN W and Z V6 only)	39
Introduction	1
Maintenance schedule	2
Manual transmission lubricant change	31
Manual transmission lubricant level check	20
Positive Crankcase Ventilation (PCV) valve check and replacement	35
Power steering fluid level check	7
Seat belt and restraint system check	16
SERVICE ENGINE SOON light	See Chapter 6
Spark plug replacement	38
Spark plug wire, distributor cap and rotor check and replacement	37
Starter safety switch check	17
Suspension and steering check	14
Tire and tire pressure checks	5
Tire rotation (every 7500 miles or 6 months)	19
Transfer case lubricant change (4WD models)	33
Transfer case lubricant level check	22
Tune-up general information	3
Underhood hose check and replacement	11
Wiper blade inspection and replacement	12

Specifications

General

Radiator cap pressure rating	15 psi
Brake pad wear limit	1/8 inch

Recommended lubricants and fluids

Note: *Listed here are manufacturer recommendations at the time this manual was written. Manufacturers occasionally upgrade their fluid and lubricant specifications, so check with your local auto parts store for current recommendations.*

Engine oil	
Type	API "certified for gasoline engines"
Viscosity	See accompanying chart
Manual transmission lubricant	
Borg Warner T-5	Dexron III® Automatic Transmission Fluid
New Venture NV3500	GM 2345349 Synchromesh transmission lubricant
New Venture NV1500	GM 9985648 Synchromesh transmission lubricant and 12380672 friction modifier
Automatic transmission fluid	Dexron III® Automatic Transmission Fluid
Transfer case lubricant	Dexron III® Automatic Transmission Fluid or GM Auto-Trak II fluid
Engine coolant	50/50 mixture of water and the specified ethylene glycol-based (green color) antifreeze or DEX-COOL® silicate-free (orange-color) coolant - DO NOT mix the two types (refer to Sections 4, 10 and 34)
Brake and clutch fluid	DOT 3 fluid
Power steering fluid	GM power steering fluid or equivalent
Differential	80W-90 (GL-5) gear oil

Capacities*

Engine oil (with filter change)	4.5 qts
Fuel tank	20 gals
Cooling system	
2.2L four cylinder engine	
through 2000	11.5 qts
2001	9.9 qts

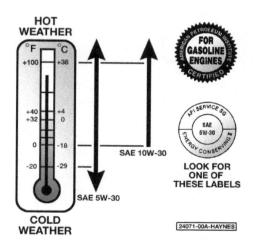

Engine oil viscosity chart - for best fuel economy and cold starting, select the lowest SAE viscosity grade for the expected temperature range

Capacities* (continued)

4.3L V6 engine
    through 2000 .... 11.9 qts
    2001 and later
        Automatic transmission... 13.8 qts
        Standard transmission .... 14.1 qts
Automatic transmission (fluid and filter replacement)............................. 5 qts
Manual transmission
    Borg Warner T-5 .... 2.9 qts
    New Venture NV3500... 2.2 qts
    New Venture NV1500... 2.9 qts
Transfer case
    Through 1998 .... 1.25 qts
    1999 and later
        NVG 133/136-NP4
            1999 and 2000 .... 2.4 qts
            2001 and later .... 2.0 qts
        NVG 233-NP1
            1999 and 2000 .... 2.3 qts
            2001 and later .... 1.1 qts
        NVG 236/246-NP8
            1999 and 2000 .... 2.4 qts
            2001 and later .... 2.0 qts
Differential
    Front .... 1.3 qts
    Rear .... 2.0 qts

*All capacities approximate. Add as necessary to bring to appropriate level.

Cylinder numbering and coil terminal location - four-cylinder engine

Ignition system

Spark plug type and gap*
    2.2L four-cylinder engine
        1994
            Type .... AC 41-908 or equivalent
            Gap .... 0.060 inch
        1995 thru 1997
            Type .... AC 41-928 or equivalent
            Gap .... 0.060 inch
        1998 and later
            Type .... AC 41-948 or equivalent
            Gap
                Through 1998 .... 0.060 inch
                1999 and later .... 0.040 inch
    4.3L V6 engine
        1994 and 1995
            Type .... AC CR43TSM or equivalent
            Gap .... 0.045 inch
        1996 and later
            Type .... AC 41-932 or equivalent
            Gap .... 0.060 inch
Firing order
    2.2L four-cylinder engine 1-3-4-2
    4.3L V6 engine 1-6-5-4-3-2
Base ignition timing (1994 and 1995 VIN W/Z V6 models only)............. 0-degrees

*The Vehicle Emission Control Label in the engine compartment, if different, supersedes this information

FIRING ORDER 1-6-5-4-3-2 with HEI ignition system

FIRING ORDER 1-6-5-4-3-2 with Enhanced Distributor Ignition (EDI) system

Cylinder numbering and distributor rotation - V6 engines

Torque specifications

	Ft-lbs
Spark plugs	
2.2L four-cylinder engine	13
4.3L V6 engine	11
Engine oil drain plug	15 to 20
Automatic transmission pan bolts	8 to 10
Wheel lug nuts	95

1 Introduction

This Chapter is designed to help the home mechanic maintain the Chevrolet S-10 and Blazer, GMC Sonoma and Jimmy, Oldsmobile Bravada and Isuzu Hombre models with the goals of maximum performance, economy, safety and reliability in mind.

Included is a master maintenance schedule, followed by procedures dealing specifically with each item on the schedule. Visual checks, adjustments, component replacement and other helpful items are included.

Refer to the accompanying illustrations of the engine compartment and the underside of the vehicle for the locations of various components.

Servicing your vehicle in accordance with the mileage/time maintenance schedule and the step-by-step procedures will result in a

Chapter 1 Tune-up and routine maintenance   1-3

Engine compartment components (four-cylinder engine)

1. PCV valve
2. Engine oil filler cap/dipstick
3. Brake fluid reservoir
4. Clutch fluid reservoir
5. Windshield washer fluid reservoir
6. Air cleaner housing
7. Radiator cap
8. Power steering fluid reservoir
9. Battery
10. Coolant reservoir

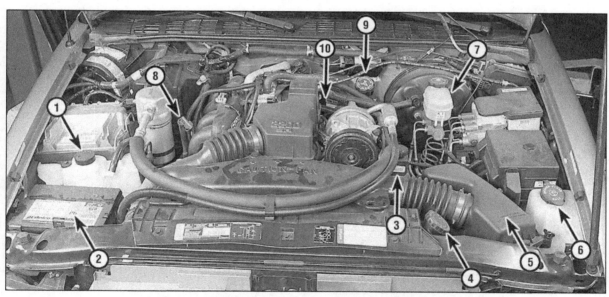

Engine compartment components (Vortec four-cylinder engine)

1. Coolant reservoir
2. Battery
3. Power steering fluid reservoir
4. Radiator cap
5. Air filter housing
6. Windshield washer reservoir
7. Brake master cylinder reservoir
8. Automatic transmission fluid dipstick location
9. Engine oil fill cap
10. PCV valve

planned maintenance program that should produce a long and reliable service life. Keep in mind that it's a comprehensive plan, so maintaining some items but not others at the specified intervals will not produce the same results.

As you service your vehicle, you'll discover that many of the procedures can - and should - be grouped together because of the nature of the particular procedure you're performing or because of the close proximity of two otherwise unrelated components.

For example, if the vehicle is raised, you should inspect the exhaust, suspension, steering and fuel systems while you're under the vehicle. When you're rotating the tires, it makes good sense to check the brakes since the wheels are already removed. Finally, let's suppose you have to borrow or rent a torque wrench. Even if you only need it to tighten the spark plugs, you might as well check the torque of as many critical fasteners as time allows.

The first step in this maintenance program is to prepare yourself before the actual work begins. Read through all the procedures you're planning to do, then gather up all the parts and tools needed. If it looks like you might run into problems during a particular job, seek advice from a mechanic or an experienced do-it-yourselfer.

1-4 Chapter 1 Tune-up and routine maintenance

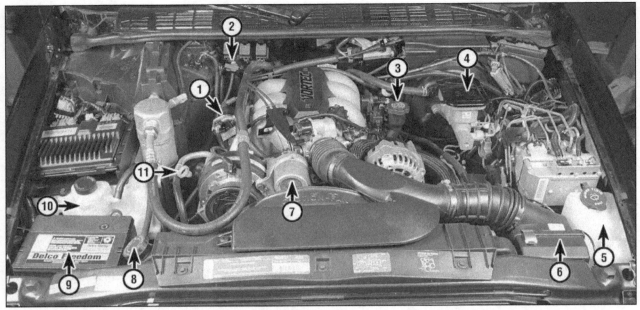

Typical engine compartment components (early V6 engine shown)

1. PCV valve
2. Automatic transmission fluid dipstick location
3. Engine oil filler cap
4. Brake fluid reservoir
5. Windshield washer fluid reservoir
6. Air cleaner housing
7. Serpentine drivebelt tensioner
8. Radiator cap
9. Battery
10. Coolant reservoir
11. Engine oil dipstick

Typical engine compartment underside components (2WD)

1. Steering grease fitting
2. Drivebelt
3. Spindle (wheel bearings)
4. Balljoint grease fitting
5. Fuel filter
6. Transmission
7. Engine oil pan drain plug
8. Exhaust system
9. Balljoint grease fitting
10. Shock absorber
11. Front disc brake

Chapter 1 Tune-up and routine maintenance

Typical engine compartment underside components (4WD)

1. Shock absorber
2. Lower radiator hose
3. Drivebelt
4. Front disc brake
5. Driveaxle boot
6. Fuel filter
7. Exhaust system
8. Transfer case
9. Front differential
10. Automatic transmission fluid pan
11. Engine oil pan

Typical rear underside components

1. Rear shock absorber
2. Exhaust system
3. Driveshaft rear universal joint
4. Fuel tank
5. Fuel filter
6. Rear drum brake

2 Maintenance schedule

The following maintenance intervals are based on the assumption that the vehicle owner will be doing the maintenance or service work, as opposed to having a dealer service department do the work. Although the time/mileage intervals are loosely based on factory recommendations, most have been shortened to ensure, for example, that such items as lubricants and fluids are checked/changed at intervals that promote maximum engine/driveline service life. Also, subject to the preference of the individual owner interested in keeping his or her vehicle in peak condition at all times, and with the vehicle's ultimate resale in mind, many of the maintenance procedures may be performed more often than recommended in the following schedule. We encourage such owner initiative.

When the vehicle is new it should be serviced initially by a factory authorized dealer service department to protect the factory warranty. In many cases the initial maintenance check is done at no cost to the owner (check with your dealer service department for more information).

Every 250 miles or weekly, whichever comes first

Check the engine oil level (Section 4)
Check the engine coolant level (Section 4)
Check the wiper/washer fluid level (Section 4)
Check the brake and clutch fluid levels (Section 4)
Check the tires and tire pressures (Section 5)

Every 3000 miles or 3 months, whichever comes first

All items listed above plus:
Check the automatic transmission fluid level (Section 6)
Check the power steering fluid level (Section 7)
Change the engine oil and filter (Section 8)
Check and service the battery (Section 9)

Every 7500 miles or 12 months, whichever comes first

All items listed above plus:
Check the cooling system (Section 10)
Inspect and replace, if necessary, all underhood hoses (Section 11)
Inspect and replace, if necessary, the wiper blades (Section 12)
Lubricate the chassis (Section 13)
Inspect the suspension and steering components (Section 14)
Inspect the exhaust system (Section 15)
Inspect the restraint system (Section 16)
Check the starter safety switch (Section 17)
Check the Brake and Transmission Shift Interlock (BTSI) system (Section 18)
Rotate the tires (Section 19)

Every 15,000 miles or 12 months, whichever comes first

Check the manual transmission lubricant level (Section 20)
Check the differential lubricant level (Section 21)
Check the transfer case lubricant level (Section 22)
Check the brakes (Section 23)*
Check the driveaxle CV joint boots (4WD models only) (Section 24)
Inspect the fuel system (Section 26)

Every 30,000 miles or 24 months, whichever comes first

All items listed above plus:
Replace the air filter (Section 26)
Replace the fuel filter (Section 27)
Check the engine drivebelt (Section 28)
Check and repack the front wheel bearings (2WD models only) (Section 29)**
Change the automatic transmission fluid and filter (Section 30)**
Change the manual transmission lubricant (Section 31)
Change the differential lubricant (Section 32)
Change the transfer case lubricant (Section 33)
Service the cooling system (drain, flush and refill) (green-colored ethylene glycol anti-freeze only) (Section 34)
Inspect and replace, if necessary, the PCV valve (Section 35)
Inspect the evaporative emissions control system (Section 36)
Inspect the spark plug wires, distributor cap and wires (Section 37)
Replace the spark plugs (conventional spark plugs) (Section 38)

Every 50,000 miles or 36 months, whichever comes first

Replace the spark plugs (platinum tipped spark plugs) (Section 38)

Every 60,000 miles or 48 months,

Check and adjust the ignition timing (1995 and earlier V6 VIN Z engines only) (Section 39)

Every 100,000 miles or 5 years, whichever comes first

Service the cooling system (drain, flush and refill) (orange-colored DEX-COOL® silicate-free coolant only) (Section 34)

If the vehicle frequently tows a trailer, is operated primarily in stop-and-go conditions or its brakes receive severe usage for any other reason, check the brakes every 3000 miles or three months.

**If operated under one or more of the following conditions, change the automatic transmission fluid and repack the front wheel bearings (2WD models) every 15,000 miles:*

In heavy city traffic where the outside temperature regularly reaches 90-degrees F (32-degrees C) or higher
In hilly or mountainous terrain
Frequent trailer pulling

Chapter 1 Tune-up and routine maintenance

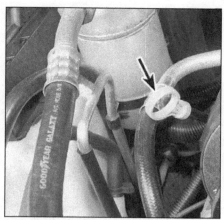

4.2 On V6 engines the engine oil dipstick (arrow) is mounted on the side of the engine - on four-cylinder engines, unscrew the oil filler cap and pull the dipstick out

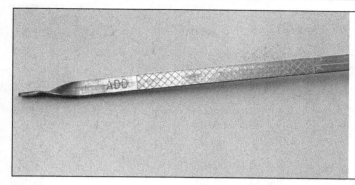

4.4 The oil level should be at or near the upper hole or cross-hatched area on the dipstick - if it's below the ADD line, add enough oil to bring the level into the upper hole or cross-hatched area

3 Tune-up general information

Caution: *On models equipped with an anti-theft audio system, be sure the lockout feature is turned off before performing any procedure which requires disconnecting the battery.*

The term tune-up is used in this manual to represent a combination of individual operations rather than one specific procedure. If, from the time the vehicle is new, the routine maintenance schedule is followed closely and frequent checks are made of fluid levels and high wear items, as suggested throughout this manual, the engine will be kept in relatively good running condition and the need for additional work due to lack of regular maintenance will be minimized. This is even more likely if a used vehicle, which has not received regular and frequent maintenance checks, is purchased. In such cases, an engine tune-up will be needed outside of the regular routine maintenance intervals.

The first step in any tune-up or diagnostic procedure to help correct a poor running engine is a cylinder compression check. A compression check (see Chapter 2, Part C) will help determine the condition of internal engine components and should be used as a guide for tune-up and repair procedures. If, for instance, a compression check indicates serious internal engine wear, a conventional tune-up won't improve the performance of the engine and would be a waste of time and money. Because of its importance, the compression check should be done by someone with the right equipment and the knowledge to use it properly.

The following procedures are those most often needed to bring a generally poor running engine back into a proper state of tune.

Minor tune-up

Check all engine related fluids (Section 4)
Clean, inspect and test the battery (Section 9)
Check the drivebelt and adjuster (Section 28)
Replace the spark plugs (Section 38)
Inspect the distributor cap, rotor and spark plug wires (Section 37)
Check the PCV valve (Section 35)
Check the air filter (Section 26)
Check the cooling system (Section 10)
Check all underhood hoses (Section 11)

Major tune-up

All items listed under Minor tune-up plus . . .
Check the EGR system (Chapter 6)
Check the ignition timing (Section 39)
Check the charging system (Chapter 5)
Check the fuel system (Section 25)
Replace the air filter (Section 26)
Replace the spark plug wires (Section 38)

4 Fluid level checks (every 250 miles or weekly)

Note: *The following are fluid level checks to be done on a 250 mile or weekly basis. Additional fluid level checks can be found in specific maintenance procedures which follow. Regardless of intervals, be alert to fluid leaks under the vehicle which would indicate a problem to be corrected immediately.*

1 Fluids are an essential part of the lubrication, cooling, brake and windshield washer systems. Because the fluids gradually become depleted and/or contaminated during normal operation of the vehicle, they must be periodically replenished. See *Recommended lubricants and fluids* at the beginning of this Chapter before adding fluid to any of the following components. **Note:** *The vehicle must be on level ground when fluid levels are checked.*

Engine oil

Refer to illustrations 4.2, 4.4 and 4.6

2 The engine oil level is checked with a dipstick **(see illustration)**. The dipstick extends through a metal tube down into the oil pan. On four-cylinder engines, the dipstick is part of the oil filler cap.
3 The oil level should be checked before the vehicle has been driven, or about 15 minutes after the engine has been shut off. If the oil is checked immediately after driving the vehicle, some of the oil will remain in the upper part of the engine, resulting in an inaccurate reading on the dipstick.
4 Pull the dipstick from the tube (V6) or unscrew the filler cap and withdraw the dipstick (four-cylinder) and wipe all the oil from the end with a clean rag or paper towel. Insert the clean dipstick all the way back into the tube and pull it out again. Note the oil at the end of the dipstick. Add oil as necessary to keep the level above the ADD mark in the cross hatched area of the dipstick **(see illustration)**.
5 Do not overfill the engine by adding too much oil since this may result in oil fouled spark plugs, oil leaks or oil seal failures.
6 Oil is added to the engine after removing a twist-off cap located on the engine **(see illustration)**. A funnel may help to reduce spills.
7 Checking the oil level is an important preventive maintenance step. A consistently low oil level indicates oil leakage through damaged seals, defective gaskets or past worn rings or valve guides. If the oil looks milky in color or has water droplets in it, the cylinder head gasket may be blown or the head or block may be cracked. The engine should be checked immediately. The condition of the oil should also be checked. Whenever you check the oil level, slide your thumb and index finger up the dipstick before wiping off the oil. If you see small dirt or metal particles clinging to the dipstick, the oil should be changed (see Section 8).

4.6 The engine oil filler cap is clearly marked and threads into the tube on the engine

4.9 The coolant reservoir is located in the right (passenger side) front corner of the engine compartment

4.14 Flip the windshield washer fluid cap up to add fluid

4.16 Remove the cell caps to check the water level in the battery - if the level is low, add distilled water only

Engine coolant

Refer to illustration 4.9

Warning: *Do not allow antifreeze to come in contact with your skin or painted surfaces of the vehicle. Flush contaminated areas immediately with plenty of water. Do not store new coolant or leave old coolant lying around where it's accessible to children or pets - they're attracted by its sweet smell. Ingestion of even a small amount of coolant can be fatal! Wipe up garage floor and drip pan coolant spills immediately. Keep antifreeze containers covered and repair leaks in the cooling system immediately.*

Caution: *Never mix green-colored ethylene glycol anti-freeze and orange-colored DEX-COOL® silicate-free coolant because doing so will destroy the efficiency of the DEX-COOL® coolant which is designed to last for 100,000 miles or five years.*

8   All vehicles covered by this manual are equipped with a pressurized coolant recovery system. A plastic coolant reservoir located at the front of the engine compartment is connected by a hose to the radiator filler neck. As the engine warms up and the coolant expands, it escapes through a valve in the radiator cap and travels through the hose into the reservoir. As the engine cools, the coolant is automatically drawn back into the cooling system to maintain the correct level.

9   The coolant level in the reservoir should be checked regularly. **Warning:** *Do not remove the radiator cap to check the coolant level when the engine is warm.* The level in the reservoir varies with the temperature of the engine. When the engine is cold, the coolant level should be at or slightly above the ADD mark on the reservoir **(see illustration)**. Once the engine has warmed up, the level should be at or near the FULL HOT mark. If it isn't, allow the engine to cool, then unscrew the cap from the reservoir and add a 50/50 mixture of ethylene glycol based green-colored antifreeze or orange-colored DEX-COOL® silicate-free coolant and water (see **Caution** above).

10   Drive the vehicle and recheck the coolant level. If only a small amount of coolant is required to bring the system up to the proper level, water can be used. However, repeated additions of water will dilute the antifreeze and water solution. In order to maintain the proper ratio of antifreeze and water, always top up the coolant level with the correct mixture. An empty plastic milk jug or bleach bottle makes an excellent container for mixing coolant. Do not use rust inhibitors or additives.

11   If the coolant level drops consistently, there may be a leak in the system. Inspect the radiator, hoses, filler cap, drain plugs and water pump (see Section 10). If no leaks are noted, have the radiator cap pressure tested by a service station.

12   If you have to remove the radiator cap, wait until the engine has cooled completely, then wrap a thick cloth around the cap and turn it to the first stop. If coolant or steam escapes, let the engine cool down longer, then remove the cap.

13   Check the condition of the coolant as well. It should be relatively clear. If it is brown or rust colored, the system should be drained, flushed and refilled. Even if the coolant appears to be normal, the corrosion inhibitors wear out, so it must be replaced at the specified intervals.

Windshield washer fluid

Refer to illustration 4.14

14   Fluid for the windshield washer system is located in a plastic reservoir on the left side of the engine compartment **(see illustration)**. In milder climates, plain water can be used in the reservoir, but it should be kept no more than two-thirds full to allow for expansion if the water freezes. In colder climates, use windshield washer system antifreeze, available at any auto parts store, to lower the freezing point of the fluid. Mix the antifreeze with water in accordance with the manufacturer's directions on the container. **Caution:** *Do not use cooling system antifreeze - it will damage the vehicle's paint.*

15   To help prevent icing in cold weather, warm the windshield with the defroster before using the washer.

Battery electrolyte

Refer to illustration 4.16

16   All vehicles covered by this manual are equipped with a battery which is permanently sealed (except for vent holes) and has no filler caps. Water does not have to be added to these batteries at any time; however, if a maintenance-type battery has been installed on the vehicle since it was new, remove all the cell caps on top of the battery (usually there are two caps that cover three cells each). If the electrolyte level is low, add distilled water until the level is above the plates. There is usually a split-ring indicator in each cell to help you judge when enough water has been added **(see illustration)**. Add water until the electrolyte level is just up to the bottom of the split ring indicator. Do not overfill the battery or it will spew out electrolyte when it is charging.

Brake and clutch fluid

Refer to illustrations 4.17 and 4.18

17   The brake master cylinder is located on the front of the power booster unit in the engine compartment **(see illustration)**. The clutch master cylinder is mounted on the inner fender to the left of the brake master cylinder.

18   The brake reservoir can be checked visually to make sure the level is above the MIN marks of the plastic windows in the reservoir. The fluid level in the clutch reservoir should be near the top of the reservoir **(see illustration)**.

19   When adding fluid, pour it carefully into the reservoir to avoid spilling it on surrounding painted surfaces. Be sure the specified fluid is used, since mixing different types of brake fluid can cause damage to the system. See *Recommended lubricants and fluids* at the front of this Chapter or your owner's manual. **Warning:** *Brake fluid can harm your eyes and damage painted surfaces, so use extreme caution when handling or pouring it. Do not use brake fluid that has been standing open or is more than one year old. Brake fluid absorbs moisture from the air. Excess moisture can cause a dangerous loss of braking effectiveness.*

Chapter 1 Tune-up and routine maintenance   1-9

4.17 The brake fluid level is easily checked by looking through the windows in the reservoir - the level must never drop below the MIN marks

4.18 The clutch fluid level should be about half an inch from the top of the reservoir

5.2 Use a tire tread depth gauge to monitor tire wear - they are available at auto parts stores and service stations and cost very little

20 At this time the fluid and master cylinder can be inspected for contamination. The system should be drained and refilled if deposits, dirt particles or water droplets are seen in the fluid.
21 After filling the reservoir to the proper level, make sure the cap is on tight to prevent fluid leakage.
22 The brake fluid level in the master cylinder will drop slightly as the pads at each wheel wear down during normal operation. If the master cylinder requires repeated replenishing to keep it at the proper level, this is an indication of leakage in the brake system, which should be corrected immediately.

Check all brake lines and connections (see Section 23 for more information).
23 If, when checking the master cylinder fluid level, you discover one or both reservoirs empty or nearly empty, the brake system should be bled (see Chapter 9).

5 Tire and tire pressure checks (every 250 miles or weekly)

Refer to illustrations 5.2, 5.3, 5.4a, 5.4b and 5.8

1 Periodic inspection of the tires may spare you the inconvenience of being stranded with a flat tire. It can also provide you with vital information regarding possible problems in the steering and suspension systems before major damage occurs.
2 The original tires on this vehicle are equipped with 1/2-inch wide bands that appear when tread depth reaches 1/16-inch, indicating the tires are worn out. Tread wear can be monitored with a simple, inexpensive device known as a tread depth indicator **(see illustration)**.
3 Note any abnormal tread wear **(see illustration)**. Tread pattern irregularities such as cupping, flat spots and more wear on one side than the other are indications of front

UNDERINFLATION

INCORRECT TOE-IN OR EXTREME CAMBER

CUPPING

Cupping may be caused by:
- Underinflation and/or mechanical irregularities such as out-of-balance condition of wheel and/or tire, and bent or damaged wheel.
- Loose or worn steering tie-rod or steering idler arm.
- Loose, damaged or worn front suspension parts.

5.3 This chart will help you determine the condition of the tires, the probable cause(s) of abnormal wear and the corrective action necessary

OVERINFLATION

FEATHERING DUE TO MISALIGNMENT

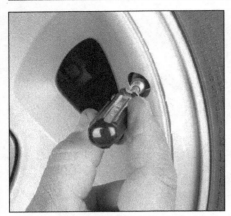

5.4a If a tire looses air on a steady basis, check the valve core first to make sure it's snug (special inexpensive wrenches are commonly available at auto parts stores)

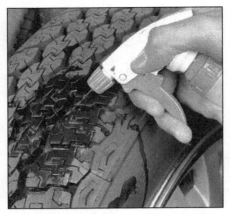

5.4b If the valve core is tight, raise the corner of the vehicle with the low tire and spray a soapy water solution onto the tread as the tire is turned slowly - leaks will cause small bubbles to appear

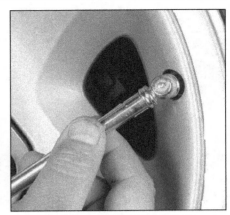

5.8 To extend the life of the tires, check the air pressure at least once a week with an accurate gauge (don't forget the spare)

6.3 The automatic transmission fluid dipstick (arrow) is located at the rear of the engine compartment

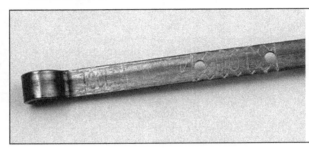

6.6 The automatic transmission fluid level must be maintained within the cross-hatched area on the dipstick, between the upper and lower holes when hot

end alignment and/or balance problems. If any of these conditions are noted, take the vehicle to a tire shop or service station to correct the problem.

4 Look closely for cuts, punctures and embedded nails or tacks. Sometimes a tire will hold air pressure for short time or leak down very slowly after a nail has embedded itself in the tread. If a slow leak persists, check the valve stem core to make sure it's tight **(see illustration)**. Examine the tread for an object that may have embedded itself in the tire or for a "plug" that may have begun to leak (radial tire punctures are repaired with a plug that's installed in a puncture). If a puncture is suspected, it can be easily verified by spraying a solution of soapy water onto the suspected area **(see illustration)**. The soapy solution will bubble if there's a leak. Unless the puncture is unusually large, a tire shop or service station can usually repair the tire.

5 Carefully inspect the inner sidewall of each tire for evidence of brake fluid. If you see any, inspect the brakes immediately.

6 Correct air pressure adds miles to the lifespan of the tires, improves mileage and enhances overall ride quality. Tire pressure cannot be accurately estimated by looking at a tire, especially if it's a radial. A tire pressure gauge is essential. Keep an accurate gauge in the vehicle. The pressure gauges attached to the nozzles of air hoses at gas stations are often inaccurate.

7 Always check tire pressure when the tires are cold. Cold, in this case, means the vehicle has not been driven over a mile in the three hours preceding a tire pressure check. A pressure rise of four to eight pounds is not uncommon once the tires are warm.

8 Unscrew the valve cap protruding from the wheel or hubcap and push the gauge firmly onto the valve stem **(see illustration)**. Note the reading on the gauge and compare the figure to the recommended tire pressure shown on the label attached to the inside of the glove compartment door. Be sure to reinstall the valve cap to keep dirt and moisture out of the valve stem mechanism. Check all four tires and, if necessary, add enough air to bring them up to the recommended pressure.

9 Don't forget to keep the spare tire inflated to the specified pressure (refer to your owner's manual or the tire sidewall).

6 Automatic transmission fluid level check (every 3000 miles or 3 months)

Refer to illustrations 6.3 and 6.6

1 The automatic transmission fluid level should be carefully maintained. Low fluid level can lead to slipping or loss of drive, while overfilling can cause foaming and loss of fluid.

2 With the parking brake set, start the engine, then move the shift lever through all the gear ranges, ending in Park. The fluid level must be checked with the vehicle level and the engine running at idle. **Note:** *Incorrect fluid level readings will result if the vehicle has just been driven at high speeds for an extended period, in hot weather in city traffic, or if it has been pulling a trailer. If any of these conditions apply, wait until the fluid has cooled (about 30 minutes).*

3 With the transmission at normal operating temperature, remove the dipstick from the filler tube. The dipstick is located at the rear of the engine compartment **(see illustration)**.

4 Carefully touch the fluid at the end of the dipstick to determine if the fluid is cool, warm or hot. Wipe the fluid from the dipstick with a clean rag and push it back into the filler tube until the cap seats.

5 Pull the dipstick out again and note the fluid level.

6 If the fluid felt cool, the level should be within the lower marks on the dipstick **(see illustration)**. If it felt warm or hot, the level should be within the cross-hatched upper areas on the dipstick. If additional fluid is required, pour it directly into the tube using a funnel. It takes about one pint to raise the level from the lower mark to the upper edge of the cross-hatched area with a hot transmission, so add the fluid a little at a time and keep checking the level until it's correct.

Chapter 1 Tune-up and routine maintenance

7.2 The power steering fluid reservoir is located near the front (drivebelt end) of the engine; turn the cap counterclockwise for removal

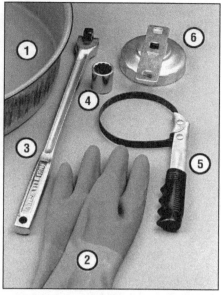

8.2 These tools are required when changing the engine oil and filter

1. **Drain pan** - It should be fairly shallow in depth, but wide to prevent spills
2. **Rubber gloves** - When removing the drain plug and filter, you will get oil on your hands (the gloves will prevent burns)
3. **Breaker bar** - Sometimes the oil drain plug is tight, and a long breaker bar is needed to loosen it
4. **Socket** - To be used with the breaker bar or a ratchet (must be the correct size to fit the drain plug)
5. **Filter wrench** - This is a metal band-type wrench, which requires clearance around the filter to be effective
6. **Filter wrench** - This type fits on the bottom of the filter and can be turned with a ratchet or breaker bar (different-size wrenches are available for different types of filters)

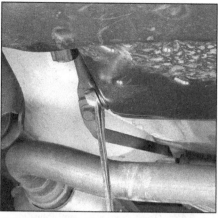

8.7 The engine oil drain plug is located at the rear of the oil pan - it is usually very tight, so use a box-end wrench to avoid rounding off the hex

7 The condition of the fluid should also be checked along with the level. If the fluid at the end of the dipstick is a dark reddish-brown color, or if the fluid has a burned smell, the fluid should be changed. If you're in doubt about the condition of the fluid, purchase some new fluid and compare the two for color and smell.

7 Power steering fluid level check (every 3000 miles or 3 months)

Refer to illustration 7.2

1 Unlike manual steering, the power steering system relies on fluid which may, over a period of time, require replenishing.
2 The fluid reservoir for the power steering pump is located at the front of the engine **(see illustration)**.
3 For the check, the front wheels should be pointed straight ahead and the engine should be off.
4 Use a clean rag to wipe off the reservoir cap and the area around the cap. This will help prevent any foreign matter from entering the reservoir during the check.
5 Twist off the cap and check the temperature of the fluid at the end of the dipstick with your finger.
6 Wipe off the fluid with a clean rag, reinsert it, then withdraw it and read the fluid level. The level should be at the HOT mark if the fluid was hot to the touch. It should be at the COLD mark if the fluid was cool to the touch.
7 If additional fluid is required, pour the specified type directly into the reservoir, using a funnel to prevent spills.
8 If the reservoir requires frequent fluid additions, all power steering hoses, hose connections, the power steering pump and the rack and pinion assembly should be carefully checked for leaks.

8 Engine oil and filter change (every 3000 miles or 3 months)

Refer to illustrations 8.2, 8.7, 8.12a, 8.12b and 8.14

1 Frequent oil changes are the best preventive maintenance the home mechanic can give the engine, because aging oil becomes diluted and contaminated, which leads to premature engine wear.
2 Make sure you have all the necessary tools before you begin this procedure **(see illustration)**. You should also have plenty of rags or newspapers handy for mopping up any spills.
3 Access to the underside of the vehicle is greatly improved if the vehicle can be lifted on a hoist, driven onto ramps or supported by jackstands. **Warning:** *Do not work under a vehicle which is supported only by a hydraulic or scissors-type jack.*
4 If this is your first oil change, get under the vehicle and familiarize yourself with the locations of the oil drain plug and the oil filter. The engine and exhaust components will be warm during the actual work, so try to anticipate any potential problems before the engine and accessories are hot.
5 Park the vehicle on a level spot. Start the engine and allow it to reach its normal operating temperature. Warm oil and sludge will flow out more easily. Turn off the engine when it's warmed up. Remove the filler cap from the valve cover.
6 Raise the vehicle and support it securely on jackstands. **Warning:** *Never get beneath the vehicle when it is supported only by a jack. The jack provided with your vehicle is designed solely for raising the vehicle to remove and replace the wheels. Always use jackstands to support the vehicle when it becomes necessary to place your body underneath the vehicle.*
7 Being careful not to touch the hot exhaust components, place the drain pan under the drain plug in the bottom of the pan and remove the plug **(see illustration)**. You may want to wear gloves while unscrewing the plug the final few turns if the engine is hot.
8 Allow the old oil to drain into the pan. It may be necessary to move the pan farther under the engine as the oil flow slows to a trickle. Inspect the old oil for the presence of metal shavings and chips.
9 After all the oil has drained, wipe off the drain plug with a clean rag. Even minute metal particles clinging to the plug would immediately contaminate the new oil.
10 Clean the area around the drain plug opening, reinstall the plug and tighten it securely, but do not strip the threads.
11 Move the drain pan into position under the oil filter.

Chapter 1 Tune-up and routine maintenance

8.12a On V6 engines the oil filter is located remote from the engine, under a cover

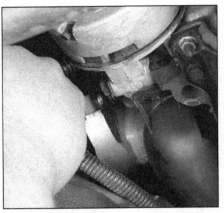

8.12b The oil filter is usually on very tight as well and will require a special wrench for removal - DO NOT use the wrench to tighten the new filter!

8.14 Lubricate the oil filter gasket with clean engine oil before installing the filter on the engine

12 Loosen the oil filter **(see illustrations)** by turning it counterclockwise with the filter wrench. Any standard filter wrench should work. Once the filter is loose, use your hands to unscrew it from the block. Just as the filter is detached from the block, immediately tilt the open end up to prevent the oil inside the filter from spilling out. **Warning:** *The exhaust system may still be hot, so be careful.*
13 With a clean rag, wipe off the oil filter mounting surface. Also make sure that the old gasket doesn't remains stuck to the mounting surface. It can be removed with a scraper if necessary.
14 Compare the old filter with the new one to make sure they are the same type. Smear some clean engine oil on the rubber gasket of the new filter and screw it into place **(see illustration)**. Because overtightening the filter will damage the gasket, do not use a filter wrench to tighten the filter. Tighten it by hand until the gasket contacts the seating surface. Then seat the filter by giving it an additional 3/4-turn.
15 Remove all tools, rags, etc. from under the vehicle, being careful not to spill the oil in the drain pan, then lower the vehicle.
16 Add new oil to the engine through the oil filler cap in the valve cover. Use a funnel, if necessary, to prevent oil from spilling onto the top of the engine. Pour three quarts of fresh oil into the engine. Wait a few minutes to allow the oil to drain into the pan, then check the level on the oil dipstick (see Section 4 if necessary). If the oil level is at or near the upper hole on the dipstick, install the filler cap hand tight, start the engine and allow the new oil to circulate.
17 Allow the engine to run for about a minute. While the engine is running, look under the vehicle and check for leaks at the oil pan drain plug and around the oil filter. If either is leaking, stop the engine and tighten the plug or filter.
18 Wait a few minutes to allow the oil to trickle down into the pan, then recheck the level on the dipstick and, if necessary, add enough oil to bring the level to the upper hole.

19 During the first few trips after an oil change, make it a point to check frequently for leaks and proper oil level.
20 The old oil drained from the engine cannot be re-used in its present state and should be discarded. Check with your local refuse disposal company, disposal facility or environmental agency to see whether they will accept the oil for recycling. Don't pour used oil into drains or onto the ground. After the oil has cooled, it can be drained into a suitable container (capped plastic jugs, topped bottles, milk cartons, etc.) for transport to one of these disposal sites.

9 Battery check, maintenance and charging (every 7500 miles or 6 months)

Refer to illustrations 9.1, 9.4, 9.5a, 9.5b and 9.5c

Warning: *Hydrogen gas is produced by the battery, so keep open flames and lighted tobacco away from it at all times. Always wear eye protection when working around the battery. Rinse off spilled electrolyte immediately with large amounts of water. When removing the battery cables, always detach the negative cable first and hook it up last!*
Caution: *On models equipped with a Delco Loc II audio system, be sure the lockout feature is turned off before performing any procedure which requires disconnecting the battery.*

1 Battery maintenance is an important procedure which will help ensure you aren't stranded because of a dead battery. Several tools are required for this procedure **(see illustration)**.
2 A sealed battery is standard equipment on all vehicles covered by this manual. Although this type of battery has many advantages over the older, capped cell type, and never requires the addition of water, it should still be routinely maintained according to the procedures which follow.

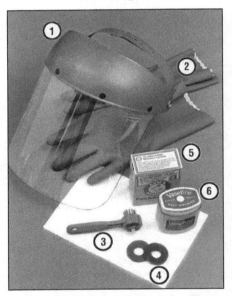

9.1 Tools and materials required for battery maintenance

1 **Face shield/safety goggles** - *When removing corrosion with a brush, the acidic particles can easily fly up into your eyes*
2 **Rubber gloves** - *Another safety item to consider when servicing the battery - remember that's acid inside the battery!*
3 **Battery terminal/cable cleaner** - *This wire brush cleaning tool will remove all traces of corrosion from the battery and cable*
4 **Treated felt washers** - *Placing one of these on each terminal, directly under the cable end, will help prevent corrosion (be sure to get the correct type for side-terminal batteries)*
5 **Baking soda** - *A solution of baking soda and water can be used to neutralize corrosion*
6 **Petroleum jelly** - *A layer of this on the battery terminal bolts will help prevent corrosion*

Chapter 1 Tune-up and routine maintenance

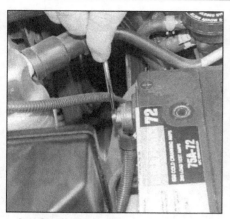

9.4 Check the tightness of the battery cable terminal bolts

9.5a A tool like this one (available at auto parts stores) is used to clean the side terminal type battery contact area

9.5b Use the brush to finish the cleaning job

Check

3   The battery is located in the right front corner of the engine compartment. The exterior of the battery should be inspected periodically for damage such as a cracked case or cover.
4   Check the tightness of the battery cable terminals and connections to ensure good electrical connections and check the entire length of each cable for cracks and frayed conductors **(see illustration)**.
5   If corrosion (visible as white, fluffy deposits) is evident, remove the cables from the terminals, clean them with a battery brush and reinstall the cables **(see illustrations)**. Corrosion can be kept to a minimum by using special treated fiber washers available at auto parts stores or by applying a layer of petroleum jelly to the terminals and cables after they are assembled.
6   Make sure that the battery tray is in good condition and the hold-down clamp bolt are tight. If the battery is removed from the tray, make sure no parts remain in the bottom of the tray when the battery is reinstalled. When reinstalling the hold-down clamp bolt, do not overtighten it.
7   Information on removing and installing the battery can be found in Chapter 5. Information on jump starting can be found at the front of this manual. For more detailed battery checking procedures, refer to the Haynes *Automotive Electrical Manual*.

Cleaning

8   Corrosion on the hold-down components, battery case and surrounding areas can be removed with a solution of water and baking soda. Thoroughly rinse all cleaned areas with plain water.
9   Any metal parts of the vehicle damaged by corrosion should be covered with a zinc-based primer, then painted.

Charging

Warning: *When batteries are being charged, hydrogen gas, which is very explosive and flammable, is produced. Do not smoke or allow open flames near a charging or a recently charged battery. Wear eye protection when near the battery during charging. Also, make sure the charger is unplugged before connecting or disconnecting the battery from the charger.*
10   Slow-rate charging is the best way to restore a battery that's discharged to the point where it will not start the engine. It's also a good way to maintain the battery charge in a vehicle that's only driven a few miles between starts. Maintaining the battery charge is particularly important in the winter when the battery must work harder to start the engine and electrical accessories that drain the battery are in greater use.
11   It's best to use a one or two-amp battery charger (sometimes called a "trickle" charger). They are the safest and put the least strain on the battery. They are also the least expensive. For a faster charge, you can use a higher amperage charger, but don't use one rated more than 1/10th the amp/hour rating of the battery. Rapid boost charges that claim to restore the power of the battery in one to two hours are hardest on the battery and can damage batteries that aren't in good condition. This type of charging should only be used in emergency situations.
12   The average time necessary to charge a battery should be listed in the instructions that come with the charger. As a general rule, a trickle charger will charge a battery in 12 to 16 hours.
13   Remove all of the cell caps (if equipped) and cover the holes with a clean cloth to prevent spattering electrolyte. Disconnect the negative battery cable and hook the battery charger leads to the battery posts (positive to positive, negative to negative), then plug in the charger. Make sure it is set at 12-volts if it has a selector switch.
14   If you're using a charger with a rate higher than two amps, check the battery regularly during charging to make sure it doesn't overheat. If you're using a trickle charger, you can safely let the battery charge overnight after you've checked it regularly for the first couple of hours.

9.5c The result should be a clean, shiny terminal area

15   If the battery has removable cell caps, measure the specific gravity with a hydrometer every hour during the last few hours of the charging cycle. Hydrometers are available inexpensively from auto parts stores - follow the instructions that come with the hydrometer. Consider the battery charged when there's no change in the specific gravity reading for two hours and the electrolyte in the cells is gassing (bubbling) freely. The specific gravity reading from each cell should be very close to the others. If not, the battery probably has a bad cell(s).
16   Some batteries with sealed tops have built-in hydrometers on the top that indicate the state of charge by the color displayed in the hydrometer window. Normally, a bright-colored hydrometer indicates a full charge and a dark hydrometer indicates the battery still needs charging. Check the battery manufacturer's instructions to be sure you know what the colors mean.
17   If the battery has a sealed top and no built-in hydrometer, you can hook up a digital voltmeter across the battery terminals to check the charge. A fully charged battery should read 12.5-volts or higher.

Check for a chafed area that could fail prematurely.

Check for a soft area indicating the hose has deteriorated inside.

Overtightening the clamp on a hardened hose will damage the hose and cause a leak.

Check each hose for swelling and oil-soaked ends. Cracks and breaks can be located by squeezing the hose.

10.4 Hoses, like drivebelts, have a habit of failing at the worst possible time - to prevent the inconvenience of a blown radiator or heater hose, inspect them carefully as shown here

10 Cooling system check (every 7500 miles or 6 months)

Refer to illustration 10.4

Caution: *Never mix green-colored ethylene glycol anti-freeze and orange-colored DEX-COOL® silicate-free coolant because doing so will destroy the efficiency of the DEX-COOL® coolant which is designed to last for 100,000 miles or five years.*

1   Many major engine failures can be attributed to a faulty cooling system. If the vehicle is equipped with an automatic transmission, the cooling system also cools the transmission fluid and plays an important role in prolonging transmission life.
2   The cooling system should be checked with the engine cold. Do this before the vehicle is driven for the day or after the engine has been shut off for at least three hours.
3   Remove the radiator cap by turning it to the left until it reaches a stop. If you hear any hissing sounds (indicating there is still pressure in the system), wait until it stops. Now press down on the cap with the palm of your hand and continue turning to the left until the cap can be removed. Thoroughly clean the cap, inside and out, with clean water. Also clean the filler neck on the radiator. All traces of corrosion should be removed. The coolant inside the radiator should be relatively transparent. If it is rust colored, the system should be drained and refilled (see Section 34). If the coolant level is not up to the top, add additional antifreeze/coolant mixture (see Section 4).
4   Carefully check the large upper and lower radiator hoses along with any smaller diameter heater hoses which run from the engine to the firewall. Inspect each hose along its entire length, replacing any hose which is cracked, swollen or shows signs of deterioration. Cracks may become more apparent if the hose is squeezed **(see illustration)**.
5   Make sure all hose connections are tight. A leak in the cooling system will usually show up as white or rust colored deposits on the areas adjoining the leak. If wire-type clamps are used at the ends of the hoses, it may be wise to replace them with more secure screw-type clamps.
6   Use compressed air or a soft brush to remove bugs, leaves, etc. from the front of the radiator or air conditioning condenser. Be careful not to damage the delicate cooling fins or cut yourself on them.
7   Every other inspection, or at the first indication of cooling system problems, have the cap and system pressure tested. If you don't have a pressure tester, most gas stations and repair shops will do this for a minimal charge.

11 Underhood hose check and replacement (every 7500 miles or 6 months)

General

1   **Warning:** *Replacement of air conditioning hoses must be left to a dealer service department or air conditioning shop that has the equipment to depressurize the system safely. Never remove air conditioning components or hoses until the system has been depressurized.*
2   High temperatures under the hood can cause the deterioration of the rubber and plastic hoses used for engine, accessory and emission systems operation. Periodic inspection should be made for cracks, loose clamps, material hardening and leaks. Information specific to the cooling system hoses can be found in Section 10.
3   Some, but not all, hoses are secured to the fittings with clamps. Where clamps are used, check to be sure they haven't lost their tension, allowing the hose to leak. If clamps aren't used, make sure the hose hasn't expanded and/or hardened where it slips over the fitting, allowing it to leak.

Vacuum hoses

4   It's quite common for vacuum hoses, especially those in the emissions system, to be color coded or identified by colored stripes molded into each hose. Various systems require hoses with different wall thickness, collapse resistance and temperature resistance. When replacing hoses, be sure the new ones are made of the same material.
5   Often the only effective way to check a hose is to remove it completely from the vehicle. If more than one hose is removed, be sure to label the hoses and fittings to ensure correct installation.
6   When checking vacuum hoses, be sure to include any plastic T-fittings in the check. Inspect the fittings for cracks and the hose where it fits over the fitting for distortion, which could cause leakage.
7   A small piece of vacuum hose (1/4-inch inside diameter) can be used as a stethoscope to detect vacuum leaks. Hold one end of the hose to your ear and probe around vacuum hoses and fittings, listening for the "hissing" sound characteristic of a vacuum leak. **Warning:** *When probing with the vacuum hose stethoscope, be careful not to allow your body or the hose to come into contact with moving engine components such as the drivebelt, cooling fan, etc.*

Fuel hose

Warning: *Gasoline is extremely flammable, so take extra precautions when you work on any part of the fuel system. Don't smoke or allow open flames or bare light bulbs near the work area, and don't work in a garage where a natural gas-type appliance (such as a water heater or clothes dryer) with a pilot light is present. If you spill any fuel on your skin, rinse it off immediately with soap and water. When you perform any kind of work on the fuel system, wear safety glasses and have a Class B type fire extinguisher on hand. The fuel system is under pressure, so if any lines must be disconnected, the pressure in the system must be relieved first (see Chapter 4 for more information).*

8   Check all rubber fuel lines for deterioration and chafing. Check especially for cracks in areas where the hose bends and just before fittings, such as where a hose attaches to the fuel filter and fuel injection unit.
9   High-quality fuel line, specifically designed for high-pressure fuel-injection applications, must be used for fuel line replacement. Never, under any circumstances, use unreinforced vacuum line, clear plastic tubing or water hose for fuel lines.
10   Spring-type clamps are commonly used on fuel lines. These clamps often lose their tension over a period of time, and can be "sprung" during the removal process. As a result spring-type clamps should be replaced with screw-type clamps whenever a hose is replaced.

Chapter 1 Tune-up and routine maintenance

1-15

12.5 Depress the release lever, then slide the blade assembly out of the hook in the end of the wiper arm

12.7 Squeeze the end of the windshield wiper element to free it from the bridge claw, then slide the element out

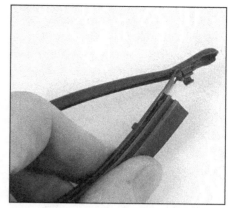

12.12 On rear wipers, pull the element out of the end, the slide it from the frame

Metal lines

11 Sections of steel tubing often used for fuel line between the fuel pump and fuel injection unit. Check carefully for cracks, kinks and flat spots in the line.
12 If a section of metal fuel line must be replaced, only seamless steel tubing should be used, since copper and aluminum tubing do not have the strength necessary to withstand normal engine vibration.
13 Check the metal brake lines where they enter the master cylinder and brake proportioning unit (if used) for cracks in the lines and loose fittings. Any sign of brake fluid leakage calls for an immediate thorough inspection of the brake system.

12 Wiper blade inspection and replacement (every 7500 miles or 6 months)

Refer to illustrations 12.5, 12.7 and 12.12

1 The windshield and rear wiper and blade assemblies should be inspected periodically for damage, loose components and cracked or worn blade elements.
2 Road film can build up on the wiper blades and affect their efficiency, so they should be washed regularly with a mild detergent solution.
3 The action of the wiping mechanism can loosen the bolts, nuts and fasteners, so they should be checked and tightened, as necessary, at the same time the wiper blades are checked.
4 If the wiper blade elements (sometimes called inserts) are cracked, worn or warped, they should be replaced with new ones.

Windshield wiper

5 Remove the wiper blade assembly from the wiper arm by depressing the release lever while pulling on the blade to release it **(see illustration)**.
6 With the blade removed from the vehicle, you can remove the rubber element from the blade.
7 Grasp the end of the wiper bridge securely with one hand and the element with the other. Detach the end of the element from the bridge claw and slide to free it, then slide the element out **(see illustration)**.
8 Compare the new element with the old for length, design, etc.
9 Slide the new element into the claw into place, notched end last and secure the claw into the notches.
10 Reinstall the blade assembly on the arm, wet the windshield and test for proper operation.

Rear wiper

11 Remove the nut and detach the wiper arm assembly.
12 Grasp the of the wiper bridge securely with one hand pull the end of the element out of the bridge claw and slide to free it, then slide the element out **(see illustration)**.
13 Installation is the reverse of removal.

13 Chassis lubrication (every 7500 miles or 6 months)

Refer to illustrations 13.1, 13.6a, 13.6b, 13.6c, 13.8a and 13.8b

1 Refer to *Recommended lubricants and fluids* at the front of this Chapter to obtain the necessary grease, etc. You'll also need a grease gun **(see illustration)**. Occasionally plugs will be installed rather than grease fittings. If so, grease fittings will have to be purchased and installed.
2 Look under the vehicle and see if grease fittings or plugs are installed. If there are plugs, remove them and buy grease fittings, which will thread into the component. A dealer or auto parts store will be able to supply the correct fittings. Straight, as well as angled, fittings are available.
3 For easier access under the vehicle, raise it with a jack and place jackstands under the frame. Make sure it's safely supported by the stands. If the wheels are to be

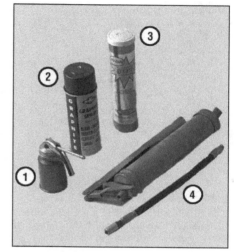

13.1 Materials required for chassis and body lubrication

1 **Engine oil** - Light engine oil in a can like this can be used for door and hood hinges
2 **Graphite spray** - Used to lubricate lock cylinders
3 **Grease** - Grease, in a variety of types and weights, is available for use in a grease gun. Check the Specification for your requirements
4 **Grease gun** - A common grease gun, shown here with a detachable hose and nozzle, is needed for chassis lubrication. After use, clean it thoroughly!

removed at this interval for tire rotation or brake inspection, loosen the lug nuts slightly while the vehicle is still on the ground.
4 Before beginning, force a little grease out of the nozzle to remove any dirt from the end of the gun. Wipe the nozzle clean with a rag.
5 With the grease gun and plenty of clean rags, crawl under the vehicle and begin lubricating the components.
6 Wipe one of the grease fittings clean

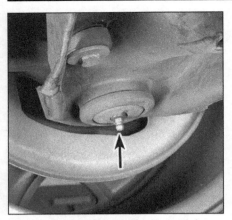

13.6a Location of the lower balljoint grease fitting (arrow)

13.6b The upper balljoint grease fitting (arrow) is easier to reach if the front wheel is removed

13.6c Be sure to grease each steering arm balljoint (arrow)

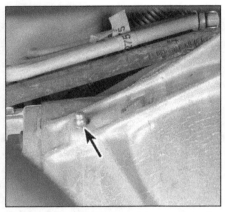

13.8a Be careful not to pump too much grease into the clutch shaft fitting (arrow) - one or two pumps should do

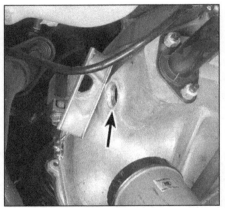

13.8b Use a mirror and flashlight to look into the inspection hole (arrow) and apply a small dab of grease in the clutch lever and release cylinder pushrod contact area

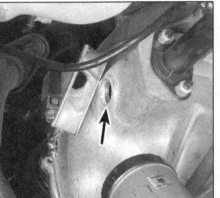

14.4 Check the suspension arm and stabilizer bar connections (arrows)

and push the nozzle firmly over it (see illustrations). Pump the gun until the balljoint rubber seal is firm to the touch. Do not pump too much grease into the fitting as it could rupture the seal.

7 Wipe the excess grease from the components and the grease fitting. Repeat the procedure for the remaining fittings

8 On manual transmission-equipped models, lubricate the clutch linkage pivot points with clean engine oil. Some models have a grease fitting for the internal clutch fork ballstud located on the clutch housing (see illustration). Be careful not to pump too much grease into this fitting because it could cause the clutch to malfunction. On some models its possible to unscrew the access plug and apply a small dab on grease on the contact area between the clutch release lever and release cylinder pushrod (see illustration). Again, be careful not to use too much grease.

9 Clean and lubricate the parking brake cable, along with the cable guides and levers. This can be done by smearing some of the chassis grease onto the cable and its related parts with your fingers.

10 Open the hood and smear a little chassis grease on the hood latch mechanism. Have an assistant pull the hood release lever from inside the vehicle as you lubricate the cable at the latch.

11 Lubricate all the hinges (door, hood, etc.) with engine oil to keep them in proper working order.

12 The key lock cylinders can be lubricated with spray graphite or silicone lubricant, which is available at auto parts stores.

13 Lubricate the door weatherstripping with silicone spray. This will reduce chafing and retard wear.

14 Suspension and steering check (every 7500 miles or 6 months)

Refer to illustrations 14.4 and 14.5

1 Indications of a fault in these systems are excessive play in the steering wheel before the front wheels react, excessive sway around corners, body movement over rough roads or binding at some point as the steering wheel is turned.

2 Raise the front of the vehicle periodically and visually check the suspension and steering components for wear. Because of the work to be done, make sure the vehicle cannot fall from the stands.

3 Check the wheel bearings. Do this by spinning the front wheels. Listen for any abnormal noises and watch to make sure the wheel spins true (doesn't wobble). Grab the top and bottom of the tire and pull in-and-out on it. Any movement would indicate a loose wheel bearing assembly. If the bearings are suspect, refer to Section 29 and Chapter 10 for more information.

4 From under the vehicle check for loose bolts, broken or disconnected parts and deteriorated rubber bushings on all suspension and steering components (see illustration). Check the steering gear seals for leaks. Check the power steering hoses and connections for leaks.

5 Check the shock absorbers or leaking fluid or damage (see illustration).

6 Have an assistant turn the steering wheel from side-to-side and check the steering components for free movement, chafing and binding. If the steering doesn't react with the movement of the steering wheel, try to determine where the slack is located.

Chapter 1 Tune-up and routine maintenance

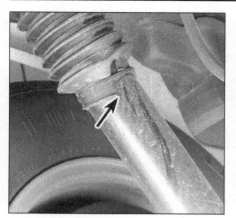

14.5 Check for signs of oil leaks in this area of each shock absorber (arrow)

15.2a Check the exhaust system rubber hangers for cracks and damage

15.2b Check the exhaust system connections to the chassis

15 Exhaust system check (every 7500 miles or 6 months)

Refer to illustrations 15.2a and 15.2b

1 With the engine cold (at least three hours after the vehicle has been driven), check the complete exhaust system from the engine to the end of the tailpipe. Ideally, the inspection should be done with the vehicle on a hoist to permit unrestricted access. If a hoist is not available, raise the vehicle and support it securely on jackstands.
2 Check the exhaust pipes and connections for evidence of leaks, severe corrosion and damage. Make sure that all brackets and hangers are in good condition and tight **(see illustrations)**.
3 At the same time, inspect the underside of the body for holes, corrosion, open seams, etc. which may allow exhaust gases to enter the interior. Seal all body openings with silicone or body putty.
4 Rattles and other noises can often be traced to the exhaust system, especially the mounts and hangers. Try to move the pipes, muffler and catalytic converter. If the components can come in contact with the body or suspension parts, secure the exhaust system with new mounts.

16 Seat belt and restraint system check (every 7500 miles or 6 months)

1 Check the seat belts, buckles, latch plates and guide loops for obvious damage and signs of wear.
2 See if the seat belt reminder light comes on when the key is turned to the Run or Start position. A chime should also sound.
3 The seat belts are designed to lock up during a sudden stop or impact, yet allow free movement during normal driving. Make sure the retractors return the belt against your chest while driving and rewind the belt fully when the buckle is unlatched.
4 If any of the above checks reveal problems with the seat belt system, replace parts as necessary.
5 On air bag-equipped models, inspect the air bag cover for damage or warping.

17 Starter safety switch check (every 7500 miles or 6 months)

Warning: *During the following checks there's a chance the vehicle could lunge forward, possibly causing damage or injuries. Allow plenty of room around the vehicle, apply the parking brake and hold-down the regular brake pedal during the checks.*
1 Try to start the engine in each gear. The engine should crank only in Park or Neutral.
2 Make sure the steering column lock allows the key to go into the Lock position only when the shift lever is in Park.
3 The ignition key should come out only in the Lock position.

18 Brake and Transmission Shift Interlock (BTSI) system check (1996 and later models with automatic transmission) (every 7500 miles or 6 months)

Warning: *During the following check there's a chance the vehicle could lunge forward, possibly causing damage or injuries. Allow plenty of room around the vehicle and be ready to hold-down the regular brake pedal during the check.*
1 Park the vehicle on a level surface with the engine off.
2 Be prepared to apply the vehicle brakes if necessary and, with the parking brake securely applied, turn the ignition key to the Run position, then try to move the shift lever out of Park.
3 If the lever will not move out of Park, the BTSI system is operating properly. If lever does move out of Park, take the vehicle to your dealer to have the BTSI system checked.

19 Tire rotation (every 7500 miles or 6 months)

Refer to illustration 19.2

1 The tires should be rotated at the specified intervals and whenever uneven wear is noticed.
2 Refer to the **accompanying illustration** for the preferred tire rotation pattern.
3 Refer to the information in *Jacking and towing* at the front of this manual for the proper procedures to follow when raising the vehicle and changing a tire. If the brakes are going to be checked, don't apply the parking brake as stated. Make sure the tires are blocked to prevent the vehicle from rolling as it's raised.
4 The entire vehicle should be raised at the same time. This can be done on a hoist or by jacking up each corner and then lowering the vehicle onto jackstands placed under the frame rails. Always use four jackstands and make sure the vehicle is safely supported.
5 After rotation, check and adjust the tire pressures as necessary and be sure to check the lug nut tightness.

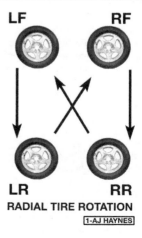

19.2 Tire rotation diagram

1-18 Chapter 1 Tune-up and routine maintenance

20.1 Locations of the manual transmission check and drain plugs (arrows)

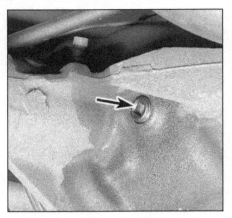

21.2a Rear differential fill/check plug location (arrow)

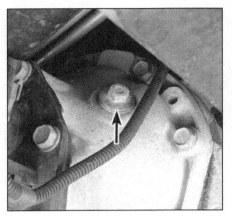

21.2b Front differential fill/check plug location (arrow)

20 Manual transmission lubricant level check (every 15,000 miles or 12 months)

Refer to illustration 20.1

1 The manual transmission has a fill plug which must be removed to check the lubricant level **(see illustration)**. If the vehicle is raised to gain access to the plug, be sure to support it safely on jackstands - DO NOT crawl under a vehicle which is supported only by a jack!
2 Remove the plug from the transmission and use your little finger to reach inside the housing to feel the lubricant level. The level should be at or near the bottom of the plug hole.
3 If it isn't, add the recommended lubricant through the plug hole with a syringe or squeeze bottle.
4 Install and tighten the plug and check for leaks after the first few miles of driving.

21 Differential lubricant level check (every 15,000 miles or 12 months)

Refer to illustrations 21.2a and 21.2b

1 The differential has a filler plug which must be removed to check the lubricant level. If the vehicle is raised to gain access to the plug, be sure to support it safely on jackstands - DO NOT crawl under the vehicle when it's supported only by the jack.
2 Remove the filler plug from the side of the differential **(see illustrations)**.
3 The lubricant level should be at the bottom of the plug opening. If not, use a syringe to add the recommended lubricant until it just starts to run out of the opening. On some models a tag is located in the area of the plug which gives information regarding lubricant type, particularly on models equipped with a limited slip differential.
4 Install the plug and tighten it securely.

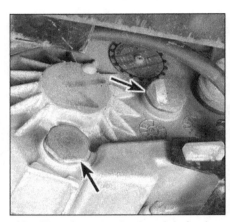

22.1 Transfer case check and drain plug locations (arrows)

22 Transfer case lubricant level check (every 15,000 miles or 12 months)

Refer to illustration 22.1

1 The transfer case lubricant level is checked by removing the upper plug located in the back of the case **(see illustration)**.
2 After removing the plug, reach inside the hole. The lubricant level should be just at the bottom of the hole. If not, add the appropriate lubricant through the opening.

23 Brake check (every 15,000 miles or 12 months)

Warning: *Brake system dust is hazardous to your health. DO NOT blow it out with compressed air or inhale it. DO NOT use gasoline or solvents to remove the dust. Use brake system cleaner or denatured alcohol only.*
Note: *For detailed photographs of the brake system, refer to Chapter 9.*

1 In addition to the specified intervals, the brakes should be inspected every time the wheels are removed or whenever a defect is

23.5 Look through each end of the caliper to check the thickness of the pad lining, which rubs against the disc (arrow)

suspected. Raise the vehicle and place it securely on jackstands. Remove the wheels (see *Jacking and towing* at the front of this manual, if necessary).

Disc brakes (front)

Refer to illustrations 23.5

2 Disc brakes can be checked without removing any parts except the wheels. Extensive disc damage can occur if the pads are not replaced when needed.
3 The disc brake pads have built-in wear indicators which make a high-pitched squealing sound when the pads are worn. **Caution:** *Expensive damage to the disc can result if the pads are not replaced soon after the wear indicators start squealing.*
4 The disc brake calipers, which contain the pads, are now visible. There is an outer pad and an inner pad in each caliper. All pads should be inspected.
5 Check the material thickness by inspecting the end of the pad or by looking into the "window" in the caliper **(see illustration)**. If the pad material has worn to about 1/8-inch thick or less, the pads should be replaced.
6 If you're unsure about the exact thick-

Chapter 1 Tune-up and routine maintenance

1-19

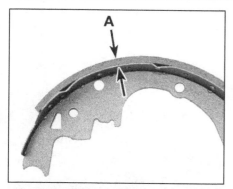

23.14 If the lining is bonded to the brake shoe, measure the lining thickness from the outer surface to the metal shoe, as shown here (A); if the lining is riveted to the shoe, measure from the lining outer surface to the rivet head

ness of the remaining lining material, remove the pads for further inspection or replacement (refer to Chapter 9).
7 Before installing the wheels, check for leakage and/or damage at the brake hoses and connections. Replace the hose or fittings as necessary, referring to Chapter 9.
8 Check the condition of the brake rotor. Look for score marks, deep scratches and overheated areas (they will appear blue or discolored). If damage or wear is noted, the disc can be removed and resurfaced by an automotive machine shop or replaced with a new one. Refer to Chapter 9 for more detailed inspection and repair procedures.

Drum brakes (rear)

Refer to illustration 23.14

9 Raise the vehicle and support it securely on jackstands. Block the front tires to prevent the vehicle from rolling; however, don't apply the parking brake or it will lock the drums in place.
10 Remove the wheels, referring to *Jacking and towing* at the front of this manual if necessary.
11 Mark the hub so it can be reinstalled in the same position. Use a scribe, chalk, etc. on the drum, hub and backing plate.
12 Remove the brake drum.
13 With the drum removed, carefully clean the brake assembly with brake system cleaner. **Warning:** *Don't blow the dust out with compressed air and don't inhale any of it (it may contain asbestos, which is harmful to your health).*
14 Note the thickness of the lining material on both front and rear brake shoes. If the material has worn away to within 1/8-inch of the recessed rivets or metal backing, the shoes should be replaced **(see illustration)**. The shoes should also be replaced if they're cracked, glazed (shiny areas), or covered with brake fluid.
15 Make sure all the brake assembly springs are connected and in good condition.
16 Check the brake components for signs of fluid leakage. With your finger or a small screwdriver, carefully pry back the rubber cups on the wheel cylinder located at the top of the brake shoes. Any leakage here is an indication that the wheel cylinders should be overhauled immediately (see Chapter 9). Also, check all hoses and connections for signs of leakage.
17 Wipe the inside of the drum with a clean rag and denatured alcohol or brake cleaner. Again, be careful not to breathe the dangerous brake dust.
18 Check the inside of the drum for cracks, score marks, deep scratches and "hard spots" which will appear as small discolored areas. If imperfections cannot be removed with fine emery cloth, the drum must be taken to an automotive machine shop for resurfacing.
19 Repeat the procedure for the remaining wheel. If the inspection reveals that all parts are in good condition, reinstall the brake drums, install the wheels and lower the vehicle to the ground.

Disc brakes (rear)

20 Some models have optional rear disc brakes. Follow the inspection procedures in Steps 2 through 8 as shown for front disc brakes.

Parking brake

21 The parking brake is operated by a hand lever and locks the rear brake system. The easiest, and perhaps most obvious, method of periodically checking the operation of the parking brake assembly is to park the vehicle on a steep hill with the parking brake set and the transmission in Neutral (be sure to stay in the vehicle during this check!). If the parking brake cannot prevent the vehicle from rolling, it needs service (see Chapter 9).

24 Driveaxle boot check (4WD models) (every 15,000 miles or 12 months)

Refer to illustration 24.2

1 The driveaxle boots are very important because they prevent dirt, water and foreign material from entering and damaging the constant velocity (CV) joints. Oil and grease can cause the boot material to deteriorate prematurely, so it's a good idea to wash the boots with soap and water.
2 Inspect the boots for tears and cracks as well as loose clamps **(see illustration)**. If there is any evidence of cracks or leaking grease, they must be replaced as described in Chapter 8.

25 Fuel system check (every 15,000 miles or 12 months)

Warning: *Gasoline is extremely flammable, so take extra precautions when you work on any part of the fuel system. Don't smoke or allow*

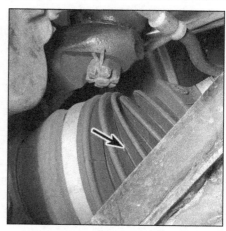

24.2 Push on the boot and check for cracks or leaking grease in the area shown (arrow)

open flames or bare light bulbs near the work area, and don't work in a garage where a gas-type appliance (such as a water heater or clothes dryer) is present. Since gasoline is carcinogenic, wear latex gloves when there's a possibility of being exposed to fuel, and, if you spill any fuel on your skin, rinse it off immediately with soap and water. Mop up any spills immediately and do not store fuel-soaked rags where they could ignite. The fuel system is under constant pressure, so, if any fuel lines are to be disconnected, the fuel pressure in the system must be relieved first (see Chapter 4 for more information). When you perform any kind of work on the fuel system, wear safety glasses and have a Class B type fire extinguisher on hand.

1 The fuel system is most easily checked with the vehicle raised on a hoist so the components underneath the vehicle are readily visible and accessible.
2 If the smell of gasoline is noticed while driving or after the vehicle has been in the sun, the system should be thoroughly inspected immediately.
3 Remove the gas tank cap and check for damage, corrosion and an unbroken sealing imprint on the gasket. Replace the cap with a new one if necessary.
4 With the vehicle raised, inspect the gas tank and filler neck for cracks and other damage. The connection between the filler neck and tank is especially critical. Sometimes a filler neck will leak due to cracks, problems a home mechanic can't repair. **Warning:** *Do not, under any circumstances, try to repair a fuel tank yourself (except rubber components). A welding torch or any open flame can easily cause the fuel vapors to explode if the proper precautions are not taken.*
5 Carefully check all rubber hoses and metal lines leading away from the fuel tank. Check for loose connections, deteriorated hoses, crimped lines and other damage. Follow the lines to the front of the vehicle, carefully inspecting them all the way. Repair or replace damaged sections as necessary.

26.8a Detach the air cleaner housing clips

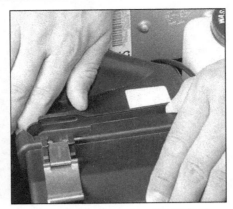

26.8b Separate the housing halves

26.8c Lift the filter element out of the housing

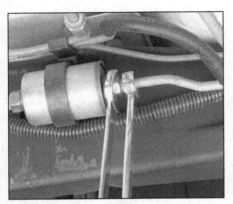

27.4a Use two wrenches to loosen the fuel lines and detach if from the filter (if available, use a flare nut wrench on the fuel line nut fitting)

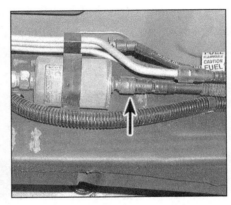

27.4b On 1997 and later models one end of the fuel filter has a quick disconnect fitting - to remove the line squeeze the plastic clips on the fitting together and pull back on the line

28.7 The serpentine drivebelt routing diagram is located on the radiator shroud

26 Air filter replacement (every 30,000 miles or 24 months)

Refer to illustrations 26.8a, 26.8b and 26.8c

1 At the specified intervals, the air filter should be replaced with a new one. The filter should be inspected between changes.

TBI models

2 The air filter is located inside the air cleaner housing mounted on top of the engine
3 Remove the wing nuts and lift off the top plate of the air cleaner housing. While the top plate is off, be careful not to drop anything down into the TBI unit.
4 Lift the air cleaner element out of the housing.
5 Wipe out the inside of the air cleaner housing with a clean rag.
6 Place the new filter into the housing, make sure it seats properly then install the top plate and wing nuts.

Central Port Injection and Multiport Fuel Injection

7 The air filter is located inside the air cleaner housing mounted in the left front corner of the engine compartment.

8 Detach the two clips, separate housing halves and lift the filter out (see illustrations).
9 Wipe out the inside of the air cleaner housing with a clean rag.
10 Place the new filter in the air cleaner housing. Make sure it seats properly, seat the two halves together and secure them with the clips.

27 Fuel filter replacement (every 30,000 miles or 24 months)

Refer to illustration 27.4a and 27.4b
Warning: *Gasoline is extremely flammable, so take extra precautions when you work on any part of the fuel system. Don't smoke or allow open flames or bare light bulbs near the work area, and don't work in a garage where a gas-type appliance (such as a water heater or clothes dryer) is present. Since gasoline is carcinogenic, wear latex gloves when there's a possibility of being exposed to fuel, and, if you spill any fuel on your skin, rinse it off immediately with soap and water. Mop up any spills immediately and do not store fuel-soaked rags where they could ignite. The fuel system is under constant pressure, so, if any fuel lines are to be disconnected, the fuel*

pressure in the system must be relieved first (see Chapter 4 for more information). When you perform any kind of work on the fuel system, wear safety glasses and have a Class B type fire extinguisher on hand.

1 Relieve the fuel system pressure (see Chapter 4).
2 Raise the vehicle and support it securely on jackstands.
3 Place a container, newspapers or rags under the fuel filter.
4 Use two wrenches, one to steady the filter and the other to unscrew the fuel line nut, then separate the connections (see illustrations). It's a good idea to use a flarenut wrench because the wrap-around design gives it a better grip on the fuel line nut.
5 Detach the filter from the bracket.
6 Install the new filter and connect the fuel lines securely.

28 Drivebelt check and replacement (every 30,000 miles or 24 months)

Refer to illustration 28.7

1 A single serpentine drivebelt is located at the front of the engine and plays an important role in the overall operation of the engine and its components. Due to its function and material make up, the belt is prone to wear and should be periodically inspected. The

Chapter 1 Tune-up and routine maintenance

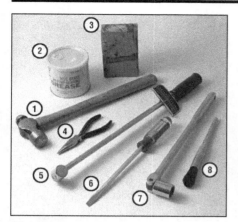

29.1 Tools and materials needed for front wheel bearing maintenance

1 **Hammer** - A common hammer will do just fine
2 **Grease** - High-temperature grease that is formulated for front wheel bearings should be used
3 **Wood block** - If you have a scrap piece of 2x4, it can be used to drive the new seal into the hub
4 **Needle-nose pliers** - Used to straighten and remove the cotter pin in the spindle
5 **Torque wrench** - This is very important in this procedure; if the bearing is too tight, the wheel won't turn freely - if it's too loose, the wheel will "wobble" on the spindle. Either way, it could mean extensive damage
6 **Screwdriver** - Used to remove the seal from the hub (a long screwdriver is preferred)
7 **Socket/breaker bar** - Needed to loosen the nut on the spindle if it's extremely tight
8 **Brush** - Together with some clean solvent, this will be used to remove old grease from the hub and spindle

serpentine belt drives the alternator, power steering pump, water pump and air conditioning compressor.

2   With the engine off, open the hood and use your fingers (and a flashlight, if necessary), to move along the belt checking for cracks and separation of the belt plies. Also check for fraying and glazing, which gives the belt a shiny appearance. Both sides of the belt should be inspected, which means you will have to twist the belt to check the underside.

3   Check the ribs on the underside of the belt. They should all be the same depth, with none of the surface uneven.

4   The tension of the belt is maintained by the tensioner assembly and isn't adjustable. The belt should be replaced at the mileage specified in the maintenance schedule at the front of this Chapter, or if it is damaged or worn.

5   To replace the belt, rotate the tensioner counterclockwise to release belt tension.

6   Remove the belt from the auxiliary com-

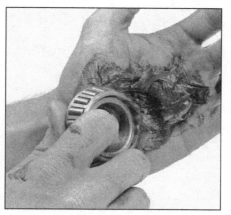

29.15 Work the grease completely into the bearing rollers - if you don't like getting greasy, special bearing packing tools that work with a common grease gun are available inexpensively from auto parts stores

ponents and slowly release the tensioner.

7   Route the new belt over the various pulleys, again rotating the tensioner to allow the belt to be installed, then release the belt tensioner. **Note:** *These models have a drivebelt routing decal on the radiator shroud to help during drivebelt installation* **(see illustration)**.

29 Front wheel bearing check, repack and adjustment (2WD models) (every 30,000 miles)

Refer to illustrations 29.1 and 29.15

1   In most cases the front wheel bearings will not need servicing until the brake pads are changed. However, the bearings should be checked whenever the front of the vehicle is raised for any reason. Several items, including a torque wrench and special grease, are required for this procedure **(see illustration)**.

2   With the vehicle securely supported on jackstands, spin each wheel and check for noise, rolling resistance and freeplay.

3   Grasp the top of each tire with one hand and the bottom with the other. Move the wheel in-and-out on the spindle. If there's any noticeable movement, the bearings should be checked and then repacked with grease or replaced if necessary.

4   Remove the wheel.

5   Remove the brake caliper (see Chapter 9) and hang it out of the way on a piece of wire. A wood block can be slid between the brake pads to keep them separated, if necessary.

6   Pry the dust cap out of the hub using a screwdriver or hammer and chisel.

7   Straighten the bent ends of the cotter pin, then pull the cotter pin out of the nut lock. Discard the cotter pin and use a new one during reassembly.

8   Remove the nut lock, nut and washer from the end of the spindle.

9   Pull the hub/disc assembly out slightly, then push it back into its original position. This should force the outer bearing off the spindle enough so it can be removed.

10   Pull the hub/disc assembly off the spindle.

11   Use a screwdriver to pry the seal out of the rear of the hub. As this is done, note how the seal is installed.

12   Remove the inner wheel bearing from the hub.

13   Use solvent to remove all traces of the old grease from the bearings, hub and spindle. A small brush may prove helpful; however make sure no bristles from the brush embed themselves inside the bearing rollers. Allow the parts to air dry.

14   Carefully inspect the bearings for cracks, heat discoloration, worn rollers, etc. Check the bearing races inside the hub for wear and damage. If the bearing races are defective, the hubs should be taken to a machine shop with the facilities to remove the old races and press new ones in. Note that the bearings and races come as matched sets and old bearings should never be installed on new races.

15   Use high-temperature front wheel bearing grease to pack the bearings. Work the grease completely into the bearings, forcing it between the rollers, cone and cage from the back side **(see illustration)**.

16   Apply a thin coat of grease to the spindle at the outer bearing seat, inner bearing seat, shoulder and seal seat.

17   Put a small quantity of grease inboard of each bearing race inside the hub. Using your finger, form a dam at these points to provide extra grease availability and to keep thinned grease from flowing out of the bearing.

18   Place the grease-packed inner bearing into the rear of the hub and put a little more grease outboard of the bearing.

19   Place a new seal over the inner bearing and tap the seal evenly into place with a hammer and blunt punch until it's flush with the hub.

20   Carefully place the hub assembly onto the spindle and push the grease-packed outer bearing into position.

21   Install the washer and spindle nut. Tighten the nut only slightly (no more than 12 ft-lbs of torque).

22   Spin the hub in a forward direction while tightening the spindle nut to approximately 20 ft-lbs to seat the bearings and remove any grease or burrs which could cause excessive bearing play later.

23   Loosen the spindle nut 1/4-turn, then using your hand (not a wrench of any kind), tighten the nut until it's snug. Install the nut lock and a new cotter pin through the hole in the spindle and the slots in the nut lock. If the nut lock slots don't line up, remove the nut lock and turn it slightly until they do.

24   Bend the ends of the cotter pin until they're flat against the nut. Cut off any extra length which could interfere with the dust cap.

25   Install the dust cap, tapping it into place with a hammer.

1-22   Chapter 1   Tune-up and routine maintenance

30.6 With the rear bolts in place but loose, pull the front of the pan down to drain the transmission fluid

30.9 Rotate the filter out of the retaining clip, then lower it from the transmission

30.10 If necessary, use a screwdriver to remove the seal from the transmission - be careful not to gouge the aluminum housing

26 Place the brake caliper near the rotor and carefully remove the wood spacer. Install the caliper (see Chapter 9).
27 Install the wheel on the hub and tighten the lug nuts.
28 Grasp the top and bottom of the tire and check the bearings in the manner described earlier in this Section.
29 Lower the vehicle.

30 Automatic transmission fluid and filter change (every 30,000 miles or 24 months)

Refer to illustrations 30.6, 30.9 and 30.10

1 At the specified intervals, the transmission fluid should be drained and replaced. Since the fluid will remain hot long after driving, perform this procedure only after the engine has cooled down completely.
2 Before beginning work, purchase the specified transmission fluid (see *Recommended lubricants and fluids* at the front of this Chapter) and a new filter.
3 Other tools necessary for this job include a floor jack, jackstands to support the vehicle in a raised position, a drain pan capable of holding at least five quarts, newspapers and clean rags.
4 Raise the vehicle and support it securely on jackstands.
5 Place the drain pan underneath the transmission pan. Remove the front and side pan mounting bolts, but only loosen the rear pan bolts approximately four turns.
6 Carefully pry the transmission pan loose with a screwdriver, allowing the fluid to drain **(see illustration)**.
7 Remove the remaining bolts, pan and gasket. Carefully clean the gasket surface of the transmission to remove all traces of the old gasket and sealant.
8 Drain the fluid from the transmission pan, clean it with solvent and dry it with compressed air.
9 Remove the filter from the mount inside the transmission **(see illustration)**.
10 If the seal did not come out with the filter, remove it from the transmission **(see illustration)**. The new seal can be driven squarely into place using a socket of the appropriate size. Install the new filter.
11 Make sure the gasket surface on the transmission pan is clean, then install a new gasket on the pan. Put the pan in place against the transmission and, working around the pan, tighten each bolt a little at a time until the final torque figure is reached.
12 Lower the vehicle and add approximately three and one-half quarts of the specified type of automatic transmission fluid through the filler tube (see Section 6).
13 With the transmission in Park and the parking brake set, run the engine at a fast idle, but don't race it.
14 Move the gear selector through each range and back to Park. Check the fluid level. It will probably be low. Add enough fluid to bring the level up to top of the COLD range on the dipstick.
15 Check under the vehicle for leaks during the first few trips.

31 Manual transmission lubricant change (every 30,000 miles or 24 months)

1 Raise the vehicle and support it securely on jackstands.
2 Move a drain pan, rags, newspapers and wrenches under the transmission.
3 Remove the transmission drain plug at the bottom of the case and allow the lubricant to drain into the pan **(see illustration 20.1)**.
4 After the lubricant has drained completely, reinstall the plug and tighten it securely.
5 Remove the fill plug from the side of the transmission case. Using a hand pump, syringe or funnel, fill the transmission with the specified lubricant until it begins to leak out through the hole. Reinstall the fill plug and tighten it securely.
6 Lower the vehicle.
7 Drive the vehicle for a short distance, then check the drain and fill plugs for leakage.

32 Differential lubricant change (every 30,000 miles or 24 months)

Refer to illustrations 32.7a, 32.7b, 32.7c and 32.9

Draining

1 This procedure should be performed after the vehicle has been driven so the lubricant will be warm and therefore will flow out of the differential more easily.
2 Raise the vehicle and support it securely on jackstands.
3 The easiest way to drain the differential is to remove the lubricant through the filler plug hole with a suction pump. In fact, this is the only way to remove the front differential lubricant because it has no drain plug or removable cover. If the rear differential's bolt-on cover gasket is leaking, it will be necessary to remove the cover to drain the lubricant (which will also allow you to inspect the differential).

Changing the lubricant with a suction pump

4 Remove the fill plug from the differential (see Section 20).
5 Insert the flexible hose. Work the hose down to the bottom of the differential housing and pump the lubricant out.

Changing the lubricant by removing the cover

6 Move a drain pan, rags, newspapers and wrenches under the vehicle.
7 Remove the bolts on the lower half of the plate **(see illustration)**. Loosen the bolts on the upper half and use them to keep the cover loosely attached **(see illustration)**. Allow the oil to drain into the pan, then completely remove the cover **(see illustration)**.
8 Using a lint-free rag, clean the inside of the cover and the accessible areas of the differential housing. As this is done, check for chipped gears and metal particles in the lubricant, indicating that the differential should be more thoroughly inspected and/or repaired.

Chapter 1 Tune-up and routine maintenance

1-23

32.7a Remove the bolts from the lower edge of the cover . . .

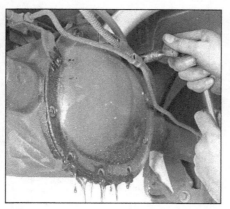

32.7b . . . then loosen the top bolts and let the lubricant drain out

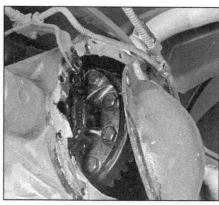

32.7c After the lubricant has drained, remove the bolts and the cover

32.9 Carefully scrape the old gasket material off to ensure a leak-free seal

34.6 The drain plug is located at the lower corner of the radiator (arrow)

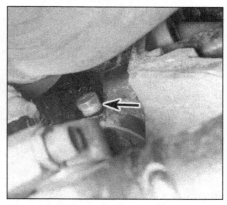

34.7 The block drain plug(s) (arrow) are located below the exhaust manifold(s) (V6 shown)

9 Thoroughly clean the gasket mating surfaces of the differential housing and the cover plate. Use a gasket scraper or putty knife to remove all traces of the old gasket (see illustration).
10 Apply a thin layer of RTV sealant to the cover flange, then press a new gasket into position on the cover. Make sure the bolt holes align properly.

Refilling

11 Use a hand pump, syringe or funnel to fill the differential housing with the specified lubricant until it's level with the bottom of the plug hole. Install the fill plug.

33 Transfer case lubricant change (4WD models) (every 30,000 miles or 24 months)

1 Drive the vehicle for at least 15 minutes in stop and go traffic to warm the lubricant in the case. Perform this warm-up procedure in 4WD. Use all gears, including Reverse, to ensure the lubricant is sufficiently warm to drain completely.
2 Raise the vehicle and support it securely on jackstands.
3 Remove the filler plug from the case (see illustration 22.1).
4 Remove the drain plug from the lower

part of the case and allow the old lubricant to drain completely.
5 Carefully clean and install the drain plug after the case is completely drained. Tighten the plug securely.
6 Fill the case with the specified lubricant until it is level with the lower edge of the filler hole.
7 Install the filler plug and tighten it securely.
8 Drive the vehicle for a short distance and recheck the lubricant level. In some instances a small amount of additional lubricant will have to be added.

34 Cooling system servicing (draining, flushing and refilling) (every 30,000 miles or 24 months)

Refer to illustrations 34.6 and 34.7
Caution: *Never mix green-colored ethylene glycol anti-freeze and orange-colored DEX-COOL® silicate-free coolant because doing so will destroy the efficiency of the DEX-COOL® coolant which is designed to last for 100,000 miles or five years.*
1 Periodically, the cooling system should be drained, flushed and refilled to replenish the antifreeze mixture and prevent formation of rust and corrosion, which can impair the performance of the cooling system and

cause engine damage.
2 At the same time the cooling system is serviced, all hoses and the radiator cap should be inspected and replaced if defective (see Section 10).
3 Since antifreeze is a corrosive and poisonous solution, be careful not to spill any of the coolant mixture on the vehicle's paint or your skin. If this happens, rinse it off immediately with plenty of clean water. Consult local authorities about the dumping of antifreeze before draining the cooling system. In many areas, reclamation centers have been set up to collect automobile oil and drained antifreeze/water mixtures, rather than allowing them to be added to the sewage system.
4 With the engine cold, remove the radiator cap.
5 Move a large container under the radiator to catch the coolant as it's drained.
6 Drain the radiator by opening the drain plug at the bottom on the left side (see illustration). If the drain plug is corroded and can't be turned easily, or if the radiator isn't equipped with a plug, disconnect the lower radiator hose to allow the coolant to drain. Be careful not to get antifreeze on your skin or in your eyes. **Note:** *On later models, the radiator drain plug flows to the side rather than down. Attach a length of hose to the fitting on the drain to lessen spillage.*

35.1a On V6 engines the PCV valve is located in the valve cover - to check it, pull it out and feel for suction and shake it to make sure it rattles

35.1b On four-cylinder engines the PCV valve (arrow) is located in the valve cover

37.9a Use a Torx head tool to remove the distributor cap screws

7   After the coolant stops flowing out of the radiator, move the container under the engine block drain plug(s), then remove the plug(s) from the block **(see illustration)**. The four-cylinder engine has one block drain plug, while the V6 engine has two (one on each cylinder bank).
8   Disconnect the hose from the coolant reservoir and remove the reservoir (see Chapter 3). Flush it out with clean water.
9   Place a garden hose in the radiator filler neck and flush the system until the water runs clear at all drain points.
10   In severe cases of contamination or clogging of the radiator, remove it (see Chapter 3) and reverse flush it. This involves inserting the hose in the bottom radiator outlet to allow the water to run against the normal flow, draining through the top. A radiator repair shop should be consulted if further cleaning or repair is necessary.
11   When the coolant is regularly drained and the system refilled with the correct antifreeze/water mixture, there should be no need to use chemical cleaners or descalers.
12   To refill the system, install the block plug, reconnect any radiator hoses and install the reservoir and the overflow hose.
13   Make sure to use the proper coolant (see **Caution** above). The manufacturer recommends adding GM cooling system sealer part number 3634621 any time the coolant is changed. Slowly fill the radiator with the recommended mixture of antifreeze and water to the base of the filler neck. Wait two minutes and recheck the coolant level, adding if necessary, then install the radiator cap. Add more coolant to the reservoir until it reaches the lower mark.
14   Keep a close watch on the coolant level and the cooling system hoses during the first few miles of driving. Tighten the hose clamps and/or add more coolant as necessary. The coolant level should be a little above the HOT mark on the reservoir with the engine at normal operating temperature.

35 Positive Crankcase Ventilation (PCV) valve check and replacement (every 30,000 miles or 24 months)

Refer to illustrations 35.1a and 35.1b

1   With the engine idling at normal operating temperature, pull the valve (with hose attached) out of the rubber grommet in the valve cover **(see illustrations)**.
2   Place your finger over the end of the valve. If there is no vacuum at the valve, check for a plugged hose, manifold port, or the valve itself. Replace any plugged or deteriorated hoses.
3   Turn off the engine and shake the PCV valve, listening for a rattle. If the valve doesn't rattle, replace it with a new one.
4   To replace the valve, pull it out of the end of the hose, noting its installed position and direction.
5   When purchasing a replacement PCV valve, make sure it's for your particular vehicle, model year and engine size. Compare the old valve with the new one to make sure they are the same.
6   Push the valve into the end of the hose until it's seated.
7   Inspect the rubber grommet for damage and replace it with a new one if necessary.
8   Push the PCV valve and hose securely into position in the valve cover.

36 Evaporative emissions control system check (every 30,000 miles or 24 months)

1   The function of the evaporative emissions control system is to draw fuel vapors from the gas tank and fuel system, store them in a charcoal canister and then burn them during normal engine operation.
2   The most common symptom of a fault in the evaporative emissions system is a strong fuel odor in the engine compartment. If a fuel odor is detected, inspect the charcoal canister. The charcoal canister is mounted on a bracket above the rear axle and behind the gas tank on pick-up models. On sport-utility models, it's located in the quarter panel area behind the left rear wheel. Check the canister and all hoses for damage and deterioration.
3   The evaporative emissions control system is explained in more detail in Chapter 6.

37 Spark plug wire, distributor cap and rotor check and replacement (every 30,000 miles or 24 months)

Refer to illustrations 37.9a, 37.9b, 37.10a and 37.10b
Note: *Four-cylinder engines are equipped with a distributorless ignition system. The spark plug wires are connected directly to the ignition coils. The distributor used on the V6 engine is mounted at the rear of the block.*

1   The spark plug wires should be checked at the recommended intervals and whenever new spark plugs are installed in the engine.
2   The wires should be inspected one at a time to prevent mixing up the order, which is essential for proper engine operation.
3   Disconnect the plug wire from the spark plug. To do this, grab the rubber boot, twist slightly and pull the wire off. Do not pull on the wire itself, only on the rubber boot.
4   Check inside the boot for corrosion, which will look like a white crusty powder. Push the wire and boot back onto the end of the spark plug. It should be a tight fit on the plug. If it isn't, remove the wire and use pliers to carefully crimp the metal connector inside the boot until it fits securely on the end of the spark plug.
5   Using a clean rag, wipe the entire length of the wire to remove any built-up dirt and grease. Once the wire is clean, check for burns, cracks and other damage. Do not bend the wire excessively or pull the wire lengthwise - the conductor inside might break.
6   On four-cylinder engines, disconnect the wire from the coil pack. Again, pull only on the rubber boot. On V6 engines disconnect the wire from the distributor at the rear of the engine. Check for corrosion and a tight fit in the same manner as the spark plug end.

Chapter 1 Tune-up and routine maintenance

37.9b Inspect the inside of the cap for corrosion, carbon tracks and wear

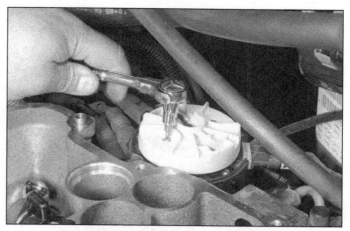

37.10a Use a Torx head screwdriver to remove the distributor rotor

37.10b Check the distributor rotor terminals for wear and burn marks

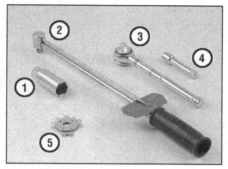

38.2 Tools required for changing spark plugs

1   **Spark plug socket** - This will have special padding inside to protect the spark plug's porcelain insulator
2   **Torque wrench** - Although not mandatory, using this tool is the best way to ensure the plugs are tightened properly
3   **Ratchet** - Standard hand tool to fit the spark plug socket
4   **Extension** - Depending on model and accessories, you may need special extensions and universal joints to reach one or more of the plugs
5   **Spark plug gap gauge** - This gauge for checking the gap comes in a variety of styles. Make sure the gap for your engine is included

38.5a Spark plug manufacturers recommend using a wire type gauge when checking the gap - if the wire does not slide between the electrodes with a slight drag, adjustment is required

Replace the wire at the coil pack or distributor.
7   Check the remaining spark plug wires one at a time, making sure they are securely fastened at the ignition coil or distributor and the spark plug when the check is complete.
8   If new spark plug wires are required, purchase a set for your specific engine model. Wire sets are available pre-cut, with the rubber boots already installed. Remove and replace the wires one at a time to avoid mix-ups in the firing order.
9   Detach the distributor cap by removing the cap retaining screws. Look inside it for cracks, carbon tracks and worn, burned or loose contacts **(see illustrations)**.
10   Remove the retaining screws and pull the rotor off the distributor shaft (it may be necessary to use a small screwdriver to gently pry off the rotor). Examine it for cracks and carbon tracks **(see illustrations)**. Replace the cap and rotor if any damage or defects are noted.
11   It is common practice to install a new cap and rotor whenever new spark plug wires are installed. When installing a new cap, remove the wires from the old cap one at a time and attach them to the new cap in the exact same location - do not simultaneously remove all the wires from the old cap or firing order mix-ups may occur.

38 Spark plug replacement (every 50,000 miles or 36 months)

Refer to illustrations 38.2, 38.5a, 38.5b, 38.6a, 38.6b, 38.8, 38.9 and 38.10

1   The spark plugs are located at the sides of the engine.
2   In most cases, the tools necessary for spark plug replacement include a spark plug socket which fits onto a ratchet (spark plug sockets are padded inside to prevent damage to the porcelain insulators on the new plugs), various extensions and a gap gauge to check and adjust the gaps on the new plugs **(see illustration)**. A special plug wire removal tool is available for separating the wire boots from the spark plugs, and is a good idea on these models because the boots fit very tightly. A torque wrench should be used to tighten the new plugs. It is a good idea to allow the engine to cool before removing or installing the spark plugs.
3   The best approach when replacing the spark plugs is to purchase the new ones in advance, adjust them to the proper gap and replace the plugs one at a time. When buying the new spark plugs, be sure to obtain the correct plug type for your particular engine. The plug type can be found in the Specifications at the front of this Chapter and on the Emission Control Information label located under the hood. If these two sources list different plug types, consider the emission control label correct.
4   Allow the engine to cool completely before attempting to remove any of the plugs. While you are waiting for the engine to cool, check the new plugs for defects and adjust the gaps.
5   Check the gap by inserting the proper

1-26 Chapter 1 Tune-up and routine maintenance

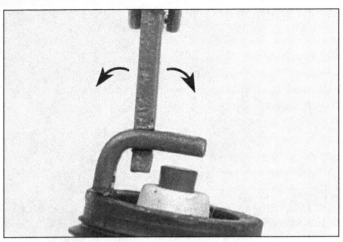

38.5b To change the gap, bend the *side* electrode only, as indicated by the arrows, and be very careful not to crack or chip the porcelain insulator surrounding the center electrode

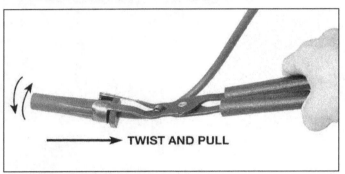

38.6b When removing the spark plug wires, grasp the boot and use a twisting, pulling motion - a tool like this one makes the job easier

38.6a When removing the spark plug wires, pull only on the boot (arrow)

38.8 A socket and extension will be required to remove the spark plugs on these models

thickness gauge between the electrodes at the tip of the plug **(see illustration)**. The gap between the electrodes should be the same as the one specified on the Emissions Control Information label. The wire should slide between the electrodes with a slight amount of drag. If the gap is incorrect, use the adjuster on the gauge body to bend the curved side electrode slightly until the proper gap is obtained **(see illustration)**. If the side electrode is not exactly over the center electrode, bend it with the adjuster until it is. Check for cracks in the porcelain insulator (if any are found, the plug should not be used).

6   With the engine cool, remove the spark plug wire from one spark plug. Pull only on the boot at the end of the wire - do not pull on the wire. A plug wire removal tool should be used if available **(see illustrations)**.

7   If compressed air is available, use it to blow any dirt or foreign material away from the spark plug hole. A common bicycle pump will also work. The idea here is to eliminate the possibility of debris falling into the cylinder as the spark plug is removed.

8   The spark plugs on these models are, for the most part, difficult to reach so a spark plug socket extension will be necessary **(see illustration)**. Place the spark plug socket over the plug and remove it from the engine

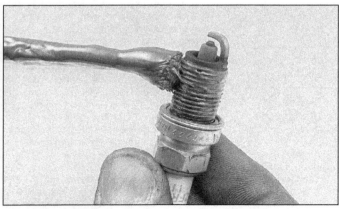

38.9 Apply a thin coat of anti-seize compound to the spark plug threads

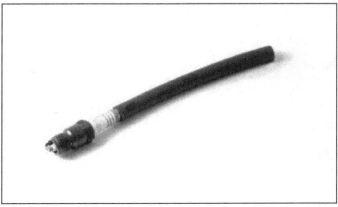

38.10 A length of 3/8-inch ID rubber hose will save time and prevent damaged threads when installing the spark plugs

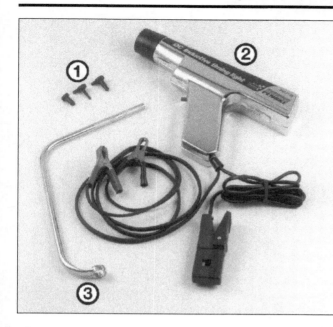

39.2 Tools needed to check and adjust the ignition timing

1. **Vacuum plugs** - Vacuum hoses will, in most cases, have to be disconnected and plugged. Molded plugs in various shapes and sizes are available for this
2. **Inductive pick-up timing light** - Flashes a bright concentrated beam of light when the number one spark plug fires. Connect the leads according to the instructions supplied with the light
3. **Distributor wrench** - On some models, the hold-down bolt for the distributor is difficult to reach and turn with a conventional wrench or socket. A special wrench like this must be used

by turning it in a counterclockwise direction.

9 Compare the spark plug with the chart shown on the inside back cover of this manual to get an indication of the general running condition of the engine. Before installing the new plugs, it is a good idea to apply a thin coat of anti-seize compound to the threads **(see illustration)**. **Caution:** *Do not get anti-seize on either the ground or center electrodes.*

10 Thread one of the new plugs into the hole until you can no longer turn it with your fingers, then tighten it with a torque wrench (if available) or the ratchet. It's a good idea to slip a short length of rubber hose over the end of the plug to use as a tool to thread it into place **(see illustration)**. The hose will grip the plug well enough to turn it, but will start to slip if the plug begins to cross-thread in the hole - this will prevent damaged threads and the accompanying repair costs.

11 Before pushing the spark plug wire onto the end of the plug, inspect it following the procedures outlined in Section 37.

12 Attach the plug wire to the new spark plug, again using a twisting motion on the boot until it's seated on the spark plug.

13 Repeat the procedure for the remaining spark plugs, replacing them one at a time to prevent mixing up the spark plug wires.

39 Ignition timing check and adjustment (1994 and 1995 VIN W and Z V6 only) (every 60,000 miles or 48 months)

Refer to illustration 39.2

Note: *Models other than those identified above do not require periodic ignition timing adjustment.*

1 All vehicles are equipped with an *Emissions Control Information* label inside the engine compartment. The label contains important ignition timing specifications and the proper timing procedure for your specific vehicle. If the information on the emissions label is different from the information included in this Section, follow the procedure on the label.

2 At the specified intervals, or when the distributor has been removed, the ignition timing must be checked and adjusted if necessary. Tools required for this procedure include an inductive pick-up timing light and a distributor wrench **(see illustration)**.

3 Apply the parking brake and block the wheels to prevent movement of the vehicle. The transmission must be in Park (automatic) or Neutral (manual).

4 If the SERVICE ENGINE SOON light is on, don't proceed with the ignition check (see Chapter 6 for more information).

5 The timing system must be bypassed prior to checking the ignition timing. Locate the single tan wire with a black stripe that's behind the glove box, in the harness to the Powertrain Control Module (PCM) and unplug the connector.

6 Before you check the timing, make sure the idle speed is correct (see Chapter 4) and the engine is at normal operating temperature.

7 With the engine off, connect a timing light in accordance with the manufacturer's instructions (an inductive timing light is preferred). **Caution:** *If an inductive pick-up timing light isn't available, don't puncture the spark plug wire to attach the timing light pick-up lead. Instead use an adapter between the spark plug and plug wire. If the insulation on the plug wire is damaged, the secondary voltage will jump to ground at the damaged point and the engine will misfire.* Connect the light to the battery and the number one spark plug wire. The number one spark plug is the one at the front of the engine on the left (driver's) side.

8 Locate the stamped-steel numbered timing scale on the front cover of the engine.

9 Locate the notched groove across the crankshaft pulley. It may be necessary to have an assistant temporarily turn the ignition on and off in short bursts without starting the engine in order to bring the groove into a position where it can easily be cleaned and marked. **Warning:** *Stay clear of all moving engine components when the engine is turned over in this manner.*

10 Use white soap-stone, chalk or paint to mark the groove in the pulley or flywheel. Also, put a mark on the timing scale corresponding to the number of degrees specified on the *Vehicle Emission Control Information* label in the engine compartment or in this Chapter's Specifications.

11 Aim the timing light at the marks, again being careful not to come into contact with moving parts. The marks made should appear stationary. If the marks are in alignment, the timing is correct. If the marks are not aligned, turn off the engine.

12 Loosen the hold-down bolt or nut at the base of the distributor. Loosen the bolt/nut only slightly, just enough to turn the distributor (see Chapter 5).

13 Now restart the engine and turn the distributor very slowly until the timing marks are aligned.

14 Shut off the engine and tighten the distributor bolt/nut, being careful not to move the distributor.

15 Start the engine and recheck the timing to make sure the marks are still in alignment.

16 Disconnect the timing light and reconnect any components which were disconnected for this procedure.

17 Reconnect the timing connector, then clear any computer trouble codes set during the ignition timing procedure (see Chapter 6).

Notes

Chapter 2 Part A
2.2L four-cylinder engine

Contents

	Section		Section
Camshaft and lifters - removal, inspection and installation	11	Intake manifold - removal and installation	6
CHECK ENGINE SOON light	See Chapter 6	Oil pan - removal and installation	12
Crankshaft front oil seal - replacement	8	Oil pump - removal and installation	13
Cylinder compression check	See Chapter 2C	Rear main oil seal - replacement	15
Cylinder head - removal and installation	10	Repair operations possible with the engine in the vehicle	2
Drivebelt check, adjustment and replacement	See Chapter 1	Rocker arms and pushrods - removal, inspection and installation	4
Engine mounts - replacement	16	Spark plug replacement	See Chapter 1
Engine oil and filter change	See Chapter 1	Timing chain cover, chain and sprockets - removal, inspection and installation	9
Engine overhaul - general information	See Chapter 2C		
Engine - removal and installation	See Chapter 2C		
Exhaust manifold - removal and installation	7	Valve cover - removal and installation	3
Flywheel/driveplate - removal and installation	14	Valve springs, retainers and seals - replacement	5
General information	1	Water pump - removal and installation	See Chapter 3

Specifications

General
Cylinder numbers (front-to-rear)	1-2-3-4
Firing order	1-3-4-2
Displacement	134 cubic inches

Camshaft
Lobe lift (intake and exhaust)	
Through 1998	0.288 inches
1999 and later	0.263 inches
Bearing journal diameter	1.868 to 1.869 inches
Bearing oil clearance	0.0015 to 0.0039 inch
Gear/thrust plate end clearance	0.0015 to 0.005 inch

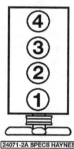

Cylinder numbering and coil terminal locations

Torque specifications
Ft-lbs (unless otherwise indicated)

Camshaft sprocket bolt	96
Camshaft thrust plate bolts	106 in-lbs
Crankshaft pulley bolts	37
Crankshaft pulley/hub-to-crankshaft bolt	77
Cylinder head bolts	
Step 1	
Long bolts	46
Short bolts	43
Step 2	Tighten an additional 90-degrees
Engine mount-to-engine bolts	39
Engine mount through-bolts	38
Engine mount bracket-to-frame bolts	33
Exhaust manifold nuts	115 in-lbs
Flywheel/driveplate-to-crankshaft bolts	55
Intake manifold-to-cylinder head nuts/bolts	
Through 1998	24
1999 and later	17
Lifter guide retainer-to-block stud	97 in-lbs
Oil pan bolts	89 in-lbs
Oil pump-to-block bolt	32
Rocker arm studs	37
Rocker-arm nuts (bolts on 1999 and later)	19 to 22
Timing chain cover bolts	97 in-lbs
Valve cover bolts	89 in-lbs

3.2 Remove the PVC valve from the rubber grommet in the valve cover (bottom arrow) - also remove the vacuum hose to the brake booster (right arrow) and the valve cover-to-air-cleaner ventilation hose (top arrow)

3.3a Remove the resonator box retaining bolt and loosen the clamp that attaches the box to the throttle body (arrows)

3.3b Remove the heater hose tube mounting bolt and pull the tube away from the valve cover (arrow)

3.3c Remove the resonator box mounting bracket bolts and remove the bracket

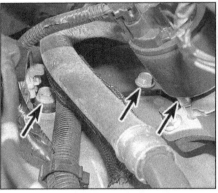

3.3d Remove the engine wiring harness bracket retaining bolts (one bolt is hidden on the rear of the head next to the EGR bracket)

1 General information

This Part of Chapter 2 is devoted to in-vehicle repair procedures for the 2.2L four-cylinder engine. Information concerning engine removal and installation, as well as engine block and cylinder head overhaul, is in Part C of this Chapter.

The following repair procedures are based on the assumption that the engine is installed in the vehicle. If the engine has been removed from the vehicle and mounted on a stand, many of the steps included in this Part of Chapter 2 will not apply.

The Specifications included in this Part of Chapter 2 apply only to the engine and procedures in this Part. The Specifications necessary for rebuilding the block and cylinder head are found in Part C.

2 Repair operations possible with the engine in the vehicle

Many major repair operations can be accomplished without removing the engine from the vehicle.

Clean the engine compartment and the exterior of the engine with some type of pressure washer before any work is done. A clean engine will make the job easier and will help keep dirt out of the internal areas of the engine.

Depending on the components involved, it may be necessary to remove the hood to improve access to the engine as repairs are performed (see Chapter 11 if necessary).

If vacuum, exhaust, oil or coolant leaks develop, indicating a need for gasket or seal replacement, the repairs can generally be made with the engine in the vehicle. The intake and exhaust manifold gaskets, valve cover gasket and cylinder head gasket are all accessible with the engine in place. Changing the oil pan gasket, however, requires virtually removing the engine to provide enough clearance for oil pan removal. Information on engine removal is in Part C.

Exterior engine components such as the intake and exhaust manifolds, the water pump, the starter motor, the alternator, the distributor and the fuel injection system can be removed for repair with the engine in place.

Since the cylinder head can be removed without pulling the engine, valve component servicing can also be accomplished with the engine in the vehicle.

3 Valve cover - removal and installation

Refer to illustrations 3.2, 3.3a, 3.3b, 3.3c, 3.3d, 3.4a, 3.4b and 3.5

Removal

1 Disconnect the negative battery cable from the battery, then remove the air intake duct assembly (see Chapter 4). **Caution:** *On models equipped with a Delco Loc II audio system, be sure the lockout feature is turned off before performing any procedure which requires disconnecting the battery.*

2 Disconnect the accelerator cable and bracket (see Chapter 4). Remove the PCV valve from the valve cover and the vacuum hose to the power brake booster (see illustration).

3 On 1998 (SFI) models:

a) Remove the air cleaner resonator box retaining bolt, and clamp, pull the box back and up to remove the PCV valve hose (see illustration).

b) Remove the heater hose tube mounting bolt (see illustration). *On some 1999 and later models, the coolant may have to be drained and the coolant pipe removed in order to access the valve cover.*

c) Remove the resonator mounting bracket bolts (see illustration).

d) Remove the engine wiring harness mounting bracket retaining bolts and bracket (see illustration).

4 Remove the valve cover bolts (see illustrations).

5 Remove the valve cover. If it sticks to the head, use a soft-face hammer or a block of wood and a hammer to dislodge it. If the cover still won't come loose, pry on it carefully at several points until the seal is broken, but don't distort the cover. **Note:** *The valve cover can be slid out toward the front of the engine without disconnecting the spark plug wires* (see illustration).

Installation

6 Prior to reinstallation, remove all dirt, oil

Chapter 2 Part A 2.2L four-cylinder engine

3.4a Remove the valve cover bolts - don't pry under the cover or you could damage the gasket sealing surfaces, the head and valve cover are both soft aluminum

3.4b Remove the valve cover mounting bolts - it is not necessary to replace the gasket unless its damaged

and old gasket material from the cover and cylinder head with a scraper. Clean the mating surfaces with lacquer thinner or acetone.

7 Place the valve cover on the cylinder head with a new gasket and install the mounting bolts. Tighten the bolts a little at a time to the torque listed in this Chapter's Specifications.

8 Complete the installation by reversing the removal procedure.

9 Start the engine and check for oil leaks at the valve cover-to-head joint.

4 Rocker arms and pushrods - removal, inspection and installation

Refer to illustrations 4.2, 4.4 and 4.5

Removal

1 Detach the valve cover from the cylinder head (see Section 3).

2 The rocker arms and pushrods should only be removed from a specific cylinder when that cylinder is at Top Dead Center, and the lifters for that cylinder are on the base circle of the camshaft. To do this, rotate the crankshaft with a socket and breaker bar on the crankshaft damper bolt, using a plastic ball-point pen in the spark plug hole **(see illustration)**. When the movement of the pen indicates that the piston is at it's highest point of travel, check the two rocker arms for that cylinder. If neither valve spring is under tension, this is roughly TDC for that particular cylinder. To find TDC for another cylinder, follow the firing order sequence. Mark the damper and front cover with chalk at number one TDC, rotate the crankshaft 180-degrees for number 3 TDC, 180-degrees further for number 4 TDC and finally 180-degrees again for number 2 TDC.

3 Beginning at the front of the cylinder head, loosen the rocker arm bolts. **Note:** *If the pushrods are the only items being removed, rotate the rocker arms to one side so the pushrods can be lifted out.*

4 Remove the rocker arm nuts, the rocker arms and the pivot balls **(see illustration)** and store them in marked containers (they must be reinstalled in their original locations). **Note:** *On 1999 and later models, the rocker arms are secured by bolts, rather than studs and nuts.*

5 Remove the pushrods and store them separately to make sure they don't get mixed up during installation **(see illustration)**.

6 If the pushrod guides must be removed for any reason, make sure they're marked so they can be reinstalled in their original locations.

3.5 The shallow valve cover can be removed toward the front of the engine without disconnecting the spark plug wires

Inspection

7 Check each rocker arm for wear, cracks and other damage, especially where the pushrods and valve stems contact the rocker arm faces.

8 Make sure the hole at the pushrod end of each rocker arm is open.

9 Check each rocker arm pivot area for wear, cracks and galling. If the rocker arms are worn or damaged, replace them with new ones and use new pivot balls as well. On 1999 and later models, the rocker arms have an integral roller bearing instead of a pivot ball. Clean the roller bearing with solvent and spin by hand to check for free operation with-

4.2 On engines with direct ignition (no distributor) and no timing marks on the crankshaft damper, a plastic pen inserted into the spark plug hole can be used to locate approximate TDC

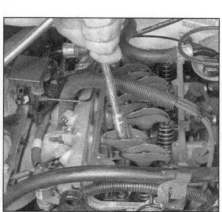

4.4 Remove the nut, ball and rocker arm and keep all components for each valve together

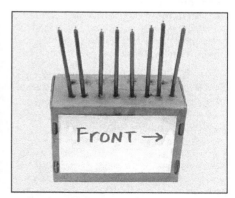

4.5 A perforated cardboard box can be used to store the pushrods to ensure they're installed in their original locations - note the label indicating the front of the engine

Chapter 2 Part A 2.2L four-cylinder engine

5.4 This is what the air hose adapter that threads into the spark plug hole looks like - they're commonly available from auto parts stores

5.7 A clamp-type spring compressor is used to compress the spring and remove the keepers to replace valve seals or springs with the head installed

5.15 Keepers don't always stay in place, so apply a small dab of grease to each one as shown here before installation - it'll hold them in place on the valve stem as the spring is released

out excessive looseness.
10 Inspect the pushrods for cracks and excessive wear at the ends. Roll each pushrod across a piece of plate glass to see if it's bent (if it wobbles, it's bent).

Installation

11 Lubricate the lower ends of the pushrods with clean engine oil or moly-base grease and install them in their original locations. Make sure each pushrod seats completely in the lifter socket.
12 Apply moly-base grease to the ends of the valve stems and the upper ends of the pushrods before positioning the rocker arms and installing the bolts.
13 Set the rocker arms in place, then install the pivot balls and bolts. Apply engine assembly lube to the pivot balls to prevent damage to the mating surfaces before engine oil pressure builds up. Tighten the rocker arm nuts to the torque listed in this Chapter's Specifications. In order for the valve spring tension to not affect the rocker nut torque reading, the nuts for a particular cylinder should be tightened when that cylinder is at TDC, and there is no valve spring load on the rocker arm (see Step 2). On 1999 and later models, align the tab on the bottom of the roller-rocker pivot with the slot in the cylinder head casting protrusion.
14 After the rocker arms are tightened, reinstall the valve cover (refer to Section 3), start the engine and check for oil leaks or noises from the valvetrain.

5 Valve springs, retainers and seals - replacement

Refer to illustrations 5.4, 5.7 and 5.15

Note: *Broken valve springs and defective valve stem seals can be replaced without removing the cylinder head. Two special tools and a compressed air source are normally required to perform this operation, so read through this Section carefully and rent or buy the tools before beginning the job.*

1 Remove the spark plug from the cylinder which has the defective component. If all of the valve stem seals are being replaced, remove all of the spark plugs.
2 If you're replacing all of the valve stem seals, begin with cylinder number one (at TDC, see Section 4, Step 2) and work on the valves for one cylinder at a time.
3 Work from cylinder-to-cylinder, at the respective TDC, following the firing order sequence (1-3-4-2).
4 Thread an adapter into the spark plug hole and connect an air hose from a compressed air source to it **(see illustration)**. Most auto parts stores can supply the air hose adapter. **Note:** *Many cylinder compression gauges utilize a screw-in fitting that may work with your air hose quick-disconnect fitting.*
5 Remove the rocker arm nut, pivot ball and rocker arm for the valve with the defective part and pull out the pushrod. If all of the valve stem seals are being replaced, remove all of the rocker arms and pushrods (see Section 4).
6 Apply compressed air to the cylinder. The valves should be held in place by the air pressure. **Warning:** *If the cylinder isn't exactly at TDC, air pressure may cause the engine to rotate. Do not leave a socket or wrench on the balancer bolt; damage or personal injury may result.*
7 Stuff shop rags into any cylinder head holes near the valves to prevent parts and tools from falling into the engine, then use a valve spring compressor to compress the spring. Remove the keepers with small needle-nose pliers or a magnet **(see illustration)**. **Note:** *A couple of different types of tools are available for compressing the valve springs with the head in place. One type utilizes the rocker arm stud/nut for leverage, while the other type, shown here, grips the lower spring coils and presses on the retainer as the knob is turned. Both types work very well, although the lever type is usually less expensive.*
8 Remove the spring retainer and valve spring, then remove the guide seal. **Note:** *If air pressure fails to hold the valve in the closed position during this operation, the valve face and/or seat is probably damaged. If so, the cylinder head will have to be removed for additional repair operations.*
9 Wrap a rubber band or tape around the top of the valve stem so the valve will not fall into the combustion chamber, then release the air pressure.
10 Inspect the valve stem for damage. Rotate the valve in the guide and check the end for eccentric movement, which would indicate that the valve is bent.
11 Move the valve up-and-down in the guide and make sure it does not bind. If the valve stem binds, either the valve is bent or the guide is damaged. In either case, the head will have to be removed for repair.
12 Reapply air pressure to the cylinder to retain the valve in the closed position, then remove the tape or rubber band from the valve stem.
13 Lubricate the valve stem with engine oil and install a new guide seal, pushing it down onto the top of the guide only until it stops. Use a deep socket to carefully tap the seal onto the guide, if necessary.
14 Install the spring and retainer in position over the valve.
15 Compress the valve spring and retainer, and position the keepers in the upper groove. Apply a small dab of grease to the inside of each keeper to hold it in place if necessary **(see illustration)**. Remove the pressure from the spring tool and make sure the keepers are seated.
16 Disconnect the air hose and remove the adapter from the spark plug hole.
17 Install the rocker arm(s) and pushrod(s) (see Section 4).
18 Install the spark plug(s) and hook up the wire(s).
19 Install the valve cover (see Section 3).
20 Start and run the engine, then check for oil leaks and unusual sounds coming from the valve cover area.

6 Intake manifold - removal and installation

Refer to illustrations 6.7, 6.8a, 6.8b and 6.13

Removal

1 Disconnect the cable from the negative

Chapter 2 Part A 2.2L four-cylinder engine

6.7 Unbolt the heater hose pipe (arrow) from the intake manifold

6.8a Remove the intake manifold nuts and pull off the manifold

6.8b Remove the EGR passage tube (arrow) and soak it in solvent to clean it thoroughly

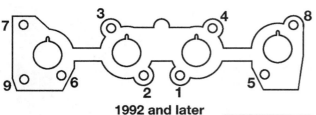

6.13 Intake manifold bolt TIGHTENING sequence

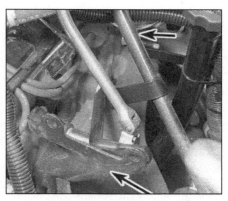

7.6 Unbolt the rear air conditioning compressor bracket (lower arrow), shown here with compressor removed for clarity), and remove the nut (upper arrow) and the oil filler tube

battery terminal. **Caution:** *On models equipped with a Delco Loc II audio system, be sure the lockout feature is turned off before performing any procedure which requires disconnecting the battery.*

2 Remove the air intake duct assembly (see Chapter 4).

3 Remove the PCV valve and hose connecting the intake manifold to the valve cover.

4 Refer to Chapter 4 and remove the upper intake manifold (1997 and earlier MFI models only). Later models have a one-piece intake manifold. Tag and disconnect all electrical connectors at the throttle body and manifold.

5 Remove the fuel lines and any wiring attached to the lower intake manifold. Remove the fuel rail bracket.

6 Remove the spark plug wires from the coil pack. Mark the spark plug wires so there's no confusion on reassembly. Pull them up through the intake manifold, then lay them over the left side of the engine, out of the way.

7 Remove the bolt holding the heater hose tube to the right rear of the manifold **(see illustration)**.

8 Remove the mounting nuts and separate the manifold from the cylinder head **(see illustration)**. Don't pry between the manifold and head, as damage to the gasket sealing surfaces may result. Also remove the EGR tube for cleaning **(see illustration)**.

9 Remove the old gasket, and if a new manifold is being installed, transfer all components still attached to the old manifold to the new one.

Installation

10 Before installing the manifold, clean the cylinder head and manifold gasket surfaces with lacquer thinner or acetone. All gasket material and sealing compound must be removed prior to installation. Also clean the carbon out of the EGR passages with a sharp tool or a length of coat-hanger wire.

11 Apply a thin coat of RTV sealant to the intake manifold and cylinder head mating surfaces. Make certain that the sealant will not spread into the ports when the manifold is installed.

12 Place a new gasket on the manifold, hold the manifold in position against the cylinder head and install the mounting nuts finger tight.

13 Tighten the mounting nuts, in several steps and in the correct sequence **(see illustration),** to the torque listed in this Chapter's Specifications.

14 Install the remaining components in the reverse order of removal.

7 Exhaust manifold - removal and installation

Refer to illustration 7.6

Removal

1 Remove the cable from the negative battery terminal. **Caution:** *On models equipped with a Delco Loc II audio system, be sure the lockout feature is turned off before performing any procedure which requires disconnecting the battery.*

2 Remove the air cleaner assembly and duct (see Chapter 4).

3 Remove the oxygen sensor from the manifold (see Chapter 6).

4 Raise the vehicle and support it securely on jackstands.

5 Disconnect the exhaust pipe from the exhaust manifold. You may have to apply penetrating oil to the fastener threads, as they are usually corroded.

6 On air-conditioned models, remove the air conditioning compressor rear mount **(see illustration)**, without disconnecting the lines. Remove the nut on the oil filler tube assembly.

7 Remove the exhaust manifold nuts and separate the exhaust manifold from the engine.

8 Remove the exhaust manifold gasket.

Installation

9 Before installing the manifold, clean the gasket mating surfaces on the cylinder head and manifold. All old gasket material and carbon deposits must be removed. Check the bolt threads for damage.

10 Place a new exhaust manifold gasket in position on the cylinder head, then place the manifold in position and tighten the mounting nuts to this Chapter's Specifications.

8.4 Remove the center hub retaining bolt

8.5 Remove the pulley-to-hub bolts

8.6 Use a puller to remove the crankshaft pulley hub

11 Install the remaining components in the reverse order of removal.
12 Start the engine and check for exhaust leaks between the manifold and cylinder head and between the manifold and exhaust pipe.

8 Crankshaft front oil seal - replacement

Refer to illustrations 8.4, 8.5 and 8.6

Crankshaft pulley and hub removal

1 Remove the cable from the negative battery terminal. **Caution:** *On models equipped with a Delco Loc II audio system, be sure the lockout feature is turned off before performing any procedure which requires disconnecting the battery.*
2 Remove the drivebelt (see Chapter 1).
3 Raise the vehicle and place it securely on jackstands.
4 Remove the hub retaining bolt **(see illustration)**.
5 Place a bar through one of the pulley holes and wedge it against the bottom of the front cover, to keep the pulley from turning while loosening the pulley bolts **(see illustration)**.
6 Remove the pulley hub. Use a puller if necessary **(see illustration)**.

Seal replacement

Refer to illustrations 8.7, 8.9a, 8.9b and 8.11
7 Note how the seal is installed - the new one must face the same direction! Carefully pry the oil seal out of the cover with a seal puller or a large screwdriver **(see illustration)**. Be very careful not to distort the cover or scratch the crankshaft!
8 Apply a thin layer of engine oil or multi-purpose grease to the oil seal lip and to the outer edge of the new seal. Install the seal in the cover with the lip (open end) facing IN. Drive the seal into place with a large socket and a hammer (if a large socket isn't available, a section of the appropriate size pipe will also work). Make sure the seal enters the bore squarely and driven in until the front face is flush with the cover.
9 The seal can also be replaced if the timing chain cover has been removed for other work. Support the cover on a block of wood and drive the seal out from the backside **(see illustration)**.
10 Install a new seal into the bore as in Step 8.
11 Reinstall the timing chain cover with a bead of RTV sealant applied to the mating

8.7 Pry the old seal out with a seal-removing tool or a screwdriver wrapped with tape to protect the crankshaft's seal surface

surface of the oil pan. **Note:** *The studs at the bottom of the front cover have a small hex at the end, and they can be removed with a female Torx socket. Removing the studs makes installing the cover with the oil pan in place much easier* **(see illustration)**.
12 Apply a small amount of RTV sealant on the crankshaft Woodruff key. Align the slot in the hub with the crankshaft Woodruff key,

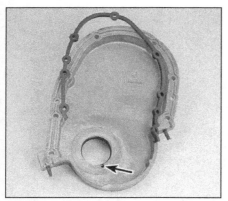

8.9a The timing chain cover uses a rubber lip seal pressed into the bore - at the bottom is a slot (arrow) to make driving the old seal out easier

8.9b A large socket and hammer can be used to squarely drive the new seal into place - stop when it is flush with the cover

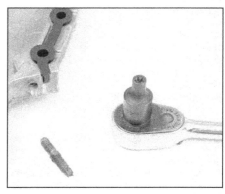

8.11 The lower studs on the front cover can be removed with a female Torx socket to make cover installation easier - be sure to install the studs once the cover is sealed in place

9.3 Remove the belt tensioner mounting bracket from under the alternator - there are three bolts; the wrench is on one and arrows indicate the other two

9.4 Remove the timing chain cover-to-block bolts

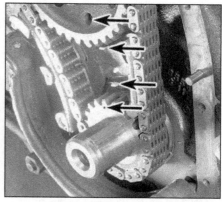

9.6 Turn the crankshaft so the marks on the sprockets line up with the two projections on the center of the tensioner (arrows)

then press the hub onto the crankshaft until it bottoms against the crankshaft timing gear. The hub retaining bolt can be used to press the hub into position.
13 Install the crankshaft pulley on the hub. Use a non-hardening thread-locking compound on the bolt threads.
14 Tighten the hub-to-crankshaft bolt to the torque listed in this Chapter's Specifications.
15 Install the drivebelt (see Chapter 1).

9 Timing chain cover, chain and sprockets - removal, inspection and installation

Refer to illustrations 9.3 and 9.4

Cover removal

1 Disconnect the negative battery cable from the battery. **Caution:** *On models equipped with a Delco Loc II audio system, be sure the lockout feature is turned off before performing any procedure which requires disconnecting the battery.*
2 Remove the upper fan shroud, drivebelt, fan/clutch assembly and water pump pulley (see Chapter 3). Remove the crankshaft pulley and hub (see Section 8).
3 Remove drivebelt tensioner bracket, which is retained with three bolts under the alternator **(see illustration)**.
4 Remove the timing chain cover-to-engine block bolts **(see illustration)**. Remove the nuts from the studs that go through the front of the oil pan. Pull the cover forward to break the gasket seal, then pull up and guide the studs out of the oil pan. **Note:** *The studs at the bottom of the front cover have a small hex at the end, and they can be removed with a female Torx socket. Removing the studs makes cover removal and installation much easier* **(see illustration 8.11)**.
5 Use a gasket scraper to remove all traces of old gasket material and sealant from the cover and engine block. Clean the gasket sealing surfaces with lacquer thinner or acetone.

Timing chain removal

Refer to illustrations 9.6, 9.7 and 9.8
6 Temporarily install the crankshaft pulley hub and bolt. Rotate the crankshaft until the timing marks on the crankshaft and camshaft sprockets align with the tabs on the chain tensioner **(see illustration)**.
7 Push the spring back on the timing chain tensioner and insert a pin into the hole to retain it in the retracted position **(see illustration)**.
8 Use a prybar against two bolts in the crankshaft balancer hub (installed temporarily) to keep the engine from turning while removing the camshaft bolt **(see illustration)**. Do not turn the camshaft in the process (if you do, realign the timing marks before the sprocket is removed).
9 Use two large screwdrivers to carefully pry the camshaft sprocket off the camshaft dowel pin (it may come off easily with no tools required), and remove the sprocket and chain.

Inspection

10 Timing chains and sprockets should be replaced in sets. If you intend to install a new timing chain, remove the crankshaft sprocket with a puller and install a new one. Be sure to align the key in the crankshaft with the keyway in the sprocket during installation.

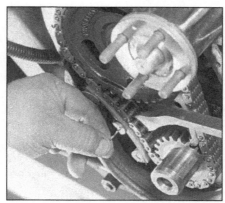

9.7 Push back the tensioner spring and insert a pin to retain it in the retracted position - then remove the tensioner

11 Clean the timing chain and sprockets with solvent and dry them with compressed air (if available). **Warning:** *Wear eye protection when using compressed air.*
12 Inspect the components for wear and damage. Look for teeth that are deformed, chipped, pitted and cracked.
13 The timing chain should be replaced with a new one if the engine has high mileage, the chain has visible damage, or total freeplay (without the tensioner) midway between the sprockets exceeds one inch. Failure to replace a worn timing chain may result in erratic engine performance, loss of power and decreased fuel mileage. Loose chains can "jump" timing and in the worst case, will result in severe engine damage. **Note:** *If the timing chain is being replaced, replace the tensioner as well.*

Installation

Refer to illustration 9.18
14 Mesh the timing chain with the camshaft sprocket, then engage it with the crankshaft sprocket. The timing marks should be aligned as shown in **illustration 9.6**. **Note:** *If the crankshaft has been disturbed, turn it until the*

9.8 Use a prybar against two bolts in the temporarily installed crankshaft hub to hold the engine while a breaker bar is used on the camshaft bolt

9.18 Apply RTV sealant to the areas where the pan and front cover meet, and along the top of the exposed portion of the oil pan gasket

10.10 Unbolt and remove this engine lifting bracket (arrow) from the rear of the cylinder head

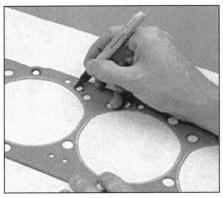

10.11 To avoid mixing up the head bolts, use the new gasket to transfer the bolt hole pattern to a piece of cardboard, then punch holes in the cardboard to accept the bolts

mark stamped on the crankshaft sprocket is pointing at the projection on the tensioner. If the camshaft was turned, install the sprocket temporarily and turn the camshaft until the timing marks align.

15 Install the camshaft sprocket bolt and tighten it to the torque listed in this Chapter's Specifications.

16 Press the timing chain against the tensioner, pull out the pin retaining the spring and release the tensioner.

17 Lubricate the chain and sprocket with clean engine oil. Rotate the engine through two complete revolutions and check the alignment of the timing marks again.

18 Apply RTV sealant to the exposed area of oil pan sealing surface **(see illustration)**. Install the timing chain cover, using a new rubber cover seal and tighten the cover bolts to the torque listed in this Chapter's Specifications.

19 The remaining installation steps are the reverse of removal.

10 Cylinder head - removal and installation

Removal

Refer to illustrations 10.10 and 10.11

1 Disconnect the negative cable from the battery. **Caution:** *On models equipped with a Delco Loc II audio system, be sure the lockout feature is turned off before performing any procedure which requires disconnecting the battery.*

2 Drain the cooling system. Remove the drivebelt, upper fan shroud, fan/clutch assembly and the water pump pulley (see Chapter 3).

3 Remove the upper radiator hose and heater hose from the thermostat housing. Disconnect the electrical connector to the coolant temperature sensor and remove the thermostat housing (see Chapter 3).

4 Remove the air cleaner assembly (see Chapter 4).

5 On models through 1998, remove the upper intake manifold (see Chapter 4), and the lower intake manifold (see Section 6). On 1999 and later models, the intake manifold is one-piece.

6 Remove the dipstick tube.

7 Remove the alternator and bracket from the right side of the engine, and the air conditioning compressor, power steering pump and their shared bracket (see Chapter 5, Chapter 3 and Section 9 of this Chapter). **Caution:** *Don't disconnect any of the air conditioning lines unless the system has been depressurized by a dealer service department or other repair shop. Lay the bracket aside with the air conditioning and power steering hoses still connected.*

8 Refer to Section 7 and remove the exhaust manifold.

9 Refer to Sections 3 and 4 and remove the valve cover, rocker arms and pushrods.

10 Unbolt and remove the engine lifting bracket from the rear of the cylinder head **(see illustration)**.

11 Using a new head gasket, outline the cylinders and bolt pattern on a piece of cardboard **(see illustration)**. Be sure to indicate the front of the engine for reference. Punch holes at the bolt locations to store the bolts as they are removed.

12 Loosen the cylinder head mounting bolts in 1/4-turn increments until they can be removed by hand. Store the bolts in the cardboard holder as they're removed - this will ensure that they are reinstalled in their original locations.

13 Lift the head off the engine. If it's stuck, pry on it only at the overhang on the thermostat end of the head. **Caution:** *If you pry on the head anywhere else, damage to the gasket surface may result.*

14 Place the head on a block of wood to prevent damage to the gasket surface. See Part C for cylinder head disassembly and valve service procedures.

Installation

Refer to illustrations 10.19 and 10.22

15 If a new cylinder head is being installed, transfer all external parts from the old cylinder head to the new one.

16 The mating surfaces of the cylinder head and block must be perfectly clean when the head is installed. It's also a good idea to have the head checked for distortion (warpage) and cracks by an automotive machine shop.

17 Use a wood or plastic gasket scraper to remove all traces of carbon and old gasket material, then clean the mating surfaces with lacquer thinner or acetone. If there's oil on the mating surfaces when the head is installed, the gasket may not seal correctly and leaks may develop. Use a vacuum cleaner to remove any debris that falls into the cylinders. **Caution:** *When cleaning the aluminum cylinder head, do not use a rotary wire brush or an abrasive disc with an electric drill; head gasket sealing problems may result.*

18 Check the block and head mating surfaces for nicks, deep scratches and other damage. If damage is slight, it can be removed with a file; if it's excessive, machining may be the only alternative. **Caution:** *The manufacturer does not advise milling this aluminum head more than 0.010-inch, nor do they recommend any welding repairs be made to this alloy.*

10.19 A die should be used to remove corrosion and sealant from the head bolt threads prior to installation

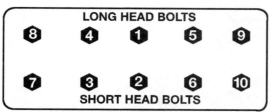

10.22 Cylinder head bolt TIGHTENING sequence

11.3 A dial indicator set up on the rocker arm over the pushrod can be used to check camshaft lobe lift

19 Use a tap of the correct size to chase the threads in the head bolt holes. Mount each bolt in a vise and run a die down the threads to remove corrosion and restore the threads **(see illustration)**. Dirt, corrosion, sealant and damaged threads will affect critical head bolt torque readings.
20 Position the new gasket over the dowel pins in the block (no sealant is used on the gasket), then carefully position the head on the block without disturbing the gasket. Most gaskets are marked "this side up" on one side.
21 Coat the threads of all cylinder head bolts with sealing compound and install the bolts finger tight. **Caution:** *The head bolts should be tightened within 15 minutes of installation, before the sealant sets up.*
22 Tighten the bolts, following the recommended sequence, to the torque listed in this Chapter's Specifications **(see illustration)**.
23 The remaining installation steps are the reverse of removal.
24 Change the oil and filter (see Chapter 1), add coolant, then run the engine and check for leaks.

11 Camshaft and lifters - removal, inspection and installation

Camshaft lobe lift check

Refer to illustration 11.3

1 To determine the extent of cam lobe wear, the lobe lift should be checked prior to camshaft removal. Position the number one piston at TDC on the compression stroke (see Section 4).
2 Remove the valve cover. Refer to Section 4 and loosen the rocker arm nuts, then pivot the rocker arms to the side for access to the pushrods. On 1999 and later models, remove the roller rocker arms to remove the pushrods (see Section 4).
3 Beginning with the number one cylinder valves, mount a dial indicator on the engine and position the plunger against the top of the first pushrod. The plunger should be directly in line with the pushrod **(see illustration)**.
4 Zero the dial indicator, then very slowly turn the crankshaft in the normal direction of rotation (clockwise when looking at the front of the engine) until the indicator needle stops and begins to move in the opposite direction. The point at which it stops indicates maximum cam lobe lift.
5 Record this figure for future reference, then reposition the piston at TDC on the compression stroke.
6 Move the dial indicator to the remaining number one cylinder pushrod and repeat the check. Be sure to record the results for each valve.
7 Repeat the check for the remaining valves. Since each piston must be at TDC on the compression stroke for this procedure, work from cylinder-to-cylinder following the firing order sequence. Make chalk marks on the damper and front cover, then turn the crankshaft 180-degrees when moving from one cylinder to the next (see Section 4, Step 2).
8 After the check is complete, compare the results to the Specifications. If camshaft lobe lift is less than specified, cam lobe wear has occurred and a new camshaft should be installed.

Lifter check and removal

Refer to illustrations 11.12a and 11.12b

9 A noisy valve lifter can be isolated when the engine is idling. Place a mechanics stethoscope or a length of hose or tubing near the location of each valve while listening at the other end. If a valve lifter is defective, it'll be evident from the noise caused by the excessive clearance in the valve train.
10 The most likely cause of a noisy valve lifter is a piece of dirt trapped between the plunger and the lifter body.
11 Removing the lifters on this engine requires that the cylinder head be removed, as the roller lifters are secured in the block with retainers under the head. Refer to Section 10 for cylinder head removal.
12 Remove the bolts holding the two lifter guides in the block and remove the lifter guides **(see illustrations)**.

11.12a Each of the two lifter guides is retained in the block with a bolt

11.12b Remove the lifter guides by pulling straight up

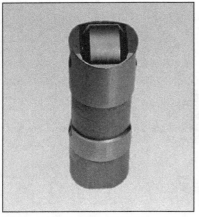

11.16 The roller on the bottom of the lifter must turn freely - check for wear and excessive play as well

11.20 Remove the camshaft thrust plate bolts (arrows) - note the orientation of the thrust plate so it can be reinstalled with the wear surface against the camshaft

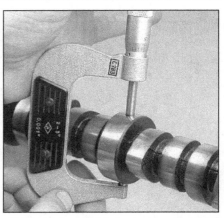

11.23a Using a micrometer, measure the camshaft journals and compare the measurements to Specifications

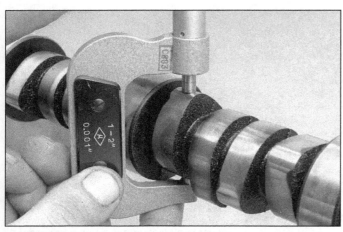

11.23b Measure the camshaft lobes at their greatest dimension and write down the measurements . . .

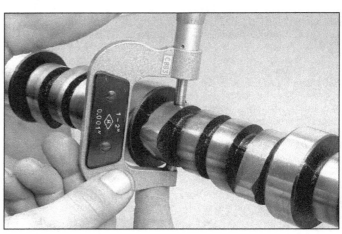

11.23c . . . and subtract the measurements of the lobe diameters at their smallest dimension to obtain the lobe lift specification

13   There are several ways to extract a lifter from its bore. A special removal tool is available, but isn't always necessary. On newer engines without a lot of varnish buildup, lifters can often be removed with a small magnet or even with your fingers. A scribe can also be used to pull the lifter out of the bore. **Caution:** *Don't use pliers of any type to remove a lifter unless you intend to replace it with a new one - they will damage the precision machined and hardened surface of the lifter, rendering it useless.*

14   Store the lifters in a clearly labeled box to insure their reinstallation in the same lifter bores.

Lifter inspection

Refer to illustration 11.16

15   Clean the lifters with solvent and dry them thoroughly without mixing them up.

16   Check each lifter wall, pushrod seat and roller for scuffing, score marks and uneven wear. Check the rollers for wear and damage. Make sure the rollers turn freely without excessive play **(see illustration)**. If the lifter walls are damaged or worn (which isn't very likely), inspect the lifter bores in the engine block as well. If the pushrod seats are worn, check the pushrod ends.

17   Even though roller lifters do not have the same cam-to-lifter wear pattern as conventional flat lifters, the manufacturer recommends replacing the cam and lifters as a set. Since the head must be removed to replace worn or noisy lifters, it's good insurance to replace the lifters if the cam is worn, or vice-versa.

Camshaft removal and inspection

Refer to illustrations 11.20, 11.23a, 11.23b and 11.23c

18   Refer to Section 9 for removal of the front cover, timing chain and sprockets.

19   Unbolt and remove the oil pump drive from the rear of the block. Disconnect and remove the camshaft position sensor (see Chapter 6).

20   Remove the two bolts and the camshaft thrust plate, noting which side faces against the block **(see illustration)**.

21   Carefully pull the camshaft out of the block. **Caution:** *To avoid damage to the camshaft bearings as the lobes pass over them, support the camshaft near the block as it's withdrawn.*

22   After the camshaft has been removed from the engine, cleaned with solvent and dried, inspect the bearing journals for uneven wear, pitting and evidence of seizure. If the journals are damaged, the bearing inserts in the block are probably damaged as well. Both the camshaft and bearings will have to be replaced. Camshaft bearing replacement is a procedure that should be done in an automotive machine shop, and requires that the engine be removed from the vehicle and stripped. Refer to Part C of this Chapter for camshaft bearing measurement procedures.

23   If the journals are in good condition, measure the bearing journals with a micrometer to determine their sizes and whether or not they're out-of-round. Also measure the camshaft lobes **(see illustrations)**.

24   Check the camshaft lobes for heat discoloration, score marks, chipped areas, pitting and uneven wear. If the lobes are in good condition and if the lobe lift measurements are as specified, the camshaft can be reused.

Chapter 2 Part A 2.2L four-cylinder engine

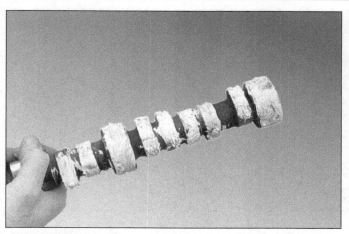

11.25 Lubricate the camshaft lobes and journals thoroughly with camshaft and lifter assembly lube before installation,

14.2 A large screwdriver can be used as shown to hold the driveplate from turning while the mounting bolts are loosened or tightened

Installation

Refer to illustration 11.25

25 Lubricate the camshaft bearing journals and cam lobes with camshaft and lifter assembly lube **(see illustration)**.
26 Slide the camshaft into the engine. Support the cam near the block and be careful not to scrape or nick the bearings.
27 Install the camshaft thrust plate. Make sure you position the same side towards the block as originally installed. Tighten the bolts to the torque listed in this Chapter's Specifications. Reinstall the oil pump drive in the block.
28 If the used lifters are being reinstalled, they must be installed in their original bores. Coat new or used lifters with camshaft and lifter assembly lube.
29 Lubricate the bearing surfaces of the lifter bores with engine oil.
30 Install the lifter(s) in the lifter bore(s). **Note:** *Make sure the rollers are oriented to rotate parallel to the camshaft lobes.*
31 Install the lifter guides, and refer to Section 10 for reinstalling the cylinder head.
32 Refer to Section 9 for installation of the timing chain and sprockets.
33 Install the pushrods, pushrod guides, rocker arms and rocker arm retaining nuts (see Section 4). **Caution:** *Make sure that each pair of lifters is on the base circle of the camshaft (that is, with both valves closed) before tightening the rocker arm bolts.*
34 Tighten the rocker arm nuts to the torque listed in this Chapter's Specifications.
35 The remaining installation steps are the reverse of removal.

12 Oil pan - removal and installation

Removal

1 The manufacturer recommends that the oil pan only be removed with the engine out of the vehicle. Refer to Chapter 2, Part C for engine removal procedures.
2 Remove the bolts and detach the oil pan. Don't pry between the block and pan or damage to the sealing surfaces may result and oil leaks could develop. If the pan is stuck, dislodge it with a block of wood and a hammer. **Note:** *Some of the fasteners are studs. Before removing them, make a mark on the pan to indicate their location for reassembly.*
3 Use a scraper to remove all traces of sealant from the pan and block, then clean the mating surfaces with lacquer thinner or acetone.

Installation

4 Install a new rubber half-circle seal to the rear main bearing cap. Apply a thin coat of RTV sealant to the ends of the rubber seal, where it fits into the main bearing cap. Apply a 3/16-inch wide by 1/8-inch thick bead of RTV sealant around the oil pan flange and the front circle of the pan, where it contacts the timing chain cover.
5 Install the oil pan and tighten the mounting bolts to the torque listed in this Chapter's Specifications. Start at the center of the pan and work out toward the ends in a spiral pattern.
6 The remainder of installation is reverse of removal. Reinstall the engine and lower the vehicle.
7 Install a new filter and add oil to the engine, then start the engine and check for leaks.

13 Oil pump - removal and installation

Removal

1 Remove the oil pan (see Section 12).
2 Remove the oil pump mounting bolt from the main bearing cap.
3 Detach the oil pump and pick-up assembly from the block.
4 If the pump is defective, replace it with a new one. If the engine is being completely overhauled, install a new oil pump - don't reuse the original or attempt to rebuild it. **Note:** *Check the oil pump pickup tube for looseness where it is pressed into the pump body. If it is loose it could cause engine oiling problems. Don't "fix" a loose one, replace the complete pump assembly.*

Installation

5 To install the pump, turn the shaft so the gear tang mates with the slot on the lower end of the oil pump driveshaft. The oil pump should slide easily into place over the oil pump driveshaft lower retainer. If it doesn't, pull it off and turn the tang until it's aligned with the pump driveshaft slot.
6 Install the pump mounting bolt and tighten it to the torque listed in this Chapter's Specifications.
7 Reinstall the oil pan (see Section 12).
8 Add oil, run the engine and check for leaks.

14 Flywheel/driveplate - removal and installation

Refer to illustration 14.2

1 Remove the transmission (see Chapter 7). If your vehicle has a manual transmission, the pressure plate and clutch will also have to be removed (see Chapter 8).
2 Jam a large screwdriver in the starter ring gear or driveplate hole to keep the crankshaft from turning, then remove the mounting bolts **(see illustration)**. Since it's fairly heavy, support the flywheel as the last bolt is removed. **Warning:** *The ring gear teeth may be sharp - wear gloves to protect your hands.*
3 Pull straight back on the flywheel/driveplate to detach it from the crankshaft. On some models there may be a spacer-shim installed between the driveplate and crankshaft, and automatic transmission-equipped models have a retainer ring

15.5 Carefully pry the oil seal out with a screwdriver - don't nick or scratch the crankshaft or the new seal may leak

15.8 Use a large-diameter section of pipe to tap the seal in, or tap around the outer edge with a blunt drift to seat it squarely in the bore

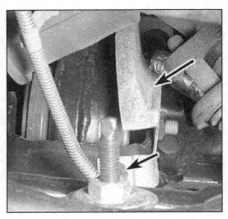

16.5 If equipped, remove the bolt (lower arrow) retaining the sheetmetal heat shield (upper arrow) to the motor mount

between the bolts and the driveplate.
4   On manual transmission equipped vehicles, check the pilot bushing and replace it if necessary (see Chapter 8).
5   Installation is the reverse of removal. Be sure to align the hole in the flywheel/driveplate with the dowel pin in the crankshaft, if used. Use non-hardening thread-locking compound on the bolt threads and tighten them to the torque listed in this Chapter's Specifications in a criss-cross pattern.

15 Rear main oil seal - replacement

Refer to illustrations 15.5 and 15.8
1   The rear main bearing oil seal can be replaced without removing the oil pan or crankshaft.
2   Remove the transmission (see Chapter 7).
3   If equipped with a manual transmission, remove the pressure plate and clutch disc (see Chapter 8).
4   Remove the flywheel or driveplate (see Section 14).
5   Using a seal removal tool or a large screwdriver, carefully pry the seal out of the block **(see illustration)**. Don't scratch or nick the crankshaft in the process.
6   Clean the bore in the block and the seal contact surface on the crankshaft. Check the crankshaft surface for scratches and nicks that could damage the new seal lip and cause oil leaks. If the crankshaft is damaged, the only alternative is a new or different crankshaft.
7   Apply a light coat of clean engine oil to the seal lip and outer edge of the new seal.

8   Press the new seal into place with GM tool no. J34924 (if available). The seal lip must face toward the front of the engine. If the special tool isn't available, carefully work the seal lip over the end of the crankshaft and tap the seal in with a hammer and blunt drift until it's seated squarely in the bore **(see illustration)**.
9   Install the flywheel or driveplate.
10 If equipped with a manual transmission, reinstall the clutch disc and pressure plate.
11 Reinstall the transmission (see Chapter 7).

16 Engine mounts - replacement

Refer to illustrations 16.5 and 16.6
Warning: *Improper lifting methods or devices are hazardous and could result in severe injury or death. DO NOT place any part of your body under the engine/transmission when it's supported only by a jack. Failure of the lifting device could result in serious injury or death.*
1   If the rubber mounts have hardened, cracked or separated from the metal backing plates, they must be replaced. This operation may be carried out with the engine/transmission still in the vehicle.
2   Disconnect the negative cable from the battery. **Caution:** *On models equipped with a Delco Loc II audio system, be sure the lock-out feature is turned off before performing any procedure which requires disconnecting the battery.*
3   Raise the front of the vehicle and support it securely on jackstands.
4   Support the engine with a jack. Position

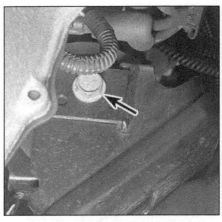

16.6 Remove the engine mount throughbolt (arrow) - right mount shown, left mount similar

a wood block between the jack head and the oil pan and raise it enough to take the tension off the mounts.
5   If equipped, remove the metal heat shield over the mount **(see illustration)**.
6   Remove the engine mount through-bolt **(see illustration)**.
7   Remove the mount-to-engine bolts and remove the mount.
8   Place the new mount in position.
9   Install the mount-to-engine bolts, lower the engine and install the through-bolt. Tighten all fasteners to the torque listed in this Chapter's Specifications.
10 Remove the jackstands and lower the vehicle.

Chapter 2 Part B
4.3L V6 engine

Contents

	Section		Section
Balance shaft - removal and installation	See Chapter 2C	General information	1
Camshaft and lifters - removal, inspection and installation	10	Intake manifold - removal and installation	6
SERVICE ENGINE SOON light	See Chapter 6	Oil pan - removal and installation	12
Crankshaft front oil seal - replacement	8	Oil pump - removal and installation	13
Cylinder compression check	See Chapter 2C	Rear main oil seal - replacement	15
Cylinder heads - removal and installation	11	Repair operations possible with the engine in the vehicle	2
Drivebelt check, adjustment and replacement	See Chapter 1	Rocker arms and pushrods - removal, inspection and installation	4
Engine mounts - check and replacement	16	Spark plug replacement	See Chapter 1
Engine oil and filter change	See Chapter 1	Timing chain cover, chain and sprockets - removal and installation	9
Engine overhaul - general information	See Chapter 2C	Valve covers - removal and installation	3
Engine - removal and installation	See Chapter 2C	Valve springs, retainers and seals - replacement	5
Exhaust manifolds - removal and installation	7	Water pump - removal and installation	See Chapter 3
Flywheel/driveplate - removal and installation	14		

Specifications

General

Cylinder numbers (front-to-rear)
  Left (driver's) side 1-3-5
  Right side 2-4-6
Firing order 1-6-5-4-3-2
Displacement .... 262 cubic inches

The blackened terminal shown on the distributor cap indicates the Number One spark plug wire position

**FIRING ORDER
1-6-5-4-3-2**
with HEI ignition system

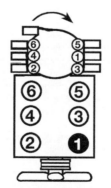

**FIRING ORDER
1-6-5-4-3-2**
with Enhanced Distributor Ignition (EDI) system

Cylinder numbering and distributor rotation - V6 engines

Camshaft

Bearing journal
    Diameter .... 1.8677 to 1.8697 inches
    Out-of-round limit .... 0.001 inch
Lobe lift
    1994 and 1995
        VIN Z
            Intake .... 0.234 inch
            Exhaust .... 0.257 inch
        VIN W
            Intake .... 0.288 inch
            Exhaust .... 0.294 inch
    1996 on
        Intake .... 0.2763 inch
        Exhaust .... 0.2855 inch
Endplay .... 0.001 to 0.009 inch

Torque specifications
Ft-lbs (unless otherwise indicated)

Valve cover bolts
    1994 and 1995 .... 90 in-lbs
    1996 on .... 106 in-lbs
Intake manifold bolts
    1994 and 1995
        VIN Z **(see illustration 6.29a)**
            All, except bolt A 35
            Bolt A 41
        VIN W
            Lower 35
            Upper 10
    1996 on
        Step 1 .... 26 in-lbs
        Step 2 .... 106 in-lbs
        Step 3 .... 132 in-lbs
Exhaust manifold bolts/stud
    Step 1 .... 132 in-lbs
    Step 2 22
Cylinder head bolts
    1994 and 1995 65
    1996 on
        Step 1 (all bolts) 22
        Step 2
            Short bolts .... Tighten an additional 55-degrees
            Medium bolts .... Tighten an additional 65-degrees
            Long bolts .... Tighten an additional 75-degrees
Timing chain cover bolts
    1994 .... 124 in-lbs
    1995 on .... 106 in-lbs
Camshaft sprocket bolts 21
Balance shaft retainer bolts .... 106 in-lbs
    Driven gear bolt
        Step 1 15
        Step 2 .... Tighten an additional 35-degrees
    Drive gear retaining stud .... 144 in-lbs
Rocker arm nuts (bolts on 2000 and later) 20
Rocker arm stud (to cylinder head) 35
Vibration damper bolt 70
Hydraulic lifter retainer bolts .... 144 in-lbs
Oil pan
    1994 and 1995
        Bolts .... 100 in-lbs
        Nuts .... 204 in-lbs
    1996 through 1998
        Bolts .... 204 in-lbs
        Nuts .... 216 in-lbs
    1999 and later (all) 18
Oil pump bolt 65
Rear main oil seal retainer bolts .... 132 in-lbs
Flywheel/driveplate bolts 75

Chapter 2 Part B 4.3L V6 engine

3.5 Disconnect the PCV valve (small arrow) and move the coil bracket (large arrow) - pull the heater hoses from their bracket at the rear

3.7 Remove the three bolts and remove the valve cover

3.10 On most models, the PCV valve (right arrow) and hose must be disconnected and moved for left valve cover removal - the oil filler tube (left arrow) easily twists out of the valve cover

1 General information

This Part of Chapter 2 is devoted to in-vehicle repair procedures for the 4.3L V6 engine. All information concerning engine removal and installation and engine block and cylinder head overhaul can be found in Part C of this Chapter.

There are three different fuel systems used on the 4.3L V6 engines covered by this manual; refer to Chapter 4 for more detailed information. Some specifications and procedures in this Chapter vary by the specific version of this same basic engine. If there's any doubt as to which version you have, refer to the Vehicle Identification Number (VIN) that is located on the forward edge of the dashboard on the driver's side. The VIN is visible from outside the vehicle, through the windshield. If the eighth position in the alpha-numeric code is a W, you have the VIN W engine. You could also have a Z code or X code.

The following repair procedures are based on the assumption that the engine is installed in the vehicle. If the engine has been removed from the vehicle and mounted on a stand, many of the steps outlined in this Part of Chapter 2 will not apply.

The Specifications included in this Part of Chapter 2 apply only to the procedures contained in this Part. Part C of Chapter 2 contains the Specifications necessary for cylinder head and engine block rebuilding.

2 Repair operations possible with the engine in the vehicle

Many major repair operations can be accomplished without removing the engine from the vehicle.

Clean the engine compartment and the exterior of the engine with some type of pressure washer before any work is done. It will make the job easier and help keep dirt out of the internal areas of the engine.

Remove the hood, if necessary, to improve access to the engine as repairs are performed (see Chapter 11 if necessary).

If vacuum, exhaust, oil or coolant leaks develop, indicating a need for gasket or seal replacement, the repairs can generally be made with the engine in the vehicle. The intake and exhaust manifold gaskets, timing chain cover gasket, oil pan gasket (on 4WD models only), crankshaft oil seals and cylinder head gaskets are all accessible with the engine in place.

Exterior engine components, such as the intake and exhaust manifolds, the oil pan and oil pump (on 4WD models only), the water pump, the starter motor, the alternator, the distributor and the fuel system components can be removed for repair with the engine in place.

Since the cylinder heads can be removed without pulling the engine, valve component servicing can also be accomplished with the engine in the vehicle. Replacement of the timing chain and sprockets is also possible with the engine in the vehicle.

3 Valve covers - removal and installation

Removal

1 Disconnect the negative cable from the battery. **Caution:** *On models equipped with a Delco Loc II audio system, be sure the lockout feature is turned off before performing any procedure which requires disconnecting the battery.*

2 Remove the air cleaner assembly (see Chapter 4).

Right side

Refer to illustrations 3.5 and 3.7

3 On 1994 pick-up models, remove the EGR controller and EGR bracket, then tag and disconnect the vacuum lines in the way.

4 On 1994 Blazer and Jimmy models, disconnect the PCV valve from the cover, disconnect the heater pipe at the intake manifold (refer to Chapter 1 for the coolant draining procedure), unclip and lay aside the wiring harness, unbolt and move the relay bracket bolted to the cowl, pull the spark plug wires from their clips, and disconnect the dipstick tube bracket at the cylinder head and wiggle the tube aside.

5 On 1995 models, disconnect the PCV valve at the valve cover, and unbolt and move the coil bracket **(see illustration)**.

6 On 1996 and later models, unclip and move the spark plug wires, disconnect the vent tube, unbolt the wiring bracket at the alternator, and remove the bolt from the brace on the dipstick tube and wiggle the dipstick tube out of the way. The heater hoses can be moved aside without disconnecting them.

7 Remove the three valve cover bolts, then detach the cover from the cylinder head **(see illustration)**. **Note:** *If the cover is stuck to the cylinder head, bump one end with a block of wood and a hammer to jar it loose. If that doesn't work, try to slip a flexible putty knife between the cylinder head and cover to break the gasket seal. Don't pry at the cover-to-head joint or damage to the sealing surfaces may occur (leading to oil leaks in the future).*

Left side

Refer to illustration 3.10

8 On 1994 and 1995 pick-up models, remove the power brake booster brace, and disconnect the large vacuum hose to the booster.

9 On 1994 Blazer and Jimmy models, disconnect the heat tube to the air cleaner system, the PCV pipe, the rear alternator bracket and remove the spark plug wires from their clips. Disconnect and move aside the large vacuum hose to the brake booster, and the fuel lines at the TBI unit (refer to Chapter 4).

10 On 1995 Blazer and Jimmy models, disconnect and move aside the PCV hose **(see illustration)**.

11 On all 1996 and later models, disconnect and move aside the PCV valve and tube. Unbolt and set aside the air conditioning compressor and bracket (refer to Chapter 3).

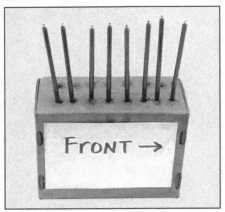

4.4 A perforated cardboard box can be used to store the pushrods to ensure that they're installed in their original positions - note the label indicating the front of the engine

4.10 Lube the ends of the pushrods and the valve stems with moly-base grease prior to installation of the rocker arms

4.11 Moly-base grease applied to the pivot balls will ensure adequate lubrication until oil pressure builds up when the engine is started

DO NOT disconnect the hoses from the compressor. **Warning:** *The air conditioning system is under high pressure. DO NOT disassemble any part of the system unless the system has been depressurized.* On 1999 and later models, the compressor does not have to be removed, but disconnect and remove the EGR tube, and disconnect the electrical connector at the coolant temperature sensor in the cylinder head.

12 Remove the three valve cover bolts, then detach the cover from the cylinder head. **Note:** *If the cover is stuck to the cylinder head, bump one end with a block of wood and a hammer to jar it loose. If that doesn't work, try to slip a flexible putty knife between the cylinder head and cover to break the gasket seal. Don't pry at the cover-to-head joint or damage to the sealing surfaces may occur (leading to oil leaks in the future).*

Installation

13 The mating surfaces of each cylinder head and valve cover must be perfectly clean when the covers are installed. Use a gasket scraper to remove all traces of sealant and old gasket material, then clean the mating surfaces with lacquer thinner or acetone. If there's sealant or oil on the mating surfaces when the cover is installed, oil leaks may develop.

14 Clean the mounting bolt threads with a die to remove any corrosion and restore damaged threads. Make sure the threaded holes in the cylinder head are clean - run a tap into them to remove corrosion and restore damaged threads.

15 The gaskets should be mated to the covers before the covers are installed. Apply a thin coat of RTV sealant to the cover flange, then position the gasket inside the cover lip and allow the sealant to set up so the gasket adheres to the cover.

16 Carefully position the cover(s) on the cylinder head and install the bolts.

17 Tighten the bolts in three or four steps to the torque listed in this Chapter's Specifications.

18 The remaining installation steps are the reverse of removal.

19 Start the engine and check carefully for oil leaks as the engine warms up.

4 Rocker arms and pushrods - removal, inspection and installation

Removal

Refer to illustration 4.4

1 Detach the valve cover(s) from the cylinder head(s) (see Section 3).

2 Beginning at the front of one cylinder head, loosen and remove the rocker arm stud nuts. Store them separately in marked containers to ensure that they will be reinstalled in their original locations. **Note 1:** *If the pushrods are the only items being removed, loosen each nut (models through 1998) just enough to allow the rocker arms to be rotated to the side so the pushrods can be lifted out.* **Note 2:** *On 2000 and later models, the roller-rocker arms are retained by a bolt, there are no pivot balls. Unlike the late four-cylinder models with roller rockers, the V6 rockers ride on long rocker arm supports that bolt against the cylinder head casting protrusions.*

3 Lift off the rocker arms and pivot balls and store them in the marked containers with the nuts (they must be reinstalled in their original locations).

4 Remove the pushrods and store them separately to make sure they don't get mixed up during installation **(see illustration)**.

Inspection

5 Check each rocker arm for wear, cracks and other damage, especially where the pushrods and valve stems contact the rocker arm faces.

6 Make sure the hole at the pushrod end of each rocker arm is open.

7 Check each rocker arm pivot area for wear, cracks and galling. If the rocker arms are worn or damaged, replace them with new ones and use new pivot balls as well. On 2000 and later models, align the roller rocker arms with the rocker arm supports.

8 Inspect the pushrods for cracks and excessive wear at the ends. Roll each pushrod across a piece of plate glass to see if it's bent (if it wobbles, it's bent).

Installation

Refer to illustrations 4.10 and 4.11

9 Lubricate the lower end of each pushrod with clean engine oil or moly-base grease and install them in their original locations. Make sure each pushrod seats completely in the lifter.

10 Apply moly-base grease to the ends of the valve stems and the upper ends of the pushrods before positioning the rocker arms over the studs **(see illustration)**.

11 Set the rocker arms in place, then install the pivot balls and nuts. Apply moly-base grease to the pivot balls to prevent damage to the mating surfaces before engine oil pressure builds up **(see illustration)**. Be sure to install each nut with the flat side against the pivot ball. Tighten the rocker-arm nuts to the torque listed in this Chapter's Specifications.

12 The remainder of the installation is the reverse of removal.

13 Start the engine and check for valve cover leaks and valvetrain noise.

Valve adjustment (1994 VIN Z engines only)

Refer to illustrations 4.14 and 4.17

Note: *On most models covered by this book, there are no provisions for valve adjustment. The rocker arm studs have a positive stop shoulder for the rocker arm nuts. After valve service, tighten the rocker arm nuts to the torque listed in this Chapter's Specifications. Unless there have been machining operations that significantly altered the valve lash, the adjustment should be correct. Some 1994 VIN Z engines have press-in rocker-arm studs, without a positive stop, and require the following valve adjustment procedure when-*

4.14 Align the timing marks on the damper and front cover to find TDC

4.17 Rotate each pushrod as the rocker-arm nut is tightened until a *slight* drag is felt (zero lash), then tighten an additional 3/4 turn

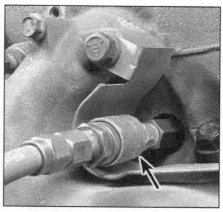

5.4 Use compressed air to hold the valve closed when the springs are removed - the air hose adapter (arrow) threads into the spark plug hole and accepts the hose from the compressor

ever the rocker arms have been loosened or removed.

14 Begin by setting the engine to TDC for piston number 1. Top Dead Center (TDC) is the highest point in the cylinder that each piston reaches as it travels up-and-down when the crankshaft turns. Each piston reaches TDC on the compression stroke and again on the exhaust stroke, but TDC generally refers to piston position on the compression stroke. The timing marks on the vibration damper installed on the front of the crankshaft are referenced to TDC **(see illustration)** and will line up at TDC for number 1 or number 4.

15 In order to bring any piston to TDC, the crankshaft must be turned using a large socket and breaker bar attached to the bolt threaded into the crankshaft damper. When looking at the front of the engine, normal crankshaft rotation is clockwise. **Warning:** *Before beginning this procedure, be sure to place the transmission in Neutral and unplug the electrical connector(s) at the distributor to disable the ignition system.*

16 Turn the engine until the mark on the damper aligns with the "0" mark on the timing tab while observing the rocker arms for the number 1 cylinder. If the valves were moving as the marks began to align, then this is TDC for number 4, and the engine must be turned one more revolution.

17 At TDC for number 1 piston, the following rocker arms can be adjusted: number 1, 5, and 6 exhaust valves, and number 1, 2 and 3 intake valves. To adjust the valves, tighten the rocker arm nut 3/4 turn past zero lash. To find zero lash, rotate the pushrod (of the valve being adjusted) between two fingers while tightening the rocker-arm nut **(see illustration)**. When a slight drag on the pushrod is just felt, all lash is removed, tighten the rocker-arm nut an additional 3/4 turn to center the hydraulic lifter plunger in its travel.

18 After the number one piston has been positioned at TDC on the compression stroke, TDC for any of the remaining cylinders can be located by turning the crankshaft 120-degrees at a time and following the firing order (see the Specifications at the beginning of this Chapter).

19 Rotate the crankshaft 360-degrees (TDC number 4) and adjust the following rocker arms: number 2, 3 and 4 exhaust valves, and number 4, 5 and 6 intake valves.

5 Valve springs, retainers and seals - replacement

Refer to illustrations 5.4, 5.7a, 5.7b, 5.8, 5.15 and 5.16

Note: *Broken valve springs and defective valve stem seals can be replaced without removing the cylinder heads. Two special tools and a compressed air source are normally required to perform this operation, so read through this Section carefully and rent or buy the tools before beginning the job.*

1 Remove the valve cover from the cylinder head (see Section 3). If all of the valve stem seals are being replaced, remove both valve covers.

2 Remove the spark plug from the cylinder which has the defective component. If all of the valve stem seals are being replaced, all of the spark plugs should be removed.

3 Turn the crankshaft until the piston in the affected cylinder is at top dead center on the compression stroke (see Section 4). If you are replacing all of the valve stem seals, begin with cylinder number 1 and work on the valves for one cylinder at a time. Move from cylinder-to-cylinder following the firing order sequence (1-6-5-4-3-2). Each cylinder in the firing order is 120-degrees of crankshaft rotation (clockwise) from the previous one.

4 Thread an adapter into the spark plug hole **(see illustration)** and connect an air hose from a compressed air source to it. Most auto parts stores can supply the air hose adapter. **Note:** *Many cylinder compression gauges utilize a screw-in fitting that may work with your air hose quick-disconnect fitting.*

5 Remove the nut, pivot ball and rocker arm for the valve with the defective part and pull out the pushrod (see Section 4). If all the valve

5.7a Once the spring is depressed, the keepers can be removed with a small magnet or needle-nose pliers (a magnet is preferred to prevent dropping the keepers

stem seals are being replaced, all of the rocker arms and pushrods should be removed.

6 Apply compressed air to the cylinder. The valves should be held in place by the air pressure. **Warning:** *If the cylinder isn't exactly at TDC, air pressure may cause the engine to rotate. Do not leave a socket or wrench on the balancer bolt; damage or personal injury may result.*

7 Stuff shop rags into the cylinder head holes above and below the valves to prevent parts and tools from falling into the engine, then use a valve spring compressor to compress the spring/damper assembly. Remove the keepers with small needle-nose pliers or a magnet **(see illustration)**. **Note:** *A couple of different types of tools are available for compressing the valve springs with the cylinder head in place. One type, shown here, grips the lower spring coils and presses on the retainer as the knob is turned, while the other type utilizes the rocker arm stud and nut for leverage* **(see illustration)**. *Both types work very well, although the lever type is usually less expensive.*

5.7b The stamped steel lever-type valve spring compressor is usually less expensive that the type that grips the spring coils

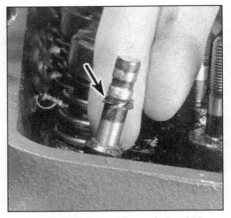

5.8 The O-ring seal (arrow) should be replaced with a new one each time the keepers and retainer are removed

5.15 Make sure the O-ring seal under the retainer is seated in the groove and not twisted before installing the keepers

8 Remove the spring retainer, oil shield and valve spring assembly (there is both an inner and outer valve spring for each valve - the inner is called a spring damper), then remove the valve stem O-ring seal and the umbrella-type guide seal (intake valves only). The O-ring seal will most likely be hardened and will probably break when removed, so plan on installing a new one each time the original is removed **(see illustration)**. *Note: If air pressure fails to hold the valve in the closed position during this operation, the valve face and/or seat is probably damaged. If so, the cylinder head will have to be removed for additional repair operations.*
9 Wrap a rubber band or tape around the top of the valve stem so the valve won't fall into the combustion chamber, then release the air pressure.
10 Inspect the valve stem for damage. Rotate the valve in the guide and check the end for eccentric movement, which would indicate that the valve is bent.
11 Move the valve up-and-down in the guide and make sure it does not bind. If the valve stem binds, either the valve is bent or the guide is damaged. In either case, the cylinder head will have to be removed for repair.
12 Reapply air pressure to the cylinder to retain the valve in the closed position, then remove the tape or rubber band from the valve stem.
13 Lubricate the valve stem with engine oil and install a new umbrella-type guide seal (intake valve only). On 1999 and later models, there are seals on all valve guides. Replacement seals for the exhaust are colored red or brown, while intake seals are black.
14 Install the spring/damper assembly and shield in position over the valve.
15 Install the valve spring retainer or rotator. Compress the valve spring assembly and carefully install the new O-ring seal in the lower groove of the valve stem. Make sure the seal isn't twisted - it must lie perfectly flat in the groove **(see illustration)**.
16 Position the keepers in the upper groove. Apply a small dab of grease to the inside of each keeper to hold it in place if necessary **(see illustration)**. Remove the pressure from the spring tool and make sure the keepers are seated.
17 Disconnect the air hose and remove the adapter from the spark plug hole.
18 Install the rocker arm(s) and pushrod(s) (see Section 4).
19 Install the valve cover(s) (see Section 3).
20 Install the spark plug(s) and hook up the wire(s).
21 Start and run the engine, then check for oil leaks and unusual sounds coming from the valve cover area.

6 Intake manifold - removal and installation

Removal

1 Disconnect the negative cable from the battery. **Caution:** *On models equipped with a Delco Loc II audio system, be sure the lockout feature is turned off before performing any procedure which requires disconnecting the battery.*

1994 and 1995 VIN Z models

2 Remove the upper fan shroud.
3 Remove the air cleaner assembly (see Chapter 4).
4 Drain the cooling system (see Chapter 1).
5 Disconnect the upper radiator and heater hoses at the front of the engine.
6 Remove the two rear braces from the drivebelt tensioner.
7 Unbolt and move the emission relays bracket.
8 Remove the spark plug wires, coil and distributor (see Chapters 1 and 5). **Note:** *The engine should be turned to TDC (see Section 4) for the number 1 piston. Mark the relationship of the distributor rotor to the distributor housing before removing the distributor.*
9 Disconnect the fuel lines at the TBI unit (see Chapter 4). **Warning:** *The fuel system is under pressure - do not disconnect the lines without following the safety procedures in Chapter 4.*
10 Remove the accelerator, cruise control and TV cables from the bracket(s) on the manifold (see Chapter 4).
11 Detach the vacuum brake booster pipe from the manifold.
12 Label and disconnect the fuel lines, vacuum hoses/pipes and wires at the manifold and TBI unit (see Chapter 4).
13 Disconnect and move the EGR hose.
14 Loosen the manifold mounting bolts in 1/4-turn increments until they can be removed by hand. **Note:** *Some of the manifold fasteners are studs and will require a deep socket to remove.* The manifold will probably be stuck to the cylinder heads and force may be required to break the gasket seal. A large prybar can be positioned under the cast-in lug near the left front mounting bolt to pry up the front of the manifold, but make sure all bolts have been removed first! **Caution:** *Don't pry between the block and manifold or the cylinder heads and manifold or damage to the gasket sealing surfaces may occur, leading to vacuum leaks.*

5.16 Apply a small dab of grease to each keeper as shown here before installation - it will hold them in place on the valve stem as the spring is released

6.15 Disconnect the upper radiator hose (lower arrow) and heater hose (upper arrow) from the intake manifold

6.19 Remove this short brace (arrow) from the air conditioning compressor to the intake manifold

6.21a There are four bolts inside the manifold (arrows) plus four bolts ahead of the plenum and four behind the plenum

1994 and 1995 VIN W and all 1996 and later models

Refer to illustrations 6.15, 6.19, 6.21a and 6.21b

15 Complete Steps 2 through 5 above **(see illustration).**
16 Refer to Chapter 4 for removal of the upper intake plenum, linkage, wiring and disconnection of the fuel lines and injector harnesses. **Warning:** *The fuel system is under pressure. Do not disconnect the lines without following the safety procedures in Chapter 4.*
Note: *On 1996 and later models, the upper intake plenum (manifold) does not have to be removed in order to remove the lower intake manifold. But, all connections must be detached from it to remove the upper and lower intake manifolds as an assembly.*
17 Unbolt the EGR valve from the lower manifold and move it out of the way (see Chapter 6).
18 Move the heater hoses from the bracket above the right valve cover. Remove the ignition coil **(see illustration 3.5).**
19 Remove the small brace from the air conditioning compressor to the intake manifold **(see illustration).**
20 Turn the engine to TDC for number 1 piston (see Section 4) and remove the distributor. **Note:** *Mark the relationship of the distributor rotor to the distributor housing before removing the distributor.*
21 The manifold will probably be stuck to the cylinder heads and force may be required to break the gasket seal. A large prybar can be positioned under the cast-in lug near the left front mounting bolt to pry up the front of the manifold, but make sure all bolts have been removed first **(see illustrations)! Caution:** *Don't pry between the block and manifold or the cylinder heads and manifold or damage to the gasket sealing surfaces may occur, leading to vacuum leaks.*

Installation (all models)

Refer to illustrations 6.22, 6.23, 6.24, 6.25, 6.27, 6.29a and 6.29b

Warning: *Composition gaskets, with a metal core and compressible material on each side,*

6.21b Pry only under the casting lug near the left front mounting bolt (arrow)

are used in the intake systems of these engines. Be careful in handling the edges of these gaskets; sharp metal edges may be exposed.
Note: *The mating surfaces of the cylinder heads, block and manifold must be perfectly clean when the manifold is installed. Gasket removal solvents in aerosol cans are available at most auto parts stores and may be helpful when removing old gasket material that's stuck to the cylinder heads and manifold (since the manifold is made of aluminum, aggressive scraping can cause damage). Be sure to follow the directions printed on the container.*
22 Use a plastic or wooden gasket scraper to remove all traces of sealant and old gasket material, then clean the mating surfaces with lacquer thinner or acetone. Do not use abrasive pads or discs to clean the gasket surfaces. If there's old sealant or oil on the mating surfaces when the manifold is installed, oil or vacuum leaks may develop. When working on the cylinder heads and block, cover the lifter valley with shop rags to keep debris out of the engine **(see illustration).** Use a vacuum cleaner to remove any gasket material that falls into the intake ports in the cylinder heads.
23 Use a tap of the correct size to chase the threads in the bolt holes, then use compressed air (if available) to remove the debris

6.22 After covering the lifter valley, use a gasket scraper to remove all traces of sealant and old gasket material from the cylinder head and manifold mating surfaces

6.23 The bolt hole threads must be clean and dry to ensure accurate torque readings when the manifold mounting bolts are installed

from the holes **(see illustration). Warning:** *Wear safety glasses or a face shield to protect your eyes when using compressed air! Remove excessive carbon deposits and corrosion from the exhaust, EGR and coolant passages in the cylinder heads and manifold.*

Chapter 2 Part B 4.3L V6 engine

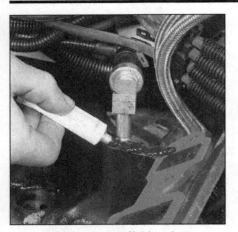

6.24 Intake manifold sealant application details

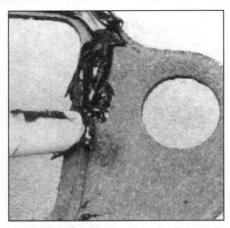

6.25 RTV sealant should be used around the coolant passage holes in the new intake manifold gaskets

6.27 Make sure the gaskets are installed with the blocked off holes (arrow) to the rear

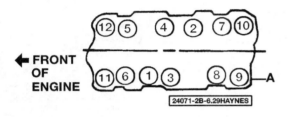

6.29a Intake manifold bolt tightening sequence (1994 and 1995 models) - on VIN Z models, bolt A must be tightened an additional amount (see this Chapter's Specifications)

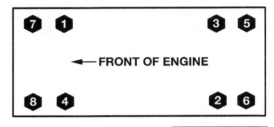

6.29b Intake manifold bolt tightening sequence (1996 and later models)

24 Apply a 3/16-inch wide bead of RTV sealant to the front and rear manifold mating surfaces of the block **(see illustration)**. Make sure the beads extend up the cylinder heads 1/2-inch on each side.

25 Apply a thin coat of RTV sealant around the coolant passage holes on the cylinder head side of the new intake manifold gaskets **(see illustration)**.

26 Position the gaskets on the cylinder heads, with the ears at each end overlapping the bead of RTV sealant on the cylinder head. The upper side of each gasket will have a THIS SIDE UP label stamped into it to ensure correct installation.

27 Make sure all intake port openings, coolant passage holes and bolt holes are aligned correctly. **Note:** *Be sure the gaskets are installed with the blocked off coolant passages at the rear of the engine (see illustration). Some gaskets may have small tabs which must be bent over until they're flush with the rear surface of each cylinder head.*

28 Carefully set the manifold in place while the sealant is still wet. **Caution:** *Don't disturb the gaskets and don't move the manifold fore-and-aft after it contacts the sealant on the block.*

29 Following the recommended sequence, install the bolts and tighten them to the torque listed in this Chapter's Specifications **(see illustrations)**. Work up to the final torque in two stages and note that there are different sequences for different models.

30 The remaining installation steps are the reverse of removal. Start the engine and check carefully for oil and coolant leaks at the intake manifold joints.

7 Exhaust manifolds - removal and installation

Removal

Refer to illustrations 7.4, 7.10, 7.13, 7.14a and 7.14b

Warning: *Use caution when working around the exhaust manifolds, the sheetmetal heat shields can be sharp on the edges. Also, the engine should be cold when this procedure is followed.*

1 Disconnect the negative battery cable from the battery. **Caution:** *On models equipped with a Delco Loc II audio system, be sure the lockout feature is turned off before performing any procedure which requires disconnecting the battery.*

2 Raise the vehicle and support it securely on jackstands.

3 Working under the vehicle, apply penetrating oil to the exhaust pipe-to-manifold studs and nuts (they're usually rusty).

4 Remove the nuts retaining the exhaust pipe(s) to the manifold(s) **(see illustration)**. **Note:** *On some later models, you may have to remove the tire for access to the exhaust nuts through the fenderwell opening.*

Right side manifold

5 On early models, remove the air cleaner assembly (see Chapter 4).

6 Remove the heat stove pipe on TBI models.

7 Detach the spark plug wires from the

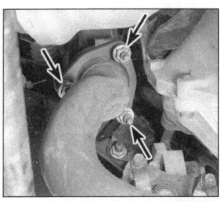

7.4 Access the exhaust pipe bolts/nuts from underneath the vehicle - on some models it may be easier to remove the wheel and work through the fenderwell opening

7.10 Remove the bolt and nut (arrows) retaining the alternator brace to the exhaust manifold stud

7.13 Use a puller like this to remove the power steering pump pulley

7.14a Remove the three power steering pump mounting bolts (arrows)

7.14b Slide the power steering pump over, with rear bracket attached, until the bracket clears the exhaust stud (arrow)

8.3 Remove the three bolts retaining the pulley to the damper, then remove the center crankshaft bolt

plugs and brackets and position them out of the way (see Chapter 1).
8 Unbolt the air conditioning compressor and its bracket and set it aside (see Chapter 3). DO NOT disconnect the hoses from the compressor. **Warning:** *The air conditioning system is under high pressure. DO NOT disassemble any part of the system unless the system has been depressurized.*
9 Remove the oil dipstick, unbolt the dipstick tube bracket and move the dipstick tube.

Left side manifold

10 On 1994 and 1995 models, unbolt and remove the rear brace from the alternator **(see illustration)**.
11 Detach the spark plug wires from the plugs and brackets and position them out of the way.
12 On 1996 and later models, disconnect the EGR pipe from the exhaust manifold.
13 The rear power steering bracket must be removed from the exhaust manifold stud, but to do so the power steering pump must be removed. Use a puller to remove the pulley from the pump **(see illustration)**.
14 With the pulley off, remove the three pump mounting bolts **(see illustration)**. Do not disconnect the power steering lines.

Remove the nut from the exhaust manifold stud, slide the bracket off the stud and set the pump aside **(see illustration)**.

Both manifolds

15 Bend the lock tabs back, then remove the mounting bolts and separate the exhaust manifold from the cylinder head. Remove the heat shields from the left side manifold after the bolts are removed. **Note:** *1994 and 1995 models use washers and locking tabs under all bolts; later models do not. Also, make note of the location of studs used on the left manifold on all models.*

Installation

16 Check the manifold for cracks and make sure the bolt threads are clean and undamaged. The manifold and cylinder head mating surfaces must be clean before the manifolds are reinstalled - use a gasket scraper to remove all carbon deposits.
17 Position the manifold, locking tabs and heat shields (if equipped) on the cylinder head and install the mounting bolts.
18 When tightening the mounting bolts, work from the center to the ends and be sure to use a torque wrench. Tighten the bolts in three equal steps to the torque listed in this Chapter's Specifications. On 1994 and 1995 models, bend the locking tabs back against the bolt heads.
19 The remaining installation steps are the reverse of removal.
20 Start the engine and check for exhaust leaks.

8 Crankshaft front oil seal - replacement

Front seal - timing chain cover in place

Refer to illustrations 8.3, 8.4, 8.5 and 8.7

1 Disconnect the negative battery cable from the battery. **Caution:** *On models equipped with a Delco Loc II audio system, be sure the lockout feature is turned off before performing any procedure which requires disconnecting the battery.*
2 Remove the upper and lower fan shrouds, the cooling fan (see Chapter 3) and the drivebelt (see Chapter 1).
3 Remove the bolts and separate the crankshaft drivebelt pulley from the vibration damper **(see illustration)**.

8.4 Use the recommended puller to remove the damper or damage to the damper may result

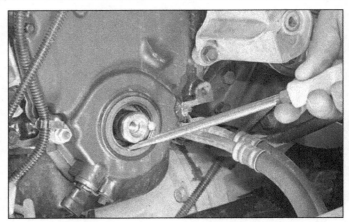

8.5 Using a seal removal tool or a screwdriver, pry the front seal out of the cover - don't distort the cover or scratch the crankshaft

8.7 The new, lubricated seal can be driven in with a large socket or section of pipe - drive it in squarely, to the depth of the original seal

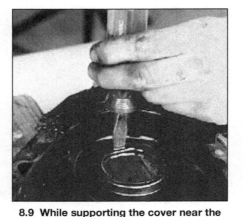

8.9 While supporting the cover near the seal bore, drive the old seal out from the inside with a hammer and punch or screwdriver

8.11 Clean the bore, then apply a small amount of oil to the outer edge of the seal and drive it squarely into the opening with a large socket and hammer - do not damage the seal in the process

4   Remove the lower flywheel cover (see Chapter 7) and have an assistant wedge a large screwdriver in the flywheel ring gear teeth to hold the engine from turning while you remove the bolt from the front of the crankshaft with a socket and breaker bar. Then use a puller to detach the vibration damper **(see illustration)**. **Caution:** *Don't use a puller with jaws that grip the outer edge of the damper. The puller must be the type shown in the illustration that applies force to the damper hub only.* Unless your puller has a tip with a short taper where it meets the crankshaft, insert a short bolt that fits the crankshaft threads, then use the puller against the bolt rather than the crankshaft, to avoid damage to the crankshaft.

5   Carefully pry the seal out of the cover with a seal removal tool or a large screwdriver **(see illustration)**. Be careful not to distort the cover or scratch the wall of the seal bore. If the engine has accumulated a lot of miles, apply penetrating oil to the seal-to-cover joint and allow it to soak in before attempting to pull the seal out.

6   Clean the bore to remove any old seal material and corrosion. Position the new seal in the bore with the open end of the seal facing IN. A small amount of grease applied to the outer edge of the new seal will make installation easier.

7   Apply a light film of grease to the lip of the seal. Drive the seal into the bore with GM tool no. J-23042A, or equivalent, until it's completely seated. If the special tool isn't available, a socket that's the same outside diameter as the seal or a section of pipe can be used **(see illustration)**.

8   Reinstall the vibration damper, aligning its groove with the keyway in the crankshaft. **Caution:** *Use the bolt to pull the damper on - do not use a hammer or the rubber-mounted inertia-weight section of the damper may be damaged.*

Front seal - timing chain cover removed

Refer to illustrations 8.9 and 8.11

Note: *1995 VIN W engines and all 1996 and later engines use a composite-material front cover, which is not reusable. It its recommended to replace it rather than try to reseal it and have it leak later. The following procedure is for 1995 and earlier models with steel front covers.*

9   Use a punch or screwdriver and hammer to drive the seal out of the cover from the back side. Support the cover as close to the seal bore as possible **(see illustration)**. Be careful not to distort the cover or scratch the wall of the seal bore. If the engine has accumulated a lot of miles, apply penetrating oil to the seal-to-cover joint on each side and allow it to soak in before attempting to drive the seal out.

10   Clean the bore to remove any old seal material and corrosion. Support the cover on blocks of wood and position the new seal in the bore with the open end of the seal facing IN. A small amount of grease applied to the outer edge of the new seal will make installation easier.

11   Drive the seal into the bore with a large socket and hammer until it's completely seated **(see illustration)**. Select a socket that's the same outside diameter as the seal (a section of pipe can be used if a socket isn't available). Install the timing chain cover.

12   Reinstall the cover (see Section 9). The remainder of the installation is the reverse of the disassembly. Start the engine and check for leaks.

Chapter 2 Part B 4.3L V6 engine

9.4 Remove the crankshaft position sensor (large arrow) and remove the nut (upper arrow) retaining the wiring harness to the cover stud

9.6 Before removing the sprockets or chain, remove the reluctor ring from the crankshaft - note the dished side faces the timing chain cover

9.7 Remove the three bolts from the end of the camshaft and remove the camshaft sprocket and chain as an assembly

9 Timing chain cover, chain and sprockets - removal and installation

Removal

Refer to illustrations 9.4, 9.6 and 9.7

1 Disconnect the negative battery cable. **Caution:** *On models equipped with a Delco Loc II audio system, be sure the lockout feature is turned off before performing any procedure which requires disconnecting the battery.*

2 Refer to Chapter 3 and remove the upper and lower fan shrouds, drivebelt, cooling fan and water pump.

3 Position the number *four* piston at TDC on the compression stroke (see Section 4). **Caution:** *Once this has been done, DO NOT turn the crankshaft until the timing chain and sprockets have been reinstalled!*

4 Refer to Section 8 and remove the drivebelt pulley and the crankshaft damper. **Note:** *On some 1995 models, and all 1996 models, remove the crankshaft position sensor and the nut retaining the wiring harness to the front cover* **(see illustration)**.

5 Loosen the front oil pan nuts, then remove the cover bolts and separate the timing chain cover from the block. It may be stuck - if so, use a putty knife or screwdriver to break the gasket seal. The cover is easily distorted, so be very careful when prying it off - because the bottom of the cover has studs going through the pan, and because there are two dowels pins into the block that are part of the cover, the cover must be pulled forward to clear the dowels, then up to clear the studs. **Note:** *1995 VIN W engines and all 1996 and later engines use a composite-material front cover, which is not reusable. It its recommended to replace it rather than try to reseal it and have it leak later.*

6 Remove the crankshaft position sensor reluctor ring just inside the front cover **(see illustration)**.

7 Remove the three bolts from the end of the camshaft, then detach the camshaft

9.11 Before installing the camshaft sprocket, make sure the timing marks on the balance shaft gears are properly aligned (arrows)

sprocket and chain as an assembly **(see illustration)**. **Note:** *The balance shaft drive gear will stay attached to the camshaft and the driven gear will stay attached to the balance shaft. The sprocket on the crankshaft can be removed with a two- or three-jaw puller, but be careful not to damage the threads in the end of the crankshaft. If the timing chain cover oil seal has been leaking, refer to Section 8 and install a new one.*

Installation

Refer to illustrations 9.11 and 9.12

8 Use a gasket scraper to remove all traces of old gasket material and sealant from the cover and engine block. Stuff a shop rag into the opening at the front of the oil pan to keep debris out of the engine. Clean the cover and block sealing surfaces with lacquer thinner or acetone.

9 Check the cover flange for distortion, particularly around the bolt holes. If necessary, place the cover on a block of wood and use a hammer to flatten and restore the gasket surface.

10 If new parts are being installed, be sure

9.12 Position the timing marks on the camshaft and crankshaft sprockets in the 6 and 12 o'clock positions, respectively. A straight line should pass through the camshaft timing mark, the center of the camshaft, the crankshaft timing mark and the center of the crankshaft - when aligned as shown the number four piston is at TDC on the compression stroke

to align the keyway in the crankshaft sprocket with the Woodruff key in the end of the crankshaft. Press the sprocket onto the crankshaft with the vibration damper bolt, a large socket and some washers; or tap it gently into place until it's completely seated. **Caution:** *If resistance is encountered, DO NOT hammer the sprocket onto the crankshaft. It may eventually move onto the shaft, but it may be cracked in the process and fail later, causing extensive engine damage.*

11 Before installing the timing chain and camshaft sprocket, align the balance shaft gears **(see illustration)**. The camshaft should be positioned with the balance shaft drive gear timing mark at 12 o'clock and the driven gear mark at 6 o'clock.

12 Position the crankshaft so the sprocket timing mark is in the 12 o'clock position. Loop the chain over the camshaft sprocket, mesh the chain with the crankshaft sprocket and position the camshaft sprocket on the

2B-12 Chapter 2 Part B 4.3L V6 engine

10.2 When checking the camshaft lobe lift, the dial indicator plunger must be positioned directly above the pushrod

10.7 The roller lifters are held in place by a retainer - remove the four retainer bolts and remove the retainer

10.9 Remove the Torx bolts and take off the camshaft retainer plate, noting which side faces the block

camshaft with the timing mark in the 6 o'clock position. When correctly installed, the marks on the sprockets will be aligned as shown **(see illustration)**. *Note: The number four piston will be at TDC on the compression stroke with the sprockets aligned as shown.*

13 Apply Loctite to the camshaft sprocket bolt threads, then install and tighten them to the torque listed in this Chapter's Specifications. Lubricate the chain with clean engine oil.

14 Check the oil pan-to-block joints to make sure that all excess sealant is removed.

15 Apply RTV sealant to the U-shaped channel on the bottom of the cover.

16 Apply a thin layer of RTV sealant to both sides of the new gasket, then position it on the engine block (the dowel pins and sealant will hold it in place). *Note: The composite covers do not have a gasket, they use sealant only. Resealing the old cover is not recommended, but rather replace it to avoid sealing problems caused by the distortion of prying the old cover off.*

17 Since the oil pan seal in the bottom of the cover must be compressed in order to position the cover over the dowel pins and thread the bolts into the block, the cover is nearly impossible to install unless the oil pan is removed first. It can be done, but it's very difficult and frustrating. If the front three bolts on each side of the oil pan are removed, the oil pan can be pried down slightly, allowing the cover to be installed. Squirt some RTV sealant into the gap between the pan and the block. If difficulty is encountered, it may be necessary to remove the oil pan (see Section 12).

18 Install the timing chain cover and tighten the bolts to the torque listed in this Chapter's Specifications. If the front of the oil pan was pried down, install the bolts and tighten them to the torque listed in this Chapter's Specifications.

19 Lubricate the oil seal contact surface of the vibration damper hub with moly-base grease or clean engine oil, then install the damper on the end of the crankshaft. The keyway in the damper must align with the Woodruff key in the crankshaft nose. If the damper cannot be seated by hand, slip a large washer over the bolt, then install the bolt and use it to force the damper into place. Remove the large washer and tighten the bolt to the torque listed in this Chapter's Specifications.

20 The remaining installation steps are the reverse of removal. Use a new O-ring when reinstalling the crankshaft position sensor.

10 Camshaft and lifters - removal, inspection and installation

Camshaft lobe lift check

Refer to illustration 10.2

1 In order to determine the extent of cam lobe wear, the lobe lift should be checked prior to camshaft removal. Remove the valve covers (see Section 3). Refer to Section 4 and loosen the rocker arm nuts, then pivot the rocker arms to the side for access to the pushrods.

2 Mount a dial indicator on the engine and position the plunger against the top of the first pushrod. The plunger should be directly in line with the pushrod **(see illustration)**. *Note: Disconnect the electrical connectors from the distributor to disable the ignition before making this test.*

3 Turn the engine slowly until the pushrod is at its lowest point; this is the base circle of the camshaft lobe. Zero the dial indicator, then very slowly turn the crankshaft in the normal direction of rotation until the indicator needle reaches its peak and begins to move in the opposite direction. The point at which it stops indicates maximum cam lobe lift.

4 Record this figure for future reference, then repeat the check for the remaining lobes. Be sure to record the results for each lobe.

5 After the check is complete, compare the results to this Chapter's Specifications. If camshaft lobe lift is 0.003 inch less than specified, cam lobe wear has occurred and a new camshaft should be installed.

Removal

Refer to illustrations 10.7, 10.9 and 10.10

6 Refer to the appropriate Sections and remove the intake manifold, the rocker arms, the pushrods and the timing chain and camshaft sprocket. The fan shrouds, fan, radiator and condenser should be removed as well (see Chapter 3).

7 Before removing the lifters, arrange to store them in a clearly labeled box to ensure that they're reinstalled in their original locations. Remove the lifter retainer **(see illustration)**. Remove the lifters and store them where they won't get dirty. DO NOT attempt to withdraw the camshaft with the lifters in place.

8 There are several ways to extract the lifters from the bores. A special tool designed to grip and remove lifters is manufactured by many tool companies and is widely available, but it may not be required in every case. On newer engines, without a lot of varnish buildup, the lifters can often be removed with a small magnet or even with your fingers. A machinist's scribe with a bent end can be used to pull the lifters out by positioning the point under the retainer ring in the top of each lifter. **Caution:** *Don't use pliers to remove the lifters unless you intend to replace them with new ones (along with the camshaft). The pliers will damage the precision machined and hardened lifters, rendering them useless.*

9 Remove the two Torx bolts and the camshaft retainer plate, noting which direction faces the block **(see illustration)**.

10 Thread three 6-inch long, 5/16-18 bolts into the camshaft sprocket bolt holes to use as a "handle" when removing the camshaft from the block **(see illustration)**.

11 Carefully pull the camshaft out. Support the cam near the block so the lobes don't nick or gouge the bearings as it's withdrawn.

Inspection

12 Visually examine the cam lobes and bearing journals for score marks, pitting, galling and evidence of overheating (blue, discolored areas). Look for flaking of the hardened surface of each lobe.

13 Using a micrometer, measure the diameter of each camshaft journal and the lift of each camshaft lobe (see Chapter 2A, Section 14). Compare your measurements with the Specifications listed at the front of this Chapter, and if the diameter of any one of these is less than specified, replace the camshaft.

Chapter 2 Part B 4.3L V6 engine

10.10 Thread three long bolts into the camshaft to use as a handle - pull the camshaft straight out, without nicking the bearings

10.15 Lubricate the camshaft journals and lobes with camshaft and lifter assembly lube before installation to provide initial lubrication

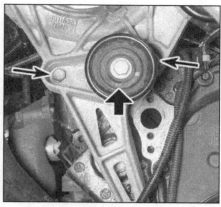

11.4 On the right cylinder head, remove the air conditioning compressor/tensioner bracket - remove the three bracket-to-block bolts (small arrows); it's necessary to remove the idler pulley (large arrow) for access to one of the bolts

Bearing replacement

14 Camshaft bearing replacement requires special tools and expertise that place it outside the scope of the home mechanic. Take the block to an automotive machine shop to ensure that the job is done correctly. Refer to Chapter 2, Part C for bearing measurement and inspection.

Installation

Refer to illustration 10.15

15 Lubricate the camshaft bearing journals and cam lobes with camshaft and lifter assembly lube **(see illustration)**.
16 Slide the camshaft into the engine. Support the cam near the block and be careful not to scrape or nick the bearings.
17 Turn the camshaft until the dowel pin is in the 3 o'clock position, and install the camshaft thrust plate, tighten the bolts to the torque listed in this Chapter's Specifications.
18 Install the balance shaft drive gear over the camshaft, aligning the dowel pin. Make sure the balance shaft timing marks are properly aligned **(see illustration 9.11)**.
19 Install the timing chain and sprockets (see Section 9).
20 Lubricate the lifters with clean engine oil and install them in the block. If the original lifters are being reinstalled, be sure to return them to their original locations. If a new camshaft is being installed, install new lifters as well.
21 The remaining installation steps are the reverse of removal.
22 Before starting and running the engine, change the oil and install a new oil filter (see Chapter 1).

11 Cylinder heads - removal and installation

Note: *The engine must be completely cool when the cylinder heads are removed. Failure to allow the engine to cool off could result in cylinder head warpage. When cool, refer to Chapter 1 and drain the cooling system.*

Removal

1 Remove the intake manifold (see Section 6), and exhaust manifolds (see Section 7).
2 Remove the valve covers (see Section 3).
3 Remove the pushrods (see Section 4).

Right cylinder head

Refer to illustration 11.4

4 On models through 1998, refer to Chapter 3 and remove the air conditioning compressor. DO NOT disconnect the hoses from the compressor. **Warning:** *The air conditioning system is under high pressure. DO NOT disassemble any part of the system unless the system has been depressurized.* Remove the compressor mounting bracket and the belt tensioner **(see illustration)**.
5 Disconnect the electrical connector from the knock sensor. On 1999 and later models, remove the alternator mounting bracket.
6 Remove the dipstick. Refer to Section 7 and unbolt and move the dipstick tube.

Left cylinder head

Refer to illustration 11.10

7 Remove the engine ground wire and electrical harness at the rear of the cylinder head.
8 Disconnect the electrical connectors from the knock sensor and the coolant temperature sensor.
9 On models through 1998, remove the alternator (see Chapter 5).
10 Unbolt and lay aside the accessory bracket retaining the power steering pump, without disconnecting the power steering hoses **(see illustration)**.

Both cylinder heads

Refer to illustration 11.11

11 Using a new cylinder head gasket, outline the cylinders and bolt pattern on a piece of cardboard **(see illustration)**. Be sure to indicate the front of the engine for reference. Punch holes at the bolt locations.
12 Loosen the cylinder head bolts in 1/4-turn increments until they can be removed by

11.10 After removing the alternator and laying it aside, remove the accessory bracket from the left cylinder head. You may leave the power steering pump attached and set it aside without disconnecting the hoses

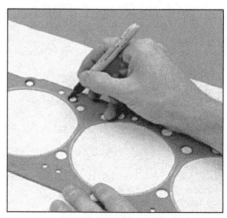

11.11 To avoid mixing up the cylinder head bolts, use a new gasket to transfer the bolt hole pattern to a piece of cardboard, then punch holes to accept the bolts

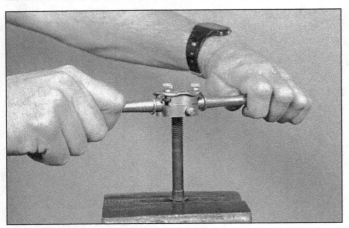

11.18 A die should be used to remove sealant and corrosion from the cylinder head bolt threads prior to installation

11.19a Locating dowels (arrows) are used to position the cylinder head gaskets on the block

hand. Work from bolt-to-bolt in a pattern that's the reverse of the tightening sequence shown in **illustration 11.22**. **Note:** *Don't overlook the row of bolts on the lower edge of each cylinder head, near the spark plug holes.* Store the bolts in the cardboard holder as they're removed; this will ensure that the bolts are reinstalled in their original holes.

13 Lift the cylinder head(s) off the engine. If resistance is felt, DO NOT pry between the cylinder head and block as damage to the mating surfaces will result. To dislodge the cylinder head, place a block of wood against the end of it and strike the wood block with a hammer. Store the cylinder heads on wood blocks to prevent damage to the gasket sealing surfaces.

14 Cylinder head disassembly and inspection procedures are covered in detail in Chapter 2, Part C.

Installation

Refer to illustrations 11.18, 11.19a, 11.19b, 11.21 and 11.22

15 The mating surfaces of the cylinder heads and block must be perfectly clean when the cylinder heads are installed.

16 Use a gasket scraper to remove all traces of carbon and old gasket material, then clean the mating surfaces with lacquer thinner or acetone. If there's oil on the mating surfaces when the cylinder heads are installed, the gaskets may not seal correctly and leaks may develop. When working on the block, cover the lifter valley with shop rags to keep debris out of the engine. Use a vacuum cleaner to remove any debris that falls into the cylinders.

17 Check the block and cylinder head mating surfaces for nicks, deep scratches and other damage. If damage is slight, it can be removed with a file - if it's excessive, machining may be the only alternative.

18 Use a tap of the correct size to chase the threads in the cylinder head bolt holes. Mount each bolt in a vise and run a die down the threads to remove corrosion and restore the threads **(see illustration)**. Dirt, corrosion, sealant and damaged threads will affect torque readings.

19 Position the new gaskets over the dowel pins in the block **(see illustration)**. **Note:** *If a steel gasket is used, apply a thin, even coat of a sealant such as K&W Copper Coat to both sides prior to installation* **(see illustration)**. **Warning:** *Composition-type gaskets are used on some engines, with a thin, sheetmetal core. Be very careful when handling because the edges may be very sharp. Composition gaskets do not require sealant.*

20 Carefully position the cylinder heads on the block without disturbing the gaskets.

21 Before installing the cylinder head bolts,

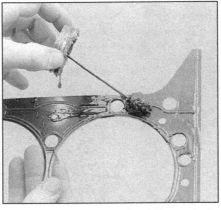

11.19b Steel gaskets should be coated with a sealant such as K&W Copper Coat before installation - composite gaskets do not use sealant

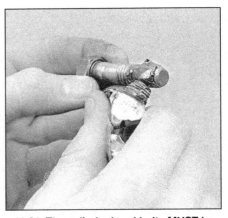

11.21 The cylinder head bolts MUST be coated with a non-hardening sealant (such as Permatex no. 2) before they're installed - coolant will leak past the bolts if this isn't done

coat the threads with a non-hardening sealant such as Permatex no. 2 **(see illustration)**.

22 Install the bolts in their original locations and tighten them finger tight. Follow the recommended sequence and tighten the bolts in several steps to the torque listed in this Chapter's Specifications **(see illustration)**.

23 The remaining installation steps are the reverse of removal.

24 Change the engine oil and filter (see Chapter 1), then start the engine and check carefully for oil and coolant leaks.

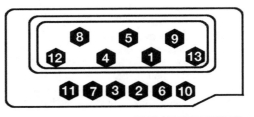

11.22 Cylinder head bolt tightening sequence

Chapter 2 Part B 4.3L V6 engine

12.3 Remove the two braces (arrows) between the bellhousing and the engine

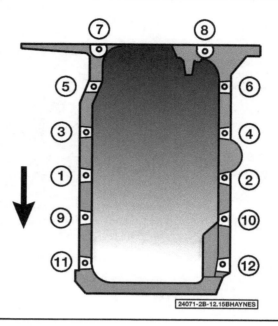

12.15 Bolt tightening sequence for the cast-aluminum oil pan

12 Oil pan - removal and installation

Note: *On 2WD models it is necessary to remove the engine from the vehicle to perform this procedure (see Chapter 2C). On 4WD models, the front differential housing must be unbolted, moved forward and supported (see Chapter 8) and the engine must be unbolted from the mounts (see Section 16) and raised with an engine hoist to provide the necessary clearance for oil pan removal.*

Removal and installation
1994 and 1995 4WD models

Refer to illustration 12.3

1 Disconnect the negative battery cable from the battery, then refer to Chapter 1 and drain the oil. **Caution:** *On models equipped with a Delco Loc II audio system, be sure the lockout feature is turned off before performing any procedure which requires disconnecting the battery.*

2 Remove the starter motor (see Chapter 5).

3 Remove the two engine to bellhousing braces and the cover from the lower part of the bellhousing **(see illustration)**.

4 Separate the exhaust pipes from the manifolds (see Section 7).

5 Remove the oil pan mounting bolts/nuts. Most models are equipped with a reinforcement strip on each side of the oil pan which may come loose after the bolts/nuts are removed. Note the location of any studs and brackets before removing them.

6 Carefully separate the pan from the block. Don't pry between the block and pan or damage to the sealing surfaces may result and oil leaks could develop. You may have to turn the crankshaft slightly to maneuver the front of the pan past the crank counterweights.

7 Clean the gasket sealing surfaces with lacquer thinner or acetone. Make sure the bolt holes in the block are clean.

8 Check the oil pan flange for distortion, particularly around the bolt holes. If necessary, place the pan on a block of wood and use a hammer to flatten and restore the gasket surface.

9 Apply a one-inch bead of RTV sealant at the corners of the block-to-front-cover joint and the rear-main-seal-to-block joint. Attach the new gasket to the pan.

10 Carefully position the pan against the block and install the bolts/nuts finger tight (don't forget the reinforcement strips, if used). Tighten the bolts/nuts in three steps to the torque listed in this Chapter's Specifications. Start at the center of the pan and work out toward the ends in a spiral pattern.

1996 and later models

Refer to illustration 12.15

11 A cast-aluminum oil pan is used on 1996 and later models that is designed as a structural reinforcement between the engine and transmission. It's alignment is critical. As with the previous models, the factory recommends removing the engine from the vehicle on 2WD models to remove the oil pan, and 4WD models can be done in-vehicle with the engine raised if the front axle is disconnected and pulled forward.

12 After removing the bolts (note the location of any studs and brackets before removal), separate the pan from the engine carefully. Do not use a pry tool between the block and pan or the soft aluminum of the pan can be gouged, leading to later sealing problems. Also, if a hammer and block of wood is used to break the gasket seal, use a rawhide or rubber mallet and only strike the pan where it is reinforced.

13 Clean the gasket sealing surfaces with lacquer thinner or acetone. Make sure the bolt holes in the block are clean and clean the pan's mounting surface, but do not damage the soft aluminum.

14 Apply a one-inch-long bead of RTV sealant at the corners of the block-to-front-cover joint and the rear-main-seal-to-block joint. Attach the new gasket to the pan, install the pan and tighten the bolts/studs finger-tight.

15 The alignment of the rear face of the aluminum pan to the rear of the block is important. When done in-vehicle (4WD models) measure with feeler gauges between the rear face of the pan and the front face of the transmission bellhousing. Clearance should ideally be flush, but up to 0.010-inch is allowable. If the clearance is OK, tighten the pan bolts/studs in sequence to the torque listed in this Chapter's Specifications **(see illustration)**. On 2WD models where the engine is out of the vehicle, use a heavy, precision steel straightedge held against the rear face of the block (bellhousing mounting surface) and measure between the straightedge and the pan for proper clearance.

16 The remaining steps are the reverse of removal. **Caution:** *Don't forget to refill the engine with oil before starting it (see Chapter 1).*

17 Start the engine and check carefully for oil leaks at the oil pan.

13 Oil pump - removal and installation

1 Remove the oil pan (see Section 12).

2 While supporting the oil pump, remove the pump-to-rear main bearing cap bolt. On some models, the oil pan baffle must be removed first, since it's also held in place by the pump mounting bolt.

3 Lower the pump and remove it along with the pump driveshaft.

4 If a new oil pump is installed, make sure the pump driveshaft is mated with the shaft inside the pump.

5 Position the pump on the engine and

make sure the slot in the upper end of the driveshaft is aligned with the tang on the lower end of the distributor shaft. The distributor drives the oil pump, so it is absolutely essential that the components mate properly.

6 Install the mounting bolt and tighten it to the torque listed in this Chapter's Specifications.

7 Install the oil pan and refill with fresh oil. The remainder of assembly is the reverse of the disassembly procedures.

14 Flywheel/driveplate - removal and installation

The flywheel replacement procedure is virtually identical to that of the 2.2L four-cylinder engine. Refer to Chapter 2, Part A for the procedure, and use the torque figures from this Chapter's Specifications.

15 Rear main oil seal - replacement

The rear main oil seal replacement procedure is virtually identical to that of the 2.2L four-cylinder engine. Refer to Chapter 2, Part A for the general procedure, and use the torque figures from this Chapter. Later model V6 engines have the rear main seal mounted in a seal retainer. The retainer has two slots on the edge where you can insert a screwdriver to more easily pry out the seal without damaging the crankshaft.

16 Engine mounts - check and replacement

1 Engine mounts seldom require attention, but broken or deteriorated mounts should be replaced immediately or the added strain placed on the driveline components may cause damage.

16.8a Engine mount through-bolt nut location (arrow) - left side (four-wheel drive model shown, with front driveshaft removed for clarity)

Check

2 During the check, the engine must be raised slightly to remove the weight from the mounts.

3 Raise the vehicle and support it securely on jackstands, then position the jack under the engine oil pan. Place a large block of wood between the jack head and the oil pan, then carefully raise the engine just enough to take the weight off the mounts.

4 Check the mounts to see if the rubber is cracked, hardened or separated from the metal plates. Sometimes the rubber will split right down the center. Rubber preservative or WD-40 should be applied to the mounts to slow deterioration.

5 Check for relative movement between the mount plates and the engine or frame (use a large screwdriver or prybar to attempt to move the mounts). If movement is noted, lower the engine and tighten the mount fasteners.

16.8b Engine mount through-bolt nut location (arrow) - right side, just above the starter motor

Replacement

Refer to illustrations 16.8a and 16.8b

6 Disconnect the negative battery cable from the battery, then raise the vehicle and support it securely on jackstands. **Caution:** *On models equipped with a Delco Loc II audio system, be sure the lockout feature is turned off before performing any procedure which requires disconnecting the battery.*

7 Attach an engine hoist to the top of the engine for lifting; do not use a jack under the oil pan or the oil pump pick-up could be damaged.

8 Raise the engine slightly, then remove the through-bolt nuts and withdraw the mount through-bolt from the frame bracket **(see illustrations)**. Raise the engine further until the engine mount can be unbolted from the block.

9 Installation is the reverse of removal. Use Loctite on the mount bolts and be sure to tighten them securely.

Chapter 2 Part C
General engine overhaul procedures

Contents

	Section		Section
Balance shaft (V6 engines only) - installation	25	Engine overhaul - general information	2
Camshaft, balance shaft and bearings - removal and inspection	13	Engine overhaul - reassembly sequence	22
		Engine rebuilding alternatives	7
Crankshaft - inspection	20	Engine removal - methods and precautions	5
Crankshaft - installation and main bearing oil clearance check	24	General information	1
Crankshaft - removal	15	Initial start-up and break-in after overhaul	29
Cylinder compression - check	3	Main and connecting rod bearings - inspection	21
Cylinder head - cleaning and inspection	10	Piston rings - installation	23
Cylinder head - disassembly	9	Piston/connecting rod assembly - inspection	19
Cylinder head - reassembly	12	Piston/connecting rod assembly - installation and bearing oil clearance check	27
Cylinder honing	18		
Driveplate - removal and installation	28	Piston/connecting rod assembly - removal	14
Engine - removal and installation	6	Rear main oil seal - installation	26
Engine block - cleaning	16	SERVICE ENGINE SOON light	See Chapter 6
Engine block - inspection	17	Vacuum gauge diagnostics - check	4
Engine overhaul - disassembly sequence	8	Valves - servicing	11

Specifications

2.2L four-cylinder engine

General
Bore and stroke	3.50 x 3.46 inches
Displacement	134 cubic inches
Oil pressure, warm, minimum	56 psi at 3000 rpm
Compression pressure	140 psi at 160 rpm

Engine block
Cylinder bore diameter	3.5036 to 3.5043 inches
Out-of-round and taper limit	0.0005 inch

Cylinder head and valvetrain
Head warpage limit	0.002 inch
Maximum head milling	0.010 inch
Valve face angle	45-degrees
Valve seat angle	46-degrees
Minimum valve margin width	1/32 inch
Valve stem-to-guide clearance	
1994	
Intake	0.001 to 0.002 inch
Exhaust	0.001 to 0.003 inch
1995 through 1998	
Intake	0.0010 to 0.0027 inch
Exhaust	0.0014 to 0.0031 inch
1999 and later	
Intake	0.0070 to 0.0020 inch
Exhaust	0.0014 to 0.0031 inch
Valve spring free length	
Through 1998	1.950 inches
1999 and later	1.910 inches
Valve spring installed height	
Through 1998	1.710 inches
1999 and later	1.600 inches

Crankshaft and connecting rods
Crankshaft endplay	0.002 to 0.007 inch
Connecting rod endplay (side clearance)	0.0039 to 0.0149 inch
Main bearing journal diameter	2.4945 to 2.4954 inches
Main bearing oil clearance	0.0006 to 0.0019 inch
Connecting rod journal diameter	1.9983 to 1.9994 inches

Crankshaft and connecting rods (continued)

Connecting rod bearing oil clearance	0.00098 to 0.0031 inch
Crankshaft journal taper/out-of-round limit	0.00019 inch

Pistons and rings

Piston-to-bore clearance	0.0007 to 0.0017 inch
Piston ring side clearance	
Compression rings	0.0019 to 0.0027 inch
Oil ring	0.0019 to 0.0082 inch
Piston ring end gap	
Compression rings	0.010 to 0.020 inch
Oil ring	
Through 1998	0.020 to 0.050 inch
1999 and later	0.010 to 0.030 inch

4.3L V6 engine

General

Bore and stroke	4.000 x 3.480 inches
Displacement	262 cubic inches
Oil pressure, hot, minimum	
at 1000 rpm	6 psi
at 2000 rpm	18 psi
at 4000 rpm	24 psi
Compression pressure	Lowest reading cylinder must be at least 70-percent of highest reading cylinder (100 psi minimum)

Engine block

Cylinder bore diameter	4.0007 to 4.0017 inches
Taper limit	0.001 inch
Out-of-round limit	0.002 inch

Cylinder head and valvetrain

Head warpage limit	0.003 inch per 6 inch span/0.006 inch overall
Valve seat angle	46-degrees
Valve face angle	45-degrees
Minimum valve margin width	1/32 inch
Valve stem-to-guide clearance	
Intake	0.001 to 0.003 inch
Exhaust	0.001 to 0.004 inch
Valve spring free length	2.03 inches
Valve spring installed height, intake and exhaust	1.670 to 1.700 inches

Crankshaft and connecting rods

Main journal	
Diameter	
No. 1 journal	2.4488 to 2.4495 inches
No. 2 and 3 journals	2.4485 to 2.4494 inches
No. 4 journal	2.4480 to 2.4489 inches
Taper limit	0.001 inch
Out-of-round limit	0.001 inch
Main bearing oil clearance	
No. 1 journal	0.0010 to 0.0020 inch
No. 2 and 3 journals	0.0010 to 0.0025 inch
No. 4 journal	0.0025 to 0.0035 inch
Connecting rod journal	
Diameter	2.2487 to 2.2497 inches
Taper limit	0.001 inch
Out-of-round limit	0.001 inch
Connecting rod bearing oil clearance	0.0013 to 0.0035 inch
Connecting rod endplay (side clearance)	
Through 1998	0.015 to 0.046 inch
1999 and later	0.006 to 0.017 inch
Crankshaft endplay	0.005 to 0.018 inch

Pistons and rings

Piston-to-cylinder bore clearance	
Standard	0.0007 to 0.0017 inch
Service limit	0.0024 inch

Piston ring-to-groove side clearance
  Compression rings
    Standard .... 0.0012 to 0.0032 inch
    Service limit .... 0.0030 inch
  Oil control ring
    Standard
      Through 1998 .... 0.002 to 0.007 inch
      1999 and later .... 0.0018 to 0.0037 inch
    Service limit
      Through 1998 .... 0.008 inch
      1999 and later .... 0.0039 inch
Piston ring end gap
  Top compression ring
    Standard .... 0.010 to 0.020 inch
    Service limit .... 0.035 inch
  Second compression ring
    Standard .... 0.010 to 0.025 inch
    Service limit .... 0.035 inch
  Oil control ring
    Standard
      Through 1998 .... 0.015 to 0.055 inch
      1999 and later .... 0.010 to 0.029 inch
    Service limit
      Through 1998 .... 0.065 inch
      1999 and later .... 0.035 inch

Camshaft
Lobe lift and journal diameters .... See Chapter 2B
End play .... 0.001 to 0.009 inch

Balance shaft
Front bearing journal diameter .... 2.1648 to 2.1654 inches
Rear bearing journal diameter .... 1.4994 to 1.5000 inches
Rear bearing journal clearance .... 0.001 to 0.0036 inch

Torque specifications   **Ft-lbs** (unless otherwise indicated)

2.2L four-cylinder engine*
Main bearing cap bolts 70
Connecting rod cap nuts
  Step 1 38
  Step 2 .... Loosen 1 turn
  Step 3 38
Oil pump to block bolts
  Through 1998 22
  1999 and later 32
Oil pick-up tube bracket nut 37
Note: Refer to Part A for additional torque specifications

4.3L V6 engine*
Main bearing cap bolts
  1994 75
  1995 81
  1996 through 1998 77
  1999 and later
    Step 1 15
    Step 2 .... turn 73 degrees
Connecting rod cap nuts
  Step 1 20
  Step 2 .... Tighten an additional 70-degrees
Oil pump-to-block bolt 65
Note: Refer to Part B for additional torque specifications

1   General information

Included in this portion of Chapter 2 are the general overhaul procedures for the cylinder head(s) and internal engine components of the 2.2L four-cylinder and the 4.3L V6. The information ranges from advice concerning preparation for an overhaul and the purchase of replacement parts to detailed, step-by-step procedures covering removal and installation of internal engine components and the inspection of parts.

The following Sections have been written based on the assumption that the engine has been removed from the vehicle. For information concerning in-vehicle engine repair, as well as removal and installation of the external components necessary for the overhaul, see Part A of Chapter 2 for the 2.2L or Part B for the 4.3L.

2.4a The oil pressure sending unit (arrow) is located on right side of engine - 2.2L OHV engine

2.4b The oil pressure sending unit (arrow) on the 4.3L V6 engine is located on top of the block behind the intake manifold

3.4 Use a compression gauge with a threaded fitting for the spark plug hole; don't use the type with a conical rubber tip - it relies on hand pressure to create a seal between the gauge and the combustion chamber, and usually gives an inaccurate reading

The Specifications included here in Part C are only those necessary for the inspection and overhaul procedures which follow. Refer to Part A or B for additional Specifications related to the two engines covered in this manual.

2 Engine overhaul - general information

Refer to illustrations 2.4a and 2.4b

It is not always easy to determine when, or if, an engine should be completely overhauled, as a number of factors must be considered.

High mileage is not necessarily an indication that an overhaul is needed while low mileage, on the other hand, does not preclude the need for an overhaul. Frequency of servicing is probably the single most important consideration. An engine that has regular (and frequent) oil and filter changes, as well as other required maintenance, will most likely give many thousands of miles of reliable service. Conversely, a neglected engine may require an overhaul very early in its life.

Excessive oil consumption is an indication that piston rings and/or valve guides are in need of attention (make sure that oil leaks are not responsible before deciding that the rings and guides are bad). Have a cylinder compression or leak-down test performed by an experienced tune-up mechanic to determine for certain the extent of the work required.

If the engine is making obvious knocking or rumbling noises, the connecting rod and/or main bearings are probably at fault. Check the oil pressure with a gauge installed in place of the oil pressure sending unit **(see illustrations)**. Compare the measured oil pressure to this Chapter's Specifications. If it is extremely low, the bearings and/or oil pump are probably worn out.

Loss of power, rough running, excessive valvetrain noise and high fuel consumption rates may also point to the need for an overhaul (especially if they are all present at the same time). If a complete tune-up does not remedy the situation, major mechanical work is the only solution.

An engine overhaul generally involves restoring the internal parts to the specifications of a new engine. During an overhaul, the piston rings are replaced and the cylinder walls are reconditioned (rebored and/or honed). If a rebore is done, then new pistons are also required. The main and connecting rod bearings are replaced with new ones and, if necessary, the crankshaft may be reground to restore the journals.

Generally, the valves are serviced as well, since they're usually in less-than-perfect condition at this point. While the engine is being overhauled, rebuild other components such as the starter and the alternator. The end result will be a like-new engine that will give as many trouble-free miles as the original.

Before beginning the engine overhaul, read through the entire procedure to familiarize yourself with the scope and requirements of the job. Overhauling an engine is not that difficult, but it is time consuming. Plan on the vehicle being tied up for a minimum of two weeks, especially if parts must be taken to an automotive machine shop for repair or reconditioning. Check on availability of parts and make sure that any necessary special tools and equipment are obtained in advance. Most work can be done with typical shop hand tools, although a number of precision measuring tools are required for inspecting parts to determine if they must be replaced. Often a reputable automotive machine shop will handle the inspection of parts and offer advice concerning reconditioning and replacement. **Note:** *Always wait until the engine has been completely disassembled and all components, especially the engine block, have been inspected before deciding what service and repair operations must be performed by an automotive machine shop.* Since the block's condition will be the major factor to consider when determining whether to overhaul the original engine or buy a rebuilt one, never purchase parts or have machine work done on other components until the block has been thoroughly inspected. As a general rule, time is the primary cost of an overhaul, so it does not pay to install worn or sub-standard parts.

As a final note, to ensure maximum life and minimum trouble from a rebuilt engine, everything must be assembled with care in a spotlessly clean environment.

3 Cylinder compression check

Refer to illustration 3.4

1   A compression check will tell you what mechanical condition the upper end (pistons, rings, valves, head gaskets) of your engine is in. Specifically, it can tell you if the compression is down due to leakage caused by worn piston rings, defective valves and seats or a blown head gasket. **Note:** *The engine must be at normal operating temperature and the battery must be fully charged for this check.*
2   Begin by cleaning the area around the spark plugs before you remove them (compressed air should be used, if available, otherwise a small brush or even a bicycle tire pump will work). The idea is to prevent dirt from getting into the cylinders as the compression check is being done. Remove all of the spark plugs from the engine (see Chapter 1).
3   Block the throttle wide open. Pull the PCM/BATT fuse from the fuse panel (see Chapter 12) to prevent the engine from starting during the test.
4   With the compression gauge installed in place of the number one spark plug **(see illustration)**, depress the accelerator pedal all the way to the floor to open the throttle valve. Crank the engine over at least four compression strokes and watch the gauge. The compression should build up quickly in a healthy engine. Low compression on the first stroke, followed by gradually increasing pressure on successive strokes, indicates worn piston rings. A low compression reading on the first stroke, which doesn't build up during successive strokes, indicates leaking valves or a blown head gasket (a cracked head could also be the cause). Record the highest gauge reading obtained.

4.4 Use a vacuum gauge connected to manifold vacuum (full vacuum available at idle) to test an engine's condition

5   Repeat the procedure for the remaining cylinders and compare the results to the Specifications.
6   Add some engine oil (about three squirts from a plunger-type oil can) to each cylinder, through the spark plug holes, and repeat the test.
7   If the compression increases after the oil is added, the piston rings are definitely worn. If the compression doesn't increase significantly, the leakage is occurring at the valves or head gasket. Leakage past the valves may be caused by burned valve seats and/or faces or warped, cracked or bent valves.
8   If two adjacent cylinders have equally low compression, there's a strong possibility that the head gasket between them is blown. The appearance of coolant in the combustion chambers or the crankcase would verify this condition.
9   If the compression is unusually high, the combustion chambers are probably coated with carbon deposits. If that's the case, the cylinder head(s) should be removed and decarbonized.
10   If compression is way down or varies greatly between cylinders, it would be a good idea to have a leak-down test performed by an automotive repair shop. This test will pinpoint exactly where the leakage is occurring and how severe it is.

4   Vacuum gauge diagnostic - checks

Refer to illustration 4.4

A vacuum gauge provides valuable information about what is going on in the engine at a low cost. You can check for worn rings or cylinder walls, leaking head or intake manifold gaskets, incorrect fuel mixtures, restricted exhaust, stuck or burned valves, weak valve springs, improper ignition or valve timing and ignition problems.

Unfortunately, vacuum gauge readings are easy to misinterpret, so they should be used in conjunction with other tests to confirm the diagnosis.

Both the gauge readings and the rate of needle movement are important for accurate interpretation. Most gauges measure vacuum in inches of mercury (in-Hg). As vacuum increases (or atmospheric pressure decreases), the reading will increase. Also, for every 1,000-foot increase in elevation above sea level, the gauge readings will decrease about one inch of mercury.

Connect the vacuum gauge directly to intake manifold vacuum **(see illustration)**. Be sure no hoses are left disconnected during the test or false readings will result.

Before you begin the test, allow the engine to warm up completely. Block the wheels and set the parking brake. With the transmission in Park, start the engine and allow it to run at normal idle speed. **Warning:** *Carefully inspect the fan blades for cracks or damage before starting the engine. Keep your hands and the vacuum tester clear of the fan and do not stand in front of the vehicle or in line with the fan when the engine is running.*

Read the vacuum gauge; an average, healthy engine should normally produce about 17 to 22 inches of vacuum with a fairly steady needle. Refer to the following vacuum gauge readings and what they indicate about the engines condition:
1   A low, steady reading usually indicates a leaking gasket between the intake manifold and throttle body, a leaky vacuum hose, late ignition timing or incorrect camshaft timing. Eliminate all other possible causes, utilizing the tests provided in this Chapter before you remove the timing chain cover to check the timing marks.
2   If the reading is three to eight inches below normal and it fluctuates at that low reading, suspect an intake manifold gasket leak at an intake port.
3   If the needle has regular drops of about two to four inches at a steady rate, the valves are probably leaking. Perform a compression or leak-down test to confirm this.
4   An irregular drop or down-flick of the needle can be caused by a sticking valve or an ignition misfire. Perform a compression or leak-down test and read the spark plugs.
5   A rapid vibration of about four inches-Hg vibration at idle combined with exhaust smoke indicates worn valve guides. Perform a leak-down test to confirm this. If the rapid vibration occurs with an increase in engine speed, check for a leaking intake manifold gasket or head gasket, weak valve springs, burned valves or ignition misfire.
6   A slight fluctuation, say one inch up and down, may mean ignition problems. Check all the usual tune-up items and, if necessary, run the engine on an ignition analyzer.
7   If there is a large fluctuation, perform a compression or leak-down test to look for a weak or dead cylinder or a blown head gasket.
8   If the needle moves slowly through a wide range, check for a clogged PCV system, incorrect idle fuel mixture, throttle body or intake manifold gasket leaks.
9   Check for a slow return after revving the engine by quickly snapping the throttle open until the engine reaches about 2,500 rpm and let it shut. Normally the reading should drop to near zero, rise above normal idle reading (about 5 in-Hg over) and then return to the previous idle reading. If the vacuum returns slowly and doesn't peak when the throttle is snapped shut, the rings may be worn. If there is a long delay, look for a restricted exhaust system (often the muffler or catalytic converter). An easy way to check this is to temporarily disconnect the exhaust ahead of the suspected part and re-test.

5   Engine removal - methods and precautions

If it has been decided that an engine must be removed for overhaul or major repair work, certain preliminary steps should be taken. Locating a suitable work area is extremely important. A shop is, of course, the most desirable place to work. Adequate work space along with storage space for the vehicle is very important. If a shop or garage is not available, at the very least a flat, level, clean work surface made of concrete or asphalt is required.

Cleaning the engine compartment and engine prior to removal will help keep tools clean and organized. An engine hoist or A-frame will also be necessary. Make sure that the equipment is rated in excess of the combined weight of the engine and its accessories. Safety is of primary importance, considering the potential hazards involved in lifting the engine out of the vehicle.

If the engine is being removed by a novice, a helper should be available. Advice and aid from someone more experienced would also be helpful. There are many instances when one person cannot simultaneously perform all of the operations required when lifting the engine out of the vehicle.

Plan the operation ahead of time. Arrange for or obtain all of the tools and equipment you will need prior to beginning the job. Some of the equipment necessary to perform engine removal and installation safely and with relative ease are (in addition to an engine hoist) a heavy-duty floor jack, complete sets of wrenches and sockets as described in the front of this manual, wooden blocks and plenty of rags and cleaning solvent for mopping up the inevitable spills. If the hoist is to be rented, make sure that you arrange for it in advance and perform beforehand all of the operations possible without it. This will save you money and time.

Plan for the vehicle to be out of use for a considerable amount of time. A machine shop will be required to perform some of the work which the do-it-yourselfer cannot accomplish due to a lack of special equipment. These shops often have a busy schedule so it would be wise to consult them before removing the engine in order to accurately estimate the amount of time required to rebuild or repair components that may need work.

2C-6   Chapter 2 Part C   General engine overhaul procedures

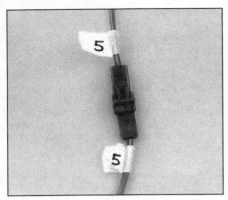

6.4 Label each wire before unplugging the connector

6.9a Unbolt the lines from the oil cooler adapter (arrow) and remove the braces (arrow) - there may be a brace on each side of the oil pan

6.9b On 4WD models, the oil filter assembly (arrow) is mounted remotely, up front under the chassis

Always use extreme caution when removing and installing the engine; serious injury can result from careless actions. Plan ahead. Take your time and a job of this nature, although major, can be accomplished successfully.

6   Engine - removal and installation

Warning 1: *1995 and later models are equipped with an airbag. Impact sensors for the airbag system are located in the dash, cowl and just in front of the radiator. The airbag could accidentally deploy if these sensors are disturbed, so be extremely careful when working in this area. Airbag system components are also located in the steering wheel and the steering column area, so be extremely careful when working in this area and don't disturb any airbag system components or wiring. You could easily be injured if an airbag accidentally deploys, and the airbag might not deploy correctly in a collision if any components or wiring in the system have been disturbed. Always turn the steering wheel to the straight ahead position, place the ignition switch in Lock and remove the key, then remove the Air Bag fuse and unplug the yellow Connector Position Assurance (CPA) connector at the base of the steering column (see Chapter 12).*
Warning 2: *Gasoline is extremely flammable, so take extra precautions when disconnecting any part of the fuel system. Don't smoke or allow open flames or bare light bulbs in or near the work area and don't work in a garage where a gas appliance (such as a clothes dryer or water heater) is installed. If you spill gasoline on your skin, rinse it off immediately. Have a fire extinguisher rated for gasoline fires handy and know how to use it!*
Warning 3: *The air-conditioning system is under high pressure. DO NOT loosen any fittings or remove any components until after the system has been discharged. Air-conditioning refrigerant should be properly discharged into an EPA-approved recovery container at a dealer service department or an automotive air-conditioning facility. Always wear eye protection when disconnecting air-conditioning system fittings.*

Caution: *On models equipped with a Delco Loc II audio system, be sure the lockout feature is turned off before performing any procedure which requires disconnecting the battery.*

Removal

Refer to illustrations 6.4, 6.9a, 6.9b and 6.19
1   Relieve the fuel system pressure (see Chapter 4), then disconnect the negative cable from the battery. See **Caution** above.
2   Cover the fenders and cowl. Special pads are available to protect the fenders, but an old bedspread or blanket will also work. It's also a good idea to remove the hood (see Chapter 11) to give yourself more room to work.
3   Remove the air intake duct assembly (see Chapter 4).
4   To ensure correct reassembly, label each vacuum line, emission system hose, wiring connector, ground strap and fuel line. Pieces of masking tape with numbers or letters written on them prevent confusion at assembly time **(see illustration)**. Or sketch the engine compartment routing of lines, hoses and wires.
5   Drain the cooling system (see Chapter 1) and label and detach all coolant hoses from the engine.
6   Disconnect the throttle and cruise control cables (see Chapter 4) and TV cable (see Chapter 7B). Disconnect the fuel lines at the intake manifold (see Chapter 4).
7   Have a dealer service department or service station evacuate the air conditioning system and recapture the refrigerant before disconnecting any of the air conditioning hoses or fittings. Disconnect the hoses from the compressor, hang them out of the way, detach the air conditioning compressor from its mounting bracket and remove it from the engine compartment (see Chapter 3).
8   Raise the vehicle and suitably support it on jackstands. While raised, perform the disassembly procedures that can be done only from underneath, such as disconnecting the exhaust, and disconnecting the transmission cooler lines where they are held by a clip to one of the oil pan bolts (automatic-transmission models) and disconnecting a wiring harness attached to the block.
9   Drain the engine oil and remove the filter (see Chapter 1). On V6 engines, disconnect the oil cooler lines from the oil cooler adapter between the oil filter and the block **(see illustrations)**. **Note:** *2WD models have the adapter between the filter and the block, while 4WD versions have the adapter on the block, with lines going forward to a remote filter.*
10   Remove the skid plate, if equipped. Remove the starter (see Chapter 5).
11   Remove the driveplate access cover. Scribe or paint an alignment mark between the driveplate and the torque converter (see Chapter 7B) then rotate the crankshaft and remove the driveplate bolts.
12   Remove the bellhousing-to-engine bolts, engine mount throughbolts (see Chapter 2, Part A or Part B), and braces (if equipped) between the engine and transmission **(see illustration 6.9a)**.
13   Lower the vehicle and place a jack under the transmission pan, then support the engine with an engine hoist from above. Remove the ground strap at the back of the engine.
14   Remove the fan shrouds, fan, water pump, radiator and air conditioning condenser (see Chapter 3).
15   Unbolt the power steering pump and bracket and tie them out of the way.
16   Remove the intake manifold plenum on V6 models, or the complete intake manifold on the 2.2L four-cylinder. Refer to Part A or B and remove the exhaust manifolds.
17   Raise the engine enough to take tension off the motor mounts and remove the motor mounts.
18   Raise the engine and pull it forward to free it from the converter snout (automatic transmissions) or transmission input shaft (standard transmissions).
19   Tie the wiring harnesses out of the way, raise the engine with the hoist, and with a combination of tilting, twisting and raising, move the engine around any obstructions and pull it out of the vehicle, raising it high enough to clear the front of the body **(see illustration)**.
20   Remove the driveplate (see Section 28) and mount the engine on an engine stand.

Installation

21 While the engine is out, check the engine mount and the transmission mounts (see Chapter 7). If they're worn or damaged, replace them.
22 Carefully lower the engine, twisting it to clear any obstructions, until the transmission input shaft or converter snout lines up and the bellhousing-to-engine bolts can be inserted. **Caution:** *DO NOT use the transmission-to-engine bolts to force the transmission and engine together. Take great care when mating the torque converter to the driveplate, following the procedure outlined in Chapter 7. Make sure the alignment marks you made on the driveplate and the torque converter during removal are lined up.*
23 Install the driveplate-to-torque converter bolts and tighten them to the torque listed in Chapter 2A or 2B Specifications.
24 Reinstall the remaining components in the reverse order of removal. Double-check to make sure everything is hooked up right, using the sketches or photos taken earlier to go by.
25 Add coolant, oil, power steering and transmission fluid as needed.
26 Run the engine and check for leaks and proper operation of all accessories, then install the hood and test drive the vehicle.

7   Engine rebuilding alternatives

The do-it-yourselfer is faced with a several options when performing an engine overhaul. The decision to replace the block, piston/connecting rod assemblies and crankshaft depends on a number of factors, with the number one consideration being the condition of the block. Other factors are cost, access to machine shop facilities, parts availability, time required to complete the project and experience.

Some of the rebuilding alternatives include:

Individual parts - If the inspection procedures reveal that the engine block and most engine components are in reusable condition, purchasing individual parts may be the most economical alternative. The block, crankshaft and piston/connecting rod assemblies should all be inspected carefully. Even if the block shows little wear, the cylinder bores should receive a finish hone; a job for an automotive machine shop.

Master kit (crankshaft kit) - This rebuild package usually consists of a reground crankshaft and a matched set of pistons and connecting rods. The pistons will already be installed on the connecting rods. Piston rings and the necessary bearings may or may not be included in the kit. These kits are commonly available for standard cylinder bores, as well as for engine blocks which have been bored to a regular oversize.

Short block - A short block consists of an engine block with a crankshaft and piston/connecting rod assemblies already installed. All new bearings are incorporated and all clearances will be correct. Depending on where the short block is purchased, a guarantee may be included. The existing camshaft, valvetrain components, cylinder head(s) and external parts can be bolted to the short block with little or no machine shop work necessary.

Long block - A long block consists of a short block plus an oil pump, oil pan, cylinder head(s), valve cover(s), camshaft and valvetrain components, timing gears or sprockets and chain and timing gear/chain cover. All components are installed with new bearings, seals and gaskets incorporated throughout. The installation of manifolds and external parts is all that is necessary. Some form of guarantee is usually included with the purchase.

Used engine assembly - While overhaul provides the best assurance of a like-new engine, used engines available from wrecking yards and importers are often a very simple and economical solution. Many used engines come with warranties, but always give any engine a thorough diagnostic check-out before purchase. Check compression, vacuum and also for signs of oil leakage. If possible, have the seller run the engine, ether in the vehicle or on a test stand so you can be sure it runs smoothly with no knocking or other noises.

Give careful thought to which alternative is best for you and discuss the situation with local automotive machine shops, auto parts stores or dealership parts department before ordering or purchasing replacement parts.

8   Engine overhaul - disassembly sequence

1 It is much easier to disassemble and work on the engine if it is mounted on a portable engine stand. These stands can often be rented for a reasonable fee from an equipment rental yard. Before the engine is mounted on a stand, the flywheel/driveplate should be removed from the engine (see Part A of this Chapter).
2 If a stand is not available, it is possible to disassemble the engine with it blocked up on a sturdy workbench or on the floor. Be extra careful not to tip or drop the engine when working without a stand.
3 If you are going to obtain a rebuilt engine, all external components must come off first in order to be transferred to the replacement engine (just as they will if you are doing a complete engine overhaul yourself). These include:

Alternator and brackets
Emissions control components
Distributor or coil pack, spark plug wires and spark plugs
Thermostat and housing cover
Water pump
Fuel injection components
Intake and exhaust manifolds
Oil filter
Engine mounts
Flywheel/driveplate

Note: *When removing the external components from the engine, pay close attention to details that may be helpful or important during installation. Note the installed position of gaskets, seals, spacers, pins, washers, bolts and other small items.*

6.19 With the accessories tied out of the way and the motor mount through bolts removed, tilt, twist and raise the engine with the hoist until it clears the engine compartment

4 If you are obtaining a short block (which consists of the engine block, crankshaft, pistons and connecting rods all assembled), then the cylinder heads, oil pan and oil pump will have to be removed also. See *Engine rebuilding alternatives* for additional information regarding the different possibilities to be considered.
5 If you are planning a complete overhaul, the engine must be disassembled and the internal components removed in the following order:

Valve cover(s)
Rocker arms and pushrods
Cylinder head(s)
Roller lifter retainers
Valve lifters
Timing chain cover
Timing chain and sprockets
Camshaft
Balance shaft (V6)
Oil pan
Oil pump
Piston/connecting rod assemblies
Crankshaft

6 Before beginning the disassembly and overhaul procedures, make sure the following items are available:

Common hand tools
Small cardboard boxes or plastic bags for storing parts
Gasket scraper
Ridge reamer
Vibration damper puller
Micrometers
Small hole gauges
Telescoping gauges
Dial indicator set
Valve spring compressor
Cylinder surfacing hone
Piston ring groove cleaning tool
Electric drill motor
Tap and die set
Wire brushes
Cleaning solvent

9.1 A small plastic bag, with an appropriate label, can be used to store the valve train components so they can be kept together and reinstalled in their original locations

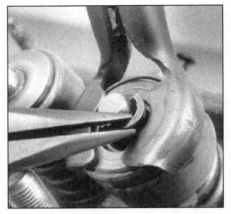

9.3a Use a valve spring compressor to compress the spring, then remove the keepers from the valve stem

9.3b If the valve won't pull through the guide, deburr the edge of the stem end and the area around the top of the keeper groove with a file or whetstone

9 Cylinder head - disassembly

Refer to illustrations 9.1, 9.3a and 9.3b

Note: *New and rebuilt cylinder heads are commonly available for most engines at dealerships and auto parts stores. Due to the fact that some specialized tools are necessary for the disassembly and inspection procedures, and replacement parts may not be readily available, it may be more practical and economical for the home mechanic to purchase a replacement head (or heads) rather than taking the time to disassemble, inspect and recondition the original head(s).*

1   Cylinder head disassembly involves removal and disassembly of the intake and exhaust valves and their related components. If they are still in place, remove the nuts or bolts and pivot balls, then separate the rocker arms from the cylinder head. Label the parts or store them separately so they can be reinstalled in their original locations **(see illustration)**.

2   Before the valves are removed, arrange to label and store them, along with their related components, so they can be kept separate and reinstalled in the same valve guides they are removed from.

3   Compress the valve spring on the first valve with a spring compressor and remove the keepers **(see illustration)**. Carefully release the valve spring compressor and remove the retainer (or rotator), the shield (if equipped), the springs, the valve guide seal and/or O-ring seal, the spring seat and the valve from the head. If the valve binds in the guide (won't pull through), push it back into the head and deburr the area around the keeper groove with a fine file or whetstone **(see illustration)**.

4   Repeat the procedure for the remaining valves. Remember to keep together all the parts for each valve so they can be reinstalled in the same locations.

5   Once the valves have been removed and safely stored, the head should be thoroughly cleaned and inspected. If a complete engine overhaul is being done, finish the engine disassembly procedures before beginning the cylinder head cleaning and inspection process.

10 Cylinder head - cleaning and inspection

1   Thorough cleaning of the cylinder head and related valvetrain components, followed by a detailed inspection, will enable you to decide how much valve service work must be done during the engine overhaul.

Cleaning

2   Scrape away all traces of old gasket material and sealing compound from the head gasket, intake manifold and exhaust manifold sealing surfaces, using a dull gasket scraper and gasket-removal solvent. **Note:** *Do not use power tools or abrasive pads to clean the block deck or cylinder heads. The manufacturer, in recent years, has machined both block and head sealing surfaces with a finer finish to accommodate composition head gaskets used without sealant, and this finish should not be disturbed.*

3   Remove any built-up scale around the coolant passages.

4   Run a stiff wire brush through the oil holes to remove any deposits that may have formed in them.

5   It is a good idea to run an appropriate size tap into each of the threaded holes to remove any corrosion and thread sealant that may be present. If compressed air is available, use it to clear the holes of debris produced by this operation.

6   Clean the exhaust and intake manifold stud threads in a similar manner with an appropriate size die. Clean the rocker-arm stud threads with a wire brush.

7   Next, clean the cylinder head with solvent and dry it thoroughly. Compressed air will speed the drying process and ensure that all holes and recessed areas are clean. **Note:** *Decarbonizing chemicals are available and may prove very useful when cleaning cylinder heads and valvetrain components. They are very caustic and should be used with caution. Be sure to follow the instructions on the container.*

8   Clean the rocker arms, pivot balls and pushrods with solvent and dry them thoroughly. Compressed air will speed the drying process and can be used to clean out the oil passages.

9   Clean all the valve springs, keepers, retainers, rotators, shields and spring seats with solvent and dry them thoroughly. Do the components from one valve at a time to avoid mixing up the parts.

10   Scrape off any heavy deposits that may have formed on the valves, then use a motorized wire brush to remove deposits from the valve heads and stems. Again, make sure the valves do not get mixed up.

Inspection

Cylinder head

Refer to illustrations 10.12 and 10.14

11   Inspect the head very carefully for cracks, evidence of coolant leakage and

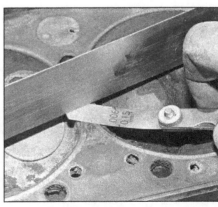

10.12 Check the cylinder head gasket surface for warpage by trying to slip a feeler gauge under the straightedge (see this Chapter's Specifications for the maximum warpage allowed and use a feeler gauge of that thickness)

Chapter 2 Part C General engine overhaul procedures

10.14 A dial indicator can be used to determine the valve stem-to-gauge clearance (move the valve stem as indicated by the arrows)

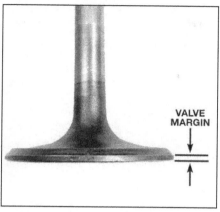

10.19 The margin width on the valve must be as specified (if no margin exists, the valve cannot be reused)

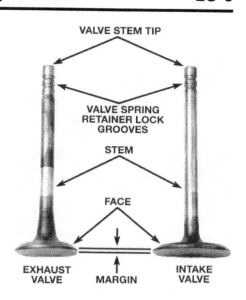

10.20 Check for valve wear at the points shown here

other damage. If cracks are found, a new cylinder head should be obtained.

12 Using a straightedge and feeler gauges, check the head gasket mating surface for warpage **(see illustration)**. If the head is warped beyond the limits given in this Chapter's Specifications, it can be resurfaced at an automotive machine shop.

13 Examine the valve seats in each of the combustion chambers. If they are pitted, cracked or burned, the head will require valve service that is beyond the scope of the home mechanic.

14 Measure the inside diameters of the valve guides at both ends and the center of each guide, with a small hole gauge and a 0-to-1-inch micrometer. Record the measurements for future reference. These measurements, along with the valve stem diameter measurements, will enable you to compute the valve stem-to-guide clearances. These clearances, when compared to the Specifications, will be one factor that will determine the extent of valve service work required. The guides are measured at the ends and at the center to determine if they are worn in a bell-mouth pattern (more wear at the ends). If they are, guide reconditioning or replacement is necessary. As an alternative, use a dial indicator to measure the lateral movement of each valve stem with the valve in the guide and approximately 1/16-inch off the seat **(see illustration)**.

Rocker arm components

15 Check the rocker arm faces (that contact the pushrod ends and valve stems) for pits, wear and rough spots. Check the pivot contact areas as well.

16 Inspect the pushrod ends for scuffing and excessive wear. Roll the pushrod on a flat surface, such as a piece of glass, to determine if it is bent.

17 Any damaged or excessively worn parts must be replaced with new ones.

Valves

Refer to illustrations 10.19 and 10.20

18 Carefully inspect each valve face for cracks, pits and burned spots. Check the valve stem and neck for cracks. Rotate the valve and check for any obvious indication that it is bent. Check the end of the stem for pits and excessive wear. The presence of any of these conditions indicates the need for valve service by a properly-equipped professional.

19 Measure the width of the valve margin **(see illustration)** on each valve and compare it to the Specifications. Any valve with a margin narrower than specified will have to be replaced with a new one.

20 Measure the valve stem diameter **(see illustration)**. By subtracting the stem diameter from the corresponding valve guide diameter, the valve stem-to-guide clearance is obtained. Compare the results to the Specifications. If the stem-to-guide clearance is greater than specified, the guides will have to be reconditioned and new valves may have to be installed, depending on the condition of the old ones.

Valve components

Refer to illustrations 10.21a and 10.21b

21 Check the ends of each valve spring for wear and pitting. Measure the free length **(see illustration)** and compare it to this Chapter's Specifications. Any springs that are shorter than specified have sagged and should not be reused. Stand the spring on a flat surface and check it for squareness **(see illustration)**.

22 Check the spring retainers and keepers for obvious wear and cracks. Any questionable parts should be replaced with new ones, as extensive damage will occur in the event of failure during engine operation.

23 If the inspection process indicates that the valve components are in generally poor condition and worn beyond the limits specified, which is usually the case in an engine that is being overhauled, reassemble the valves in the cylinder head and refer to Section 11 for valve servicing recommendations.

24 If the inspection turns up no excessively worn parts, and if the valve faces and seats are in good condition, the valvetrain compo-

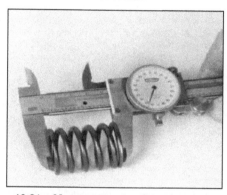

10.21a Measure the free length of each valve spring with a dial or vernier caliper

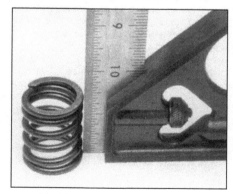

10.21b Check each valve spring for squareness

nents can be reinstalled in the cylinder head without major servicing. However, if the engine has accumulated a lot of mileage and the rest of the engine is being rebuilt to like-new specs, the heads should at least receive a fresh valve job at an automotive machine shop to ensure lasting performance. Refer to the appropriate Section for cylinder head reassembly procedures.

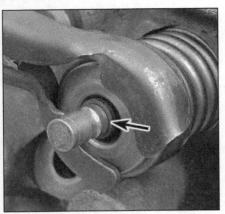

12.6a Make sure the O-ring seal is seated in the groove and not twisted before installing the keepers (V6 engine only)

12.6b Apply a small dab of grease to each keeper as shown here before installation - it'll hold them in place on the valve stem as the spring is released

12.8 Be sure to check the valve spring installed height (the distance from the top of the seat/shims to the top of the shield/bottom of the retainer)

11 Valves - servicing

1 Because of the complex nature of the job and the special tools and equipment needed, servicing of the valves, the valve seats and the valve guides (commonly known as a "valve job") is best left to a professional.

2 The home mechanic can remove and disassemble the head, do the initial cleaning and inspection, then reassemble and deliver the head to a dealer service department or a reputable automotive machine shop for the actual valve servicing.

3 The dealer service department or automotive machine shop will remove the valves and springs, recondition or replace the valves and valve seats, recondition the valve guides, check and replace the valve springs, spring retainers and keepers (as necessary), replace the valve seals with new ones, reassemble the valve components and make sure the installed spring height is correct. The cylinder head gasket surface will also be resurfaced if it is warped.

4 After the valve job has been performed by a professional, the head will be in like-new condition. When the head is returned, be sure to clean it again, very thoroughly (before installation on the engine), to remove any metal particles and abrasive grit that may still be present from the valve service or head resurfacing operations. Use compressed air, if available, to blow out all the oil holes and passages.

12 Cylinder head - reassembly

Refer to illustrations 12.6a, 12.6b and 12.8

1 Regardless of whether or not a head was sent to an automotive repair shop for valve servicing, make sure it's clean before beginning reassembly.

2 If a head was sent out for valve servicing, the valves and related components will already be in place. Begin the reassembly procedure with Step 8.

3 Beginning at one end of the head, lubricate and install the first valve. Apply engine assembly lube or clean engine oil to the valve stem.

4 Install the spring seats over each of the valve guides (four-cylinder engine only). Install new seals on the valve guides (intake valve only on the V6 engine). Using a hammer and a deep socket or seal installation tool, gently tap each seal into place until it's completely seated on the guide. Don't twist or cock the seals during installation or they won't seal properly on the valve stems.

5 Set the valve springs, oil shield (V6 engine intake valve only) and retainer (or rotator on the V6 engine exhaust valve) in place.

6 Compress the springs with a valve spring compressor and carefully install the O-ring oil seal in the lower groove of the valve stem (V6 engine only). Make sure the seal isn't twisted - it must lie perfectly flat in the groove **(see illustration)**. Position the keepers in the upper groove, then slowly release the compressor and make sure the keepers seat properly. Apply a small dab of grease to each keeper to hold it in place if necessary **(see illustration)**.

7 Repeat the procedure for the remaining valves. Be sure to return the components to their original locations - don't mix them up!

8 Check the installed valve spring height with a ruler graduated in 1/32-inch increments or a dial caliper. If the head was sent out for service work, the installed height should be correct (but don't assume that it is). The measurement is taken from the top of each spring seat or shim(s) to the top of the oil shield (or the bottom of the retainer - the two points are the same) **(see illustration)**. If the height is greater than specified, shims can be added under the springs to correct it. **Caution:** *Do not, under any circumstances, shim the springs to the point where the installed height is less than specified.*

9 Apply engine assembly lube to the rocker arm faces and the pivot balls, then install the rocker arms and pivots on the cylinder head studs. Thread the nuts on three or four turns only (when the heads are installed on the engine, the nuts will be tightened following a specific procedure, see Part A or Part B of this Chapter).

13 Camshaft, balance shaft and bearings - removal and inspection

Camshaft

1 After the camshaft has been removed from the engine (see Part A or Part B of this Chapter for camshaft removal procedures), cleaned with solvent and dried, inspect the bearing journals for uneven wear, pitting and evidence of seizure. If the journals are damaged, the bearing inserts in the block are probably damaged as well. Both the camshaft and bearings will have to be replaced with new ones. Measure the inside diameter of each camshaft bearing and record the results (take two measurements, 90-degrees apart, at each bearing). The inside diameter of each bearing can be determined with an inside micrometer or a small hole gauge and outside micrometer.

2 Subtract the bearing journal diameters (see Part A or Part B of this Chapter for camshaft measurement procedures and Specifications) from the corresponding bearing inside diameter measurement to obtain the oil clearance. If it is excessive, new bearings must be installed.

3 Camshaft bearing replacement requires special tools and expertise that place it outside the scope of the do-it-yourselfer. Take the block to an automotive machine shop to ensure that the job is done correctly.

4 Check the camshaft lobes for heat discoloration, score marks, chipped areas, pitting and uneven wear. If the lobes are in good condition and if the lobe lift measurements (see Part A or Part B of this Chapter for camshaft measurement procedures and Specifications) were as specified, the camshaft can be reused.

13.7 Remove the two Torx bolts and the balance shaft retainer on V6 engines

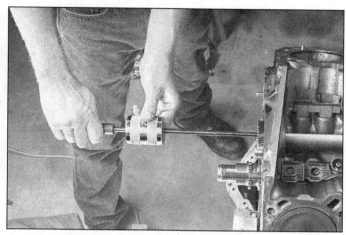

13.8 A small slide-hammer must be threaded into the front of the balance shaft to pull the balance shaft and front bearing out of the block

Lifters

5 Inspection procedures for the hydraulic roller lifters are found in Part A and Part B of this Chapter.
6 Good used roller lifters can be reinstalled with a new camshaft and the original camshaft can be used if new lifters are installed, but the manufacturer recommends replacing both lifters and camshaft together, especially if the engine has high mileage.

Balance shaft (V6 engine only)

Removal

Refer to illustrations 13.7 and 13.8

Note: *If the balance shaft is to be removed, the balance shaft gear bolt should be removed before the timing chain is removed, so that the shaft is held while the bolt is loosened.*

7 Remove the two bolts and the balance shaft retainer at the front of the block **(see illustration)**.
8 The balance shaft front bearing fits tightly into the block, use a slide-hammer threaded into the front of the balance shaft to knock it out **(see illustration)**.

Inspection

9 Inspect the balance shaft and measure the bearing journal diameters in the same manner as the camshaft. If your measurements exceeds this Chapter's Specifications, replace the balance shaft, bearing and bushing as a set. The balance shaft rides in a bearing at the front of the block and a bushing at the rear.
10 The rear bushing is similar to a camshaft bearing, pressed into the rear of the block. It must be installed by a machine shop with the proper tools, to the proper depth and with the oil hole aligned.
11 Inspect the balance shaft drive gear (the one behind the camshaft gear), and the balance shaft driven gear (bolted to the front of the balance shaft) for signs of wear, pitting, broken teeth or rough operation. As with the balance shaft and its bearings, the two balance shaft gears are serviced only as a set.

14 Piston/connecting rod assembly - removal

Refer to illustrations 14.2, 14.6 and 14.8

1 Prior to removal of the piston/connecting rod assemblies, the engine should be positioned upright.
2 Using a ridge reamer, completely remove the ridge at the top of each cylinder (follow the manufacturer's instructions provided with the ridge reaming tool) **(see illustration)**. Failure to remove the ridge before attempting to remove the piston/connecting rod assemblies will result in piston breakage.
3 After all of the cylinder wear ridges have been removed, turn the engine upside-down.
4 Before the connecting rods are removed, check the endplay as follows. Mount a dial indicator with its stem in line with the crankshaft and touching the side of the number one cylinder connecting rod cap.
5 Push the connecting rod forward, as far as possible, and zero the dial indicator. Next, push the connecting rod all the way to the rear and check the reading on the dial indicator. The distance that it moves is the endplay. If the endplay exceeds the service limit, a new connecting rod will be required. Repeat the procedure for the remaining connecting rods.
6 An alternative method is to slip feeler gauges between the connecting rod and the crankshaft throw until the play is removed **(see illustration)**. The endplay is then equal to the thickness of the feeler gauge(s).
7 Check the connecting rods and connecting rod caps for identification marks. If they are not plainly marked, identify each rod and cap using a small punch to make the appropriate number of indentations to indicate the cylinders they are associated with.
8 Loosen each of the connecting rod cap nuts approximately 1/2-turn. Remove the number one connecting rod cap and bearing insert. Do not drop the bearing insert out of the cap. Slip a short length of plastic or rubber hose over each connecting rod cap bolt (to protect the crankshaft journal and cylinder

14.2 A ridge reamer is required to remove the ridge from the top of each cylinder - do this before removing the pistons!

14.6 Check the connecting rod side clearance with a feeler gauge as shown

14.8 To prevent damage to the crankshaft journals and cylinder walls, slip sections of rubber or plastic hose over the rod bolts before removing the pistons/rods

15.3 Checking crankshaft endplay with a feeler gauge

15.4 The arrow on the main bearing cap indicates the front of the engine

wall when the piston is removed) **(see illustration)** and push the connecting rod/piston assembly out through the top of the engine. Use a wooden tool to push on the upper bearing insert in the connecting rod. If resistance is felt, double-check to make sure that all of the ridge was removed from the cylinder.

9 Repeat the procedure for the remaining cylinders. After removal, reassemble the connecting rod caps and bearing inserts in their respective connecting rods and install the cap nuts finger tight. Leaving the old bearing inserts in place until reassembly will help prevent the connecting rod bearing surfaces from being accidentally nicked or gouged.

15 Crankshaft - removal

Refer to illustrations 15.3 and 15.4

1 Before the crankshaft is removed, check the endplay as follows. Mount a dial indicator with the stem in line with the crankshaft and just touching one of the crank throws.

2 Push the crankshaft all the way to the rear and zero the dial indicator. Next, pry the crankshaft to the front as far as possible and check the reading on the dial indicator. The distance that it moves is the endplay. If it is greater than specified, check the crankshaft thrust surfaces for wear. If no wear is apparent, new main bearings should correct the endplay.

3 If a dial indicator is not available, feeler gauges can be used. Gently pry or push the crankshaft all the way to the front of the engine. Slip feeler gauges between the crankshaft and the front face of the thrust main bearing **(see illustration)** to determine the clearance (which is equivalent to crankshaft endplay).

4 Loosen each of the main bearing cap bolts 1/4 of a turn at a time, until they can be removed by hand. Check the main bearing caps to see if they are marked as to their locations. They are usually numbered consecutively (beginning with 1) from the front of the engine to the rear. If they are not, mark them with number stamping dies or a center-punch. Most main bearing caps have a cast-in arrow, which points to the front of the engine **(see illustration)**.

5 Gently tap the caps with a soft-faced hammer, then separate them from the engine block. If necessary, use the main bearing cap bolts as levers to remove the caps. Try not to drop the bearing insert if it comes out with the cap.

6 Carefully lift the crankshaft out of the engine. It is a good idea to have an assistant available, since the crankshaft is quite heavy. With the bearing inserts in place in the engine block and in the main bearing caps, return the caps to their respective locations on the engine block and tighten the bolts finger tight.

16 Engine block - cleaning

Refer to illustrations 16.1, 16.8 and 16.10

1 Remove the freeze plugs from the engine block. To do this, knock one side of the plugs into the block with a hammer and punch, then grasp them with large pliers and pull them back through the holes **(see illustration)**.

2 Using a dull gasket scraper and gasket-removal solvent, remove all traces of gasket material from the engine block. Be very careful not to nick or gouge the gasket sealing surfaces. **Note:** *Do not use power tools or abrasive pads to clean the block deck or cylinder heads. The manufacturer, in recent years, has machined both block and head sealing surfaces with a finer finish to accommodate composition head gaskets used without sealant, and this finish should not be disturbed.*

3 Remove the main bearing caps and separate the bearing inserts from the caps and the engine block. Tag the bearings according to which cylinder they were removed from (and whether they were in the cap or the block) and set them aside.

4 Using a wrench of the appropriate size, remove the threaded oil gallery plugs from the front and back of the block. In stubborn cases, the plugs may have to be drilled out and the holes re-threaded.

5 If the engine is extremely dirty, it should be taken to an automotive machine shop to be steam cleaned or hot tanked. Any bearings left in the block (such as the camshaft bearings) will be damaged by the cleaning process, so plan on having new ones installed while the block is at the machine shop.

6 After the block is returned, clean all oil holes and oil galleries one more time (brushes for cleaning oil holes and galleries are available at most auto parts stores). Flush the passages with warm water until the water runs clear, dry the block thoroughly and wipe all machined surfaces with a light, rust-preventative oil. If you have access to compressed air, use it to speed the drying process and to blow out all the oil holes and galleries.

7 If the block is not extremely dirty or sludged up, you can do an adequate cleaning job with warm soapy water and a stiff brush. Take plenty of time and do a thorough job. Regardless of the cleaning method used, be very sure to thoroughly clean all oil holes and galleries, dry the block completely and coat all machined surfaces with light oil.

16.1 Use a hammer and large punch to knock the core plugs sideways in their bores, then pull them out with pliers

Chapter 2 Part C General engine overhaul procedures

16.8 All bolt holes in the block - particularly the main bearing cap and head bolt holes - should be cleaned and restored with a tap (be sure to remove debris from the holes after this is done)

16.10 A large socket on an extension can be used to drive the new core plugs into the bores

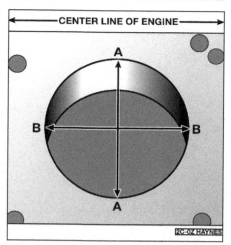

17.4a Measure the diameter of each cylinder at a right angle to engine centerline (A), and parallel to engine centerline (B) - out-of-round is the difference between A and B; taper is the difference between A and B at the top of the cylinder and A and B at the bottom of the cylinder

8 The threaded holes in the block must be clean to ensure accurate torque readings during reassembly. Run the proper size tap into each of the holes to remove any rust, corrosion, thread sealant or sludge and to restore any damaged threads (see illustration). If possible, use compressed air to clear the holes of debris produced by this operation. Now is a good time to thoroughly clean the threads on the head bolts and the main bearing cap bolts as well.
9 Reinstall the main bearing caps and tighten the bolts finger tight.
10 After coating the sealing surfaces of the new freeze plugs with a good quality gasket sealant, install them in the engine block (see illustration). Make sure they are driven in straight and seated properly or leakage could result. Special tools are available for this purpose, but equally good results can be obtained using a large socket (with an outside diameter that will just slip into the soft plug) and a large hammer.
11 On V6 engines equipped with an oil cooler, the oil cooler adapter (between the oil filter and the block on the left side) must be removed and thoroughly cleaned with solvent and compressed air. The adapter has passages inside that can collect dirt and machining particles, which could remain in the engine and circulate after the rebuilt engine is run. Professional engine rebuilders also suggest that on these models the oil cooler itself (in front of the radiator) be flushed with solvent and pressure, and some suggest that it cannot be adequately cleaned and the cooler should simply be replaced with a new one to protect the new engine. The same cautions also apply to V6 models that have a remote oil filter adapter instead of the oil cooler.
12 If the engine is not going to be reassembled right away, cover it with a large plastic trash bag to keep it clean.

17 Engine block - inspection

Refer to illustrations 17.4a, 17.4b and 17.4c
1 Thoroughly clean the engine block as described in Section 16 and double-check to make sure that the ridge at the top of each cylinder has been completely removed.

2 Visually check the block for cracks, rust and corrosion. Look for stripped threads in the threaded holes. It is also a good idea to have the block checked for hidden cracks by an automotive machine shop that has the special equipment to do this type of work. If defects are found, have the block repaired, if possible, or replaced.
3 Check the cylinder bores for scuffing and scoring.
4 Check the cylinders for taper and out-of-round conditions as follows (see illustrations):
5 Measure the diameter of each cylinder at the top (just under the ridge area), center and bottom of the cylinder bore, parallel to the crankshaft axis.
6 Next, measure each cylinder's diameter at the same three locations perpendicular to the crankshaft axis.

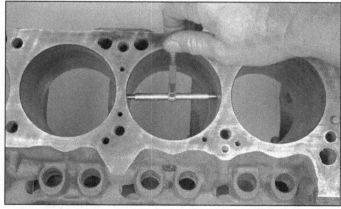

17.4b The ability to "feel" when the telescoping gauge is at the correct point will be developed over time, so work slowly and repeat the check until you're satisfied the bore measurement is accurate

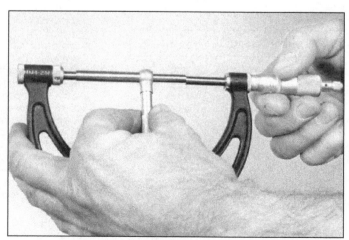

17.4c The gauge is then measured with a micrometer to determine the bore size

18.3a A "bottle brush" hone will produce better results if you're never honed cylinders before

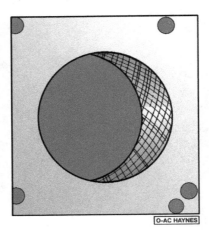

18.3b The cylinder hone should leave a smooth, crosshatch pattern with the lines intersecting at approximately a 45-60-degree angle

19.4a The piston ring grooves can be cleaned with a special tool, as shown here, . . .

7   The taper of each cylinder is the difference between the bore diameter at the top of the cylinder and the diameter at the bottom. The out-of-round specification of the cylinder bore is the difference between the parallel and perpendicular readings. Compare your results to this Chapter's Specifications.

8   If the cylinder walls are badly scuffed or scored, or if they're out-of-round or tapered beyond the limits given in this Chapter's Specifications, have the engine block rebored and honed at an automotive machine shop.

9   If a rebore is done, oversize pistons and rings will be required.

10   Using a precision straightedge and feeler gauge, check the block deck (the surface the cylinder heads mate with) for distortion as you did with the cylinder heads (see Section 10). If it's distorted beyond the specified limit, the block decks can be resurfaced by an automotive machine shop.

11   If the cylinders are in reasonably good condition and not worn to the outside of the limits, and if the piston-to-cylinder clearances can be maintained properly, they don't have to be rebored. Honing is all that's necessary (see Section 18).

18 Cylinder honing

Refer to illustrations 18.3a and 18.3b

1   Prior to engine reassembly, the cylinder bores must be honed so the new piston rings will seat correctly and provide the best possible combustion chamber seal. **Note:** *If you don't have the tools or don't want to tackle the honing operation, most automotive machine shops will do it for a reasonable fee.*

2   Before honing the cylinders, install the main bearing caps and tighten the bolts to the torque listed in this Chapter's Specifications.

3   Two types of cylinder hones are commonly available - the flex hone or "bottle brush" type and the more traditional surfacing hone with spring-loaded stones. Both will do the job, but for the less experienced mechanic the "bottle brush" hone will probably be easier to use. You'll also need some honing oil (kerosene will work if honing oil isn't available), rags and an electric drill motor. Proceed as follows **(see illustrations)**:

a) *Mount the hone in the drill motor, compress the stones and slip it into the first cylinder. Be sure to wear safety goggles or a face shield!*

b) *Lubricate the cylinder with plenty of honing oil, turn on the drill and move the hone up-and-down in the cylinder at a pace that will produce a fine crosshatch pattern on the cylinder walls, and with the drill square and centered with the bore. Ideally, the crosshatch lines should intersect at approximately a 45-60-degree angle. Be sure to use plenty of lubricant and don't take off any more material than is absolutely necessary to produce the desired finish.* **Note:** *Piston ring manufacturers may specify a different crosshatch angle - read and follow any instructions included with the new rings.*

c) *Don't withdraw the hone from the cylinder while it's running. Instead, shut off the drill and continue moving the hone up-and-down in the cylinder until it comes to a complete stop, then compress the stones and withdraw the hone. If you're using a "bottle brush" type hone, stop the drill motor, then turn the chuck in the normal direction of rotation while withdrawing the hone from the cylinder.*

d) *Wipe the oil out of the cylinder and repeat the procedure for the remaining cylinders.*

4   After the honing job is complete, chamfer the top edges of the cylinder bores with a small file so the rings won't catch when the pistons are installed. Be very careful not to nick the cylinder walls with the end of the file.

5   The entire engine block must be washed again very thoroughly with warm, soapy water to remove all traces of the abrasive grit produced during the honing operation. **Note:** *The bores can be considered clean when a lint-free white cloth - dampened with clean engine oil - used to wipe them out doesn't pick up any more honing residue, which will show up as gray areas on the cloth. Be sure to run a brush through all oil holes and galleries and flush them with running water.*

6   After rinsing, dry the block and apply a coat of light rust preventative oil to all machined surfaces. Wrap the block in a plastic trash bag to keep it clean and set it aside until reassembly.

19 Piston/connecting rod assembly - inspection

Refer to illustrations 19.4a, 19.4b, 19.10 and 19.11

1   Before the inspection process can be carried out, the piston/connecting rod assemblies must be cleaned and the original piston rings removed from the pistons. **Note:** *Always use new piston rings when the engine is reassembled.*

2   Using a piston ring installation tool, carefully remove the rings from the pistons. Do not nick or gouge the pistons in the process.

3   Scrape all traces of carbon from the top (or crown) of the piston. A hand-held wire brush, preferably with brass rather than steel bristles, or a piece of fine emery cloth can be used once the majority of the deposits have been scraped away. Do not, under any circumstances, use a steel wire brush mounted in a drill motor to remove deposits from the pistons. The piston material is soft and will be eroded away by the wire brush.

4   Use a piston ring groove-cleaning tool to remove any carbon deposits from the ring grooves. If a tool is not available, a piece broken off the old ring will do the job. Be very

19.4b ... or a section of broken ring

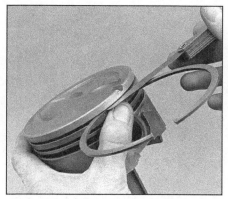

19.10 Check the ring side clearance with a feeler gauge at several points around the groove

19.11 Measure the piston diameter at a 90-degree angle to the piston pin and in line with it

careful to remove only the carbon deposits. Do not remove any metal and do not nick or scratch the sides of the ring grooves **(see illustrations)**.

5 Once the deposits have been removed, clean the piston/rod assemblies with solvent and dry them thoroughly. Make sure that the oil hole in the big end of the connecting rod and the oil return holes in the back sides of the ring grooves are clear.

6 If the pistons are not damaged or worn excessively, and if the engine block is not rebored, new pistons will not be necessary. Normal piston wear appears as even vertical wear on the piston thrust surfaces and slight looseness of the top ring in its groove. New piston rings, on the other hand, should always be used when an engine is rebuilt.

7 Carefully inspect each piston for cracks around the skirt, at the pin bosses and at the ring lands.

8 Look for scoring and scuffing on the thrust faces of the skirt, holes in the piston crown and burned areas at the edge of the crown. If the skirt is scored or scuffed, the engine may have been suffering from overheating and/or abnormal combustion, which caused excessively high operating temperatures. The cooling and lubrication systems should be checked thoroughly. A hole in the piston crown, an extreme to be sure, is an indication that abnormal combustion (preignition) was occurring. Burned areas at the edge of the piston crown are usually evidence of spark knock (detonation). If any of the above problems exist, the causes must be corrected or the damage will occur again.

9 Corrosion of the piston (evidenced by pitting) indicates that coolant is leaking into the combustion chamber and/or the crankcase. Again, the cause must be corrected or the problem may persist in the rebuilt engine.

10 Measure the piston ring side clearance by laying a new piston ring in each ring groove and slipping a feeler gauge in beside it **(see illustration)**. Check the clearance at three or four locations around each groove. Be sure to use the correct ring for each groove; they are different. If the side clearance is greater than specified, new pistons

will have to be used.

11 Check the piston-to-bore clearance by measuring the bore (see Section 17) and the piston diameter **(see illustration)**. Make sure that the pistons and bores are correctly matched. Measure the piston across the skirt, on the thrust faces (at a 90-degrees angle to the piston pin), directly in line with the center of the pin hole. Subtract the piston diameter from the bore diameter to obtain the clearance. If it is greater than specified, the block will have to be rebored and new pistons and rings installed. Check the piston-to-rod clearance by twisting the piston and rod in opposite directions. Any noticeable play indicates that there is excessive wear, which must be corrected. The piston/connecting rod assemblies should be taken to an automotive machine shop to have new piston pins installed and the pistons and connecting rods rebored.

12 If the pistons must be removed from the connecting rods, such as when new pistons must be installed, or if the piston pins have too much play in them, they should be taken to an automotive machine shop. While they are there, it would be convenient to have the connecting rods checked for bend and twist, as automotive machine shops have special equipment for this purpose. Unless new pistons or connecting rods must be installed, do not disassemble the pistons from the connecting rods.

13 Check the connecting rods for cracks and other damage. Temporarily remove the rod caps, lift out the old bearing inserts, wipe the rod and cap bearing surfaces clean and inspect them for nicks, gouges and scratches. After checking the rods, replace the old bearings, slip the caps into place and tighten the nuts finger tight.

20 Crankshaft - inspection

Refer to illustrations 20.1, 20.2, 20.5 and 20.7

1 Remove all burrs from the crankshaft oil holes with a stone, file or scraper **(see illustration)**.

2 Clean the crankshaft with solvent and dry it with compressed air (if available). **Warning:** *Wear eye protection when using compressed air.* Be sure to clean the oil holes with a stiff brush **(see illustration)** and flush them with solvent.

3 Check the main and connecting rod bearing journals for uneven wear, scoring, pits and cracks.

4 Check the rest of the crankshaft for cracks and other damage. It should be magnafluxed to reveal hidden cracks - an auto-

20.1 The oil holes should be chamfered so sharp edges don't gouge or scratch the new bearings

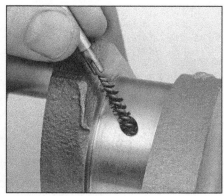

20.2 Use a wire or stiff plastic bristle brush to clean the oil passages in the crankshaft

20.5 Measure the diameter of each crankshaft journal at several points to detect taper and out-of-round conditions

20.7 If the seals have worn grooves in the crankshaft journals, or if the seal contact surfaces are nicked or scratched, the new seals will leak

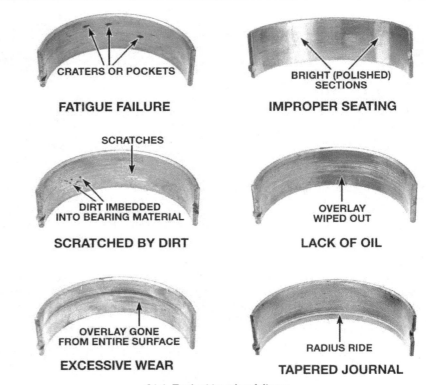

21.1 Typical bearing failures

motive machine shop will handle the procedure.

5 Using a micrometer, measure the diameter of the main and connecting rod journals and compare the results to this Chapter's Specifications **(see illustration)**. By measuring the diameter at a number of points around each journal's circumference, you'll be able to determine whether or not the journal is out-of-round. Take the measurement at each end of the journal, near the crank throws, to determine if the journal is tapered.

6 If the crankshaft journals are damaged, tapered, out-of-round or worn beyond the limits given in the Specifications, have the crankshaft reground by an automotive machine shop. Be sure to use the correct size bearing inserts if the crankshaft is reconditioned.

7 Check the oil seal journals at each end of the crankshaft for wear and damage. If the seal has worn a groove in the journal, or if it's nicked or scratched **(see illustration)**, the new seal may leak when the engine is reassembled. In some cases, an automotive machine shop may be able to repair the journal by pressing on a thin sleeve. If repair isn't feasible, a new or different crankshaft should be installed.

8 Refer to Section 21 and examine the main and rod bearing inserts.

21 Main and connecting rod bearings - inspection

Refer to illustration 21.1

1 Even though the main and connecting rod bearings should be replaced with new ones during the engine overhaul, the old bearings should be retained for close examination, as they may reveal valuable information about the condition of the engine **(see illustration)**.

2 Bearing failure occurs mainly because of lack of lubrication, the presence of dirt or other foreign particles, overloading the engine and corrosion. Regardless of the cause of bearing failure, it must be corrected before the engine is reassembled to prevent it from happening again.

3 When examining the bearings, remove them from the engine block, the main bearing caps, the connecting rods and the rod caps and lay them out on a clean surface in the same general position as their location in the engine. This will enable you to match any noted bearing problems with the corresponding crankshaft journal.

4 Dirt and other foreign particles get into the engine in a variety of ways. If may be left in the engine during assembly, or it may pass through filters or breathers. It may get into the oil, and from there into the bearings. Metal chips from machining operations and normal engine wear are often present. Abrasives are sometimes left in engine components after reconditioning, especially when parts are not thoroughly cleaned using the proper cleaning methods. Whatever the source, these foreign objects often end up embedded in the soft bearing material and are easily recognized. Large particles will not embed in the bearing and will score or gouge the bearing and shaft. The best prevention for this cause of bearing failure is to clean all parts thoroughly and keep everything spotlessly clean during engine assembly. Frequent and regular engine oil and filter changes are also recommended.

5 Lack of lubrication (or lubrication breakdown) has a number of interrelated causes. Excessive heat (which thins the oil), overloading (which squeezes the oil from the bearing face) and oil leakage or throw-off (from excessive bearing clearances, worn oil pump or high engine speeds) all contribute to lubrication breakdown. Blocked oil passages, which usually are the result of misaligned oil holes in a bearing shell, will also oil-starve a bearing and destroy it. When lack of lubrication is the cause of bearing failure, the bearing material is wiped or extruded from the steel backing of the bearing. Temperatures may increase to the point where the steel backing turns blue from overheating.

6 Driving habits can have a definite effect on bearing life. Low speed operation in too high a gear (lugging the engine) puts very high loads on bearings, which tends to squeeze out the oil film. These loads cause the bearings to flex, which produces fine cracks in the bearing face (fatigue failure). Eventually the bearing material will loosen in pieces and tear

Chapter 2 Part C General engine overhaul procedures

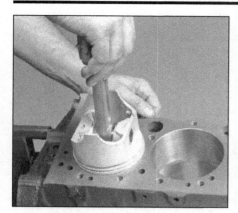

23.3a When checking piston ring end gap, the ring must be square in the cylinder bore (this is done by pushing the ring down with the top of a piston as shown)

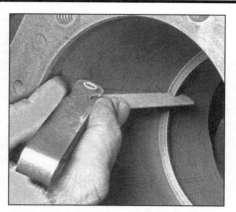

23.3b With the ring square in the cylinder, measure the end gap with a feeler gauge

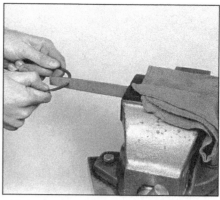

23.5 If the end gap is too small, clamp a file in a vise and file the ring ends (from the outside only) to enlarge the gap slightly

away from the steel backing. Short-trip driving leads to corrosion of bearings because insufficient engine heat is produced to drive off the condensed water and corrosive gases. These products collect in the engine oil, forming acid and sludge. As the oil is carried to the engine bearings, the acid attacks and corrodes the bearing material.

7 Incorrect bearing installation during engine assembly will lead to bearing failure as well. Tight-fitting bearings leave insufficient bearing oil clearance and will result in oil starvation. Dirt or foreign particles trapped behind a bearing insert result in high spots on the bearing which lead to failure.

22 Engine overhaul - reassembly sequence

1 Before beginning engine reassembly, make sure you have all the necessary new parts, gaskets and seals as well as the following items on hand:

Common hand tools
A 1/2-inch drive torque wrench
Piston ring installation tool
Piston ring compressor
Short lengths of rubber or plastic hose to fit over connecting rod bolts
Plastigage
Feeler gauges
A fine-tooth file
New, clean engine oil
Engine assembly lube
RTV-type gasket sealant
Anaerobic-type gasket sealant
Thread locking compound

2 In order to save time and avoid problems, engine reassembly must be done in the following order.

Crankshaft and main bearings
Rear main oil seal
Piston rings
Piston/connecting rod assemblies
Oil pump
Oil pan
Camshaft
Balance shaft (V6 engines)
Timing chain/sprockets
Timing chain cover
Valve lifters
Cylinder head(s)
Pushrods and rocker arms
Intake and exhaust manifolds
Oil filter/cooler adapter
Valve cover(s)
Water pump
Flywheel/driveplate
Thermostat and housing cover
Distributor (V6 engines), spark plug wires and spark plugs
Emissions control components
Fuel injection system components

23 Piston rings - installation

Refer to illustrations 23.3a, 23.3b, 23.5, 23.9a, 23.9b and 23.12

1 Before installing the new piston rings, the ring end gaps must be checked. It is assumed that the piston ring side clearance has been checked and verified correct (see Section 19).

2 Lay out the piston/connecting rod assemblies and the new ring sets so the ring sets will be matched with the same piston and cylinder during the end gap measurement and engine assembly.

3 Insert the top (number one) ring into the first cylinder and square it up with the cylinder walls by pushing it in with the top of the piston (see illustration). The ring should be near the bottom of the cylinder at the lower limit of ring travel. To measure the end gap, slip a feeler gauge between the ends of the ring (see illustration). Compare the measurement to the Specifications.

4 If the gap is larger or smaller than specified, double-check to make sure that you have the correct rings before proceeding.

5 If the gap is too small, it must be enlarged or the ring ends may come in contact with each other during engine operation, which can cause serious damage to the engine. The end gap can be increased by filing the ring ends very carefully with a fine file.

Mount the file in a vise equipped with soft jaws, slip the ring over the file with the ends contacting the file face and slowly move the ring to remove material from the ends. When performing this operation, file only from the outside in (see illustration). **Note:** *When you have the end-gap correct, remove any burrs from the filed ends of the rings with a whetstone.*

6 Excessive end gap is not critical unless it is greater than 0.040-inch. Again, double-check to make sure you have the correct rings for your engine.

7 Repeat the procedure for each ring that will be installed in the first cylinder and for each ring in the remaining cylinders. Remember to keep rings, pistons and cylinders matched up.

8 Once the ring end gaps have been checked/corrected, the rings can be installed on the pistons.

9 The oil control ring (lowest one on the piston) is installed first. It is composed of three separate components. Slip the spacer expander into the groove (see illustration), then install the upper side rail. Do not use a piston ring installation tool on the oil ring side rails, as they may be damaged. Instead, place one end of the side rail into the groove between the spacer expander and the ring land, hold it firmly in place and slide a finger around the piston while pushing the rail into

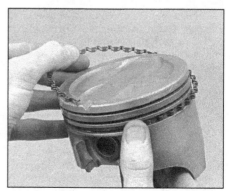

23.9a Install the spacer/expander in the oil control ring groove

23.9b DO NOT use a piston ring installation tool when installing the oil ring side rails

23.12 Install the compression rings with a ring expander - the mark (arrow) must face up

24.10 Lay the Plastigage strips (arrow) on the main bearing journals, parallel to the crankshaft centerline

the groove **(see illustration)**. Next, install the lower side rail in the same manner.
10 After the three oil ring components have been installed, check to make sure that both the upper and lower side rails can be turned smoothly in the ring groove.
11 The number two (middle) ring is installed next. It should be stamped with a mark so it can be readily distinguished from the top ring. **Note:** *Always follow the instructions printed on the package that the new rings are in - different manufacturers may require different approaches. Do not mix up the top and middle rings, as they have different cross-sections.*
12 Use a piston ring installation tool and make sure that the identification mark is facing the top of the piston, then slip the ring into the middle groove on the piston **(see illustration)**. Do not expand the ring any more than is necessary to slide it over the piston.
13 Finally, install the number one (top) ring in the same manner. Make sure the identifying mark is facing up.
14 Repeat the procedure for the remaining pistons and rings. Be careful not to confuse the number one and number two rings.

24 Crankshaft - installation and main bearing oil clearance check

Refer to illustrations 24.10 and 24.14
1 Crankshaft installation is generally one of the first steps in engine reassembly; it is assumed at this point that the engine block and crankshaft have been cleaned, inspected and repaired or reconditioned.
2 Position the engine with the bottom facing up.
3 Remove the main bearing cap bolts and lift out the caps. Lay them out in the proper order to help ensure that they are installed correctly.
4 If they are still in place, remove the old bearing inserts from the block and the main bearing caps. Wipe the main bearing surfaces of the block and caps with a clean, lint-free cloth (they must be kept spotlessly clean).

5 Clean the back sides of the new main bearing inserts and lay one bearing half in each main bearing saddle in the block. Lay the other bearing half from each bearing set in the corresponding main bearing cap. Make sure the tab on the bearing insert fits into the recess in the block or cap. Also, the oil holes in the block and cap must line up with the oil holes in the bearing insert. Do not hammer the bearing into place and do not nick or gouge the bearing faces. No lubrication should be used at this time.
6 The flanged thrust bearing must be installed in the number four cap and saddle on the four-cylinder engine, and the rear cap and saddle on the V6 engine.
7 Clean the faces of the bearings in the block and the crankshaft main bearing journals with a clean, lint-free cloth. Check or clean the oil holes in the crankshaft, as any dirt here can only go one way - straight through the new bearings.
8 Once you are certain that the crankshaft is clean, carefully lay it in position (an assistant would be very helpful here) in the main bearings with the counterweights lying sideways.
9 Before the crankshaft can be permanently installed, the main bearing oil clearance must be checked.
10 Trim several pieces of the appropriate type of Plastigage (so they are slightly shorter than the width of the main bearings) and place one piece on each crankshaft main bearing journal, parallel with the journal axis **(see illustration)**. Do not lay them across the oil holes.
11 Clean the faces of the bearings in the caps and install the caps in their respective positions (do not mix them up) with the arrows pointing toward the front of the engine. Do not disturb the Plastigage.
12 Starting with the center main and working out toward the ends, tighten the main bearing cap bolts, in three steps, to the torque listed in this Chapter's Specifications. Do not rotate the crankshaft at any time during this operation.
13 Remove the bolts and carefully lift off the main bearing caps. Keep them in order.

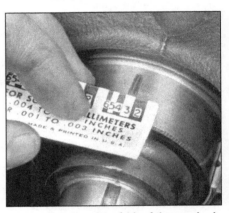

24.14 Measure the width of the crushed Plastigage to determine the main bearing oil clearance (be sure to use the correct scale - standard and metric ones are included)

Do not disturb the Plastigage or rotate the crankshaft. If any of the main bearing caps are difficult to remove, tap them gently from side-to-side with a soft-faced hammer to loosen them.
14 Compare the width of the crushed Plastigage on each journal to the scale printed on the Plastigage container to obtain the main bearing oil clearance **(see illustration)**. Check the Specifications to make sure it is correct.
15 If the clearance is not correct, double-check to make sure you have the right size bearing inserts. Also, make sure that no dirt or oil was between the bearing inserts and the main bearing caps or the block when the clearance was measured.
16 Carefully scrape all traces of the Plastigage material off the main bearing journals and/or the bearing faces. Do not nick or scratch the bearing faces.
17 Carefully lift the crankshaft out of the engine. Clean the bearing faces in the block, then apply a thin, uniform layer of clean, high-quality engine assembly lube to each of the bearing faces. Be sure to coat the thrust flange faces as well as the journal face of the thrust bearing.

Chapter 2 Part C General engine overhaul procedures   2C-19

25.2 Drive the balance shaft and front bearing into the block until the bearing retainer can be bolted in place

25.6 Position the balance shaft drive and driven gears with the timing marks aligned as shown

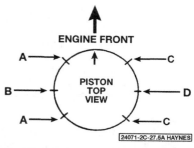

27.5a Ring end gap positions - V6 engine

- A Oil ring rail gaps
- B Second compression ring gap
- C Oil ring spacer gap (position in-between marks)
- D Top compression ring gap

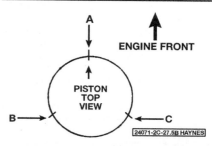

27.5b Ring end gap positions - four-cylinder engine

- A Top compression ring gap
- B Second compression ring gap
- C Oil ring assembly gap

18 Carefully reinstall the crankshaft on the block, lube the lower bearing halves in the caps, and install the caps in their respective positions with the arrows pointing toward the front of the engine.

19 On V6 engines, install the bolts and tighten them to the torque listed in this Chapter's Specifications, starting with the center main and working out toward the ends, except for the rear main (thrust bearing). Work up to the final torque in three steps. Pry the crankshaft forward and back in the block to seat the thrust bearings and then torque the rear main cap.

20 On four-cylinder engines, install all the bearing caps with the bolts finger-tight, then set the thrust bearings by hitting the crank forward and back with a rubber mallet, then torque all caps to Specifications.

21 Rotate the crankshaft a number of times by hand and check for any obvious binding.

22 Finally, check the crankshaft endplay. This can be done with a feeler gauge or a dial indicator set. Refer to Section 15 for the procedure.

25 Balance shaft - installation (V6 engines only)

Refer to illustrations 25.2 and 25.6

Note: *The balance shaft is installed after the camshaft. Refer to the camshaft installation procedures in Part B of this Chapter.*

1 Lubricate the balance shaft bearing journals with clean engine oil or engine assembly lube.

2 Slide the balance shaft into the engine. Support the balance shaft near the block and be careful not to scrape or nick the rear bearing. Using a bearing driver or large socket, gently drive the shaft in to the block until the front bearing is seated **(see illustration)**.

3 Install the balance shaft retainer and two bolts and tighten them to the torque listed in Chapter 2B Specifications **(see illustration 13.10)**.

4 Install the balance shaft driven gear and tighten the bolt to the torque listed in Chapter 2B Specifications.

5 Rotate the camshaft so that, with the balance shaft drive gear temporarily installed, the timing mark is straight up at the 12 o'clock position. Remove the drive gear.

6 Rotate the balance shaft until the timing mark on the driven gear is facing straight down at the 6 o'clock position. Reinstall the balance shaft drive gear to the camshaft and make sure both balance shaft gears are aligned **(see illustration)**.

7 Install the timing chain, sprockets and cover (see Part B of this Chapter).

26 Rear main oil seal - installation

Refer to Chapter 2 Part A or B, depending on engine type, for the rear main oil seal installation procedure.

27 Piston/connecting rod assembly - installation and bearing oil clearance check

Refer to illustrations 27.5a, 27.5b, 27.10, 27.12 and 27.14

1 Before installing the piston/connecting rod assemblies, the cylinder walls must be perfectly clean, the top edge of each cylinder must be chamfered, and the crankshaft must be in place.

2 Remove the connecting rod cap from the end of the number one connecting rod. Remove the old bearing inserts and wipe the bearing surfaces of the connecting rod and cap with a clean, lint-free cloth (they must be kept spotlessly clean).

3 Clean the back side of the new upper bearing half, then lay it in place in the connecting rod. Make sure that the tab on the bearing fits into the recess in the rod. Do not hammer the bearing insert into place and be very careful not to nick or gouge the bearing face. Do not lubricate the bearing at this time.

4 Clean the back side of the other bearing insert and install it in the rod cap. Again, make sure the tab on the bearing fits into the recess in the cap, and do not apply any lubricant. It is critically important that the mating surfaces of the bearing and connecting rod are perfectly clean and oil-free when they are assembled.

5 Position the piston ring gaps as shown **(see illustrations)**, then slip a section of plastic or rubber hose over the connecting rod cap bolts.

6 Lubricate the piston and rings with clean engine oil and attach a piston ring compressor to the piston. Leave the skirt protruding about 1/4-inch to guide the piston into the cylinder. The rings must be compressed as far as possible.

7 Rotate the crankshaft until the number one connecting rod journal is as far from the number one cylinder as possible (bottom dead center), and apply a uniform coat of engine oil to the cylinder walls.

8 With the notch on top of the piston facing to the front of the engine, gently place the piston/connecting rod assembly into the number one cylinder bore and rest the bottom edge of the ring compressor on the engine block. Tap the top edge of the ring compressor to make sure it is contacting the block around its entire circumference.

9 Clean the number one connecting rod journal on the crankshaft and the bearing faces in the rod.

10 Carefully tap on the top of the piston with the end of a wooden hammer handle (see illustration) while guiding the end of the connecting rod into place on the crankshaft journal. The piston rings may try to pop out of the ring compressor just before entering the cylinder bore, so keep some downward pressure on the ring compressor. Work slowly, and if any resistance is felt as the piston enters the cylinder, stop immediately. Find out what is hanging up and fix it before proceeding. Do not, for any reason, force the piston into the cylinder, as you will break a ring and/or the piston.
11 Once the piston/connecting rod assembly is installed, the connecting rod bearing oil clearance must be checked before the rod cap is permanently bolted in place.
12 Cut a piece of the appropriate type Plastigage slightly shorter than the width of the connecting rod bearing and lay it in place on the number one connecting rod journal, parallel with the journal axis (it must not cross the oil hole in the journal) (see illustration).
13 Clean the connecting rod cap bearing face, remove the protective hoses from the connecting rod bolts and gently install the rod cap in place. Make sure the mating mark on the cap is on the same side as the mark on the connecting rod. Install the nuts and tighten them to the torque listed in this Chapter's Specifications, working up to it in three steps. Do not rotate the crankshaft at any time during this operation.
14 Remove the rod cap, being very careful not to disturb the Plastigage. Compare the width of the crushed Plastigage to the scale printed on the Plastigage container to obtain the oil clearance (see illustration). Compare it to the Specifications to make sure the clearance is correct. If the clearance is not correct, double-check to make sure that you have the correct size bearing inserts. Also, recheck the crankshaft connecting rod journal diameter and make sure that no dirt or oil was between the bearing inserts and the connecting rod or cap when the clearance was measured.
15 Carefully scrape all traces of the Plastigage material off the rod journal and/or bearing face (be very careful not to scratch the bearing - use your fingernail or a piece of hardwood). Make sure the bearing faces are perfectly clean, then apply a uniform layer of clean, high-quality engine assembly lube to both of them. You will have to push the piston into the cylinder to expose the face of the bearing insert in the connecting rod; be sure to slip the protective hoses over the rod bolts first.
16 Slide the connecting rod back into place on the journal, remove the protective hoses from the rod cap bolts, install the rod cap and tighten the nuts to the torque listed in this Chapter's Specifications. Again, work up to the torque in three steps.
17 Repeat the entire procedure for the remaining piston/connecting rod assemblies. Keep the back sides of the bearing inserts and the inside of the connecting rod and cap perfectly clean when assembling them. Make sure

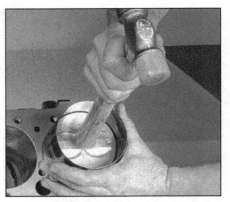

27.10 Drive the piston into the cylinder bore with the end of a wooden or plastic hammer handle

27.12 Lay the Plastigage strips on each rod bearing journal, parallel to the crankshaft centerline

you have the correct piston for the cylinder and that the notch, arrow or F on the piston faces to the front of the engine when the piston is installed. Remember, use plenty of oil to lubricate the piston before installing the ring compressor and cap. Also, when installing the rod caps for the final time, be sure to lubricate the bearing faces adequately.
18 After all the piston/connecting rod assemblies have been properly installed, rotate the crankshaft a number of times by hand and check for any obvious binding.
19 Finally, check the connecting rod endplay. Refer to Section 14 for the procedure. Compare the measured endplay to this Chapter's Specifications to make sure it is correct.

28 Driveplate - removal and installation

Refer to Chapter 2 Part A or B, depending on engine type, for flywheel/driveplate removal and installation procedures and torque Specifications.

29 Initial start-up and break-in after overhaul

1 Once the engine has been properly installed in the vehicle, double-check the engine oil and coolant levels.
2 Remove the fuel injection fuse from the fuse block.
3 With the spark plugs out of the engine and the PCM/BATT fuse removed from the fuse panel (see Chapter 12), crank the engine over until oil pressure registers on the gauge (if so equipped) or until the oil light goes off.
4 Install the spark plugs, hook up the plug wires and reinstall the PCM/BATT fuse.
5 Make sure the throttle is closed, then start the engine. It may take a few moments for the fuel pump pressure to build up, but the engine should start without a great deal of effort.

6 As soon as the engine starts, it should be set at a fast idle (to ensure proper oil circulation) and allowed to warm up to normal operating temperature. While the engine is warming up, make a thorough check for oil and coolant leaks.
7 Shut the engine off and recheck the engine oil and coolant levels.
8 Drive the vehicle to an area with no traffic, accelerate sharply from 30 to 50 mph, then allow the vehicle to slow to 30 mph with the throttle closed. Repeat the procedure 10 or 12 times. This will load the piston rings and cause them to seat properly against the cylinder walls. Check again for oil and coolant leaks.
9 Drive the vehicle gently for the first 500 miles (no sustained high speeds) and keep a constant check on the oil level. It is not unusual for an engine to use oil during the break-in period.
10 At approximately 500 to 600 miles, change the oil and filter.
11 For the next few hundred miles, drive the vehicle normally, neither pampering or abusing it.
12 After 2000 miles, change the oil and filter again and consider the engine fully broken in.

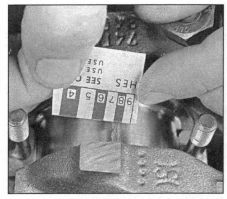

27.14 Measure the width of the crushed Plastigage to determine the rod bearing oil clearance (be sure to use the correct scale - standard and metric ones are included)

Chapter 3
Cooling, heating and air conditioning systems

Contents

	Section		Section
Air conditioning accumulator - removal and installation	15	Engine cooling fan and clutch - check, removal and installation	4
Air conditioning and heating system - check and maintenance	13	General information	1
Air conditioning compressor - removal and installation	14	Heater and air conditioning control assembly - removal and installation	12
Air conditioning condenser - removal and installation	16	Heater core - removal and installation	11
Air conditioning evaporator - removal and installation	17	Radiator and coolant reservoir tank - removal and installation	5
Antifreeze/coolant - general information	2	Thermostat - check and replacement	3
Blower motor - removal and installation	10	Water pump - check	6
Blower motor and circuit - check	9	Water pump - removal and installation	7
Coolant temperature sending unit - check and replacement	8		

Specifications

General

Coolant capacity	See Chapter 1
Drivebelt tension	See Chapter 1
Radiator pressure cap rating	15 psi
Thermostat opening temperature	195-degrees F

Torque specifications

Ft-lbs (unless otherwise indicated)

Thermostat housing nuts, four-cylinder engine	89 in-lbs
Thermostat housing bolts, V6 engine	18
Water pump attaching bolts	
Four-cylinder engine	18
V6 engine	33
Fan assembly-to-water pump nut	17
Fan-to-fan clutch bolts	24
Oil cooler line-to-adapter bolt	26

1 General information

The cooling system consists of a radiator and coolant reserve system, a radiator pressure cap, a thermostat, an engine-driven fan with a thermostatically-controlled viscous-drive clutch, and a water pump.

The system is pressurized by a spring-loaded radiator cap, which, by maintaining pressure, increases the boiling point of the coolant. If the coolant temperature goes above this increased boiling point, the extra pressure in the system forces the radiator cap valve off its seat and exposes the overflow pipe or hose. The overflow pipe/hose leads to a coolant recovery system. This consists of a plastic reservoir into which the coolant that normally escapes due to expansion is retained. When the engine cools, the excess coolant is drawn back into the radiator by the vacuum created as the system cools, maintaining the system at full capacity. This is a continuous process and provided the level in the reservoir is correctly maintained, it is not necessary to add coolant to the radiator.

Coolant in the right tank of the radiator is drawn by the water pump, which forces it through the water passages in the cylinder block. The coolant then travels up into the cylinder head(s), circulates around the combustion chambers and valve seats, travels out of the cylinder head(s) past the open thermostat into the upper radiator hose and back into the radiator.

When the engine is cold, the thermostat restricts the circulation of coolant to the engine. When the minimum operating temperature is reached, the thermostat begins to open, allowing coolant to return to the radiator.

Automatic transmission-equipped models have a cooler element incorporated into the right tank of the radiator to cool the transmission fluid, and on some models, an optional engine oil cooler is connected to the left side of the radiator.

The heating system works by directing air through the heater core mounted behind the dash and then to the interior of the vehicle by a system of ducts. Temperature is controlled by mixing heated air with fresh air, using a system of flapper doors in the ducts, and a blower motor.

Air conditioning is an optional accessory, consisting of an evaporator core located in a plastic module on the engine side of the firewall, a condenser in front of the radiator, an accumulator in the engine compartment and a belt-driven compressor mounted at the front of the engine.

2 Antifreeze/coolant - general information

Refer to illustration 2.6

Warning: *Do not allow antifreeze to come in contact with your skin or painted surfaces of the vehicle. Rinse off spills immediately with plenty of water. Antifreeze is highly toxic if ingested. Never leave antifreeze lying around in an open container or in puddles on the floor; children and pets are attracted by it's sweet smell and may drink it. Check with local authorities about disposing of used antifreeze. Many communities have collection centers which will see that antifreeze is disposed of safely. Never dump used antifreeze on the ground or pour it into drains.*

Note: *Non-toxic antifreeze is now manufactured and available at local auto parts stores, but even these types should be disposed of properly.*

2.6 An inexpensive hydrometer can be used to test the condition of your coolant

3.9a On models with four-cylinder engines, remove the three nuts and lift off the thermostat cover (arrow)

3.9b Models with V6 engines use two bolts to secure the thermostat cover (arrow)

The cooling system should be filled with a water/ethylene glycol based antifreeze solution which will prevent freezing down to at least -20-degrees F (even lower in cold climates). It also provides protection against corrosion and increases the coolant boiling point.

The cooling system should be drained, flushed and refilled at least every other year (see Chapter 1). The use of antifreeze solutions for periods of longer than two years is likely to cause damage and encourage the formation of rust and scale in the system. However, 1996 models are filled with a new long-life coolant called "DEX-COOL®," which the factory claims is good for five years, or 100,000 miles. This new coolant is identified by its orange color, in contrast to conventional coolants that are green. It's best not to mix conventional coolants with DEX-COOL®, since this will diminish its corrosion resistance.

Before adding antifreeze to the system, check all hose connections. Antifreeze can leak through very minute openings.

The exact mixture of antifreeze to water which you should use depends on the relative weather conditions. The mixture should contain at least 50-percent antifreeze, but should never contain more than 70-percent antifreeze. Consult the mixture ratio chart on the antifreeze container before adding coolant. Hydrometers are available at most auto parts stores to test the coolant **(see illustration)**. Use antifreeze which meets the vehicle manufacturer's specifications.

3 Thermostat - check and replacement

Refer to illustrations 3.9a, 3.9b and 3.10

Warning: *The engine must be completely cool when this procedure is performed.*

Note: *Don't drive the vehicle without a thermostat! The computer may stay in open loop and emissions and fuel economy will suffer.*

Check

1 Before condemning the thermostat, check the coolant level, drivebelt tension and temperature gauge (or light) operation.
2 If the engine takes a long time to warm up, the thermostat is probably stuck open. Replace the thermostat.
3 If the engine runs hot, check the temperature of the upper radiator hose. If the hose isn't hot, the thermostat is probably stuck shut. Replace the thermostat.
4 If the upper radiator hose is hot, it means the coolant is circulating and the thermostat is open. Refer to the *Troubleshooting* section for the cause of overheating.
5 If an engine has been overheated, you may find damage such as leaking head gaskets, scuffed pistons and warped or cracked cylinder heads.

Replacement

6 Drain coolant (about 1 gallon) from the radiator, until the coolant level is below the thermostat housing (see Chapter 1).
7 Remove the air intake duct (see Chapter 4).
8 Disconnect the radiator hose from the thermostat cover. **Note:** *Some models do not have a thermostat housing gasket, and it isn't necessary to remove the hose unless it is being replaced.*
9 Remove the bolts and lift the cover off **(see illustrations)**. It may be necessary to tap the cover with a soft-face hammer to break the gasket seal. **Note 1:** *Have a new rubber thermostat gasket handy before removing the old thermostat, the gasket may tear while pulling out the thermostat.* **Note 2:** *On later model V6 engines, the accelerator cable bracket may have to be removed from the throttle body.*
10 Note how it's installed, then remove the thermostat **(see illustration)**. Be sure to use a replacement thermostat with the correct opening temperature (see this Chapter's Specifications).
11 If a gasket was used, use a scraper or putty knife to remove all traces of old gasket material and sealant from the mating surfaces. **Caution:** *Be careful not to gouge or damage the gasket surfaces, because a leak could develop after assembly. Make sure no gasket material falls into the coolant passages; it is a good idea to stuff a rag in the passage. Wipe the mating surfaces with a rag saturated with lacquer thinner or acetone.*
12 Install the thermostat and make sure the correct end faces out - the spring is directed toward the engine **(see illustration 3.10)**.
13 Use a new rubber gasket. If no gasket was use, apply a bead of RTV to the groove in the coolant outlet sealing surface.
14 Carefully position the cover and install the bolts. Tighten them to the torque listed in this Chapter's Specifications - do not overtighten the bolts or the cover may crack or become distorted.
15 Reattach the radiator hose to the cover and tighten the clamp - now may be a good time to check and replace the hoses and

3.10 Note how it's installed (the spring end faces down), then remove the thermostat - four-cylinder engine model shown. Note how the old rubber seal has torn upon removal

Chapter 3 Cooling, heating and air conditioning systems

4.4 Position a thermometer between the radiator and fan - make sure there is clearance and that the thermometer is securely held in place with duct tape. Here a hole has been drilled in the upper fan shroud for the thermometer; just be sure the thermometer probe doesn't contact the fan!

4.9a Remove the upper fan shroud bolts along the top front edge . . .

4.9b . . . then the two bolts (arrows) at each side where the upper and lower shrouds meet (left side shown, socket is on one of the bolts)

4.10 Remove the nuts securing the fan/clutch assembly to the water pump (four-cylinder engine shown)

clamps (see Chapter 1).
16 Refer to Chapter 1 and refill the system, then run the engine and check carefully for leaks.
17 Repeat steps 1 through 5 to be sure the repairs corrected the previous problem(s).

4   Engine cooling fan and clutch - check, removal and installation

Warning 1: *The covered models are equipped with airbags. Impact sensors for the airbag system are located in the dash, cowl and just in front of the radiator. The airbag(s) could accidentally deploy if these sensors are disturbed, so be extremely careful when working in this area. Airbag system components are also located in the steering wheel, steering column and base of the steering column, so be extremely careful when working in this area and don't disturb any airbag system components or wiring. You could easily be injured if an airbag accidentally deploys, and the airbag might not deploy correctly in a collision if any components or wiring in the system have been disturbed. Refer to Chapter 12 for the procedure to disable the airbag system.*
Warning 2: *Keep hands, tools and clothing away from the fan. To avoid injury or damage DO NOT operate the engine with a damaged fan. Do not attempt to repair fan blades - replace a damaged fan with a new one.*

Check

Refer to illustration 4.4

1 Noise from the cooling fan is not necessarily an indication of a damaged or worn assembly. The fan will make some noise when a cold engine is first started, as the fluid in the clutch redistributes itself and then goes into normal disengaged mode, and will also make a noticeable change in sound when the vehicle warms up and the fan engages. *Constant* noise at all speed above 2500 rpm indicates the clutch has failed in the locked-up mode and should be replaced.
2 With a cold vehicle (engine off, key out), rotate the fan by hand. If it free-wheels very easily, with no drag, replace the clutch. On the other hand, if you can't rotate the fan at all, or it feels jerky or not consistently-smooth in operation, this also indicates a need for replacement.
3 Look carefully around the clutch (engine off, key out). If there is strong evidence of fluid leakage there, more than just a little around the bearing, replace the clutch.
4 Dynamic performance of the fan clutch can be tested by taping a thermometer with a probe to the inside of the fan shroud with duct tape, between the radiator and the fan **(see illustration)**, or by inserting it into a hole in the fan shroud. **Warning:** *Make certain there is plenty of clearance for the fan to turn and that the thermometer is securely taped in place. Also, if you drill a hole for the thermometer, do not drill into the radiator core by accident.*
5 Cover the grille in front of the vehicle with cardboard to block the airflow temporarily. Have an assistant run the engine at high idle with the air conditioning on to induce higher engine temperature. It may take ten minutes or so.
6 Watching the thermometer, note the temperature when the fan clutch engages. The fan should engage between 150 and 195 degrees F.
7 Once the fan has engaged, have your assistant turn off the air conditioning while you remove the cardboard from in front of the grille, to allow the engine to cool down at reduced engine speed. **Caution:** *Do not continue testing beyond the 195 degrees F or you risk overheating the engine. If the fan never came on, replace it.* With the engine at high temperature and turned off, the fan should be much harder to turn than when the engine is cold.

Replacement

Refer to illustrations 4.9a, 4.9b, 4.10 and 4.11
Note: *This procedure applies to either fan.*
8 Disconnect the battery cable from the negative battery terminal. **Caution:** *On models equipped with the Delco Loc II audio system, be sure the lockout feature is turned off before performing any procedure which requires disconnecting the battery.*
9 Take off the upper air intake duct (see Chapter 4), remove the upper fan shroud screws, and take off the upper fan shroud **(see illustrations)**.
10 On 1996 and earlier models, remove the nuts holding the fan/clutch assembly to the studs on the water pump **(see illustration)**. On 1997 and later models, hold the water pump pulley stationary and loosen the large nut retaining the fan/clutch assembly to the water pump hub. Remove the fan and clutch.
11 Off the vehicle, separate the fan from the clutch. There are bolts from the front on four-cylinder engine fan assemblies, and bolts

4.11 Remove these four bolts to separate the fan from the fan clutch

5.4a On the right side, disconnect the transmission cooler lines (arrows) and lower radiator hose (above bottom transmission fitting)

5.4b On the left side, if equipped, disconnect the engine oil cooler lines (arrows)

from behind on the V6 **(see illustration)**.
12 Installation is the reverse of removal.

5 Radiator and coolant reservoir tank - removal and installation

Warning 1: *The covered models are equipped with airbags. Impact sensors for the airbag system are located in the dash, cowl and just in front of the radiator. The airbag(s) could accidentally deploy if these sensors are disturbed, so be extremely careful when working in this area. Airbag system components are also located in the steering wheel, steering column and base of the steering column, so be extremely careful when working in this area and don't disturb any airbag system components or wiring. You could easily be injured if an airbag accidentally deploys, and the airbag might not deploy correctly in a collision if any components or wiring in the system have been disturbed. Refer to Chapter 12 for the procedure to disable the airbag system.*
Warning 2: *The engine must be completely cool when this procedure is performed.*

Radiator
Refer to illustrations 5.4a, 5.4b, 5.5 and 5.6

Removal
1 Disconnect the battery cable at the negative battery terminal.
Caution: *On models equipped with the Delco Loc II audio system, be sure the lockout feature is turned off before performing any procedure which requires disconnecting the battery.*
2 Drain the cooling system as described in Chapter 1. Refer to the coolant **Warning** in Section 2.
3 Remove the air intake duct (see Chapter 4) and the upper fan shroud (see Section 4).
4 If equipped with an automatic transmission or engine oil cooler, remove the cooler lines from the radiator **(see illustrations)** - be careful not to damage the lines or fittings. Plug the ends of the disconnected lines to prevent leakage and stop dirt from entering the system. Have a drip pan ready to catch any spills. **Note:** *On later models the lines are*

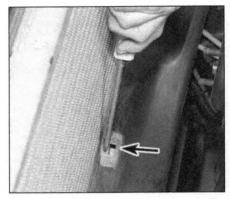

5.5 The lower fan shroud detaches easily by depressing the plastic tabs (center one shown, there are three in all) and pulling the shroud up

connected to the radiator with quick-connect fittings. Use pliers to remove the retaining spring-clips, then pull the lines straight out.
5 Disconnect the upper and lower radiator hoses. Remove the lower fan shroud by depressing the plastic tabs with a screwdriver **(see illustration)**.
6 Pull the radiator straight up and out **(see illustration)**. **Caution:** *If any coolant spills on the paint, clean it off immediately. It may damage the finish.*

Installation
7 Prior to installation of the radiator, replace any damaged hose clamps and radiator hoses.
8 If leaks have been noticed or there have been cooling problems, have the radiator cleaned and tested at a radiator shop.
9 Radiator installation is the reverse of removal. On later models, use new spring clips on the oil cooler connections. Install the clips into their slots by hooking one end, then rotating them in. With the clips in place, push the oil cooler lines straight in until they lock in place.
10 After installation, fill the system with the proper mixture of antifreeze, and also check the automatic transmission fluid level, where applicable (see Chapter 1).

5.6 Pull the radiator straight up and out

Coolant reservoir tank
Refer to illustration 5.12

Removal
11 On Blazer and Jimmy models, refer to Chapter 6 and remove the PCM (computer) from the top of the coolant recovery tank. On some pickup models, the PCM is located elsewhere and access to the recovery tank is easier.
12 Remove the overflow hose and mounting nuts from the recovery tank and remove it **(see illustration)**.

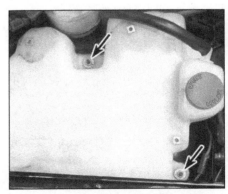

5.12 After removal of the PCM on Blazer and Jimmy models, remove these nuts (arrows) and the overflow hose

6.2 Check the weep hole (arrow) for leakage (pump removed for clarity) - V6 pump shown, four-cylinder engine similar

6.4 Check the pump for loose or rough bearings (this can be done with the fan and pulley in place)

7.4 Behind the fan and fan pulley on the four-cylinder engine is a housing and bearing (left arrow) - this is NOT the water pump - the water pump is beneath the pulley indicated by the right arrow

13 Prior to installation make sure the reservoir is clean and free of debris which could be drawn into the radiator (wash it with soapy water and a brush if necessary, then rinse thoroughly).

Installation
14 Installation is the reverse of removal.

6 Water pump - check

Refer to illustrations 6.2 and 6.4
Warning: *The covered models are equipped with airbags. Impact sensors for the airbag system are located in the dash, cowl and just in front of the radiator. The airbag(s) could accidentally deploy if these sensors are disturbed, so be extremely careful when working in this area. Airbag system components are also located in the steering wheel, steering column and base of the steering column, so be extremely careful when working in this area and don't disturb any airbag system components or wiring. You could easily be injured if an airbag accidentally deploys, and the airbag might not deploy correctly in a collision if any components or wiring in the system have been disturbed. Refer to Chapter 12 for the procedure to disable the airbag system.*

1 Water pump failure can cause overheating and serious damage to the engine. There are three ways to check the operation of the water pump while it is installed on the engine. If any one of the following quick-checks indicates water pump problems, it should be replaced immediately.

2 A seal protects the water pump impeller shaft bearing from contamination by engine coolant. If this seal fails, a weep hole in the water pump snout will leak coolant **(see illustration)** (an inspection mirror can be used to look at the underside of the pump if the hole isn't on top). If the weep hole is leaking, shaft bearing failure will follow. Replace the water pump immediately. **Note:** *A small amount of black discoloration around the weep hole is normal, but the presence of brown residue or dripping water indicates replacement is required.*

3 Besides contamination by coolant after a seal failure, the water pump impeller shaft bearing can also be prematurely worn out by an improperly-tensioned drivebelt. When the bearing wears out, it emits a high-pitched squealing sound. If such a noise is coming from the water pump during engine operation, the shaft bearing has failed - replace the water pump immediately. **Note:** *Do not confuse belt noise with bearing noise.*

4 To identify excessive bearing wear on water pumps before the bearing actually fails, grasp the water pump pulley and try to force it up-and-down or from side-to-side **(see illustration)**. If the pulley can be moved either horizontally or vertically, the bearing is nearing the end of its service life. Replace the water pump. Don't mistake drivebelt slippage, which causes a squealing sound, for water pump bearing failure.

5 Its possible for a water pump to be bad, even if it doesn't howl or leak water. Sometimes the fins on the back of the impeller can corrode away until the pump is no longer effective. The only way to check for this is to remove the pump for examination.

7 Water pump - removal and installation

Warning: *The covered models are equipped with airbags. Impact sensors for the airbag system are located in the dash, cowl and just in front of the radiator. The airbag(s) could accidentally deploy if these sensors are disturbed, so be extremely careful when working in this area. Airbag system components are also located in the steering wheel, steering column and base of the steering column, so be extremely careful when working in this area and don't disturb any airbag system components or wiring. You could easily be injured if an airbag accidentally deploys, and the airbag might not deploy correctly in a collision if any components or wiring in the system have been disturbed. Refer to Chapter 12 for the procedure to disable the airbag system.*

Removal
Refer to illustrations 7.4, 7.5, 7.6a, 7.6b, 7.6c and 7.6d

1 Disconnect the battery cable from the negative battery terminal. **Caution:** *On models equipped with the Delco Loc II audio system, be sure the lockout feature is turned off before performing any procedure which requires disconnecting the battery.*
2 Drain the coolant (see Chapter 1).
3 Remove the upper radiator shroud and fan/clutch assembly (see Section 4).
4 Remove the serpentine belt (see Chapter 1) and the water pump pulley. **Note:** *The pulley behind the fan on the four-cylinder engine engine is not the water pump pulley. The water pump is mounted to the left side of the engine and has a separate pulley* **(see illustration)**.
5 On V6 engines, detach the radiator hose and heater hose from the water pump **(see illustration)**.

7.5 Detach the radiator hose from the water pump (lower arrow), and the heater hose (top arrow)

7.6a Remove the bolts and remove the water pump (four-cylinder engine)

7.6b On air conditioning-equipped models, remove these two bolts and lift off the drivebelt tensioner brace (arrows)

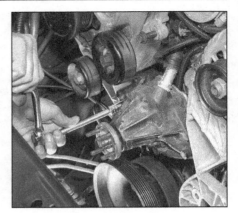

7.6c Remove the four bolts and remove the water pump (V6 models)

7.6d If necessary, tap the pump with a soft-faced hammer to break the gasket seal

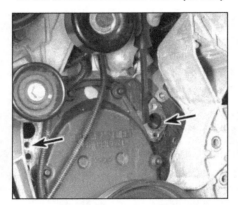

7.7 Remove all traces of old gasket material (gasket areas on V6 models are indicated by the arrows) - use care to avoid gouging the pump or sealing surfaces on the engine

8.1a The four-cylinder engine coolant-temperature gauge sending unit (arrow) is located on the bottom right side of the thermostat housing

6 Unbolt the water pump **(see illustrations)**. It may be necessary to tap the pump with a soft-face hammer to break the gasket seal. Inspect the pump's impeller blades on the backside for corrosion. If any fins are missing or badly corroded, replace the pump with a new one.

Installation

Refer to illustration 7.7

7 Clean the sealing surfaces of all gasket material on both the water pump and block **(see illustration)**. Wipe the mating surfaces with a rag saturated with lacquer thinner or acetone.
8 Apply a thin layer of RTV sealant to both sides of the new gasket and install the gasket on the water pump. On V6 engines there are two small gaskets, one at each side of the pump.
9 Place the water pump in position and install the bolts finger tight. Use caution to ensure that the gasket doesn't slip out of position. Tighten the bolts to the torque listed in this Chapter's Specifications. **Note:** *Use RTV sealant on the water pump bolt threads.*
10 The remainder of the installation procedure is the reverse of removal.
11 Add coolant to the specified level (see Chapter 1) and start the engine and check for the proper coolant level and the water pump and hoses for leaks. Bleed the cooling system of air as described in Chapter 1.

8 Coolant temperature sending unit - check and replacement

Refer to illustrations 8.1a and 8.1b
Warning: *Wait until the engine is completely cool before beginning this procedure.*

Check

1 The coolant temperature indicator system is composed of a temperature gauge mounted in the dash and a coolant temperature sending unit mounted on the engine **(see illustrations)**. **Note:** *On 1993 and 1994 V6 models, the coolant temperature sending unit is located on the right cylinder head, above the starter.*
2 If an overheating indication occurs, check the coolant level in the system and then make sure the wiring between the gauge and the sending unit is secure and all fuses are intact.
3 Remove the sending unit from the engine and clip ohmmeter leads to the sending unit body and the terminal. Place the sending unit in a pan of water on the stove,

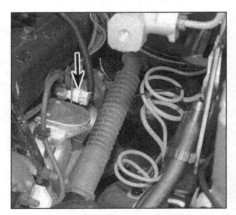

8.1b On 1995 and later V6 models, the coolant temperature sending unit (arrow) is located in the left cylinder head

with a cooking thermometer in the water. On a cold (50-80 degrees F) sender, the resistance should be 5,700 to 2,200 ohms. When it has warmed to operating temperature (170 to 200 degrees F) the resistance should drop to 200-300 ohms.
4 If the sender operated correctly, check the circuit and gauge operation by grounding the wire to the sending unit while the ignition is

Chapter 3 Cooling, heating and air conditioning systems

9.3a The blower motor (arrow) on pickup models is easily accessible in the engine compartment . . .

9.3b . . . while on Blazer and Jimmy models, it is in a plastic case (arrow) where it must be cut out of the housing for replacement

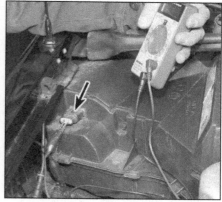

9.4 Connect a voltmeter to the heater blower motor connector (arrow) by backprobing, and check the running voltage at each blower switch position

9.7 The blower motor resistor (arrow) is located on the blower/evaporator housing, on the side toward the engine - to its right here are the blower/air conditioning relays

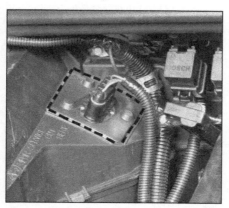

9.10 On Blazer and Jimmy models, the blower resistor must be cut out of the housing for replacement - the dotted line indicates cut lines - the replacement resistor must be glued in place with black weather-stripping adhesive

on (engine NOT running). If the gauge deflects full scale, the circuit and gauge are OK.

Replacement

5 Make sure the engine is cool before removing the defective sending unit. There will be some coolant loss as the unit is removed, so be prepared to catch it. Refer to the coolant **Warning** in Section 2.
6 Prepare the new sending unit by coating the threads with sealant. Disconnect the electrical connector and simply unscrew the old sensor from the engine and install the replacement.
7 Check the coolant level after the replacement unit has been installed and top up the system, if necessary (see Chapter 1). Check now for proper operation of the gauge and sending unit.

9 Blower motor and circuit - check

Refer to illustrations 9.3a, 9.3b, 9.4, 9.7 and 9.10

1 Check the fuse (marked HVAC) and all connections in the circuit for looseness and corrosion. Make sure the battery is fully charged.
2 With the transmission in Park, the parking brake securely set, turn the ignition switch to the Run position. It isn't necessary to start the vehicle.
3 The blower motor is mounted in a housing with the air conditioning evaporator (if equipped) on the right side of the firewall in the engine compartment **(see illustrations)**.
4 Backprobe the blower motor electrical connector with two small paper clips (straightened out) and connect a voltmeter to the blower motor connector and ground **(see illustration)**.
5 Move the blower switch through each of its positions and note the voltage readings. Changes in voltage indicate that the motor speeds will also vary as the switch is moved to the different positions.
6 If there is voltage present, but the blower motor does not operate, the blower motor is probably faulty. Disconnect the

blower motor connector and hook one side to a chassis ground and the other to a fused source of battery voltage. If the blower doesn't operate, it is faulty.
7 If there was no voltage present at the blower motor at one or more speeds, and the motor itself tested OK, check the blower motor resistor **(see illustration)**.
8 Backprobe the connector to the resistor while it is connected and look for varying voltages at each speed switch position.
9 Disconnect the electrical connector from the blower motor resistor. With the ignition on, check for voltage at each of the terminals in the connector as the blower speed switch is moved to the different positions. If the voltmeter responds correctly to the switch then the resistor is probably faulty. If there is no voltage present from the switch, then the switch, control panel or related wiring is probably faulty.
10 To change the resistor, remove the three mounting screws and pull the resistor from the housing **(see illustration 9.7)**. **Note:** *On Blazer and Jimmy models, the blower resistor is riveted to the plastic housing. The change the resistor, a marked portion of the plastic must be cut out with a sharp utility knife* **(see illustration)**. *When replacing it, it must be glued in with black weather-stripping cement.*

10 Blower motor - removal and installation

Refer to illustrations 10.1, 10.3, 10.4 and 10.7

Pick-up models

1 The pick-up models have a blower/evaporator housing that bolts to the firewall from the engine compartment side. With the ignition OFF, the blower motor can be removed easily in the engine compartment by disconnecting the electrical connector, removing the three mounting screws (refer to Section 5 for removal of the coolant recovery

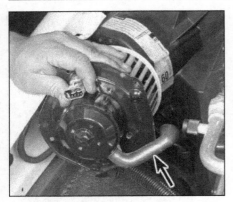

10.1 On pickup models, remove the coolant recovery tank, then disconnect the blower electrical plug (held in photo). Disconnect the drain hose (arrow), then remove the three screws to pull the blower motor out

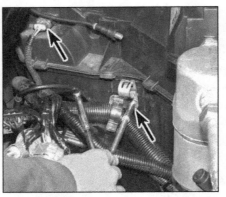

10.3 On Blazer and Jimmy models, unplug the blower electrical connector (upper arrow) and the nut (lower arrow) holding the harness to the front of the case

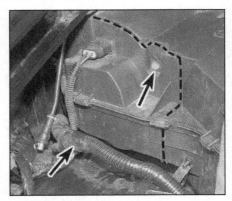

10.4 On Blazer and Jimmy models, the blower motor housing must be cut with a utility knife to access the blower motor - the dotted line indicates where to cut; arrows show the locations of the two blower motor mounting bolts

tank first) and pulling the motor and fan out (see illustration).

Blazer and Jimmy models

2   Refer to Chapter 6 and remove the PCM (computer) from the top of the coolant recovery tank, then remove the coolant recovery tank (see Section 5).

3   Disconnect the electrical plug to the blower motor and remove the nut holding the wiring harness to the front of the case (see illustration).

4   The plastic cover of the blower/evaporator housing must be cut with a sharp utility knife, using the lines molded into the cover as a guide (see illustration). There are also two clips (between the upper and lower halves of the housing) that must be removed. **Warning:** *Use extreme caution while handling the utility knife.*

5   Remove the two blower mounting screws (see illustration 10.4).

6   Disconnect the blower motor cooling tube and pull the blower up and out of the housing.

7   To remove the fan from the blower

10.7 The fan is retained to the motor by a clip on some models, but on others, like this one, the fan is pressed onto the shaft and can be removed by prying with two screwdrivers, 180 degrees apart

motor, squeeze the spring clip together and slip the fan off the shaft. **Note:** *On some models, there is no clip or nut retaining the fan to the shaft, in this case the fan can be levered off with two screwdrivers* (see illustration).

8   Install the fan onto the motor and install the blower motor into the heater housing and reinstall the mounting screws.

9   Reattach the upper housing panel that was removed, using the three mounting clips and aligning the cut edges carefully. Making a clean bead, seal the cut edges with black weather-stripping adhesive (not RTV or epoxy). Allow the adhesive to dry at least a half hour before doing any further work around the housing.

11 Heater core - removal and installation

Warning 1: *The covered models are equipped with airbags. Impact sensors for the airbag system are located in the dash, cowl and just in front of the radiator. The airbag(s) could accidentally deploy if these sensors are disturbed, so be extremely careful when working in this area. Airbag system components are also located in the steering wheel, steering column and base of the steering column, so be extremely careful when working in this area and don't disturb any airbag system components or wiring. You could easily be injured if an airbag accidentally deploys, and the airbag might not deploy correctly in a collision if any components or wiring in the system have been disturbed. Refer to Chapter 12 for the procedure to disable the airbag system.*

Warning 2: *The air conditioning system is under high pressure. DO NOT loosen any fittings or remove any components until after the system has been discharged. Air conditioning refrigerant should be properly discharged into an EPA-approved container at a dealership service department or an automotive air conditioning facility. Always wear eye protection when disconnecting air conditioning system fittings.*

Removal

1997 and earlier models

Refer to illustrations 11.3, 11.4, 11.5, 11.6, 11.7, 11.8, 11.9, 11.10, 11.12, 11.13a and 11.13b

Note: *The dash removal procedure required to access the heater core housing behind the dash is a difficult procedure because of the number of hidden screws and the many wiring connectors that must be tagged and disconnected. Once the dash is removed, however, the remainder of the heater core removal procedure is straightforward.*

1   Disconnect the battery cable at the negative battery terminal. **Caution:** *On models equipped with the Delco Loc II audio system, be sure the lockout feature is turned off before performing any procedure which requires disconnecting the battery.*

2   Drain the cooling system (see Chapter 1).

3   Disconnect the heater hoses at the heater core inlet and outlet on the engine side of the firewall and plug the open fittings (see illustration). **Caution:** *If too much force is used on twisting off the heater hoses, the heater core pipes can be damaged. If the hoses are on too tight to twist off, cut them from the pipes.*

4   The heater core housing is behind the dash, attached to the firewall. The housing is in two halves, with the frontmost portion being bolted to the firewall and the rear housing (cover) attached to it with screws. Only the rear cover housing need be removed to access the heater core. Begin the dash removal by taking out the left and right lower dash panels (see illustration). Refer to Chapter 12 for removal of the radio so that the antenna cable can be disconnected.

5   Under the dash at either end, remove the bolts (one at each end) holding the pivot bracket, then remove the two Torx screws near each bracket (see illustration).

6   Remove the four nuts holding the steering column to the dash and firewall bracing, then lower the column down (see illustration).

Chapter 3 Cooling, heating and air conditioning systems

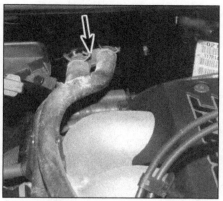

11.3 Loosen the hose clamps (arrow) and disconnect the heater hoses from the heater core tubes at the firewall

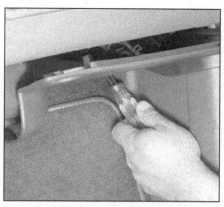

11.4 Begin the dash removal process by taking out the lower dash panels (right shown) - each panel has two clips and two screws

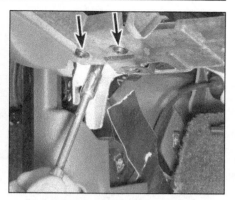

11.5 At each end of the dash on the bottom, remove the vertical bolts through the silver pivot bracket (shown here with a socket on it), then remove the two Torx bolts (arrows)

11.6 Remove the four nuts holding the steering column (two shown with arrows), then drop the column down

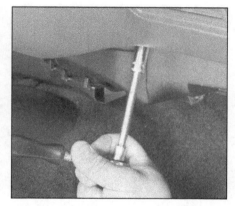

11.7 Remove this one bolt from below the dash, just below the vent that is right of center

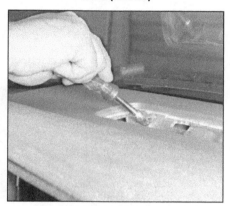

11.8 Under the speaker grilles and defroster grille (shown) are several dashboard bolts

11.9 When the dash is pivoted rearward on its lower brackets, there is room to reach your hand ahead of it to access any wires to be disconnected

11.10 Remove the central bolt (arrow) holding the main electrical plug to the right side of the dash (shown here looking through the windshield

11.12 Remove the one bolt holding the pencil brace (arrow), then remove the six heater core cover screws and remove the cover

7   Remove one bolt from the bottom of the dash, just to the right of center. It connects to a steel "pencil" brace behind the dashboard **(see illustration)**.
8   Pry up the upper dashboard speaker grilles and the defroster outlet grille to access the hidden dashboard bolts underneath **(see illustration)**.
9   When the last fasteners are removed,

hold on to the dash and it will pivot rearward, allowing access to any of the cables behind the dash that need to be disconnected **(see illustration)**.
10   On the right side of the dash, with the dash leaned backward on its pivots, reach around to the front side of the dash and disconnect the main electrical harness plug **(see illustration)**.

11   If the radio antenna lead has not been disconnected already, disconnect it now and pull the entire dashboard assembly out of the vehicle.
12   The cover for the heater core is now exposed. Unbolt and remove the pencil brace over the cover, then remove the six screws holding the cover to the heater core housing **(see illustration)**.

11.13a Remove the rear heater housing cover screws and pull it off, exposing the core - remove the four mounting strap screws (arrows) and the straps

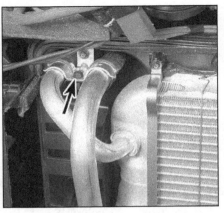

11.13b Remove this screw (arrow) and the heater core can be pulled out

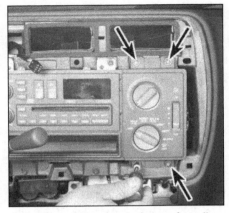

12.4 Remove the screws (arrows) to allow removal of the heater/air conditioning control panel

13 With the cover off, remove the four screws holding the straps, then the screw holding the bracket over the tubes at top left. Pull the core out, being careful not to bend or kink the heater core tubes **(see illustrations)**.

1998 and later models

14 On these models, the entire heating/ventilation module must be removed from the firewall. Follow the general procedures in Steps 1 through 11 for dash removal first.
15 Tag and disconnect all electrical and vacuum connectors at the heating/ventilation module, under the right side of the dash.
16 Remove the blower motor resistor on the engine side of the firewall (see Section 9).
17 Using a flashlight and a small ratchet and extension, reach through the blower resistor hole to remove the one engine-side bolt securing the heating/ventilation module. Mark this bolthead with paint, so you use only this bolt here for reassembly. A longer bolt will interfere with mode door functions.
18 Inside the vehicle, remove the heating/ventilation module mounting nuts and bolts. **Note:** *The nut/stud does not come all the way off. Just loosen it until it comes free of the firewall.*
19 Pull the heating/ventilation module from the vehicle.
20 To remove the heater core, remove the two screws on top of the module (over the heater core), remove the heater core cover, and slide the heater core up and out of the heating/ventilation module.

Installation

21 Installation is the reverse of removal. **Note:** *When reinstalling the heater core, make sure any original insulating/sealing materials are in place around the heater core pipes and around the core.*
22 Refill the cooling system (see Chapter 1).
23 Start the engine and check for leaks and proper operation of the heating system.

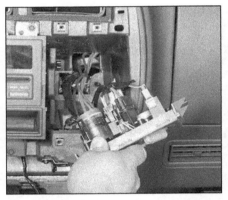

12.5a Pull the panel forward - for complete removal, detach the wiring and vacuum lines

12 Heater and air conditioning control assembly - removal and installation

Warning: *The covered models are equipped with airbags. Impact sensors for the airbag system are located in the dash, cowl and just in front of the radiator. The airbag(s) could accidentally deploy if these sensors are disturbed, so be extremely careful when working in this area. Airbag system components are also located in the steering wheel, steering column and base of the steering column, so be extremely careful when working in this area and don't disturb any airbag system components or wiring. You could easily be injured if an airbag accidentally deploys, and the airbag might not deploy correctly in a collision if any components or wiring in the system have been disturbed. Refer to Chapter 12 for the procedure to disable the airbag system.*

1994 through 1997 models

Removal

Refer to illustrations 12.4, 12.5a and 12.5b
1 Disconnect the battery cable from the negative battery terminal. **Caution:** *On models equipped with the Delco Loc II audio sys-*

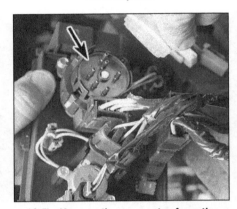

12.5b Unsnap the connector from the vacuum block (arrow) and lines from behind the mode-control switch

tem, be sure the lockout feature is turned off before performing any procedure which requires disconnecting the battery.
2 Refer to Chapter 11 and remove the instrument panel bezel.
3 Removal of the bezel allows access to the heat/air conditioning control mounting screws (see Chapter 11).
4 Remove the four screws holding the control unit to the dash **(see illustration)**. The control unit is linked to the radio panel to its left, so the screws on the radio may have to be loosened to allow removing the control unit.
5 Pull the unit from the dash. It can be pulled out just far enough to allow disconnecting the control panel light, blower speed switch and vacuum lines (on air conditioning-equipped models) from the control head **(see illustrations)**.

Installation

6 To install the control assembly, reverse the removal procedure. **Caution:** *The individual vacuum lines should not be disconnected from the white plug that attaches to the control unit. If a vacuum line must be replaced, do not use any lubricant to make it slip on easier; it can affect vacuum operation. If necessary, use a drop of plain water to make reconnection easier.*

Chapter 3 Cooling, heating and air conditioning systems

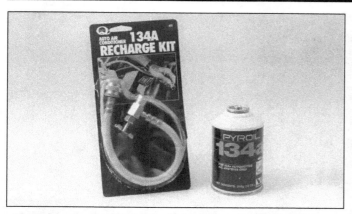

13.10 A basic charging kit for 134a systems is available at most auto parts stores - it must say 134a (not R-12) and so must the 12-ounce can of refrigerant

13.13 Add R-134a refrigerant to the low-side port only - the procedure is easier if you wrap the can with a warm, wet towel to prevent icing

1998 and later models

7 Remove the knee bolster trim plate attaching screws. Remove the two instrument panel trim plate lower screws and the single screw at the top center of the cluster lens.
8 Gently pry the instrument panel trim plate away from the dash. Disconnect all electrical harnesses from the trim plate.
9 Remove the trim plate.
10 Remove the control attaching screws and pull it out from the dash. Disconnect the vacuum and electrical connections.
11 Installation it the reverse of removal.

13 Air conditioning and heating system - check and maintenance

Air conditioning system

Warning: *The air conditioning system is under high pressure. DO NOT loosen any fittings or remove any components until after the system has been discharged. Air conditioning refrigerant should be properly discharged into an EPA-approved recovery container at a dealership service department or an automotive air conditioning repair facility. Always wear eye protection when disconnecting air conditioning system fittings.*
1 The following maintenance steps should be performed on a regular basis to ensure that the air conditioner continues to operate at peak efficiency:

a) Check the tension of the drivebelt and adjust if necessary (see Chapter 1).
b) Check the condition of the hoses. Look for cracks, hardening and deterioration. Look at potential leak areas (hoses and fittings) for signs of refrigerant oil leaking out. **Warning:** *Do not replace air conditioning hoses until the system has been discharged by a dealership or air conditioning repair facility.*
c) Check the fins of the condenser for leaves, bugs and other foreign material. A soft brush and compressed air can be used to remove them.
d) Check the wire harness for correct routing, broken wires, damaged insulation,

etc. Make sure the electrical connectors are clean and tight.
e) Maintain the correct refrigerant charge.

2 The system should be run for about 10 minutes at least once a month. This is particularly important during the winter months because long-term non-use can cause hardening of the internal seals.
3 Because of the complexity of the air conditioning system and the special equipment required to effectively work on it, accurate troubleshooting of the system should be left to a certified air conditioning technician.
4 If the air conditioning system doesn't operate at all, check the fuse panel. Check the HVAC fuse and the air conditioning compressor relay.
5 The most common cause of poor cooling is simply a low system refrigerant charge. If a noticeable drop in cool air output occurs, the following quick check will help you determine if the refrigerant level is low. For more complete information on the air conditioning system, refer to the *Haynes Automotive Heating and Air Conditioning Manual*.

Checking the refrigerant charge

6 Warm the engine up to normal operating temperature.
7 Place the air conditioning temperature selector at the coldest setting and the blower at the highest setting. Open the doors (to make sure the air conditioning system doesn't cycle off as soon as it cools the passenger compartment).
8 With the compressor engaged (the clutch will make an audible click and the center of the clutch will rotate), touch the accumulator surface and the evaporator inlet pipe. If there's no perceptible temperature difference between the inlet pipe and the accumulator, the system is properly charged. If there's a difference, there's something wrong with the system. It might be a low charge, but it might be something else. To be sure, take the vehicle to a dealer service department or other qualified shop for further diagnosis.
9 Place a thermometer in the dashboard vent nearest the evaporator and add refrigerant to the system until the indicated temperature is

around 40 to 45-degrees F. If the ambient (outside) air temperature is very high, say 110 degrees F, the duct air temperature may be as high as 60 degrees F, but generally the air conditioning is 30-50 degrees F cooler than the ambient air. **Note:** *Humidity of the ambient air also affects the cooling capacity of the system. Higher ambient humidity lowers the effectiveness of the air conditioning system.*

Adding refrigerant

Refer to illustrations 13.10 and 13.13
10 Buy an automotive charging kit at an auto parts store. A charging kit includes a 14-ounce can of refrigerant, a tap valve and a short section of hose that can be attached between the tap valve and the system low side service valve **(see illustration)**. Because one can of refrigerant may not be sufficient to bring the system charge up to the proper level, it's a good idea to buy an additional can. Make sure that one of the cans contains red refrigerant dye. If the system is leaking, the red dye will leak out with the refrigerant and help you pinpoint the location of the leak. **Caution:** *There are two types of refrigerant: R-12, used on vehicles up to 1992 (and some models until 1994), and the more environmentally-friendly R-134a used in all the models covered by this book. These two refrigerant (and their appropriate refrigerant oils) are not compatible and must never be mixed or components will be damaged. Use only R-134a refrigerant in the models covered by this book.* **Warning:** *Never add more than two cans of refrigerant to the system.*
11 Hook up the charging kit by following the manufacturer's instructions. **Warning:** *DO NOT hook the charging kit hose to the system high side!* The fittings on the charging kit are designed to fit **only** on the low side of the system.
12 Back off the valve handle on the charging kit and screw the kit onto the refrigerant can, making sure first that the O-ring or rubber seal inside the threaded portion of the kit is in place. **Warning:** *Wear protective eye wear when dealing with pressurized refrigerant cans.*
13 Remove the dust cap from the low-side charging connection and attach the quick-connect fitting on the kit hose **(see illustration)**.

14.4a Disconnect the electrical connector (arrow) from the air conditioning compressor (V6 model shown)

14.4b On four-cylinder engine models, there is one electrical connector (arrow) at the front of the compressor, and . . .

14.4c . . . two at the rear

14 Warm up the engine and turn on the air conditioner. Keep the charging kit hose away from the fan and other moving parts. **Note:** *The charging process requires the compressor to be running. Your compressor may cycle off if the pressure is low due to a low charge. If the clutch cycles off, you can pull the low-pressure cycling switch plug from the accumulator and attach a jumper wire. This will keep the compressor ON.*

15 Turn the valve handle on the kit until the stem pierces the can, then back the handle out to release the refrigerant. You should be able to hear the rush of gas. Add refrigerant to the low side of the system until both the accumulator surface and the evaporator inlet pipe feel about the same temperature. Allow stabilization time between each addition.

16 If you have an accurate thermometer, you can place it in the center air conditioning duct inside the vehicle and keep track of the "conditioned" air temperature at the dash. A charged system that is working properly should put out air that is 40 degrees F. If the ambient (outside) air temperature is very high, say 110 degrees F, the duct air temperature may be as high as 60 degrees F, but generally the air conditioning is 30-50 degrees F cooler than the ambient air.

17 When the can is empty, turn the valve handle to the closed position and release the connection from the low-side port. Replace the dust cap.

18 Remove the charging kit from the can and store the kit for future use with the piercing valve in the UP position, to prevent inadvertently piercing the can on the next use.

Heating system

19 If the carpet under the heater core is damp, or if antifreeze vapor or steam is coming through the vents, the heater core is leaking. Remove it (see Section 11) and install a new unit (most radiator shops will not repair a leaking heater core).

20 If the air coming out of the heater vents isn't hot, the problem could stem from any of the following causes:

a) *The thermostat is stuck open, preventing the engine coolant from warming up enough to carry heat to the heater core. Replace the thermostat (see Section 3).*
b) *A heater hose is blocked, preventing the flow of coolant through the heater core. Feel both heater hoses at the firewall. They should be hot. If one of them is cold, there is an obstruction in one of the hoses or in the heater core. Detach the hoses and back flush the heater core with a water hose. If the heater core is clear but circulation is impeded, remove the two hoses and flush them out with a water hose.*
c) *If flushing fails to remove the blockage from the heater core, the core must be replaced (see Section 11).*

Eliminating air conditioning odors

21 Unpleasant odors that often develop in air conditioning systems are caused by the growth of a fungus, usually on the surface of the evaporator core. The warm, humid environment there is a perfect breeding ground for mildew to develop.

22 The evaporator core on most vehicles is difficult to access, and factory dealerships have a lengthy, expensive process for eliminating the fungus by opening up the evaporator case and using a powerful disinfectant and rinse on the core until the fungus is gone. You can service your own system at home, but it takes something much stronger than basic household germ-killers or deodorizers.

23 Aerosol disinfectants for automotive air conditioning systems are available in most auto parts stores, but remember when shopping for them that the most effective treatments are also the most expensive. The basic procedure for using these sprays is to start by running the system in the RECIRC mode for ten minutes with the blower on its highest speed. Use the highest heat mode to dry out the system and keep the compressor from engaging by disconnecting the wiring connector at the compressor (see Section 14).

24 The disinfectant can usually comes with a long spray hose. Drill a 1/8-inch hole in the evaporator case, just deep enough to go through the plastic, between the blower motor and the evaporator core, point the nozzle inside the hole and to the left towards the evaporator core, and spray according to the manufacturer's recommendations. Try to cover the whole surface of the evaporator core, by aiming the spray up, down and sideways. Follow the manufacturer's recommendations for the length of spray and waiting time between applications.

25 Once the evaporator has been cleaned, the best way to prevent the mildew from coming back again is to make sure your evaporator housing drain tube is clear.

Automatic heating and air conditioning systems

26 Some models are equipped with an optional automatic climate control system. This system has its own computer that receives inputs from various sensors in the heating and air conditioning system. This computer, like the PCM, has self diagnostic capabilities to help pinpoint problems or faults within the system. Vehicles equipped with automatic heating and air conditioning systems are very complex and considered beyond the scope of the home mechanic. Vehicles equipped with automatic heating and air conditioning systems should be taken to dealer service department or other qualified facility for repair.

14 Air conditioning compressor - removal and installation

Refer to illustrations 14.4a, 14.4b, 14.4c, 14.5 and 14.6

Removal

Warning: *The air conditioning system is under high pressure. DO NOT loosen any fittings or remove any components until after the system has been discharged. Air conditioning refrigerant should be properly dis-*

Chapter 3 Cooling, heating and air conditioning systems

14.5 Remove the refrigerant line fitting bolt from the rear of the compressor - pull the lines away and seal them to prevent entry of dirt

14.6 The front compressor mount can remain on the engine - remove the bolts and lift off the compressor

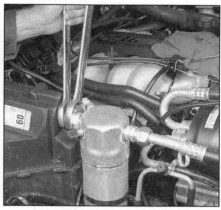

15.2 Use backup wrenches when disconnecting the fittings at the accumulator - on four-cylinder engine models, this is a pair of lines, while on V6 models the accumulator attaches directly to the evaporator

charged into an EPA-approved container at a dealership service department or an automotive air conditioning repair facility. Always wear eye protection when disconnecting air conditioning system fittings.

Note: *The accumulator (see Section 15) should be replaced whenever the compressor is replaced.*

1 Have the air conditioning system discharged (see **Warning** above). Disconnect the battery cable at the negative battery terminal.
Caution: *On models equipped with the Delco Loc II audio system, be sure the lockout feature is turned off before performing any procedure which requires disconnecting the battery.*
2 Clean the compressor thoroughly around the refrigerant line fittings.
3 Remove the serpentine drivebelt (refer to Chapter 1).
4 Disconnect the electrical connector from the air conditioning compressor **(see illustrations)**. Raise the vehicle and securely support it on stands.
5 Disconnect the suction and discharge lines from the compressor **(see illustration)**. Both lines are mounted to the back of the compressor with one bolt on 2.2L engines. On 4.3L engines, they are mounted on top of the compressor. Plug the open fittings to prevent the entry of dirt and moisture, then discard the seals between the plate and the compressor.
6 Remove the compressor-to-front-bracket bolts and nuts and pull the compressor up from the engine compartment **(see illustration)**. On four-cylinder pickup models, remove the rear compressor mount bolts.

Installation

7 If a new compressor is being installed, pour the oil from the old compressor into a graduated container and add that exact amount of new refrigerant oil to the new compressor. Also follow any directions included with the new compressor. **Note:** *Some replacement compressors come with new refrigerant oil in them. Drain the new compressor before adding the new amount of fresh oil.*

Caution: *The oil used must be labeled as compatible with R-134a refrigerant systems.*
8 Installation is basically the reverse of the disassembly. When installing the line fitting bolt to the compressor, use new seals (seals must be compatible with R-134a refrigerant) and tighten the bolt securely.
9 Reconnect the battery cable to the negative battery terminal.
10 Have the system evacuated, recharged and leak tested by a dealership service department or an automotive-air conditioning repair facility.

15 Air conditioning accumulator - removal and installation

Removal

Refer to illustration 15.2

Warning: *The air conditioning system is under high pressure. DO NOT loosen any fittings or remove any components until after the system has been discharged. Air conditioning refrigerant should be properly discharged into an EPA-approved container at a dealership service department or an automotive air conditioning repair facility. Always wear eye protection when disconnecting air conditioning system fittings.*

1 Have the air conditioning system discharged (see **Warning** above). Disconnect the battery cable at the negative battery terminal. **Caution:** *On models equipped with the Delco Loc II audio system, be sure the lockout feature is turned off before performing any procedure which requires disconnecting the battery.*
2 Disconnect the refrigerant inlet and outlet lines **(see illustration)**, using backup wrenches. Cap or plug the open lines immediately to prevent the entry of dirt or moisture.
3 Loosen the clamp bolt on the mounting bracket and slide the accumulator up and out of the compartment. **Note:** *On some Blazer and Jimmy models, the screw through the retaining bracket is installed in such a way as to require removal of the coolant recovery tank (see Section 5) to access it.*

Installation

Refer to illustration 15.5

4 If you are replacing the accumulator with a new one, add one ounce of fresh refrigerant oil to the new unit (oil must be R-134a compatible). On 1999 and later models, add two ounces to the accumulator.
5 Place the new accumulator into position in the bracket. If you reverse the mounting clamp and put the screw in from the engine side, future removal will not require removing the computer or recovery tank **(see illustration)**.
6 Install the inlet and outlet lines, using clean refrigerant oil on the new O-rings (O-rings and oil must be R-134a compatible). Tighten the mounting bolt securely.
7 Connect the cable to the negative terminal of the battery.
8 Have the system evacuated, recharged and leak tested by a dealership service department or an automotive air conditioning repair facility.

15.5 As factory installed, this mounting screw is accessible only after removing the coolant recovery tank - reverse the clamp and screw when reassembling it for easier future work

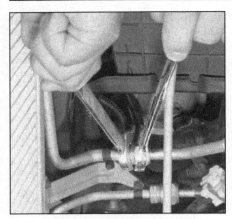

16.2 Near the lower right side of the radiator, disconnect the two lines connected to the condenser - use backup wrenches to avoid damaging the fittings

16.4 On each side of the condenser, remove the three screws (arrows) holding the plastic end panels in place

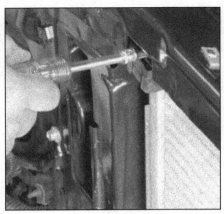

16.5a Remove the three bolts that hold the insulator brackets in place

16 Air conditioning condenser - removal and installation

Warning 1: *The covered models are equipped with airbags. Impact sensors for the airbag system are located in the dash, cowl and just in front of the radiator. The airbag(s) could accidentally deploy if these sensors are disturbed, so be extremely careful when working in this area. Airbag system components are also located in the steering wheel, steering column and base of the steering column, so be extremely careful when working in this area and don't disturb any airbag system components or wiring. You could easily be injured if an airbag accidentally deploys, and the airbag might not deploy correctly in a collision if any components or wiring in the system have been disturbed. Refer to Chapter 12 for the procedure to disable the airbag system.*

Warning 2: *The air conditioning system is under high pressure. DO NOT loosen any fittings or remove any components until after the system has been discharged. Air conditioning refrigerant should be properly discharged into an EPA-approved container at a dealership service department or an automotive air conditioning repair facility. Always wear eye protection when disconnecting air conditioning system fittings.*

Removal

Refer to illustrations 16.2, 16.4, 16.5a and 16.5b

1 Have the air conditioning system discharged (see **Warning** above). Disconnect the battery cable at the negative battery terminal. **Caution:** *On models equipped with the Delco Loc II audio system, be sure the lockout feature is turned off before performing any procedure which requires disconnecting the battery.*

2 Disconnect the refrigerant line fittings from the right side of the condenser and cap the open fittings to prevent the entry of dirt and moisture. One bolt secures the condenser-to-accumulator line to the top of the condenser, while the compressor-to-condenser line is attached with fittings to the bottom of the condenser **(see illustration)**.

3 Remove the upper radiator shroud and radiator (see Section 7).

4 On each side of the condenser opening, remove the screws holding the vertical plastic panels in place **(see illustration)**.

5 Remove the screws holding the upper insulator brackets, remove the brackets and insulators, and pull the condenser from the vehicle **(see illustrations)**. **Caution:** *The condenser is made of aluminum - be careful not to damage it during removal.*

Installation

6 Installation is the reverse of removal. Be sure to use new, R-134a-compatible O-rings on the refrigerant line fittings (lubricate the O-rings with clean refrigerant oil, compatible with R-134a refrigerant). If a new condenser is installed, add 1 oz of new refrigerant oil to the system.

7 Have the system evacuated, recharged and leak tested by a dealership service department or an automotive air conditioning repair facility.

17 Air conditioning evaporator - removal and installation

Warning 1: *The covered models are equipped with airbags. Impact sensors for the airbag system are located in the dash, cowl and just in front of the radiator. The airbag(s) could accidentally deploy if these sensors are disturbed, so be extremely careful when working in this area. Airbag system components are also located in the steering wheel, steering column and base of the steering column, so be extremely careful when working in this area and don't disturb any airbag system components or wiring. You could easily be injured if an airbag accidentally deploys, and the airbag might not deploy correctly in a collision if any components or wiring in the system have*

16.5b Pull the condenser up and out - be careful not to damage the fins

been disturbed. Refer to Chapter 12 for the procedure to disable the airbag system.

Warning 2: *The air conditioning system is under high pressure. DO NOT loosen any fittings or remove any components until after the system has been discharged. Air conditioning refrigerant should be properly discharged into an EPA-approved container at a dealership service department or an automotive air conditioning repair facility. Always wear eye protection when disconnecting air conditioning system fittings.*

Removal

Refer to illustrations 17.4 and 17.12

1 Have the air conditioning system discharged (see **Warning** above). Disconnect the battery cable at the negative battery terminal. **Note:** *This procedure is easier if the hood is removed. On 1999 and later models, also remove the front bumper and right front fender (see Chapter 11).* **Caution:** *On models equipped with the Delco Loc II audio system, be sure the lockout feature is turned off before performing any procedure which requires disconnecting the battery.*

Chapter 3 Cooling, heating and air conditioning systems

17.4 Pry this wiring bundle clip from the stud in the upper left corner of the housing and remove the electrical connector (arrow), then remove the nut from the stud - there are two more nuts/studs near the bottom of the housing

17.12 Performance of the condenser can sometimes be improved by straightening bent fins with a "fin comb" - the tool has a head for various fins-per-inch spacings

2   Drain the cooling system (see Chapter 1).
3   Disconnect the air conditioning lines from the blower/evaporator housing at the passenger's-side of the firewall. On the four-cylinder engine models, the accumulator is attached with a hose, while V6 models have the accumulator attached directly to the evaporator in front of the housing (see Section 15).
4   Remove the wiring connector at the blower motor resistor (see Section 9), then pull off the large wiring bundle clip from the mounting stud at the upper left corner of the blower/evaporator housing **(see illustration)**.
5   Remove the computer (Blazer and Jimmy models) and coolant recovery tank (see Section 5).
6   Unbolt and move the relay mounting bracket out of the way (see Chapter 12).
7   Remove the two housing mounting nuts from the studs, one at the lower left corner of the housing, one below the housing, near the split line that denotes where the evaporator is in the housing.
8   Remove the remaining bolts holding the housing to the firewall. **Note:** *The bolt in the upper right is well-hidden by the fender on Blazer and Jimmy models. It will require several extensions and a flex-joint/socket to reach this one. Have a magnetic retrieval tool handy before doing this - chances are the bolt will fall out where you can't reach it by hand.*
9   With all the fasteners removed that hold the housing to the firewall, pull the housing forward to break the seal of the gasket, then pull the housing forward and out.
10   Remove the clips and screws and separate the upper and lower halves of the housing.
11   Remove the retaining clip and pull the condenser out of the housing.
12   Check the core over carefully for signs of leaks. The efficiency can be improved slightly if any bent fins are straightened with a "fin comb" **(see illustration)**.
13   If a new evaporator core is to be installed, save all of the sealing gaskets from the original unit and transfer them.

Installation

14   Installation is the reverse of the removal procedure. Be sure to re-fill the cooling system (see Chapter 1). Use new R-134a-compatible O-rings on all connections, lubricated by R-134a-compatible refrigerant oil.
15   When reinstalling the blower/evaporator housing to the firewall, make sure the foam gasket is in place all around the housing-to-firewall juncture.
16   If a new evaporator has been installed, add 1 ounce of refrigerant oil. Have the system evacuated, recharged and leak tested by a dealership service department or an automotive air conditioning repair facility.

Notes

Chapter 4
Fuel and exhaust systems

Contents

	Section
Accelerator cable - removal and installation	10
Air cleaner assembly - removal and installation	9
Air filter replacement	See Chapter 1
Central Multiport Fuel Injection (CMFI) and Central Sequential Electronic Fuel Injection (CSEFI) - component check and replacement	15
Exhaust system check	See Chapter 1
Exhaust system servicing - general information	16
Fuel filter replacement	See Chapter 1
Fuel injection system - check	12
Fuel injection system - general information	11
Fuel level sending unit - check and replacement	8

	Section
Fuel lines and fittings - repair and replacement	4
Fuel pressure relief procedure	2
Fuel pump - removal and installation	7
Fuel pump/fuel system pressure - check	3
Fuel tank - removal and installation	5
Fuel tank cleaning and repair - general information	6
General information	1
Model 220 Throttle Body Injection (TBI) - removal, overhaul and installation	13
Multiport Fuel Injection (MFI) - component check and replacement	14
SERVICE ENGINE SOON light	See Chapter 6

Specifications

General

Fuel pressure
  Throttle Body Injection (TBI) system
    Key On, engine not running .... Approximately 9 to 13 psi
    Engine running (at idle) .... Pressure should decrease by 3 to 10 psi
  Multiport Fuel Injection (MFI) system
    Key On, engine not running .... Approximately 41 to 47 psi
    Engine running (at idle) .... Pressure should decrease by 3 to 10 psi
  Central Multiport Fuel Injection (CMFI) system
    Key On, engine not running .... Approximately 58 to 64 psi
    Engine running (at idle) .... Pressure should decrease by 3 to 10 psi
  Central Sequential Electronic Fuel Injection (CSEFI) system
    VIN X
      Key On, engine not running .... Approximately 55 to 61 psi
      Engine running (at idle) .... Pressure should decrease by 3 to 10 psi
    VIN W
      Key On, engine not running .... 60 to 66 psi
      Engine running (at idle) .... Pressure should decrease by 3 to 10 psi
Injector resistance
  TBI system .... 1.16 to 1.36 ohms
  MFI system .... 11.6 to 12.4 ohms
  CMFI system .... 1.37 to 1.77 ohms
  CSEFI system .... 11 to 14 ohms

Torque specifications

Ft-lbs (unless otherwise indicated)

Exhaust pipe-to-manifold nuts .... 15 to 22
Air intake plenum mounting bolts/nuts
  2.2L engines (through 1998) 22
  4.3L engines with CMPFI .... 124 in-lbs
  4.3L engines with CSEFI
    Step 1 .... 44 in-lbs
    Step 2 .... 80 in-lbs
Fuel rail bracket bolts (2.2L engine) .... 31 in-lbs
Throttle body bolts/nuts
  2.2L engines .... 89 in-lbs
  4.3L engines .... 144 in-lbs
IAC valve screws
  2.2L four-cylinder MFI engine .... 27 in-lbs
  4.3L V6 TBI engine .... 156 in-lbs
  4.3L V6 CMFI engine .... 27 in-lbs
  4.3L V6 CSEFI engine .... 28 in-lbs

Chapter 4 Fuel and exhaust systems

1 General information

These models are equipped with either the Throttle Body Injection (TBI) system (4.3L V6 engine, OBD I), Multiport Fuel Injection (MFI) system (2.2L four-cylinder engine), Central Multiport Fuel Injection (CMFI) system (4.3L V6 engine, OBD II) or the Central Sequential Electronic Fuel Injection (CSEFI) system (1996 and later 4.3L V6 engine, OBD II).

The fuel system consists of a fuel tank, an electric fuel pump and fuel pump relay, an air cleaner assembly and a Sequential Electronic Fuel Injection (SEFI) system.

Throttle Body Injection (TBI) system

Some 4.3L V6 engines are equipped with a Throttle Body Injection (TBI) system. The throttle body system utilizes two injectors, centrally mounted in a carburetor-like housing. The injector is an electrical solenoid; fuel is delivered to the injector at a constant pressure level. To maintain the fuel pressure at a constant level, fuel is supplied at a greater pressure than required and controlled by a fuel pressure regulator, which returns excess fuel to the fuel tank.

A signal from the PCM opens the solenoid, allowing fuel to spray through the injector into the throttle body. The amount of time the injector is held open by the PCM determines the fuel/air mixture ratio.

Multiport Fuel Injection (MFI) system

2.2L four-cylinder engines are equipped with the Multiport Fuel Injection (MFI) system. This system utilizes injectors of a different type than the throttle body injection system. Instead of an injector mounted in a centrally located throttle body, one injector is installed above each intake port. The throttle body on the MFI system serves only to control the amount of air passing into the system. Because each cylinder is equipped with an injector mounted immediately adjacent to the intake valve, much better control of the fuel/air mixture ratio is possible.

Central Multiport Fuel Injection (CMFI) and Central Sequential Electronic Fuel Injection (CSEFI) systems

Some 4.3L engines are equipped with the Central Multiport Fuel Injection system or another similar system called the Central Sequential Electronic Fuel Injection (CSEFI) system (1996 and later models). The CMFI system incorporates a centrally mounted injector assembly housed in the air intake plenum which is mounted to the intake manifold. The CMFI has a low gain fuel pressure regulator that maintains fuel pressure at the injector. When the PCM energizes the injector solenoid, pressurized fuel flows through the fuel tubes into each poppet nozzle. The injector opens and closes for each cylinder firing signal.

1996 and later models are equipped with the CSEFI fuel injection system. This system uses a centrally mounted fuel meter body that incorporates six separate injectors that sequentially distribute the fuel to each cylinder. The fuel pressure regulator is mounted directly to the fuel meter body. The PCM pulses the injectors once per crankshaft revolution. Fuel injection provides optimum mixture ratios at all stages of combustion. Combined with its immediate response characteristics, sequential fuel injection permits the engine to run on the leanest possible air/fuel mixture, which greatly reduces exhaust gas emissions.

Both systems use a Powertrain Control Module (PCM), which automatically adjusts the air/fuel mixture in accordance with engine load and performance.

Fuel pump and lines

Fuel is circulated from the fuel tank to the fuel injection system, and back to the fuel tank, through a pair of lines running along the underside of the vehicle. An electric fuel pump is attached to the fuel sending unit inside the fuel tank. A return system routes all vapors and excess fuel back to the fuel tank through separate return lines.

Exhaust system

The exhaust system includes an exhaust manifold fitted with an exhaust oxygen sensor, a catalytic converter, an exhaust pipe, and a muffler. The catalytic converter is an emission control device added to the exhaust system to reduce pollutants. A single-bed converter is used in combination with a three-way (reduction) catalyst. Refer to Chapter 6 for more information regarding the catalytic converter.

2 Fuel pressure relief procedure

Warning: *Gasoline is extremely flammable, so take extra precautions when you work on any part of the fuel system. Don't smoke or allow open flames or bare light bulbs near the work area, and don't work in a garage where a gas-type appliance (such as a water heater or a clothes dryer) is present. Since gasoline is carcinogenic, wear latex gloves when there's a possibility of being exposed to fuel, and, if you spill any fuel on your skin, rinse it off immediately with soap and water. Mop up any spills immediately and do not store fuel-soaked rags where they could ignite. The fuel system is under constant pressure, so, if any fuel lines are to be disconnected, the fuel pressure in the system must be relieved first. When you perform any kind of work on the fuel system, wear safety glasses and have a Class B type fire extinguisher on hand.*
Note: *After the fuel pressure has been relieved, it's a good idea to lay a shop towel over any fuel connection to be disassembled, to absorb the residual fuel that may leak out when servicing the fuel system.*

1 Before servicing any fuel system component, you must relieve the fuel pressure to minimize the risk of fire or personal injury.
2 Remove the fuel filler cap - this will relieve any pressure built up in the tank.

TBI systems

Note: *Although the Model 220 TBI units have an automatic internal bleed system, follow the procedure described below to relieve the fuel pressure on all TBI engines.*

3 TBI systems have an internal bleed feature that allows the fuel system to automatically depressurize while not running (ignition key OFF). It is a good idea to make sure all the fuel pressure has been bled out of the fuel lines before working on the fuel system. Position a metal pan under the fuel filter. Carefully place shop rags or towels near the inlet fuel line at the fuel filter (see Chapter 1) and then disconnect the fuel line to allow any residual fuel pressure to dissipate.

MFI systems

4 It's a good idea to ensure all the fuel pressure has been bled out of the fuel lines before working on the fuel system. Position a drain pan under the fuel filter. Place wrenches on the fuel filter and fuel line outlet fitting. Cover the wrenches and fitting with shop rags or towels (see Chapter 1), then slowly loosen the fuel line to allow any residual fuel pressure to dissipate. Dispose of the rags in a covered, marked container.
5 Unless this procedure is followed before servicing fuel lines or connections, fuel spray (and possible injury) may occur.

CMFI and CSEFI systems

Refer to illustration 2.6

6 Use one of the two following methods:
a) Attach a fuel pressure gauge to the Schrader valve on the fuel rail **(see illustration)**. *Place the gauge bleeder hose in an approved fuel container. Open the valve on the gauge to relieve pressure, then disconnect the cable from the negative terminal of the battery.*
b) Locate the fuel pressure test port and carefully place several shop towels around the test port. Remove the cap and, using the tip of a screwdriver, depress the Schrader valve and let the fuel drain into the shop towels. Be careful to catch any fuel that might spray upward by using another shop towel.

7 Unless this procedure is followed before servicing fuel lines or connections, fuel spray (and possible injury) may occur.

2.6 Attach a fuel pressure gauge to the test port and open the valve to drain the excess fuel out of the drain tube and into an approved fuel container

Chapter 4 Fuel and exhaust systems

3.2a Remove the fuel filter and install a T fitting with the fuel pressure gauge attached (2.2L MFI engine shown)

3.2b Attach a fuel pressure gauge to the test port on the fuel rail (arrow), turn the ignition key ON (engine not running) and check the fuel pressure (4.3L CMFI engine shown)

3.3 The fuel pump can be activated by attaching a jumper wire to the battery positive post (+) and touch the jumper to the fuel pump test terminal (red wire) in the driver's side corner of the engine compartment

3 Fuel pump/fuel system pressure - check

Warning: *Gasoline is extremely flammable, so take extra precautions when you work on any part of the fuel system. Don't smoke or allow open flames or bare light bulbs near the work area, and don't work in a garage where a gas-type appliance (such as a water heater or a clothes dryer) is present. Since gasoline i1s carcinogenic, wear latex gloves when there's a possibility of being exposed to fuel, and, if you spill any fuel on your skin, rinse it off immediately with soap and water. Mop up any spills immediately and do not store fuel-soaked rags where they could ignite. The fuel system is under constant pressure, so, if any fuel lines are to be disconnected, the fuel pressure in the system must be relieved first (see Section 2). When you perform any kind of work on the fuel system, wear safety glasses and have a Class B type fire extinguisher on hand.*

Note 1: *The On Board Diagnostic (OBD) system may set a trouble code 44 or 45. Refer to the code extracting procedure in Chapter 6 for additional fuel system diagnostic information.*

Note 2: *The following checks assume the fuel filter is in good condition. If you doubt its condition, install a new one (see Chapter 1).*

1 Check that there is adequate fuel in the fuel tank. Relieve the fuel pressure (see Section 2).

Fuel pump output and pressure check

Refer to illustrations 3.2a, 3.2b, 3.3, 3.5 and 3.6

2 Install a fuel pressure gauge on the fuel rail. Each type of system has a different location on the fuel line for the fuel gauge test connection (see illustrations):

TBI systems - Between the fuel inlet line and throttle body

MFI system - Between the fuel filter inlet and outlet lines

3.5 Connect a vacuum pump to the fuel pressure regulator, apply vacuum to the fuel pressure regulator and check the fuel pressure - the fuel pressure should decrease as the vacuum increases

CMFI system - Schrader valve on the fuel inlet connector

CSEFI system - Schrader valve on the fuel inlet connector

3 Turn the ignition switch ON (engine not running) with the air conditioning turned off. The fuel pump should run for about two seconds - note the reading on the gauge. After the pump stops running the pressure should hold steady. It should be within the range listed in this Chapter's Specifications. **Note:** *If there is no response from the fuel pump, use a jumper wire attached to the positive terminal of the battery and apply battery voltage to the test terminal (red wire) located on the driver's side of the engine compartment* **(see illustration)**.

4 Start the engine and let it idle at normal operating temperature. The pressure should be lower by 3 to 10 psi. If all the pressure readings are within the limits listed in this Chapter's Specifications, the system is operating properly.

3.6 Checking vacuum to the fuel pressure regulator on a 2.2L engine

5 If the pressure did not drop by 3 to 10 psi after starting the engine, apply 12 to 14 inches of vacuum to the pressure regulator **(see illustration)**. If the pressure drops, repair the vacuum source to the regulator. If the pressure does not drop, replace the regulator. **Note:** *This test will work only on 2.2L MFI engines where the fuel pressure regulator is easily accessible. On the other types of fuel systems, have the fuel pressure regulator diagnosed by a dealer service department or other qualified repair shop.*

6 If the fuel pressure is not within specifications, check the following: **Note:** *The next two tests will work only on 2.2L MFI engines where the fuel pressure regulator is easily accessible. On the other types of fuel systems, have the fuel pressure regulator diagnosed by a dealer service department.*

  a) If the pressure is higher than specified, check for vacuum to the fuel pressure regulator **(see illustration)**. *Vacuum must fluctuate with the increase or decrease in the engine rpm. If vacuum is present, check for a pinched or clogged fuel return hose or pipe. If the return line is OK, replace the regulator.*

3.12 Checking for battery voltage on the fuel pump relay connector

b) *If the pressure is lower than specified, change the fuel filter to rule out the possibility of a clogged filter. If the pressure is still low, install a fuel line shut-off adapter (J-37287) between the pressure regulator and the return line. With the valve open, start the engine (if possible) and slowly close the valve. If the pressure rises above 47 psi, replace the regulator (see Section 13).* **Warning:** *Don't allow the fuel pressure to exceed 60 psi. Also, don't attempt to restrict the return line by pinching it, as the nylon fuel line will be damaged.*

c) *If the pressure is still low with the fuel return line restricted, an injector (or injectors) may be leaking (see Section 13) or the fuel pump may be faulty.*

7 After the testing is done, relieve the fuel pressure (see Section 2) and remove the fuel pressure gauge.

8 If there are no problems with any of the above-listed components, check the fuel pump electrical circuits (see below).

Fuel pump electrical circuit check

Refer to illustration 3.12

Note: *Refer to Chapter 12 for additional wiring schematics that detail the fuel pump relay and circuit.*

9 If you suspect a problem with the fuel pump, verify the pump actually runs. Have an assistant turn the ignition switch to ON - you should hear a brief whirring noise as the pump comes on and pressurizes the system. Have the assistant start the engine. This time you should hear a constant whirring sound from the pump (but it's more difficult to hear with the engine running).

10 If the pump does not come on (makes no sound), proceed to the next Step.

11 Check the fuel pump fuse. **Note:** *These models are equipped with an ECM BAT fuse in the fuse center which governs the fuel pump relay, the fuel pump, the injectors, the ignition system and the PCM. If the fuse is blown, replace the fuse and see if the pump works. If the pump still does not work, go to the next Step.*

12 Check the fuel pump relay circuit. If the pump does not run, check for an open circuit between the relay and the fuel pump. With the ignition key ON (engine not running), check for battery voltage at the relay connector **(see illustration)**. The fuel pump relay on most models is located in the passenger side glove box inside the relay control box. **Note 1:** *On 1994 pick-ups, the fuel pump relay is located under the instrument panel, forward of the center console. On later models, it's located in the underhood fuse/relay box.* **Note 2:** *If oil pressure drops below the specified pressure level, the oil pressure switch will act as a fuel pressure cut-off device. Be sure to check the oil pressure switch and circuit in the event of a difficult problem diagnosing the fuel pump circuit.*

13 If battery voltage exists, replace the relay with a known good relay and retest. If necessary, have the relay checked by a qualified automotive parts store.

14 If the fuel pump does not activate, check for power to the fuel pump at the fuel tank. Access to the fuel pump is difficult but it is possible to check for battery voltage at the electrical connector near the tank.

4 Fuel lines and fittings - repair and replacement

Refer to illustrations 4.10, 4.12 and 4.13

Warning: *Gasoline is extremely flammable, so take extra precautions when you work on any part of the fuel system. Don't smoke or allow open flames or bare light bulbs near the work area, and don't work in a garage where a gas-type appliance (such as a water heater or a clothes dryer) is present. Since gasoline is carcinogenic, wear latex gloves when there's a possibility of being exposed to fuel, and, if you spill any fuel on your skin, rinse it off immediately with soap and water. Mop up any spills immediately and do not store fuel-soaked rags where they could ignite. The fuel system is under constant pressure, so, if any fuel lines are to be disconnected, the fuel pressure in the system must be relieved first (see Section 2). When you perform any kind of work on the fuel system, wear safety glasses and have a Class B type fire extinguisher on hand.*

1 Always relieve the fuel pressure before servicing fuel lines or fittings on fuel-injected vehicles (see Section 2).

2 The fuel feed, return and vapor lines extend from the fuel tank to the engine compartment. The lines are secured to the underbody with clip and screw assemblies. These lines must be occasionally inspected for leaks, kinks and dents.

3 If evidence of dirt is found in the system or fuel filter during disassembly, the line should be disconnected and blown out. Check the fuel strainer on the fuel level sending unit (see Section 8) for damage and deterioration.

Steel and nylon tubing

4 Because fuel lines used on fuel-injected vehicles are under high pressure, they require special consideration.

5 If replacement of a metal fuel line or emission line is called for, use welded steel tubing meeting GM specification 124-M or its equivalent. Don't use copper or aluminum tubing to replace steel tubing. These materials cannot withstand normal vehicle vibration.

6 If is becomes necessary to replace a section of nylon fuel line, replace it only with the correct part number - don't use any substitutes.

7 Most fuel lines have threaded fittings with O-rings. Any time the fittings are loosened to service or replace components:

a) *Use a backup wrench while loosening and tightening the fittings.*
b) *Check all O-rings for cuts, cracks and deterioration. Replace any that appear worn or damaged.*
c) *If the lines are replaced, always use original equipment parts, or parts that meet the GM standards specified in this Section.*

Rubber hose

Warning: *These models are equipped with electronic fuel injection, use only original equipment replacement hoses or their equivalent. Others may fail from the high pressures of this system.*

8 High-quality fuel line, specifically designed for high-pressure fuel-injection applications, must be used for fuel line replacement. Never, under any circumstances, use unreinforced vacuum line, clear plastic tubing or water hose for fuel lines. Hose(s) not clearly marked like this could fail prematurely and could fail to meet Federal emission standards. Hose inside diameter must match line outside diameter. **Warning:** *Don't substitute rubber hose for metal line on high-pressure (MFI, CMFI and CSEFI) systems. Use only genuine factory replacement lines or lines meeting factory specifications.*

9 Don't use rubber hose within four inches of any part of the exhaust system or within ten inches of the catalytic converter. Metal lines and rubber hoses must never be allowed to chafe against the frame. A minimum of 1/4-inch clearance must be maintained around a line or hose to prevent contact with the frame.

Removal and installation

Note: *The following procedure and accompanying illustrations are typical for vehicles covered by this manual. On quick-disconnect (non-threaded) fittings, clean off the fittings before disconnection to prevent dirt from getting in the fittings. After disconnection, clean the fittings with compressed air and apply a few drops of oil.*

10 Relieve the fuel pressure (see Section 2) and disconnect the fuel feed, return or vapor line at the fuel tank **(see illustration)**. **Note:** *Some fuel line connections may be threaded. Be sure to use a back-up wrench when separating the connections.*

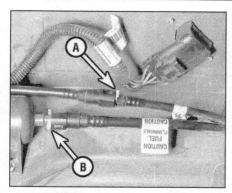

4.10 Some models are equipped with fuel lines that can be disconnected by pinching the tabs and separating each connector
- A Fuel return line
- B Fuel feed line

11 Remove all fasteners attaching the lines to the vehicle body.
12 Detach the fitting(s) that attach the fuel hoses to the engine compartment metal lines **(see illustration)**. Twisting them back and forth will allow them to separate more easily.
13 Installation is the reverse of removal. Be sure to use new O-rings at the threaded fittings, if equipped **(see illustration)**.

5 Fuel tank - removal and installation

Refer to illustrations 5.5, 5.6, 5.7, 5.8 and 5.10

Warning: *Gasoline is extremely flammable, so take extra precautions when you work on any part of the fuel system. Don't smoke or allow open flames or bare light bulbs near the work area, and don't work in a garage where a gas-type appliance (such as a water heater or a clothes dryer) is present. Since gasoline is carcinogenic, wear latex gloves when there's a possibility of being exposed to fuel, and, if you spill any fuel on your skin, rinse it off immediately with soap and water. Mop up any spills immediately and do not store fuel-soaked rags where they could ignite. The fuel system is under constant pressure, so, if any fuel lines are to be disconnected, the fuel pressure in the system must be relieved first (see Section 2). When you perform any kind of work on the fuel system, wear safety glasses and have a Class B type fire extinguisher on hand.*

Note: *Don't begin this procedure until the fuel gauge indicates the tank is empty or nearly empty. If the tank must be removed when it isn't empty (for example, if the fuel pump malfunctions), siphon any remaining fuel from the tank prior to removal.*

1 Unless the vehicle has been driven far enough to completely empty the tank, it's a good idea to siphon the residual fuel out before removing the tank from the vehicle. **Warning:** *DO NOT start the siphoning action by mouth! Use a siphoning kit, available at most auto parts stores.* **Note:** *On pickup models, do not siphon from the filler pipe, which has screens at the tank end.*

2 Relieve the fuel system pressure (see Section 2).
3 Detach the cable from the negative terminal of the battery. **Caution:** *On models equipped with a Delco Loc II audio system, be sure the lockout feature is turned off before performing any procedure which requires disconnecting the battery.*

4.12 On quick-connect fuel lines, use a special tool and push them into the connector (arrows) to separate the fuel lines

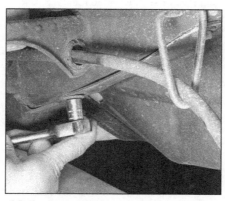

5.5 Remove the bolts (arrows) that retain the fuel tank shield to the frame

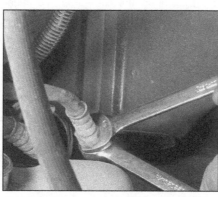

5.7 Use two wrenches to disconnect the fuel line - one to prevent the stationary fitting from turning and one to unscrew the tube fitting

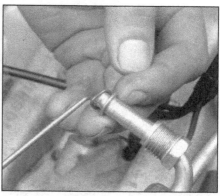

4.13 Always replace the fuel line O-rings (if equipped)

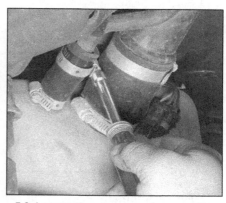

5.6 Loosen the clamps and remove the fuel inlet and vapor return hoses

4 Raise the vehicle and support it securely on jackstands placed underneath the jacking points.
5 Remove the fuel tank shield from the chassis **(see illustration)**.
6 Disconnect the fuel inlet and vapor lines from the top of the fuel tank **(see illustration)**.
7 Disconnect the fuel feed and return lines and the vapor return line from the fuel pump **(see illustration)**. **Note:** *Disconnect the fuel feed line at the fuel filter* (see Chapter 1).
8 Disconnect the vapor lines from the fuel tank rollover valve **(see illustration)**.

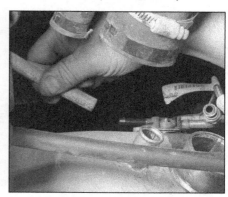

5.8 Remove the line from the fuel tank rollover valve that is located on top of the fuel pump/fuel level sending unit assembly

5.10 Support the fuel tank with a jack and remove the fuel tank straps or strap brackets

7.4 Carefully tap the lock ring counterclockwise until the locking tabs align with the slots in the fuel tank. Use a brass punch or wood dowel to prevent sparks

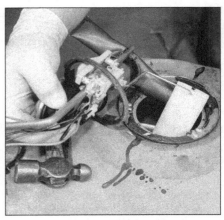

7.5 Lift the fuel pump assembly from the fuel tank and make sure the float and the sock filter are not damaged by carefully angling the assembly out of the fuel tank opening

9 Support the fuel tank with a floor jack.
10 Disconnect both fuel tank retaining straps **(see illustration)**.
11 Lower the tank enough to disconnect the wires and ground strap from the fuel pump/fuel level sending unit.
12 Remove the tank from the vehicle.
13 Installation is the reverse of removal.

6 Fuel tank cleaning and repair - general information

1 All repairs to the fuel tank or filler neck should be carried out by a professional who has experience in this critical and potentially dangerous work. Even after cleaning and flushing of the fuel system, explosive fumes can remain and ignite during repair of the tank.
2 If the fuel tank is removed from the vehicle, it should not be placed in an area where sparks or open flames could ignite the fumes coming out of the tank. Be especially careful inside garages where a gas-type appliance is located.

7 Fuel pump - removal and installation

Refer to illustrations 7.4, 7.5, 7.7, 7.8 and 7.9

Warning: *Gasoline is extremely flammable, so take extra precautions when you work on any part of the fuel system. Don't smoke or allow open flames or bare light bulbs near the work area, and don't work in a garage where a gas-type appliance (such as a water heater or a clothes dryer) is present. Since gasoline is carcinogenic, wear fuel-resistant gloves when there's a possibility of being exposed to fuel, and, if you spill any fuel on your skin, rinse it off immediately with soap and water. Mop up any spills immediately and do not store fuel-soaked rags where they could ignite. The fuel system is under constant pressure, so, if any fuel lines are to be disconnected, the fuel pressure in the system must be relieved first (see Section 2). When you perform any kind of work on the fuel system, wear safety glasses and have a Class B type fire extinguisher on hand.*

1 Relieve the fuel system pressure (see Section 2).
2 Disconnect the cable from the negative battery terminal. **Caution:** *On models equipped with a Delco Loc II audio system, be sure the lockout feature is turned off before performing any procedure which requires disconnecting the battery.*
3 Remove the fuel tank (see Section 5).
4 On some models, the fuel pump is retained by a lock ring; turn the lock ring in a counterclockwise direction until the locking tabs align with the slots in the fuel tank **(see illustration)**. On later models, the fuel pump is retained by a snap-ring; push down on the top of the pump, then remove the snap-ring with a pick or screwdriver.
5 Lift the fuel pump/sending unit assembly from the fuel tank **(see illustration)**. **Caution:** *The fuel level float and sending unit are delicate. Do not bump them against the tank during removal or the accuracy of the sending unit may be affected.* Inspect the condition of the rubber gasket around the opening of the tank. If it is dried, cracked or deteriorated, replace it.
6 On 1996 and later models, the electric fuel pump is not serviced separately. In the event of failure, the complete assembly must be replaced. Transfer the fuel pressure sensor and fuel level sending unit to the new fuel

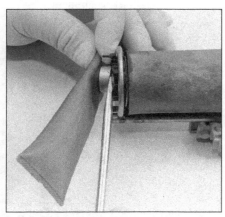

7.7 Pry on the collar to detach the strainer (filter) from the fuel pump assembly

7.8 Lift the tab to release the electrical connector from the fuel pump

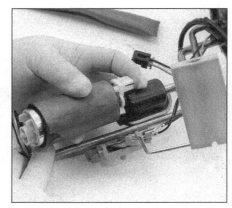

7.9 Remove the fuel pump from the bracket

Chapter 4 Fuel and exhaust systems

8.2a On 1996 and earlier models, connect an ohmmeter to the signal and ground terminals of the connector - measure the fuel level sensor resistance with the float lowered (empty), then raised (full)

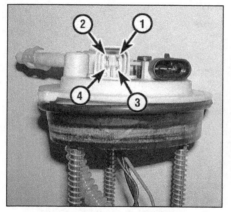

8.2b Fuel pump module connector terminal identification - 1997 and later models

1 Fuel level sending unit signal
2 Fuel pump 12-volt supply (from fuel pump relay)
3 Ground
4 Fuel level sending unit ground

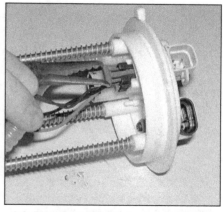

8.9 Disconnect the fuel pump/fuel level sending unit electrical connector from the fuel pump module

pump module assembly, if necessary (see Section 8). On 1995 and earlier models, replace the fuel pump as follows.
7 Remove the strainer from the lower end of the fuel pump (see illustration). If it is dirty, clean it with a suitable solvent and blow it out with compressed air. If it is too dirty to be cleaned, replace it.
8 Disconnect the electrical connector from the fuel pump (see illustration).
9 Separate the fuel pump from the assembly (see illustration).
10 Installation is the reverse of removal. Reassemble the fuel pump to the sending unit bracket assembly and insert the assembly into the fuel tank.
11 Install the fuel tank (see Section 5).

8 Fuel level sending unit - check and replacement

Warning: *Gasoline is extremely flammable, so take extra precautions when you work on any part of the fuel system. Don't smoke or allow open flames or bare light bulbs near the work area, and don't work in a garage where a gas-type appliance (such as a water heater or a clothes dryer) is present. Since gasoline is carcinogenic, wear latex gloves when there's a possibility of being exposed to fuel, and, if you spill any fuel on your skin, rinse it off immediately with soap and water. Mop up any spills immediately and do not store fuel-soaked rags where they could ignite. The fuel system is under constant pressure, so, if any fuel lines are to be disconnected, the fuel pressure in the system must be relieved first (see Section 2). When you perform any kind of work on the fuel system, wear safety glasses and have a Class B type fire extinguisher on hand.*

Check
Refer to illustrations 8.2a and 8.2b
1 Remove the fuel tank and the fuel pump

(see Sections 5 and 7).
2 Connect the probes of an ohmmeter to the fuel level sensor signal and ground terminals of the fuel pump module electrical connector (see illustrations).
3 Position the float in the down (empty) position and note the reading on the ohmmeter. The resistance should be low (approximately 2 ohms).
4 Move the float up to the full position while watching the meter. At the full position, the resistance should be higher (approximately 90 ohms).
5 If the fuel level sending unit resistance values are incorrect or if the resistance does not change smoothly as the float travels from empty to full, replace the fuel level sending unit assembly.

Replacement
Refer to illustrations 8.9, 8.10 and 8.11
6 Remove the fuel tank and the fuel pump/fuel level sending unit assembly (see Sections 5 and 7).

1995 and earlier models
7 The fuel level sending unit is part of the tank unit assembly, only the fuel pump is serviced separately. Transfer the fuel pump to the new unit (see Section 7).
8 Installation is the reverse of removal.

1996 and later models
Four-cylinder models
9 Disconnect the fuel level sending unit electrical connector from the module cover (see illustration).
10 Remove the sending unit retaining clip (see illustration).
11 Pinch the tabs together and slide the fuel level sending unit off the module (see illustration). Note the routing of the wiring for installation.
12 Installation is the reverse of removal.

Six-cylinder models
13 The fuel level sending unit is part of the fuel pump/fuel level sending unit assembly and not serviced separately. If the fuel level sending unit is defective, replace the fuel pump/fuel level sending unit assembly (see Section 7).

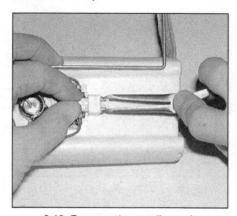

8.10 Remove the sending unit retaining clip

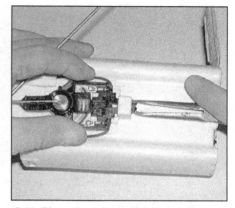

8.11 Pinch the tabs together and remove the fuel level sending unit from module

4-8 Chapter 4 Fuel and exhaust systems

9.4 Removing the air intake ducts, resonator and air cleaner assembly on a 4.3L CMFI engine

10.2 Slide the accelerator cable through the slot in the pedal arm

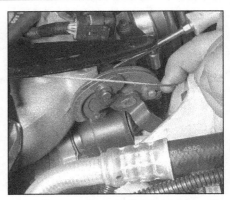

10.4 Remove the accelerator cable end through the slot in the throttle lever (4.3L CMFI engine shown)

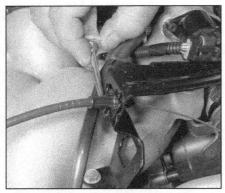

10.5 Lift the accelerator cable retaining tabs with a flat-bladed screwdriver and pull the assembly through the bracket housing

9 Air cleaner assembly - removal and installation

Refer to illustration 9.4

Note: *TBI systems are equipped with the conventional (round) style air cleaner housing mounted directly on top of the throttle body. Refer to Chapter 1 for additional illustrations.*

1 Detach the cable from the negative terminal of the battery. **Caution:** *On models equipped with a Delco Loc II audio system, be sure the lockout feature is turned off before performing any procedure which requires disconnecting the battery.*
2 Disconnect the electrical connector to the intake air temperature (IAT) sensor (see Chapter 6).
3 Remove the clamps that retain the air intake assembly to the throttle body.
4 Lift the air cleaner, air intake ducts and resonator chamber from the engine compartment as a complete assembly **(see illustration)**.
5 Installation is the reverse of removal.

10 Accelerator cable - removal and installation

Refer to illustrations 10.2, 10.4 and 10.5

Removal

1 Detach the screws and the clip retaining the lower instrument panel trim and lower the trim (if necessary) (see Chapter 11).
2 Detach the accelerator cable from the accelerator pedal **(see illustration)**.
3 Squeeze the accelerator cable cover tangs and push the cable through the firewall into the engine compartment.
4 Rotate the throttle lever and detach the accelerator cable from the throttle lever **(see illustration)**.
5 Squeeze the accelerator cable retaining tangs and push the cable through the accelerator cable bracket **(see illustration)**.

Installation

6 Installation is the reverse of removal. **Note:** *To prevent possible interference, flexible components (hoses, wires, etc.) must not be routed within two inches of moving parts, unless routing is controlled.*
7 Operate the accelerator pedal and check for any binding condition by completely opening and closing the throttle.
8 At the engine compartment side of the firewall, apply sealant around the accelerator cable.

11 Fuel injection system - general information

General Information

Refer to illustrations 11.6a, 11.6b, 11.11 and 11.12

These models are equipped with either the Throttle Body Injection (TBI) system (4.3L V6 engine, OBD I), Multiport Fuel Injection (MFI) system (2.2L four-cylinder engine, OBD I

11.6a Fuel injection component locations on the 2.2L MFI engine

1	Air intake plenum	4	Fuel pump test terminal	6	Coolant temperature sensor (in cylinder head below duct)
2	IAC valve				
3	EGR valve	5	Air intake duct		

[1994 and 1995] or OBD II [beginning in 1996]), Sequential electronic Fuel Injection (SFI) system (1998 2.2L four-cylinder engine, OBDII), Central Multiport Fuel Injection (CMFI) system (4.3L V6 engine, OBD II) or the Central Sequential Electronic Fuel Injection (CSEFI) system (4.3L V6 engine, OBD II, beginning in 1996).

The fuel system consists of a fuel tank, an electric fuel pump and fuel pump relay, an air cleaner assembly and a Sequential Electronic Fuel Injection (SEFI) system.

Throttle Body Injection (TBI) systems

The main component of the TBI system is the Throttle Body Injection (TBI) unit, which is mounted on the intake manifold just like a carburetor. The TBI unit is made up of two major assemblies: the throttle body and the fuel metering assembly.

The throttle body contains a single throttle valve, controlled by the accelerator pedal, similar to a carburetor. Attached to the exterior of the body are the Throttle Position Sensor (TPS), which sends throttle position information to the PCM, and the Idle Air Control (IAC) assembly, which is used by the PCM to maintain a constant idle speed during normal engine operation.

The fuel metering assembly contains the fuel pressure regulator and the single fuel injector. The regulator dampens the pulsations of the fuel pump and maintains a steady pressure at the injector. The fuel injector is controlled by the PCM through an electrically operated solenoid. The amount of fuel injected into the intake manifold is varied by the length of time the injector plunger is held open.

Multiport Fuel Injection (MFI)

Multiport Fuel Injection (MFI) consists of an air intake manifold, the throttle body, the injectors, the fuel rail assembly, an electric fuel pump and associated plumbing (see illustrations).

Air is drawn through the air cleaner and throttle body. A Manifold Absolute Pressure (MAP) sensor informs the PCM of temperature and pressure variations.

While the engine is running, the fuel constantly circulates through the fuel rail, which removes vapors and keeps the fuel cool while maintaining sufficient pressure to the injectors under all running conditions.

As with TBI, the operation of the injection system is controlled by the PCM so that it works in conjunction with the rest of the vehicle functions to provide optimum driveability and emissions control.

Because the MFI/SFI system meters fuel and air precisely, it is important to the proper operation of the vehicle that the fuel and air filters be changed at the specified intervals.

Central Multiport Fuel Injection (CMFI) and Central Sequential Electronic Fuel Injection (CSEFI) systems

Some 4.3L V6 engines are equipped with the Central Multiport Fuel Injection system or another similar system called the Central Sequential Electronic Fuel Injection (CSEFI) system. The CMFI system incorporates a centrally mounted injector assembly housed in the air intake plenum which is mounted to the intake manifold (see illustration). The CMFI

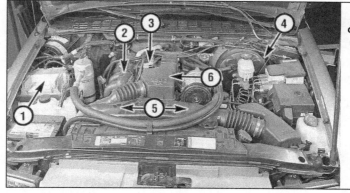

11.6b Fuel injection component locations on the 1998 2.2L SFI engine

1. Powertrain Control Module
2. IAC valve
3. EGR valve
4. Fuel pump test terminal
5. Air intake ducts
6. Air cleaner outlet resonator

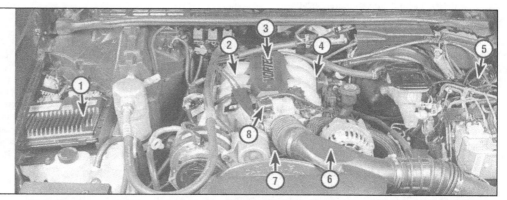

11.11 Fuel injection component locations on the 4.3L CMFI engine

1. Powertrain Control Module
2. Air Intake plenum
3. Plenum cover
4. Fuel pressure gauge test port
5. Fuel pump test terminal
6. Air intake duct
7. Coolant temperature sensor (in air intake plenum below duct)
8. IAC valve

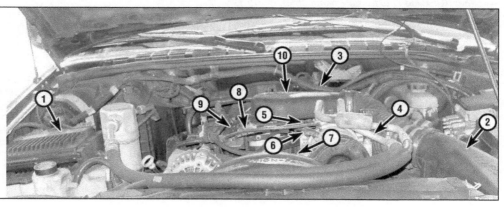

11.12 Fuel injection component location on the 4.3L CSEFI engine

1. PCM
2. Air intake
3. Fuel pressure gauge test port
4. Coolant temperature sensor
5. Throttle position sensor
6. Idle air control motor
7. Exhaust gas recirculation valve
8. Manifold Absolute pressure sensor
9. Purge solenoid
10. Camshaft position sensor (inside distributor housing)

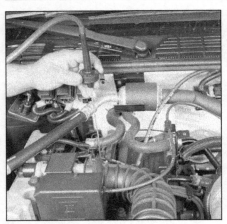

12.11 Use a stethoscope or screwdriver to determine if the injectors are working properly - they should make a steady clicking sound that rises and falls with engine speed changes (2.2L MFI system shown)

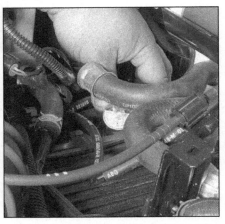

12.13a Install the "noid" light into each injector electrical connector and confirm that it blinks when the engine is cranking or running (2.2L MFI engine shown)

12.13b Disconnect the injector harness connector near the oil filler tube and install the "noid" light into the harness on the computer side (4.3L CMFI engine shown)

has a low gain fuel pressure regulator that maintains fuel pressure at the injector. When the PCM energizes the injector solenoid, pressurized fuel flows through the fuel tubes into each poppet nozzle.

1996 and later models are equipped with the CSEFI fuel injection system. This system uses a centrally mounted fuel meter body that incorporates six separate injectors that sequentially distribute the fuel to each cylinder. The fuel pressure regulator is mounted directly on the fuel meter body.

Both systems use a Powertrain Control Module (PCM), which automatically adjusts the air/fuel mixture in accordance with engine load and performance.

The PCM has a learning capability for certain performance conditions. If the battery is disconnected, part of the PCM memory is erased, which makes it necessary to "re-learn" the computer. This is done by thoroughly warming up the engine and operating the vehicle at part throttle, stop and go and idle conditions.

12 Fuel injection system - check

Warning: *Gasoline is extremely flammable, so take extra precautions when you work on any part of the fuel system. Don't smoke or allow open flames or bare light bulbs near the work area, and don't work in a garage where a gas-type appliance (such as a water heater or a clothes dryer) is present. Since gasoline is carcinogenic, wear latex gloves when there's a possibility of being exposed to fuel, and, if you spill any fuel on your skin, rinse it off immediately with soap and water. Mop up any spills immediately and do not store fuel-soaked rags where they could ignite. The fuel system is under constant pressure, so, if any fuel lines are to be disconnected, the fuel pressure in the system must be relieved first* (see Section 2). *When you perform any kind of work on the fuel system, wear safety glasses and have a Class B type fire extinguisher on hand.*

Note: *The following procedure is based on the assumption that the fuel pump is working and the fuel pressure is adequate* (see Section 3).

Preliminary checks

1 Check all electrical connectors that are related to the system. Loose electrical connectors and poor grounds can cause many problems that resemble more serious malfunctions.
2 Check to see that the battery is fully charged, as the control unit and sensors depend on an accurate supply voltage in order to properly meter the fuel.
3 Check the air filter element - a dirty or partially blocked filter will severely impede performance and economy (see Chapter 1).
4 If a blown fuse is found, replace it and see if it blows again. If it does, search for a grounded wire in the harness to the fuel pump.

System checks

Refer to illustrations 12.11, 12.13a and 12.13b

5 Check the ground wire connections on the intake manifold for tightness. Check all electrical connectors that are related to the system. Loose connectors and poor grounds can cause many problems that resemble more serious malfunctions.
6 Check to see that the battery is fully charged, as the control unit and sensors depend on an accurate supply voltage in order to properly meter the fuel.
7 Check the air filter element - a dirty or partially blocked filter will severely impede performance and economy (see Chapter 1).
8 If a blown fuse is found, replace it and see if it blows again. If it does, search for a grounded wire in the harness to the fuel pump.
9 Check the air intake duct to the intake manifold for leaks, which will result in an excessively lean mixture. Also check the condition of all vacuum hoses connected to the intake manifold.
10 Remove the air intake duct from the throttle body and check for dirt, carbon or other residue build-up. If it's dirty, clean it with carburetor cleaner and a toothbrush.
11 With the engine running, place an automotive stethoscope against each injector, one at a time, and listen for a clicking sound, indicating operation **(see illustration)**. If you don't have a stethoscope, place the tip of a screwdriver against the injector and listen through the handle. **Note:** *The injector(s) on the 4.3L CMFI and CSEFI are located under the air intake plenum, making this test impossible. Therefore, it will be necessary to check for voltage to the injector(s)* (see Step 13) *and fuel pressure* (see Section 3) *to draw the necessary conclusions for proper fuel system diagnosis. Further fuel system testing can be confirmed by disassembling the air intake plenum and checking the Central Port Injection unit* (see Section 15).
12 Unplug the injector electrical connector(s) and test the resistance of each injector. Compare the values to the Specifications listed in this Chapter. **Note:** *On CMFI systems, disconnect the harness connector by the oil fill tube and measure the resistance on the injector side. On CSEFI systems, it will be difficult to check resistance, therefore it will be necessary to take the vehicle to a dealer service department for diagnosis.*
13 Install an injector test light ("noid" light) into each injector electrical connector, one at a time **(see illustrations)**. Crank the engine over. Confirm that the light flashes evenly on each connector. This will test for voltage to the injectors.
14 The remainder of the system checks can be found in the following Sections.

Chapter 4 Fuel and exhaust systems

13.6 Carefully peel away the old fuel meter outlet passage gasket and fuel meter cover gasket with a razor blade

13.7 Never remove the four pressure regulator screws (arrows) from the fuel meter cover

13.14 To remove either injector electrical connector, depress the two tabs on the front and rear of each connector and lift straight up

13 Model 220 Throttle Body Injection (TBI) - removal, overhaul and installation

Warning: *Gasoline is extremely flammable, so take extra precautions when you work on any part of the fuel system. Don't smoke or allow open flames or bare light bulbs near the work area, and don't work in a garage where a gas-type appliance (such as a water heater or a clothes dryer) is present. Since gasoline is carcinogenic, wear latex gloves when there's a possibility of being exposed to fuel, and, if you spill any fuel on your skin, rinse it off immediately with soap and water. Mop up any spills immediately and do not store fuel-soaked rags where they could ignite. The fuel system is under constant pressure, so, if any fuel lines are to be disconnected, the fuel pressure in the system must be relieved first (see Section 2). When you perform any kind of work on the fuel system, wear safety glasses and have a Class B type fire extinguisher on hand.*

Note: *Because of its relative simplicity, the throttle body assembly does not need to be removed from the intake manifold or disassembled for component replacement. However, for the sake of clarity, the following procedures are shown with the TBI assembly removed from the vehicle.*

1 Relieve system fuel pressure (see Section 2).
2 Detach the cable from the negative terminal of the battery.
3 Remove the air cleaner housing assembly, adapter and gaskets.

Fuel meter cover/fuel pressure regulator assembly

Refer to illustrations 13.6 and 13.7

Note: *The fuel pressure regulator is housed in the fuel meter cover. Whether you are replacing the meter cover or the regulator itself, the entire assembly must be replaced. The regulator must not be removed from the cover.*

4 Unplug the electrical connectors to the fuel injectors.
5 Remove the long and short fuel meter cover screws and remove the fuel meter cover.
6 Remove the fuel meter outlet passage gasket, cover gasket and pressure regulator seal. Carefully remove any old gasket material that is stuck with a razor blade **(see illustration)**. **Caution:** *Do not attempt to re-use either of these gaskets.*
7 Inspect the cover for dirt, foreign material and casting warpage. If it is dirty, clean it with a clean shop rag soaked in solvent. Do not immerse the fuel meter cover in cleaning solvent - it could damage the pressure regulator diaphragm and gasket. **Warning:** *Do not remove the four screws* **(see illustration)** *securing the pressure regulator to the fuel meter cover. The regulator contains a large spring under compression which, if accidentally released, could cause injury. Disassembly might also result in a fuel leak between the diaphragm and the regulator housing. The new fuel meter cover assembly will include a new pressure regulator.*
8 Install the new pressure regulator seal, fuel meter outlet passage gasket and cover gasket.
9 Install the fuel meter cover using Loctite 262 or equivalent on the screws. **Note:** *The short screws go next to the injectors.*
10 Attach the electrical connectors to both injectors.
11 Attach the cable to the negative terminal of the battery.
12 With the engine off and the ignition on, check for leaks around the gasket and fuel line couplings.
13 Install the air cleaner, adapter and gaskets.

Fuel injector(s)

Refer to illustrations 13.14, 13.16, 13.21, 13.22, 13.23, 13.24 and 13.25

Note: *When replacing a fuel injector, be sure to use one having the identical part number. Injectors from other models are calibrated with different flow rates but can be interchanged with other Model 220 TBI units.*

13.16 To remove an injector, slip the tip of a flat-bladed screwdriver under the lip of the lug on top of the injector and, using another screwdriver as a fulcrum, carefully pry the injector up and out

Check with a dealer parts department for correct identification.

14 To unplug the electrical connectors from the fuel injectors, squeeze the plastic tabs and pull straight up **(see illustration)**.
15 Remove the fuel meter cover/pressure regulator assembly. **Note:** *Do not remove the fuel meter cover assembly gasket - leave it in place to protect the casting from damage during injector removal.*
16 Use two screwdrivers **(see illustration)** to pry out the injector(s).
17 Remove the upper (larger) and lower (smaller) O-rings and filter from the injector(s).
18 Remove the steel backup washer from the top of each injector cavity.
19 Inspect the fuel injector filters for evidence of dirt and contamination. If present, check for the presence of dirt in the fuel lines and fuel tank.
20 Be sure to replace the fuel injector with an identical part. Injectors from other models can fit in the Model 220 TBI assembly but are calibrated for different flow rates.
21 Slide the new filter into place on the

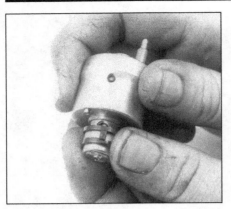

13.21 Slide the new filter onto the nozzle of the fuel injector

13.22 Lubricate the lower O-ring with transmission fluid then place it on the shoulder in the bottom of the injector cavity

13.23 Place the steel back-up washer on the shoulder near the top of the injector cavity

nozzle of the injector **(see illustration)**.
22 Lubricate the new lower (smaller) O-ring with automatic transmission fluid and place it on the small shoulder at the bottom of the fuel injector cavity in the fuel meter body **(see illustration)**.
23 Install the steel back-up washer in the injector cavity **(see illustration)**.
24 Lubricate the new upper (larger) O-ring with automatic transmission fluid and install it on top of the steel back-up washer **(see illustration)**. *Note: The backup washer and the large O-ring must be installed before the injector. If they aren't, improper seating of the large O-ring could cause fuel leakage.*
25 To install an injector, align the raised lug on the injector base with the notch in the fuel meter body cavity **(see illustration)**. Push down on the injector until it is fully seated in the fuel meter body. *Note: The electrical terminals should be parallel with the throttle shaft.*
26 Install the fuel meter cover assembly and gasket.
27 Attach the cable to the negative terminal of the battery.
28 With the engine off and the ignition on, check for fuel leaks.
29 Attach the electrical connectors to the fuel injectors.

30 Install the air cleaner housing assembly, adapter and gaskets.

Throttle Position Sensor (TPS)

31 Remove the two TPS attaching screws and retainers and remove the TPS from the throttle body.
32 If you intend to re-use the same TPS, do not attempt to clean it by soaking it in any liquid cleaner or solvent. The TPS is a delicate electrical component and can be damaged by solvents.
33 Install the TPS on the throttle body while lining up the TPS lever with the TPS drive lever.
34 Install the two TPS attaching screws and retainers.
35 Install the air cleaner housing assembly, adapter and gaskets.
36 Attach the cable to the negative terminal of the battery.

Idle Air Control (IAC) valve

Refer to illustration 13.37

37 Unplug the electrical connector from the IAC valve and remove the IAC valve **(see illustration)**.

38 Remove and discard the old IAC valve gasket. Clean any old gasket material from the surface of the throttle body assembly to insure proper sealing of the new gasket.
39 All pintles in IAC valves on Model 220 TBI units have the same dual taper. However, the pintles on some units have a 12 mm diameter and the pintles on others have a 10 mm diameter. A replacement IAC valve must have the appropriate pintle taper and diameter for proper seating of the valve in the throttle body.
40 Measure the distance between the tip of the pintle and the housing mounting surface. If dimension "A" is greater than 1-1/8 inches, it must be reduced to prevent damage to the valve.
41 To adjust the pintle of an IAC valve, grasp the valve and exert firm pressure on the pintle with the thumb. Use a slight side-to-side movement on the pintle as you press it in with your thumb.
42 Install the IAC valve and tighten it to the specified torque. Attach the electrical connector.
43 Install the air cleaner housing assembly, adapter and gaskets.
44 Attach the cable to the negative terminal of the battery.

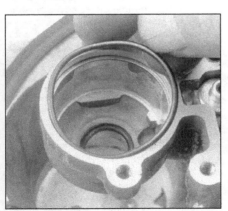

13.24 Lubricate the upper O-ring with transmission fluid then install it on top of the steel washer

13.25 Make sure that the lug is aligned with the groove in the bottom of the fuel injector cavity

13.37 The IAC valve can be removed with an adjustable wrench (shown) or a 1-1/4 inch wrench

Chapter 4 Fuel and exhaust systems

13.50 Remove the fuel inlet and outlet nuts from the fuel meter body

13.52 Once the fuel inlet and outlet nuts are off, pull the fuel meter body straight up to separate it from the throttle body

13.65 When disconnecting the fuel feed and return lines from the fuel inlet and outlet nuts, be sure to use a backup wrench to prevent damage to the lines

45 Start the engine and allow it to reach operating temperature, then turn it off. No adjustment of the IAC valve is required after installation. The IAC valve is reset by the PCM when the engine is turned off.

Fuel meter body assembly

Refer to illustrations 13.50 and 13.52

46 Unplug the electrical connectors from the fuel injectors.
47 Remove the fuel meter cover/pressure regulator assembly, fuel meter cover gasket, fuel meter outlet gasket and pressure regulator seal.
48 Remove the fuel injectors.
49 Unscrew the fuel inlet and return line threaded fittings, detach the lines and remove the O-rings.
50 Remove the fuel inlet and outlet nuts and gaskets from the fuel meter body assembly **(see illustration)**. Note the locations of the nuts to ensure proper reassembly. The inlet nut has a larger passage than the outlet nut.
51 Remove the gasket from the inner end of each fuel nut.
52 Remove the fuel meter body-to-throttle body attaching screws and remove the fuel meter body from the throttle body **(see illustration)**.
53 Install the new throttle body-to-fuel meter body gasket. Match the cut-out portions in the gasket with the openings in the throttle body.
54 Install the fuel meter body on the throttle body. Coat the fuel meter body-to-throttle body attaching screws with thread locking compound before installing them.
55 Install the fuel inlet and outlet nuts, with new gaskets, in the fuel meter body and tighten the nuts to the specified torque. Install the fuel inlet and return line threaded fittings with new O-rings. Use a backup wrench to prevent the nuts from turning.
56 Install the fuel injectors.
57 Install the fuel meter cover/pressure regulator assembly.
58 Attach the cable to the negative terminal of the battery.
59 Attach the electrical connectors to the fuel injectors.
60 With the engine off and the ignition on, check for leaks around the fuel meter body, the gasket and around the fuel line nuts and threaded fittings.
61 Install the air cleaner housing assembly, adapters and gaskets.

Throttle body assembly

Refer to illustrations 13.65 and 13.66

62 Unplug all electrical connectors - the IAC valve, TPS and fuel injectors. Detach the grommet with the wires from the throttle body.
63 Detach the throttle linkage, return spring(s), transmission control cable (automatics) and, if equipped, cruise control.
64 Clearly label, then detach, all vacuum hoses.
65 Using a backup wrench, detach the inlet and outlet fuel line nuts **(see illustration)**. Remove the fuel line O-rings from the nuts and discard them.
66 Remove the TBI mounting bolts **(see illustration)** and lift the TBI unit from the intake manifold. Remove and discard the TBI manifold gasket.
67 Place the TBI unit on a holding fixture.
Note: *If you don't have a holding fixture, and decide to place the TBI directly on a work bench surface, be extremely careful when servicing it. The throttle valve can be easily damaged.*
68 Remove the fuel meter body-to-throttle body attaching screws and separate the fuel meter body from the throttle body.
69 Remove the throttle body-to-fuel meter body gasket and discard it.
70 Remove the TPS.
71 Invert the throttle body on a flat surface for greater stability and remove the IAC valve.
72 Clean the throttle body assembly in a cold immersion cleaner. Clean the metal parts thoroughly and blow dry with compressed air. Be sure that all fuel and air passages are free of dirt or burrs. **Caution:** *Do not place the TPS, IAC valve, pressure regulator diaphragm, fuel injectors or other com-

13.66 To remove the Model 220 throttle body from the intake manifold, remove the three nuts (arrows) (third nut not visible)

ponents containing rubber in the solvent or cleaning bath. If the throttle body requires cleaning, soaking time in the cleaner should be kept to a minimum. Some models have throttle shaft dust seals that could lose their effectiveness by extended soaking.*
73 Inspect the mating surfaces for damage that could affect gasket sealing. Inspect the throttle lever and valve for dirt, binds, nicks and other damage.
74 Invert the throttle body on a flat surface for stability and install the IAC valve and the TPS.
75 Install a new throttle body-to-fuel meter body gasket and place the fuel meter body assembly on the throttle body assembly. Coat the fuel meter body-to-throttle body attaching screws with thread locking compound and tighten them securely.
76 Install the TBI unit and tighten the mounting bolts to the specified torque. Use a new TBI-to-manifold gasket.
77 Install new O-rings on the fuel line nuts. Install the fuel line and outlet nuts by hand to prevent stripping the threads. Using a backup wrench, tighten the nuts to the specified torque once they have been correctly threaded into the TBI unit.

Chapter 4 Fuel and exhaust systems

14.2 Clean the throttle body with carburetor cleaner to remove sludge deposits

14.4 Remove the four throttle body mounting bolts (arrows)

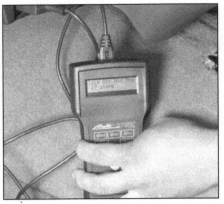

14.7 Connect a SCAN tool to the ALDL and observe the position of the IAC valve by observing the steps (counts) as the throttle is opened then closed

78 Attach the vacuum hoses, throttle linkage, return spring(s), transmission control cable (automatics) and, if equipped, cruise control cable. Attach the grommet, with wire harness, to the throttle body.
79 Plug in all electrical connectors, making sure that the connectors are fully seated and latched.
80 Check to see if the accelerator pedal is free by depressing the pedal to the floor and releasing it with the engine off.
81 Connect the negative battery cable, and, with the engine off and the ignition on, check for leaks around the fuel line nuts.
82 Check the TPS output (see Chapter 6).
83 Install the air cleaner housing assembly, adapter and gaskets.

14 Multiport Fuel Injection (MFI/SFI) - component check and replacement

Warning: *Gasoline is extremely flammable, so take extra precautions when you work on any part of the fuel system. Don't smoke or allow open flames or bare light bulbs near the work area, and don't work in a garage where a gas-type appliance (such as a water heater or a clothes dryer) is present. Since gasoline is carcinogenic, wear latex gloves when there's a possibility of being exposed to fuel, and, if you spill any fuel on your skin, rinse it off immediately with soap and water. Mop up any spills immediately and do not store fuel-soaked rags where they could ignite. The fuel system is under constant pressure, so, if any fuel lines are to be disconnected, the fuel pressure in the system must be relieved first (see Section 2). When you perform any kind of work on the fuel system, wear safety glasses and have a Class B type fire extinguisher on hand.*

Throttle body
Check
Refer to illustration 14.2
1 Detach the air intake duct from the throttle body and move the duct out of the way.
2 Have an assistant depress the throttle pedal while you watch the throttle valve. Check that the throttle valve moves smoothly when the throttle is moved from closed (idle position) to fully open (wide open throttle). **Note:** *Spray carburetor cleaner into the throttle body, especially around the shaft area* **(see illustration)** *to free-up any binding caused by the accumulation of carbon deposits or sludge buildup.*

Removal
Refer to illustration 14.4
3 On MFI systems (1997 and earlier), the throttle body is incorporated into the air intake plenum as a single unit. Follow the removal and installation procedures of the air intake plenum as described in Steps 19 through 31.
4 On SFI systems (1998 and later) the throttle body is mounted to the intake manifold. To remove the throttle body unplug the electrical connections to the (IAC) and (TPS) and disconnect the vacuum lines and throttle cable. Remove the four throttle body mounting bolts **(see illustration)**.

Idle Air Control (IAC) valve
Check
Refer to illustrations 14.7, 14.8a and 14.8b
5 The idle air control valve (IAC) controls the engine idle speed. This output actuator is mounted on the throttle body **(see illustration 14.27 for valve location)** and is controlled by voltage pulses sent from the PCM (computer). The IAC valve pintle moves in or out allowing more or less intake air into the system according to the engine conditions. To increase idle speed, the PCM retracts the IAC valve pintle away from the seat and allows more air to bypass the throttle bore. To decrease idle speed, the PCM extends the IAC valve pintle towards the seat, reducing the air flow.
6 To check the IAC valve, unplug the electrical connector and, using an ohmmeter, measure the resistance across terminals A and B, then terminals C and D. Each resistance check should indicate 40 to 80 ohms **(see illustration 15.51)**. If not, replace the

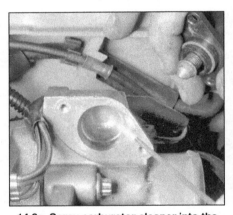

14.8a Spray carburetor cleaner into the IAC valve housing and check for clogged air passages in the air intake plenum

IAC valve.
7 There is an alternate method for testing the IAC valve. Various SCAN tools are available from auto parts stores and specialty tool companies that can be plugged into the ALDL (diagnostic connector) for the purpose of monitoring the sensors. Connect the SCAN tool and switch to the Idle Air Valve Position mode and monitor the steps (valve winding position) **(see illustration)**. The SCAN tool should indicate between 10 to 200 steps depending upon the rpm range. Allow the engine to idle for several minutes and while observing the count reading, snap the throttle to achieve high rpm (under 3,500). Repeat the procedure several times and observe the SCAN tool steps (counts) when the engine goes back to idle. The readings should be within 5 to 10 steps each time. If the readings fluctuate greatly, replace the IAC valve. **Note:** *When the IAC valve electrical connector is disconnected for testing, the PCM will have to "relearn" its idle mode. In other words, it will take a certain amount of time before the idle valve resets for the correct idle speed. Make sure the idle is smooth and not misfiring before plugging in the SCAN tool.*

Chapter 4 Fuel and exhaust systems

4-15

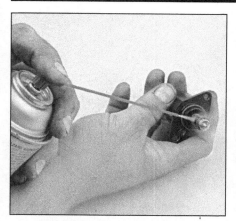

14.8b Clean the IAC valve pintle with carburetor cleaner to remove carbon deposits

14.25 Pry the retaining tang out and lift the vacuum harness assembly from the throttle body

14.27 Remove the bolts (lower arrows) from the air intake plenum (not all bolts visible in this photo) - upper arrow points to the Idle Air Control (IAC) valve

8 Next, remove the valve (see Step 10) and inspect it:
  a) Check the pintle for excessive carbon deposits. If necessary, clean it with carburetor cleaner spray **(see illustration)**. Also clean the IAC valve housing to remove any deposits **(see illustration)**.
  b) Check the IAC valve electrical connections. Make sure the pins are not bent and make good contact with the connector terminals.

Removal

9 Unplug the electrical connector from the Idle Air Control (IAC) valve.
10 Remove the two IAC valve attaching screws and withdraw the valve **(see illustration 14.27 for valve location)**.
11 Check the condition of the rubber O-ring. If it's hardened or deteriorated, replace it. On models equipped with a gasket, remove the gasket.
12 Clean the sealing surface and the bore of the idle air/vacuum signal housing assembly to ensure a good seal. **Caution:** *The IAC valve itself is an electrical component and must not be soaked in any liquid cleaner, as damage may result.*
13 Before installing the IAC valve, the position of the pintle must be checked. If the pintle is extended too far, damage to the assembly may occur.

Installation

14 Measure the distance from the flange or gasket mounting surface of the IAC valve to the tip of the pintle. If the distance is greater than 1-1/8 inch, reduce the distance by applying firm pressure onto the pintle to retract it. Try some side-to-side motion in the event the pintle binds.
15 Position the new O-ring or gasket on the IAC valve. Lubricate the O-ring with a light film of engine oil. If the IAC valve is the screw-in type, apply a light film of RTV sealant to the threads of the valve. Install the IAC valve and tighten the valve or the mounting screws securely.
16 Plug in the electrical connector at the IAC valve assembly. **Note:** *No adjustment is made to the IAC assembly after reinstallation. The IAC resetting is controlled by the PCM when the engine is started.*

Throttle Position Sensor (TPS)

Check

17 Check for stored trouble codes in the PCM using the On Board Diagnosis system (see Chapter 6).
18 To check the operation and replacement of the TPS, refer to the *Information sensors* in Chapter 6.

Air intake plenum

Removal

Refer to illustrations 14.25 and 14.27
19 Disconnect the cable from the negative terminal of the battery. **Caution:** *On models equipped with a Delco Loc II audio system, be sure the lockout feature is turned off before performing any procedure which requires disconnecting the battery.*
20 Detach the air intake duct from the throttle body.
21 Drain the coolant (see Chapter 1).
22 Disconnect the accelerator cable, transmission control cable and cruise control cable (if equipped) from the throttle lever (see Section 10). Unbolt the accelerator cable bracket and position the bracket and cables aside.
23 Detach the fuel injector electrical connectors from each injector and the fuel rail assembly.
24 Detach any hoses and electrical connectors from the throttle body and plenum such as the IAC valve, TPS, MAP and EVAP canister control electrical connectors. If necessary, mark them with pieces of numbered tape to avoid confusion during reassembly.
25 Disconnect the vacuum harness assembly from the top of the throttle body **(see illustration)**.
26 Remove the EGR pipe from the manifold (see Chapter 6).
27 Remove the plenum bolts and lift the plenum from the intake manifold **(see illustration)**. If the plenum sticks, use a block of wood and a hammer to dislodge it. Don't pry between the sealing flanges, as this will damage the machined surfaces and could cause vacuum leaks to develop.
28 Remove all traces of old gasket material from the plenum and intake manifold mating surfaces. It's a good idea to stuff rags into the intake manifold openings to prevent debris from falling in.

Installation

29 Install the new gasket(s) and set the plenum into position.
30 Install the plenum bolts and tighten them to the torque listed in this Chapter's Specifications in a criss-cross pattern.
31 The remainder of installation is the reverse or removal.

Fuel injectors

Refer to illustrations 14.36, 14.37, 14.38 and 14.39
Warning: *Before any work is performed on the fuel lines, fuel rail or injectors, the fuel system pressure must be relieved* (see Section 2).
Note: *Refer to Section 12 for the injector checking procedure.*
32 Detach the cable from the negative terminal of the battery. **Caution:** *On models equipped with a Delco Loc II audio system, be sure the lockout feature is turned off before performing any procedure which requires disconnecting the battery.*
33 Remove the fuel pressure regulator (see Steps 45 through 48).
34 Label and unplug the injector electrical connectors.
35 Remove the air intake plenum (see Steps 19 through 31).
36 Remove the fuel injector bracket retaining bolts **(see illustration)**.
37 Remove the fuel injector bracket, leaving the fuel injectors inside the lower intake manifold **(see illustration)**.
38 To remove the fuel injectors, carefully pry them out of the intake manifold using a

14.36 Remove the fuel rail bracket bolts using a Torx driver

14.37 Lift the fuel injector bracket from the intake manifold/fuel rail assembly

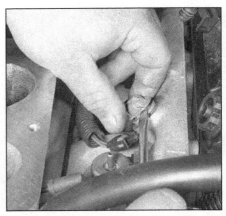

14.38 Carefully pry the fuel injectors out of the manifold

flat-bladed screwdriver **(see illustration)**.
39   Remove the injector O-rings **(see illustration)**. **Caution:** *Make sure the lower O-ring is not stuck inside the intake manifold bore after the injector is removed. These O-rings must NOT be reused once the injector has been removed.*
40   Install the new O-ring seal(s), as required, on the injector(s) and lubricate them with a light film of engine oil.
41   Install the injectors into the intake manifold/fuel rail assembly.
42   Secure the injectors with the fuel rail bracket. **Caution:** *Because these injectors are the bottom feed type, it is important to make sure that they are properly seated inside the intake manifold before starting the engine. If they are not seating correctly, fuel pressure inside the fuel rail portion of the intake manifold will force excessive amounts of fuel into the cylinders and possibly damage the engine. To check the seal, refer to Section 3 and check the fuel pressure with the ignition key ON (engine not running). If the fuel system holds fuel pressure (gauge remains steady), the fuel system is sealed properly and the engine can be started safely.*
43   Installation is the reverse of the removal procedure.

Fuel pressure regulator

Check

44   Refer to Section 3 for the fuel pressure checking procedure.

Replacement

Refer to illustration 14.48
45   Relieve the fuel system pressure (see Section 2).
46   Disconnect the cable from the negative terminal of the battery. **Caution:** *On models equipped with a Delco Loc II audio system, be sure the lockout feature is turned off before performing any procedure which requires disconnecting the battery.*
47   Disconnect the vacuum hose from the fuel pressure regulator.
48   Remove the fuel return pipe clamp from the fuel pressure regulator **(see illustration)**.

14.39 Remove the upper O-ring using a curved pick or screwdriver (the lower O-ring has already been removed)

49   Remove the Torx drive bolts from the fuel pressure regulator and slide the regulator out. Clean the housing with carburetor cleaner. When reassembling, lightly lubricate the O-ring with clean engine oil.
50   Reassembly is the reverse of disassembly. Be sure to replace the seal, otherwise a dangerous fuel leak may develop. When installing the seal, lubricate it with a light film of engine oil.

15   Central Multiport Fuel Injection (CMFI) and Central Sequential Electronic Fuel Injection (CSEFI) systems - component check and replacement

General information

1   The function of the Central Multiport Fuel Injection (CMFI) unit or the Central Sequential Electronic Fuel Injection (CSEFI) unit is to control fuel delivery to the engine, from a centrally mounted injector (CMFI) or injectors (CSEFI). Each unit is controlled by the Powertrain Control Module (PCM) (see Chapter 6).
2   The PCM monitors voltage from several

14.48 Disconnect the fuel return line from the fuel pressure regulator

sensors to determine how much fuel the engine needs. When the key is first turned "ON", the PCM turns on the fuel pump relay for two seconds and the fuel pump builds up pressure to the CMFI or CSEFI unit. The PCM monitors the Coolant Temperature Sensor (CTS), the Intake Air Temperature (IAT) Sensor, Throttle Position Sensor (TPS) and Manifold Absolute Pressure (MAP) sensor and then determines the proper air/fuel ratio for starting.
3   The fuel control system has an electric fuel pump, located in the fuel tank on the fuel gauge sending unit. The pump provides pressure above the regulated pressure needed by the CMFI or CSEFI injector(s).
4   The intake manifold is designed with an upper and lower manifold assembly. The upper manifold is a variable-tuned split-plenum design that also includes an intake manifold tuning valve, MAP sensor and a throttle valve attached to the plenum.
5   The throttle valve is used to control air flow into the engine and consequently engine output. During engine idle, the throttle valve is almost completely closed and air flow control is handled by the Idle Air Control (IAC) valve.
6   Most of the CMFI or CSEFI components are housed directly under the air intake plenum. In order to service the CMFI or CSEFI unit, it will be necessary to remove the air intake plenum from the intake manifold. The

Chapter 4 Fuel and exhaust systems

4-17

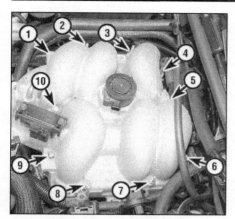

15.18 Remove the bolts (arrows) from the air intake plenum (the numbers indicate the bolt tightening sequence)

15.19a Lift the air intake plenum from the intake manifold

15.19b Lift the gasket off the intake manifold

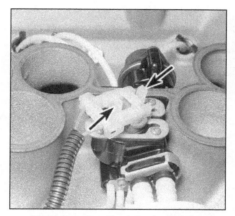

15.22 Squeeze the tabs (arrows) to disconnect the injector electrical connector

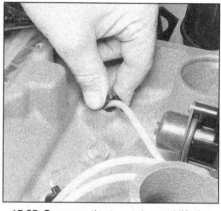

15.25 Squeeze the two tabs and lift the assembly to remove the poppet nozzle from the intake manifold

pressure regulator assembly consists of a fuel meter body, gasket seal, fuel pressure regulator, fuel injector(s) and six poppet nozzles with fuel tubes. The CMFI or CSEFI unit is not repairable and must be replaced as an assembly if found to be defective. Be sure to contact a dealer service department for any service contracts or warranties that might cover the repair of the fuel system before attempting it on your own.

7 The CMFI or CSEFI system has a fuel pressure regulator to maintain pressure at the fuel injector through a range of fuel recirculation rates from the in-tank fuel pump. With the ignition "ON" and the engine NOT running, the fuel pressure should be between 54 to 64 psi depending upon the particular system. Fuel enters the fuel meter body through the inlet line and flows directly into the injector cavity. When the PCM de-energizes the injector solenoid, fuel is recirculated through the pressure regulator. Fuel pressure applied to the regulator diaphragm acts against the spring force and opens the valve from its seat. This allows fuel to return to the fuel tank through the fuel meter body outlet and the return line. When the PCM energizes the injector solenoid, the armature lifts off the six fuel tube seats and delivers fuel through the fuel meter body out to the six poppet nozzles.

8 When the PCM energizes the injector solenoid, pressurized fuel flows through the fuel tubes to each poppet nozzle. There are six poppet nozzles (one for each cylinder). An increase in fuel pressure will cause the poppet nozzle ball to lift from its seat against spring force and spray fuel at approximately 52 psi. De-energizing the injector solenoid closes the armature and reduces the fuel pressure on the poppet nozzle ball.

CMFI unit removal and installation

Refer to illustrations 15.18, 15.19a, 15.19b, 15.22 and 15.25

Note: *Do not attempt to disassemble the CMFI unit. It is a non-serviceable part.*

Removal

9 Read the Warning at the beginning of Section 14, then relieve the fuel pressure (see Section 2). Remove the Torx screws and lift off the plastic cover from the air intake plenum.
10 Disconnect the electrical connectors on the TPS, IAC valve, MAP sensor and intake manifold tuning valve assembly (see Chapter 6).
11 Disconnect the throttle valve (TV) cable, cruise control cable and accelerator cable (see Section 10) from the throttle valve assembly.
12 Remove the bolts that retain the coil/module bracket to the air intake plenum (see Chapter 5)
13 Remove the intake air ducts from the plenum and the fan shroud (see Section 9).
14 Remove the ignition coil.
15 Disconnect the PCV hose from the plenum and the valve cover.
16 Disconnect the vacuum lines from the front and rear of the air intake plenum.
17 Remove the ignition wire and harness bracket from the plenum. Be sure to mark the position of each stud and nut.
18 Remove the bolts from the air intake plenum. Start on the front bolt and work your way counterclockwise (standing directly in front of the engine compartment) around the plenum **(see illustration)**.
19 Lift the air intake plenum from the intake manifold **(see illustration)**. Remove the gasket from the intake manifold **(see illustration)**.
20 Inspect the gasket surface on the plenum and the intake manifold for any chips, burrs or cracks. If the plenum is damaged, replace it with a new unit.
21 Clean the gasket surface on the plenum and the intake manifold with a soft cloth and solvent.
22 Detach the electrical connector from the CMFI assembly **(see illustration)**.
23 Disconnect the fuel fitting clip and dispose it.
24 Disconnect the fuel inlet and outlet lines from the CMFI assembly. Remove the O-ring seals and dispose them. Use new seals for assembly.
25 Squeeze the poppet nozzle locking tabs **(see illustration)** while lifting the nozzle out of the intake manifold.
26 After disconnecting the six poppet nozzles, lift the CMFI assembly out of the intake manifold as a single unit.

Chapter 4 Fuel and exhaust systems

15.36 Remove the retaining clip and disconnect the electrical connector from the fuel meter body

15.37a Disconnect the fuel supply and return lines from the fittings (arrows)

15.37b Remove the fuel line nuts and retainers (arrows) and remove the fuel lines from the fuel meter body

Installation

27 Align the CMFI assembly grommet with the casting grommet slots and push down until it is seated at the bottom of the guide hole.

28 Push the poppet nozzles into the casting sockets. **Caution:** *Be sure the poppet nozzles are seated and secured in the casting sockets before installing the plenum. Check by pulling each poppet nozzle firmly until it is in its correct place. This will prevent fuel leaks and consequently any fire danger.*

29 Connect the fuel inlet and outlet lines to the CMFI assembly. Be sure to install new O-rings. Coat the new O-rings with clean engine oil.

30 Install the new fuel fitting clip.

31 Be sure the fuel pump relay is connected and pressurize the fuel pump by turning the ignition key "ON". Do not start the engine. Check all the fittings for fuel leaks before installing the air intake plenum.

32 Install a new gasket on the intake plenum. Be sure to keep the green stripe facing UP.

33 Install the air intake plenum and tighten the bolts in two or three passes to the torque listed in this Chapter's Specifications, following the recommended torque sequence **(see illustration 15.18)**.

CSEFI unit removal and installation

Warning: See the **Warning** in Section 14.
Note: *When replacing components of the fuel meter body/injector assembly, refer to the identification numbers on the fuel meter body and injectors. Fuel injectors are calibrated with different flow rates and must not be interchanged with injectors from a different application.*

Removal

Refer to illustrations 15.36, 15.37a, 15.37b, 15.38, 15.40a, 15.40b, 15.41 and 15.42

34 Relieve the fuel system pressure (see Section 2).

35 Disconnect the cable from the negative battery terminal. **Caution:** *On models equipped with a Delco Loc II audio system, be sure the lockout feature is turned off before performing any procedure which requires disconnecting the battery.*

36 Remove the retaining clip and disconnect the electrical connector from the fuel meter body **(see illustration)**.

37 Disconnect the fuel supply and return lines from the fittings at the rear of the engine **(see illustration)**. Remove the bracket bolt. Loosen the nuts attaching the fuel lines to the fuel meter body and remove the lines **(see illustration)**.

38 Remove the throttle body and the accelerator control cable bracket. Remove the ignition coil/module assembly. Remove the EVAP purge valve. Remove the remaining mounting bolts and carefully remove the air intake plenum **(see illustration)**.

39 Detach the poppet nozzles by squeezing the tabs together and pulling the nozzle straight out of the intake manifold **(see illustration 15.25)**. **Note:** *Apply a numbered tag to each nozzle or line with the corresponding cylinder number.*

40 Pry the bracket locking tabs away from the fuel meter body and pull the assembly off the bracket **(see illustrations)**. Place the assembly on a clean work bench.

41 Remove the fuel injector hold-down plate nuts and remove the plate from the fuel meter body **(see illustration)**. Remove the retaining clip and withdraw the fuel pressure

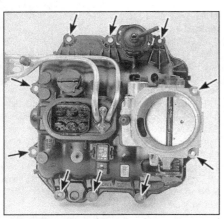

15.38 Air intake plenum bolt locations (arrows) (air intake plenum removed for clarity)

15.40a Using two tools (arrows), pry the locking tabs away from the fuel meter body . . .

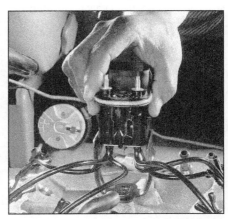

15.40b . . . and remove the fuel meter body along with the fuel lines and poppet nozzles from the intake manifold

Chapter 4 Fuel and exhaust systems

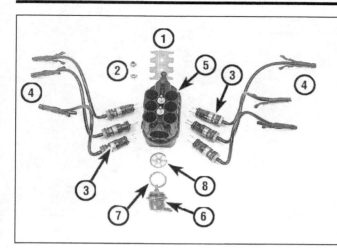

15.41 CSEFI fuel meter body components

1 Fuel injector hold-down plate
2 Nuts
3 Fuel injectors
4 Poppet nozzles
5 Fuel meter body
6 Fuel pressure regulator
7 O-ring
8 Filter screen

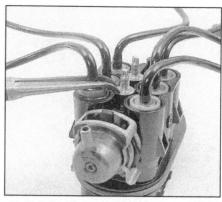

15.42 To remove an injector from the fuel meter body, pull on the tube fitting while pushing the injector out

regulator from the fuel meter body.

42 While pulling down on the injector tube fitting, push the injector out of the fuel meter body with a dull screwdriver **(see illustration)**. Be careful not to damage the electrical terminals. **Caution:** *Do not attempt to remove the fuel line or poppet nozzle from the injector. They are serviced as a complete assembly.*

Installation

43 Replace the injector O-rings. Apply a light coat of clean engine oil to the O-rings and press the injector into the fuel meter body until seated. Make sure the electrical terminals are properly aligned and the fuel tubes and nozzles are properly routed.

44 Install the injector hold-down plate and nuts. Replace the fuel pressure regulator O-rings and install the fuel pressure regulator and retaining clip.

45 Install the fuel meter body onto the intake manifold bracket. Install the poppet nozzles into the intake manifold, snapping them into place. Gently pull up on the fuel tube to ensure the nozzles are properly seated.

46 Inspect the fuel meter body and upper intake manifold seals for damage. Install new seals, if necessary. Install the air intake plenum. Apply thread locking compound to the air intake plenum bolts and tighten the bolts to the torque listed in this Chapter's Specifications. Install the EVAP purge valve and ignition coil/module assembly.

47 Inspect the fuel line O-rings and retainers for damage. Replace the O-rings and retainers, if necessary. Install the fuel lines onto the fuel meter body. Apply thread locking compound to the fuel line bracket bolt and install the bracket.

48 Inspect the throttle body seal for damage. Replace the seal, if necessary and install the throttle body.

49 The remainder of installation is the reverse of removal.

Idle Air Control (IAC) valve

Check

Refer to illustrations 15.51

50 The idle air control (IAC) valve controls the engine idle speed. This output actuator is mounted on the throttle body and is controlled by voltage pulses sent from the PCM (computer). The IAC valve pintle moves in or out allowing more or less intake air into the system according to the engine conditions. To increase idle speed, the PCM retracts the IAC valve pintle away from the seat and allows more air to bypass the throttle bore. To decrease idle speed, the PCM extends the IAC valve pintle towards the seat, reducing the air flow.

51 To check the IAC valve, unplug the electrical connector and, using an ohmmeter, measure the resistance across terminals A and B of the valve, then terminals C and D. Each resistance check should indicate 40 to 80 ohms **(see illustration)**. If not, replace the IAC valve.

52 There is an alternate method for testing the IAC valve. Various SCAN tools are available from auto parts stores and specialty tool companies that can be plugged into the ALDL (diagnostic connector) for the purpose of monitoring the sensors. Connect the SCAN tool and switch to the Idle Air Valve Position mode and monitor the steps (valve winding position) **(see illustration 14.7)**. The SCAN tool should indicate between 10 to 200 steps depending upon the rpm range. Allow the engine to idle for several minutes and while observing the count reading, snap the throttle to achieve high rpm (under 3,500). Repeat the procedure several times and observe the SCAN tool steps (counts) when the engine goes back to idle. The readings should be within 5 to 10 steps each time. If the readings fluctuate greatly, replace the IAC valve. **Note:** *When the IAC valve electrical connector is disconnected for testing, the PCM will have to "relearn" its idle mode. In other words, it will take a certain amount of time before the idle valve resets for the correct idle speed. Make sure the idle is smooth and not misfiring before plugging in the SCAN tool.*

53 Next, remove the valve (see Step 38) and inspect it:

a) Check the pintle for excessive carbon deposits. If necessary, clean it with carburetor cleaner spray. Also clean the IAC valve housing to remove any deposits.

b) Check the IAC valve electrical connections. Make sure the pins are not bent and make good contact with the connector terminals.

Removal

Refer to illustration 15.55

54 Unplug the electrical connector from the

15.51 First measure the resistance across terminals D and C, then across terminals A and B

15.55 Remove the bolts from the IAC valve (arrows)

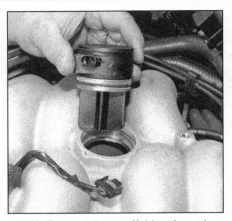

15.70 Remove the manifold tuning valve

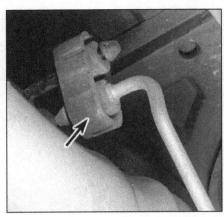

16.1 Remove the exhaust hanger

Idle Air Control (IAC) valve.
55 Unscrew the valve or remove the two IAC valve attaching screws and withdraw the valve **(see illustration)**.
56 Check the condition of the rubber O-ring. If it's hardened or deteriorated, replace it. On models equipped with a gasket, remove the gasket.
57 Clean the sealing surface and the bore of the idle air/vacuum signal housing assembly to ensure a good seal. **Caution:** *The IAC valve itself is an electrical component and must not be soaked in any liquid cleaner, as damage may result.*
58 Before installing the IAC valve, the position of the pintle must be checked. If the pintle is extended too far, damage to the assembly may occur.

Installation
59 Measure the distance from the flange or gasket mounting surface of the IAC valve to the tip of the pintle. If the distance is greater than 1-1/8 inch, reduce the distance by applying firm pressure onto the pintle to retract it. Try some side-to-side motion in the event the pintle binds.
60 Position the new O-ring or gasket on the IAC valve. Lubricate the O-ring with a light film of engine oil. If the IAC valve is the screw-in type, apply a light film of RTV sealant to the threads of the valve. Install the IAC valve and tighten the valve or the mounting screws securely.
61 Plug in the electrical connector at the IAC valve assembly. **Note:** *No adjustment is made to the IAC assembly after reinstallation. The IAC resetting is controlled by the PCM when the engine is started.*

Throttle Position Sensor (TPS)
Check
62 Check for stored trouble codes in the PCM using the On Board Diagnosis system (see Chapter 6).
63 To check the operation and replacement of the TPS, refer to the *Information sensors* in Chapter 6.

Manifold tuning valve (models through 1995)
Check
64 The manifold tuning valve is controlled by the PCM for the purpose of switching the passageway of the air intake charge for optimum torque and peak horsepower depending upon the speed and rpm range of the engine. During low and high end of the rpm range, the computer (PCM) de-energizes the valve and enables a "split" plenum condition thereby shortening or lengthening the distance the air charge travels before combustion. During midrange rpm, the PCM closes the valve and allows only the "single" passageway for the intake charge.
65 Disconnect the intake manifold tuning valve relay and using a voltmeter, check for battery voltage with the ignition key ON (engine not running).
66 If voltage exists, have the relay diagnosed by a qualified repair shop or parts department.
67 Check for battery voltage to the manifold tuning valve. There should be voltage present with the engine running.
68 Remove the valve and check it for proper movement and free from binding. Replace it if necessary.

Replacement
Refer to illustration 15.70
69 Remove the Torx bolts from the tuning valve.
70 Lift the manifold tuning valve from the plenum **(see illustration)**.
71 Installation is the reverse of removal. Be sure to install a new O-ring if it is cracked, hardened or otherwise deteriorated.

16 Exhaust system servicing - general information

Refer to illustrations 16.1
Warning: *The vehicle's exhaust system generates very high temperatures and must be allowed to cool down completely before any of the components are touched. Be especially careful around the catalytic converter, where the highest temperatures are generated.*

General information
1 Replacement of exhaust system components is basically a matter of removing the heat shields, disconnecting the component and installing a new one **(see illustration)**. The heat shields and exhaust system hangers must be reinstalled in the original locations or damage could result. Due to the high temperatures and exposed locations of the exhaust system components, rust and corrosion can seize parts together. Penetrating oils are available to help loosen frozen fasteners. However, in some cases it may be necessary to cut the pieces apart with a hacksaw or cutting torch. The latter method should be employed only by persons experienced in this work.

Crossover pipe
2 Remove the bolts or nuts securing the crossover pipe to the exhaust manifolds. Remove the crossover pipe.
3 Installation is the reverse of removal. Tighten the fasteners evenly and securely.

Chapter 5
Engine electrical systems

Contents

	Section		Section
Alternator - removal and installation	13	Ignition module - check and replacement	9
Battery cables - check and replacement	4	Ignition pick-up coil (HEI systems) - check and replacement	10
Battery - emergency jump starting	2	Ignition system - check	6
Battery - removal and installation	3	Ignition system - general information	5
Charging system - check	12	SERVICE ENGINE SOON light	See Chapter 6
Charging system - general information and precautions	11	Starter motor - removal and installation	16
Distributor (HEI and EDI systems) - removal and installation	8	Starter motor - testing in vehicle	15
General Information	1	Starter solenoid - removal and installation	17
Ignition coil - check, removal and installation	7	Starting system - general information	14

Specifications

Ignition coil resistance

2.2L engine with DIS
    Primary resistance .... 0.35 to 1.50 ohms
    Secondary resistance .... 5,000 to 10,000 ohms
4.3L engine with HEI
    Primary resistance .... 0.2 to 1.5 ohms
    Secondary resistance .... 5,000 to 25,000 ohms
4.3L engine with EDI
  1995 models
    Primary resistance .... 0.3 to 1.3 ohms
    Secondary resistance .... 5,000 to 10,000 ohms
  1996 and later models
    Primary resistance .... 0.2 to 0.5 ohms
    Secondary resistance .... 5,000 to 25,000 ohms

1 General information

Warning: *Because of the very high voltage generated by the ignition system, extreme care should be taken whenever an operation involving ignition components is performed. This not only includes the distributor, coil(s), module and spark plug wires, but related items that are connected to the systems as well, such as the plug connections, tachometer and testing equipment.*

4.3L V6 engines equipped with Throttle Body Injection (OBD I) are equipped with the High Energy Ignition (HEI) system. 2.2L four-cylinder engines are equipped with the Distributorless Ignition System (DIS) and the 4.3L CMFI or CSEFI (OBD II) engines are equipped with the Enhanced Distributor Ignition (EDI) system.

Although these systems use different ways to generate the ignition signals, they share certain similar components, such as the ignition switch, battery, coil, primary (low tension) and secondary (high tension) wiring circuits and spark plugs.

The charging system consists of a belt-driven alternator with an integral voltage regulator and the battery. These components work together to supply electrical power for the ignition system, the lights and all accessories.

Both engines are equipped with the CS-130D (100 amp) alternator. All types use a conventional pulley and fan.

2 Battery - emergency jump starting

Refer to the *Booster battery (jump) starting* procedure at the front of this manual.

3 Battery - removal and installation

Refer to illustration 3.3

Warning: *Hydrogen gas is produced by the battery, so keep open flames and lighted cigarettes away from it at all times. Always wear eye protection when working around a battery. Rinse off spilled electrolyte immediately with large amounts of water.*

Removal

1 The battery is located at the left front of the engine compartment.
2 Detach the cables from the negative and positive terminals of the battery. **Warning:** *To prevent arcing, disconnect the negative (-) cable first, then remove the positive (+) cable.* **Caution:** *On models equipped with a Delco Loc II audio system, be sure the lockout feature is turned off before performing any proce-*

3.3 Remove the battery hold-down bolt from the battery carrier

dure which requires disconnecting the battery.
3   Remove the hold-down clamp bolt and the clamp from the battery carrier **(see illustration)**.
4   Carefully lift the battery from the carrier. **Warning:** *Always keep the battery in an upright position to reduce the likelihood of electrolyte spillage. If you spill electrolyte on your skin, rinse it off immediately with large amounts of water.*

Installation

Note: *The battery carrier and hold-down clamp should be clean and free from corrosion before installing the battery. Make certain that there are no parts in the carrier before installing the battery.*
5   Set the battery in position in its carrier. Don't tilt it.
6   Install the hold-down clamp and bolt. The bolt should be snug, but overtightening it may damage the battery case.
7   Install both battery cables, positive first, then the negative. **Note:** *The battery terminals and cable ends should be cleaned prior to connection* (see Chapter 1).

4   Battery cables - check and replacement

1   Periodically inspect the entire length of each battery cable for damage, cracked or burned insulation and corrosion. Poor battery cable connections can cause starting problems and decreased engine performance.
2   Check the cable-to-terminal connections at the ends of the cables for cracks, loose wire strands and corrosion. The presence of white, fluffy deposits under the insulation at the cable terminal connection is a sign the cable is corroded and should be replaced. Check the terminals for distortion, missing mounting bolts or nuts and corrosion.
3   If only the positive cable is to be replaced, be sure to disconnect the negative cable from the battery first. **Caution:** *On models equipped with a Delco Loc II audio system, be sure the lockout feature is turned off before performing any procedure which requires disconnecting the battery.*
4   Disconnect and remove the cable. Make sure the replacement cable is the same length and diameter.
5   Clean the threads of the starter or ground connection with a wire brush to remove rust and corrosion. Apply a light coat of petroleum jelly to the threads to ease installation and prevent future corrosion.
6   Attach the cable to the starter or ground connection and tighten the mounting nut securely.
7   Before connecting the new cable to the battery, make sure it reaches the terminals without having to be stretched.
8   Connect the positive cable first, followed by the negative cable. Tighten the nuts and apply a thin coat of petroleum jelly to the terminal and cable connection.

5   Ignition system - general information

Warning: *Because of the very high voltage generated by the ignition system, extreme care should be taken whenever an operation involving ignition components is performed. This not only includes the distributor, coil(s), module and spark plug wires, but related items that are connected to the systems as well, such as the plug connections, tachometer and testing equipment.*

2.2L MFI engines are equipped with the Distributorless Ignition System (DIS), the 4.3L CMFI or CSEFI (OBD II) engine is equipped with the Enhanced Distributor Ignition (EDI) system and the 4.3L TBI (OBD I) engine is equipped with the High Energy Ignition (HEI) system.

High Energy Ignition (HEI) system

4.3L TBI OBD I engines use a special HEI distributor that is mounted at the front of the engine. HEI distributors covered by this manual are equipped with separate coils. All spark timing changes are carried out by the Powertrain Control Module which monitors data from various engine sensors, computes the desired spark timing and signals the distributor to change the timing accordingly. No vacuum or mechanical advance is used.

Enhanced Distributor Ignition (EDI) system

4.3L CMFI and CSEFI OBD II engines are equipped with an Enhanced Distributor Ignition (EDI) system. The EDI ignition system consists of the camshaft sensor (distributor housing), cap and rotor, distributor drive shaft, an ignition module, an ignition coil/coil driver, knock sensors, primary and secondary wiring, spark plugs and the necessary control circuits (wiring harness) for the entire system. This ignition system is enhanced to operate in the close tolerances of the OBD II system. The PCM monitors the information from various engine sensors and computes the correct spark timing via a direct line from the computer to the coil driver (IC).

The camshaft sensor detects camshaft position and sends this information to the computer (PCM). The sole purpose of the camshaft sensor is to provide the PCM with the information for any misfire trouble codes. In the event of failure or damage, the camshaft sensor is not directly involved with driveability.

Distributorless Ignition System (DIS)

The 2.2L engine uses a distributorless ignition system called DIS. This system uses a "waste spark" method of spark distribution. Each cylinder is paired with its opposing cylinder in the firing order, 1-4, 2-3, so that one cylinder on compression fires simultaneously with its opposing cylinder on the exhaust stroke. Since the cylinder on the exhaust stroke requires very little of the available voltage to fire its plug, most of the voltage is used to fire the cylinder on the compression stroke.

The DIS system includes two coil packs, an ignition control module (ICM), a crankshaft reluctor ring, a magnetic crankshaft sensor and the PCM. The ignition module is located under the coil packs and is connected to the PCM.

The crankshaft sensor is located on the side of the engine block near the coil packs. The magnetic crankshaft sensor protrudes through the engine block, within about 0.050-inch of the crankshaft reluctor ring. The reluctor ring is a special disc cast into the crankshaft, which acts as a signal generator for the ignition timing.

The system uses control wires from the PCM, just like conventional distributor systems. The PCM controls timing using crankshaft position, engine rpm, engine temperature and manifold absolute pressure (MAP) sensing.

6   Ignition system - check

Warning: *Because of the very high voltage generated by the ignition system, extreme care should be taken whenever an operation is performed involving ignition components. This not only includes the coils, control module and spark plug wires, but related items connected to the system as well, such as the plug connections, tachometer and any test equipment.*

General checks

Refer to illustration 6.3
1   Check all ignition wiring connections for tightness, cuts, corrosion or any other signs of a bad connection. A faulty or poor connec-

Chapter 5 Engine electrical systems

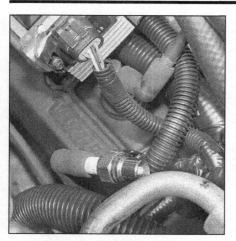

6.3 To use a calibrated ignition tester (available at most auto parts stores), simply disconnect a spark plug wire, attach the wire to the tester, clip the tester to a convenient ground and operate the starter - if there's enough power to fire the plug, sparks will be visible between the electrode tip and the tester body

tion at a spark plug could also result in a misfire. Also check for carbon deposits inside the spark plug boots. Remove the spark plugs, if necessary, and check for fouling.

2 Check for ignition and battery supply to the PCM. Check the ignition fuses (see Chapter 12). **Note:** *The ECM BAT fuse controls the computer and the fuel injection system while the IGN fuse controls power to the ignition system.*

3 Use a calibrated ignition tester to verify adequate available secondary voltage (25,000 volts) at the spark plug **(see illustration)**. Using an ohmmeter, check the resistance of the spark plug wires). Each wire should measure less than 30,000 ohms.

4 Check to see if the fuel pump and relay are operating properly (see Chapter 4). The fuel pump should activate for two seconds when the ignition key is cycled ON. Install an injector test light and monitor the blinks as the injector voltage signal pulses (see Chapter 4, *Fuel injection system - check*).

High Energy Ignition (HEI) systems

Refer to illustration 6.5

5 Refer to the illustration for terminal pin designations for testing the ignition module **(see illustration)**.

6 Check for spark at the spark plugs checking at least two or more spark plug wires **(see illustration 6.3)**. On a NO START condition, check the fuel pump relay and fuel pump systems for fuel delivery problems in the event the ignition spark is available.

7 Unplug the four-wire connector from the ignition module at the distributor and check for spark at the coil wire. If the fuel system is working properly and the ignition system will not start the engine, a spark indicates that the problem must be the distributor cap or rotor. **Note:** *A few sparks followed by no spark is the same condition as no spark at all.*

8 Check the ignition coil (see Section 7). Next, disconnect the two terminal plug on the module and check for voltage on the "C" and the "+" terminals. Normally, there should be battery voltage at the "C" and "+" terminals. Low voltage indicates an open or high resistance circuit from the distributor to the coil or ignition switch. If the "C" terminal voltage is low, but the "+" terminal voltage is 10 volts or more, the circuit from "C" terminal to the ignition coil or ignition coil primary winding is open.

9 Reconnect the "C+" connector onto the ignition module and with the ignition key ON (engine not running), check for voltage at the TACH terminal. The TACH terminal is taped back in the harness near the distributor. If there is less than 10 volts available, check the TACH circuit from the distributor. This test, checks for a shorted module or a grounded circuit from the ignition coil to the module. The distributor module should be turned off, so normal voltage should be about 12 volts. If the module is turned ON, the voltage will be low, but above one volt. This could cause the ignition coil to fail from excessive heat.

10 Check the module itself. Construct a battery pack that is more than 1.5 volts and less than 8 volts (penlight batteries). Disconnect the ignition module four-wire connector from the distributor. Disconnect the pick-up coil two-wire connector and remove the distributor cap and rotor. Connect a voltmeter from TACH to ground. Connect a test light from the battery pack (1.5 to 8.0 volts) and touch the test light probe to terminal P on the ignition module and watch the voltmeter. Voltage should drop. Applying a voltage (1.5V) to module terminal "P" should turn the module on and the TACH terminal voltage should drop to about 7 to 9 volts. **Note:** *Remember, the TACH terminal is usually taped to the harness near the distributor.*

11 This test will determine whether the module or coil is faulty or if the pick-up coil is not generating the proper signal to turn the module on. This test can be performed by using a DC battery with a rating of 1.5 volts. Monitor the change in voltage with a volt-

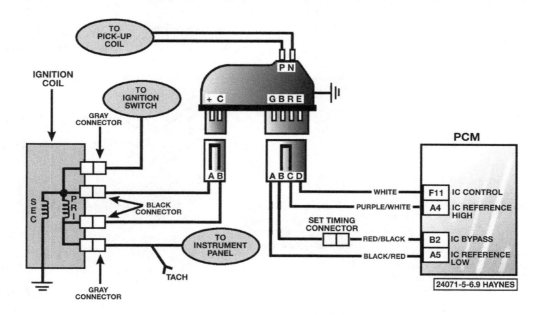

6.5 Schematic of the High Energy Ignition (HEI) system

meter connected to the TACH terminal. Some digital multimeters can also be used to trigger the module by selecting ohms, usually the diode position. In this position, the meter may have a voltage across its terminals which can be used to trigger the module. The voltage in the ohms position can be checked by using a second meter or by checking the manufacturer's specifications for the tool being used.

12 Install a spark testing tool **(see illustration 6.3)**. Repeat the previous test but instead observe spark at the spark plug when the test light tip is removed from terminal P. There should be a spark. This test checks the ignition module's ability to cause a spark. If no spark occurs, the fault is most likely in the ignition coil or pick-up coil because most module problems would have been found before this point in the procedure. A GM HEI module tester can determine which is at fault. If you cannot obtain the module tester, take the vehicle to a dealer service department or other qualified repair shop at this point. Also, check the resistance of the pick-up coil and if the specifications are incorrect, replace it with a new part.

Distributorless Ignition System (DIS)

Engine runs but misfires

13 First, check the cylinder compression for any severe variations between cylinders (see Chapter 2C). Disconnect the IAC valve (see Chapter 4) and with the engine idling, disconnect each spark plug wire, one-by-one, using insulated pliers (be sure to ground the disconnected wire). Listen carefully for a drop in engine rpm when the cylinder is removed from the firing order. If the engine rpm does not drop then the cylinder is not firing correctly. If the engine has a NO START condition, proceed to the NO START diagnostic procedure.

14 With the ignition key OFF, install a spark tester onto the plug lead which did NOT exhibit a drop in engine rpm **(see illustration 6.3)**. Have an assistant crank the engine and observe the spark. Check for spark at the spark plugs checking at least two or more spark plug wires. On a NO START condition, check the fuel pump relay and fuel systems for fuel delivery problems in the event the ignition spark is available (see Chapter 4).

15 Ground the opposite plug wire of the misfiring coil at the plug wire end using a jumper wire and observe the spark tester on the original plug. The spark should jump the tester gap easily. If the spark jumps the test gap after grounding the opposite plug wire, it indicates excessive resistance in the plug which was bypassed. A faulty or poor connection at that plug could also result in a misfire condition. Also check for carbon deposits inside the spark plug boot. **Note:** *Remember the companion cylinders are 1-4, 2-3.*

16 Check the resistance of the ignition wire of the misfiring cylinder. Wire resistance should be 30,000 ohms or less. If necessary, remove the coil pack and check for carbon tracking. If carbon tracking is evident, replace the coil and be sure that the plug wires relating to that coil are clean and tight. Excessive wire resistance or faulty connections could cause damage to the coil.

17 If the ignition wires are defective, switch the defective coil pack with a known good coil pack and check the spark using a spark tester. If a no spark condition disappears when the coil is switched for another coil, the original coil is faulty. If not, the ignition module is the cause of the no spark condition. This test can also be performed by substituting a known good coil for the one causing the no spark condition.

18 Test the ignition coil and the ignition module.

No start condition

19 Make sure each plug wire is connected to the coil towers on the coil packs then, disconnect one-by-one and install a spark tester onto the plug ends and check for a well defined, blue spark when the engine is cranked over by an assistant. This test verifies the ability of the system to produce at least 25,000 volts **(see illustration 6.3)**.

20 Check the resistance of each ignition wire and the resistance of each coil pack (see Section 7). No spark on one cylinder may be caused by an open plug wire or secondary winding. Both wires related to a coil and the secondary winding resistance should therefore be checked. Resistance readings over the upper limit, but not infinite, will probably not cause a no-start but may cause an engine miss under certain conditions.

21 Next, check for a trigger signal from the ignition module. Remove the coil pack from the ignition module to expose the module terminals. Connect a test light to each of the module terminals, one at a time, and have an assistant crank the engine over. This test checks the triggering circuit in the ignition module. A blinking test light indicates the module is triggering. The test light should blink quickly and constantly as each coil pack is triggered to fire by the switching signal from the ignition module.

22 If there is no blinking light, the ignition module is most likely the problem but not always. The PCM should relay reference signals to fire the fuel injectors if all the information is being received from the camshaft and crankshaft sensors. It is best to connect a SCAN tool and monitor the PCM activity to differentiate the ignition system problems (see Chapter 6).

23 A slowly blinking light, at this point, indicates the PCM is not seeing a crank sensor signal (see Chapter 6). At this point, the problem is in the camshaft or crankshaft sensor(s), sensor circuits or the ignition module. **Note:** *Refer to Chapter 6 for additional information and testing procedures for the camshaft sensor and the crankshaft sensors. It will be necessary to verify that the crankshaft sensor and camshaft sensor are operating correctly before changing the ignition module. A defective ignition module can only be diagnosed by process of elimination.*

24 If the trigger signals all are correct, check the fuel pump and circuit. Turn the ignition ON (engine not running) and listen for the fuel pump within the first two seconds. If the fuel pump runs, the fuse is okay.

7 Ignition coil - check, removal and installation

High Energy Ignition (HEI) systems

Refer to illustration 7.2

1 Disconnect the cable from the negative terminal of the battery.

Check

2 Check the coil for opens and grounds by performing the following three tests with an ohmmeter **(see illustration)**.

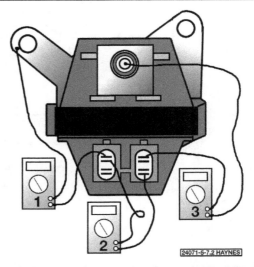

7.2 Check the coil for grounds (1), primary resistance (2) and secondary resistance (3)

Chapter 5 Engine electrical systems

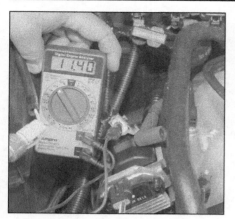

7.10 Unplug the coil electrical connector and check for battery voltage (EDI system)

7.11 Checking for coil-to-body shorts. Resistance should be infinite (EDI system)

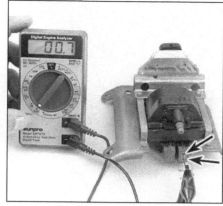

7.12 Checking the coil primary resistance (EDI system)

3 Using the ohmmeter's high scale, hook up the ohmmeter leads as illustrated **(see test 1 in illustration 7.2)**. The ohmmeter should indicate a very high, or infinite, resistance value. If it doesn't, replace the coil.
4 Check the coil primary resistance. Using the low scale, hook up the leads as illustrated **(see test 2 in illustration 7.2)**. The ohmmeter should indicate a very low, or zero, resistance value. If it doesn't, replace the coil. Check the Specifications listed in the beginning of this Chapter.
5 Check the coil secondary resistance. Using the high scale, hook up the leads as illustrated **(see test 3 in illustration 7.2)**. The ohmmeter should not indicate an infinite resistance. If it does, replace the coil.

Removal

6 Unplug the coil high tension wire and both electrical leads from the coil.
7 Remove both mounting nuts and remove the coil from the engine.

Installation

8 Installation of the coil is the reverse of the removal procedure.

Enhanced Distributor Ignition (EDI) systems

Refer to illustrations 7.10, 7.11, 7.12, 7.13 and 7.16

Check

9 Detach the primary electrical connector from the coil.
10 Check to make sure the coil is receiving battery voltage with the ignition key ON (engine not running) **(see illustration)**.
11 Using the ohmmeter's low scale, hook up the negative probe (-) of the ohmmeter to the coil body and the positive probe (+) of the ohmmeter to the coil primary terminal on the ignition coil **(see illustration)**. The ohmmeter should indicate infinite resistance. If it doesn't, replace the coil. Refer to the Specifications listed in the beginning of this Chapter.
12 Using the ohmmeter's low scale, hook up the ohmmeter leads to the primary termi-

7.13 Checking the coil secondary resistance (EDI system)

nals on the ignition coil **(see illustration)**. The ohmmeter should indicate a very low resistance value. If it doesn't, replace the coil. Refer to the Specifications listed in the beginning of this Chapter.
13 Using the high scale, hook up one lead to the ignition coil primary terminal and the other lead to the secondary terminal **(see illustration)**. Refer to the Specifications listed in the beginning of this Chapter. The ohmmeter should not indicate an infinite resistance. If it does, replace the coil.

Removal

14 Detach the cable from the negative terminal of the battery. **Caution:** *On models equipped with a Delco Loc II audio system, be sure the lockout feature is turned off before performing any procedure which requires disconnecting the battery.*
15 Unplug the coil high tension wire and both electrical leads from the coil.
16 Remove both mounting nuts and remove the coil from the engine **(see illustration)**.

Installation

17 Installation of the coil is the reverse of the removal procedure.

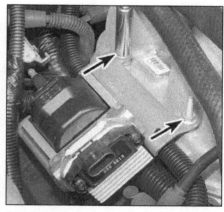

7.16 Remove the coil nuts (arrows) and lift the unit from the engine

Distributorless Ignition System (DIS)

Check

Refer to illustrations 7.18, 7.19 and 7.20

18 Check to make sure the coil is receiving battery voltage with the ignition key ON (engine not running) **(see illustration)**.

7.18 Remove the two-wire connector from the ignition module and check for battery voltage on the pink wire

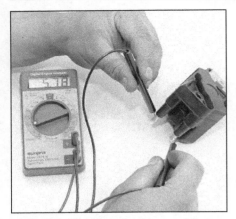

7.19 Checking the secondary resistance of a coil pack

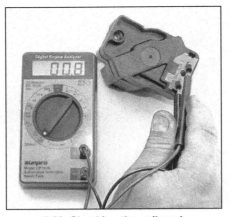

7.20 Checking the coil pack primary resistance

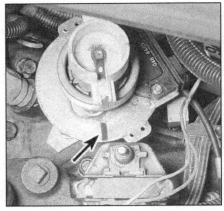

8.6a Make a mark on the distributor body (arrow) to show the direction the rotor is pointing before removing the distributor (HEI system)

19 Use an ohmmeter and check secondary resistance for each coil pack **(see illustration)**. Refer to the Specifications listed in this Chapter for the correct amount of resistance.
20 Next, check the primary resistance for each coil pack **(see illustration)**. Refer to the Specifications listed in this Chapter for the correct amount of resistance.

Removal

21 Detach the cable from the negative terminal of the battery. **Caution:** *On models equipped with a Delco Loc II audio system, be sure the lockout feature is turned off before performing any procedure which requires disconnecting the battery.*
22 Unplug the electrical connectors from the module.
23 If the plug wires are not numbered, label them and detach the plug wires at the coil assembly.
24 Remove the module/coil assembly mounting bolts and lift the assembly from the vehicle.

Installation

25 Installation is the reverse of removal.
26 When installing the coils, make sure they are connected properly to the module and all plug wires are fully seated.

8 Distributor (HEI and EDI systems) - removal and installation

Removal

Refer to illustrations 8.6a, 8.6b and 8.7
1 Disconnect the cable from the negative battery terminal. **Caution:** *On models equipped with a Delco Loc II audio system, be sure the lockout feature is turned off before performing any procedure which requires disconnecting the battery.*
2 On HEI systems, remove the air cleaner assembly (see Chapter 4).
3 On EDI systems, remove the air intake plenum (see Chapter 4).
4 Unplug the distributor electrical connectors from the side of the base.

8.6b Mark the position of the rotor on the body of the distributor (EDI system)

5 Remove the distributor cap (see Chapter 1).
6 Paint a mark on the distributor body to show the direction the rotor is pointing **(see illustrations)**.
7 Mark the position of the distributor body in relation to the engine block near the clamp **(see illustration)**. As long as the engine is not cranked over, the engine timing will remain correct in relation to the timing marks.
8 Remove the distributor hold-down bolt and clamp.
9 Remove the distributor from the engine.

Installation (crankshaft not turned after distributor removal)

HEI Ignition

10 Position the rotor in the exact location it was in when the distributor was removed.
11 Lower the distributor into the engine. To mesh the gears at the bottom of the distributor it may be necessary to turn the rotor slightly. It is possible that the distributor may not seat down fully against the block because the lower part of the distributor shaft has not properly engaged the oil pump shaft. Make sure the distributor and rotor are properly

8.7 Mark the position of the distributor in relation to the engine (arrow) before loosening the distributor hold-down clamp (HEI system)

aligned with the marks made earlier, then use a large socket and breaker bar on the crankshaft bolt to turn the engine in the normal direction of rotation until the two shafts engage and the distributor drops down against the block.
12 With the base of the distributor seated against the engine block turn the distributor housing to align the marks made on the distributor base and the engine block. Make sure the rotor is pointing to its mark.
13 Place the hold-down clamp in position and loosely install the hold-down bolt.
14 Reconnect the ignition wiring harness.
15 Install the distributor cap.
16 Reconnect the coil connector.
17 With the distributor in its original position, tighten the hold-down bolt.
18 Check the ignition timing (see Chapter 1).

Installation (crankshaft turned after distributor removal)

19 Remove the number one spark plug.
20 Place your finger over the spark plug hole while turning the crankshaft with a wrench on the pulley bolt at the front of the engine.
21 When you feel compression, continue

turning the crankshaft slowly until the timing mark on the vibration damper is aligned with the "0" on the engine timing indicator.
22 Position the rotor between the number one and six spark plug terminals on the cap.
23 Lower the distributor into the engine. To mesh the gears at the bottom of the distributor, it may be necessary to turn the rotor slightly. If the distributor does not drop down flush against the block it is because the distributor shaft has not mated to the oil pump shaft. Place a large socket and breaker bar on the crankshaft bolt and turn the engine over in the normal direction of rotation until the two shafts engage properly, allowing the distributor to seat flush against the block.
24 With the base of the distributor properly seated against the engine block, turn the distributor housing to align the marks made on the distributor base and the engine block.
25 Place the hold-down clamp in position and loosely install the hold-down bolt.
26 Reconnect the ignition wiring harness.
27 Install the distributor cap. If the secondary wiring harness was removed from the cap, reinstall it.
28 Reconnect the coil connector.
29 With the distributor in its original position, tighten the hold-down bolt and check the ignition timing.

EDI ignition

Refer to illustration 8.32

Note: *This distributor cannot be rotated to "set the timing", instead, it must be aligned correctly upon installation.*

30 Remove the number one spark plug.
31 Place your finger over the spark plug hole while turning the crankshaft with a wrench on the pulley bolt at the front of the engine. When you feel the compression, continue turning the crankshaft slowly, until the timing mark on the balancer is aligned with the "V" notch on the front cover. **Note:** *Some models have two notches on the crankshaft balancer – on these models, the proper notch to use is the notch on the left (as you are looking at the balancer).*
32 The distributor has a small number "6" molded into the housing. Mark this number with a dab of paint. Some distributors have a "6" and an "8" and are used in either engine application **(see illustration)**.
33 Install the distributor into the engine. If it does not seat correctly, it may be necessary to rotate the oil pump driveshaft by inserting a long screwdriver into the shaft and rotating it the necessary amount to allow the distributor to fit flush into the intake manifold.
34 Once the distributor is seated, check the rotor alignment. The rotor should point, within a few degrees to the "6" on the distributor housing. If it is not aligned, the distributor may be off a tooth and need to be adjusted in the direction necessary to point the rotor at the "6".
35 Tighten the hold-down bolt and install the camshaft sensor connector, cap and wires as necessary.
36 If the "SERVICE ENGINE" light comes on after installing the distributor, and has a

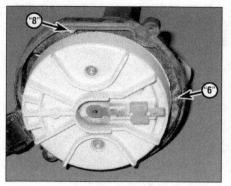

8.32 Some EDI distributors are marked "6" and "8." Be sure to align the rotor with the "6" mark

trouble code P1345, the distributor has been incorrectly installed. Repeat the installation procedure.

9 Ignition module - check and replacement

High Energy Ignition (HEI) systems

Note 1: *It is not necessary to remove the distributor to check or replace the module.*
Note 2: *Refer to illustration 6.5 for the terminal designations on the ignition module.*

Check

1 Disconnect the four-terminal connector from the distributor.
2 Check for a spark at the coil using a spark tester connected to the coil wire (see Section 6). If there is spark, check the cap, rotor and coil wire for opens or other damage.
3 If there is no spark, remove the distributor cap. Reconnect the four-terminal connector to the ignition module (distributor).
4 Unplug the two-wire connector from the distributor. With the ignition switch turned ON, check for voltage at the module positive (+) terminal of the two-wire connector.
5 If the reading is less than ten volts, there is a fault in the wire between the module positive (+) terminal and the ignition coil positive connector or the ignition coil and primary circuit-to-ignition switch.
6 If the reading is ten volts or more, check the "C" terminal on the module (two wire connector).
7 If the reading is less than one volt, there is an open or grounded lead in the distributor-to-coil "C" terminal connection or ignition coil or an open primary circuit in the coil itself.
8 If the reading is one to ten volts, replace the module with a new one and check for a spark (see Section 6). If there is a spark the module was faulty and the system is now operating properly. If there is no spark, there is a fault in the ignition coil.
9 If the reading in Step 4 is 10 volts or more, unplug the pick-up coil connector (two wire) from the module (terminals P and N).

9.14 Unplug the electrical connectors from the ignition module

Locate the TACH harness terminal (taped to the harness near the coil) and install a voltmeter positive probe (+). With the ignition key ON (engine not running), use a 1.5 volt power source (flashlight battery) and jumper wires and carefully touch the positive probe to terminal P on the module **(see illustration 6.9)**. Voltage at TACH should momentarily drop then return to normal.
10 If there is no drop in voltage, check the module ground and, if it is good, replace the module with a new one.
11 If the voltage drops, check for spark at the coil wire (spark tester hooked to coil tower wire) as the test light is removed from the module terminal. If there is no spark, the module is faulty and should be replaced with a new one. If there is a spark, the pick-up coil or connections are faulty or not grounded.

Replacement

Refer to illustration 9.14

12 Detach the cable from the negative terminal of the battery. **Caution:** *If the vehicle is equipped with a Delco Loc II audio system, make sure you have the correct activation code before disconnecting the battery. See the information at the front of this manual for the radio re-activation procedure.*
13 Remove the distributor cap and rotor (see Chapter 1).
14 Disconnect the electrical connectors from the module **(see illustration)**. Note that the connectors cannot be interchanged.
15 Remove both module attaching screws and lift the module up and away from the distributor.
16 Do not wipe the grease from the module or the distributor base if the same module is to be reinstalled. If a new module is to be installed, a package of silicone grease will be included with it. Wipe the distributor base and the new module clean, then apply the silicone grease on the face of the module and on the distributor base where the module seats. This grease is necessary for heat dissipation.
17 Install the module and attach both electrical leads.
18 Install the distributor rotor and cap (see Chapter 1).
19 Attach the cable to the negative terminal of the battery.

Chapter 5 Engine electrical systems

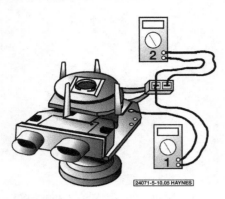

10.5 Pick-up coil test connections on the HEI distributor

10.7 Remove the spring clip from the distributor shaft

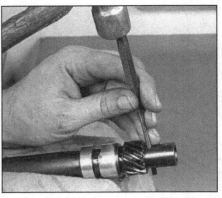

10.9a Mount the distributor shaft in a soft-jawed vise and using a drift punch and hammer, knock out the roll pin

Enhanced Distributor Ignition (EDI) system

Check

20 Refer to Steps 13 through 24 of Section 6 and follow the procedures before performing the module checks.
21 Check the crankshaft sensor reference voltage and resistance. Also check for any problems with the crankshaft sensor circuit (see Chapter 6 and the wiring diagrams at the end of Chapter 12).

Replacement

22 Detach the cable from the negative terminal of the battery. **Caution:** *On models equipped with a Delco Loc II audio system, be sure the lockout feature is turned off before performing any procedure which requires disconnecting the battery.*
23 Disconnect the ignition module electrical connector. **Note:** *The ignition module on EDI systems is located on the ignition coil.*
24 Remove both module attaching screws and lift the module up and away from its mount.
25 Install the module and tighten the screws securely.
26 Plug in the electrical connector.
27 Attach the cable to the negative terminal of the battery.

Distributorless Ignition Systems (DIS)

Check

28 First, perform the ignition system checks detailed in Section 6 (see Steps 13 through 24).
29 Disconnect the two wire electrical connector from ignition module and check for battery voltage on the pink wire.
30 If no voltage is present, check the circuit from the ignition module to the ignition key (refer to the wiring diagrams at the end of Chapter 12). **Note:** *If battery voltage is not present, check the IGNITION fuse. This might be a simple solution to a NO START condition.*
31 Next, using an ohmmeter, check the resistance of the crankshaft sensor and circuit. Probe the crankshaft sensor terminals (yellow and purple wires) on the sensor side of the

module connector. It should read between 900 and 1,200 ohms (see Chapter 6).
32 Finally, check the output voltage signal from the crankshaft sensor. Switch the voltmeter to AC scale, backprobe the crankshaft sensor electrical terminal, crank the engine over and confirm that the voltage output is greater than 0.1 volt (100 millivolts). **Note:** *Refer to Chapter 6, Section 4 for additional information and testing procedures for the camshaft sensor and crankshaft sensor.*
33 If all these tests are correct and there is no spark output at any of the coils, the ignition module is defective. In the event any of the system check findings are incorrect, diagnose the individual circuits and components.

Ignition module replacement

34 Detach the cable from the negative terminal of the battery. **Caution:** *On models equipped with a Delco Loc II audio system, be sure the lockout feature is turned off before performing any procedure which requires disconnecting the battery.*
35 Clearly label, then disconnect, all spark plug wires from the DIS assembly.
36 Unplug the electrical connector at the module.
37 Unbolt and remove the DIS and support bracket assembly.
38 Remove the coil-to-module attaching screws.
39 Separate the coil and module assemblies.
40 Open the coil and module halves. Label, then detach, the wires between the module and the coil assemblies from the spade terminals on the underside of the coils.
41 Unbolt the module from the support bracket.
42 Installation is the reverse of removal. Be sure to attach the wires of the new module to the coil assembly spade terminals in exactly the same order in which they were removed.

10 Ignition pick-up coil (HEI systems) - check and replacement

Refer to illustrations 10.5, 10.7, 10.9a, 10.9b and 10.10

1 Detach the cable from the negative terminal of the battery.

10.9b Remove the driven gear and spacer washers from the end of the shaft, making sure to note the order in which you remove any spacers

2 Remove the distributor cap and rotor.
3 Remove the distributor from the engine (only do this if you have difficulty in accessing the terminals of the pick-up coil) (see Section 8).
4 Detach the pick-up coil leads from the module.

Check

5 Connect one lead of an ohmmeter to the terminal of the pick-up coil lead and the other to ground as shown in test 1 **(see illustration)**. Flex the leads by hand to check for intermittent opens. The ohmmeter should indicate infinite resistance at all times. If it doesn't, the pick-up coil is defective and must be replaced.
6 Connect the ohmmeter leads to both terminals of the pick-up coil lead (see test 2 in **illustration 10.5**). Flex the leads by hand to check for intermittent opens. The ohmmeter should read one steady value between 500 and 1,500 ohms as the leads are flexed by hand. If it doesn't, the pick-up coil is defective and must be replaced.

Replacement

7 Remove the distributor, if not already done. Remove the spring clip from the distributor shaft **(see illustration)**.

Chapter 5 Engine electrical systems

8 Mark the distributor tang drive and shaft so they can be reassembled in the same position.
9 Carefully set the distributor in a vise with a shop rag to protect the shaft. Using a hammer and punch, remove the roll pin from the distributor shaft and gear **(see illustrations)**.
10 Remove the distributor shaft **(see illustration)**.
11 Lift the pick-up coil assembly straight up and remove it from the distributor. Note the order in which you remove the pieces.
12 Reassembly is the reverse of disassembly.

11 Charging system - general information and precautions

Caution: *On models equipped with a Delco Loc II audio system, be sure the lockout feature is turned off before performing any procedure which requires disconnecting the battery.*

The charging system consists of a belt-driven alternator with an integral voltage regulator and the battery. These components work together to supply electrical power for the ignition system, the lights and all accessories.

These models are equipped with the CS-130D (100 amp) alternator. All types use a conventional pulley and fan. These alternators should be considered non-serviceable and, if found to be faulty, should be exchanged as cores for new or rebuilt units. These alternators cannot be disassembled and repaired. Because of the expense and the limited availability of parts, no alternator overhaul information is included in this manual.

The purpose of the voltage regulator is to limit the alternator's voltage to a preset value. This prevents power surges, circuit overloads, etc., during peak voltage output. On all models with which this manual is concerned, the voltage regulator is contained within the alternator housing.

The charging system does not ordinarily require periodic maintenance. The drivebelt, electrical wiring and connections should, however, be inspected at the intervals suggested in Chapter 1.

Take extreme care when making circuit connections to a vehicle equipped with an alternator and note the following. When making connections to the alternator from a battery, always match correct polarity. Before using arc welding equipment to repair any part of the vehicle, disconnect the wires from the alternator and the battery terminals. Never start the engine with a battery charger connected. Always disconnect both battery leads before using a battery charger.

The charging indicator light on the dash lights up when the ignition switch is turned on and goes out when the engine starts. If the light stays on or comes on once the engine is running, a charging system problem has occurred. See Section 12 for the proper diagnosis procedure for each type of alternator.

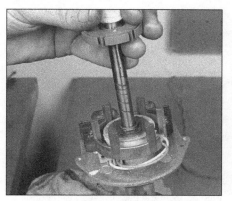

10.10 Remove the shaft from the distributor

12 Charging system - check

1 If a malfunction occurs in the charging circuit, do not immediately assume that the alternator is causing the problem.
2 First, check the following items:
   a) *Make sure the battery cable connections at the battery are clean and tight.*
   b) *The battery electrolyte specific gravity (if possible). If it is low, charge the battery.*
   c) *Check the external alternator wiring and connections. They must be in good condition.*
   d) *Check the drivebelt condition and tension (see Chapter 1).*
   e) *Make sure the alternator mounting bolts are tight.*
   f) *Run the engine and check the alternator for abnormal noise (may be caused by a loose drive pulley, loose mounting bolts, worn or dirty bearings, defective diode or defective stator).*
3 Check the charge light bulb and circuit. With the ignition key ON and the engine not running, the lamp should be ON. If not detach the wiring harness at the alternator.
   a) *Install a fused jumper wire (5 amp) to ground and connect the other end to the lead that was removed from the L terminal on the alternator.*
   b) *If the charging lamp on the dash comes ON, the alternator is defective. Replace the alternator.*
   c) *If the charging lamp on the dash remains OFF, locate the open circuit between the alternator and the bulb on the dash. First check the bulb to make sure it is not blown.*
   d) *With the ignition key ON and the engine running, the lamp should be OFF. If it remains ON while running, stop the engine and remove the lead from the L terminal on the alternator*
   e) *If the lamp on the dash goes OFF, the alternator is defective.*
   f) *If the lamp on the dash remains ON, there is a grounded L terminal in the wiring harness*

13.4 Remove the bolts (arrows) that hold the alternator to its bracket (4.3L engine)

4 Using a voltmeter, check the battery voltage with the engine off. It should be at least 12.6 volts.
5 Start the engine and check the battery voltage again. It should now be approximately 13.5 to 15 volts.

13 Alternator - removal and installation

Refer to illustration 13.4

1 Detach the cable from the negative terminal of the battery. **Caution:** *On models equipped with a Delco Loc II audio system, be sure the lockout feature is turned off before performing any procedure which requires disconnecting the battery.*
2 Clearly label, if necessary, then unplug and unbolt the electrical connectors from the alternator.
3 Remove the serpentine drivebelt (see Chapter 1). On 4.3L V6 engines, remove the heater hose bracket from the rear of the alternator.
4 Remove the alternator mounting bolts **(see illustration)** and remove the alternator.
5 Installation is the reverse of removal.

14 Starting system - general information

The function of the starting system is to crank the engine quickly enough for it to start. The starting system is composed of a starter motor, solenoid, ignition switch and battery. The battery supplies the electrical energy to the solenoid, which then completes the circuit to the starting motor, which does the actual work of cranking the engine.

The solenoid and starter motor are mounted together at the lower front side of the engine. No periodic lubrication or maintenance is required.

The electrical circuitry of the vehicle is arranged so that the starter motor can only be operated when the clutch pedal is

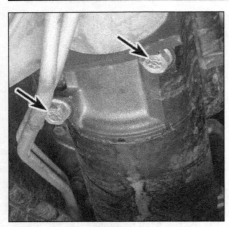

16.18 Remove the starter bolts (arrows) on the 4.3L engine

depressed (manual transmission) or the transmission selector lever is in Park or Neutral (automatic transmission).

Never operate the starter motor for more than 15 seconds at a time without pausing to allow it to cool for at least two minutes.

Excessive cranking can cause overheating, which can seriously damage the starter.

There are two types of starters equipped in these models. 2.2L engines are equipped with the SD-210 and the 4.3L engines are equipped with the SD-260. Both are the direct drive system with the pinion (gear and shaft) driven by the armature shaft. All types are not serviceable and in the event of failure, they must be replaced as a single unit as a rebuilt (exchange) or new part.

15 Starter motor - testing in vehicle

1   If the starter motor does not turn at all when the switch is operated, make sure that the shift lever is in Neutral or Park (automatic transmission) or that the clutch pedal is depressed (manual transmission).
2   Make sure that the battery is charged and that all cables, both at the battery and starter solenoid terminals, are secure.
3   If the starter motor spins but the engine is not cranking, the overrunning clutch in the starter motor is slipping and the motor must be removed from the engine for replacement.
4   If, when the switch is actuated, the starter motor does not operate at all but the solenoid clicks, then the problem lies with either the battery, the main solenoid contacts or the starter motor itself. **Note:** *Before diagnosing starter problems, make sure the battery is fully charged.*
5   If the solenoid plunger cannot be heard when the switch is actuated, the solenoid itself is defective or the solenoid circuit is open.
6   To check the solenoid, connect a jumper lead between the battery (+) and the "S" terminal on the solenoid. If the starter motor now operates, the solenoid is OK and the problem is in the ignition switch, neutral start switch or in the wiring.

7   If the starter motor still does not operate, remove the starter/solenoid assembly for disassembly, testing and repair.
8   If the starter motor cranks the engine at an abnormally slow speed, first make sure that the battery is charged and that all terminal connections are clean and tight. If the engine is partially seized, or has the wrong viscosity oil in it, it will crank slowly.
9   Run the engine until normal operating temperature is reached, then stop the engine, disconnect the coil wire from the distributor cap and ground it on the engine.
10 Connect a voltmeter positive lead to the starter motor terminal of the solenoid and then connect the negative lead to ground.
11 Crank the engine and take the voltmeter readings as soon as a steady figure is indicated. Do not allow the starter motor to turn for more than 15 seconds at a time. A reading of 9 volts or more, with the starter motor turning at normal cranking speed, is normal. If the reading is 9 volts or more but the cranking speed is slow, the motor is faulty. If the reading is less than 9 volts and the cranking speed is slow, the solenoid contacts are probably burned.

16 Starter motor - removal and installation

2WD models

1   Disconnect the negative battery cable. **Caution:** *On models equipped with a Delco Loc II audio system, be sure the locked feature is turned off before performing any procedure which requires disconnecting the battery.*
2   Raise the front of the vehicle and support it securely on jackstands. On V6 models, remove the right front wheel for easier access.
3   Remove the front exhaust pipe from the exhaust manifold (see Chapter 4).
4   On 2.2L engines, remove the brace rod from the front of the engine to the bellhousing. On models through 1999, there may be a support bracket on 2.2L engines connecting the front of the starter to the block. Remove the bracket-to-block bolts.
5   On models so equipped, remove the starter heat shield.
6   From under the vehicle, disconnect the solenoid wire and battery cable from the terminals on the solenoid.
7   Remove the starter mounting bolts.
8   Remove the starter motor. Note the location of spacer shim(s), if equipped.
9   Installation is the reverse of removal. Be sure to install the spacer shim(s) in exactly the same location, if equipped (see Steps 21 through 26 for the correct shim adjustment procedure).

4WD models

10 Disconnect the negative battery cable. **Caution:** *On models equipped with a Delco*

Loc II audio system, be sure the locked feature is turned off before performing any procedure which requires disconnecting the battery.
11 Raise the front of the vehicle and support it securely on jackstands.
12 It will be helpful to remove the wheel for better access (left wheel on 2.2L models, right wheel on 4.3L V6 models).
13 Remove the skid plate (if equipped).
14 On 2.2L models, remove the front driveshaft (see Chapter 8).
15 On some early models, it may be necessary to raise the engine/transmission with a floorjack under the transmission and remove the transmission crossmember for access to the starter motor bolts (see Chapter 7).
16 On models so equipped, remove the starter heat shield.
17 From under the vehicle, disconnect the solenoid wire and battery cable from the terminals on the solenoid.
18 Remove the starter mounting bolts **(see illustration)**.
19 Remove the starter motor. Note the location of spacer shim(s), if equipped.
20 Installation is the reverse of removal. Be sure to install the spacer shim(s) in exactly the same location, if equipped (see Steps 21 through 26 for the correct shim adjustment procedure).

Shim adjustment procedure

21 Remove the flywheel housing cover (see Chapter 7).
22 Inspect the flywheel or driveplate for signs of unusual wear such as chipped or missing gear teeth.
23 Using a wire type feeler gauge (round diameter), measure the clearance between the top of the flywheel ring gear tooth and the bottom of the pinion tooth on the starter. Normal clearance should be 0.01 to 0.06 inches.
24 If the starter clearance is less than 0.02 inches and the starter whined after engagement, the starter is too close to the flywheel. Add a 0.04 shim to gain clearance. Do not use more than two shims.
25 If the starter clearance is more than 0.06 inches and the starter whines during engagement then the starter is far from the flywheel. Remove a 0.04 shim to gain clearance. Do not remove more than two shims.
26 To install a shim, loosen the inside bolt, remove the outer bolt and slide the shim between the engine and the starter without removing the starter from the engine block.

17 Starter solenoid - removal and installation

Note: *The starter solenoid is not easily removed from the main starter body without complete disassembly. It is recommended that the starter/solenoid assembly be exchanged as a complete unit in the event of failure.*

Chapter 6
Emissions and engine control systems

Contents

	Section		Section
Catalytic converter	10	Powertrain Control Module (PCM) - replacement	3
Evaporative Emissions Control (EVAP) system	6	Secondary Air Injection (AIR) system - general description and component replacement	9
Exhaust Gas Recirculation (EGR) system	5		
General information	1	Self-diagnosis system and trouble codes	2
Information sensors - check and replacement	4	Transmission Converter Clutch (TCC) system	8
Positive Crankcase Ventilation (PCV) system	7		

Specifications

Torque specifications

Crankshaft sensor bolts
    Four-cylinder MFI engine
        Through 1998 .... 108 in-lbs
        1999 .... 88 in-lbs
        2000 and later .... 71 in-lbs
    V6 CMFI or CSEFI engine
        Through 1998 .... 96 in-lbs
        1999 and later .... 180 in-lbs

1 General information

Refer to illustrations 1.5a, 1.5b and 1.5c

To prevent pollution of the atmosphere from burned and evaporating gases, a number of emissions control systems are incorporated on the vehicles covered by this manual. The combination of systems used depends on the year in which the vehicle was manufactured, the locality to which it was originally delivered and the engine type. The major systems incorporated on the vehicles with which this manual is concerned include the:

    Fuel Control System
    Exhaust Gas Recirculation (EGR) system
    Evaporative Emissions Control (EVAP) system
    Transmission Converter Clutch (TCC) system
    Positive Crankcase Ventilation (PCV) system
    Catalytic converter

All of these systems are linked, directly or indirectly, to the On Board Diagnostic (OBD) system. The Sections in this Chapter include general descriptions, checking procedures (where possible) and component replacement procedures (where applicable) for each of the systems listed above.

Before assuming that an emissions control system is malfunctioning, check the fuel and ignition systems carefully. In some cases special tools and equipment, as well as specialized training, are required to accurately diagnose the causes of a rough running or difficult to start engine. If checking and servicing become too difficult, or if a procedure is beyond the scope of the home mechanic, consult your dealer service department. This does not necessarily mean, however, that the emissions control systems are particularly difficult to maintain and repair. You can quickly and easily perform many checks and do most (if not all) of the regular maintenance at home with common tune-up and hand tools. **Note:** *The most frequent cause of emissions system problems is simply a loose or broken vacuum hose or wiring connection. Therefore, always check the hose and wiring connections first.*

Pay close attention to any special precautions outlined in this Chapter. It should be noted that the illustrations of the various systems may not exactly match the system installed on your particular vehicle due to changes made by the manufacturer during production or from year to year.

A Vehicle Emissions Control Information (VECI) label is located in the engine compartment of all vehicles with which this manual is concerned **(see illustrations)**. This label contains important emissions specifications and setting procedures, as well as a vacuum hose schematic with emissions components identified. When servicing the engine or emissions systems, the VECI label in your particular vehicle should always be checked for up-to-date information. **Note:** *Because of a federally mandated extended warranty which covers the emission control system components (and any components which have a primary purpose other than emission control but have significant effects on emissions), check with your dealer about warranty coverage before working on any emission related systems.*

The number of emissions control system components on later model fuel-injected vehicles has actually decreased due to the high efficiency of the new fuel injection and ignition systems. These models are equipped with a three way catalytic converter containing beads which are coated with a catalyst material containing platinum, palladium and rhodium to reduce the level of nitrogen oxides.

1.5a A Vehicle Emissions Control Information (VECI) label will be found in the engine compartment of all vehicles - if it's missing, obtain a new one from a dealer parts department

Chapter 6 Emissions and engine control systems

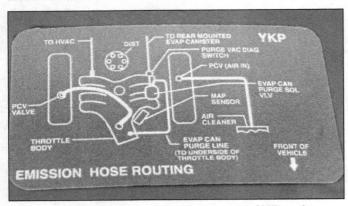

1.5b Typical vacuum schematic for the V6 CMFI engine

1.5c Typical vacuum schematic for the four-cylinder MFI engine

2 Self-diagnosis system and trouble codes

Note: *1995 and earlier models are equipped with the OBD I self diagnosis system, while 1996 and later models are equipped with the OBD II self diagnosis system. 1994 and later models require the use of a Scan tool to access trouble codes. However, many of the information sensor checks and replacement procedures do apply to both systems. Because 1994 and later systems require a special SCAN tool to access the trouble codes, have the vehicle diagnosed by a dealer service department or other qualified automotive repair facility if the proper SCAN tool is not available. 1994 V6 models and all 1996 and later models utilize five-digit trouble codes. These "generic" trouble codes listed in the following table do not include the manufacturer's specific trouble codes. Consult a dealer service department or other qualified repair shop for additional information. Refer to the troubleshooting tips in the beginning of this manual and the information described in Section 4 to gain some insight to the most likely causes of a problem.*

Diagnostic tool information

Refer to illustrations 2.1, 2.2 and 2.4

1 A digital multimeter is a necessary tool for checking fuel injection and emission related components **(see illustration)**. A digital volt-ohmmeter is preferred over the older style analog multimeter for several reasons. The analog multimeter cannot display the volts-ohms or amps measurement in hundredths and thousandths increments. When working with electronic circuits which are often very low voltage, this accurate reading is most important. Another good reason for the digital multimeter is the high impedance circuit. The digital multimeter is equipped with a high resistance internal circuitry (10 million ohms). Because a voltmeter is hooked up in parallel with the circuit when testing, it is vital that none of the voltage being measured should be allowed to travel the parallel path set up by the meter itself. This dilemma does not show itself when measuring larger amounts of voltage (9 to 12 volt circuits) but if you are measuring a low voltage circuit such as the oxygen sensor signal voltage, a fraction of a volt may be a significant amount when diagnosing a problem.

2 Hand-held scanners are the most powerful and versatile tools for analyzing engine management systems used on later model vehicles **(see illustration)**. Early model scanners handle codes and some diagnostics for many OBD I systems. Each brand scan tool must be examined carefully to match the year, make and model of the vehicle you are working on. Often interchangeable cartridges are available to access the particular manufacturer (Ford, GM, Chrysler, etc.). Some manufacturers will specify by continent; Asia, Europe, USA, etc.

3 With the arrival of the federally mandated emission control system (OBD II), a specially designed scanner has been developed. Several tool manufacturers have released OBD II scan tools for the home mechanic. Ask the parts salesperson at a local auto parts store for additional information concerning dates and costs. **Note:** *Although 1994 and 1995 OBDI and 1996 OBD II codes cannot be accessed without a Scan tool, follow the simple component checks in Section 4.*

4 Another type of code reader is available at parts stores **(see illustration)**. These tools simplify the procedure for extracting codes from the engine management computer by simply "plugging in" to the diagnostic connector on the vehicle wiring harness and are much less expensive (however, they will not work on models that require the use of a scan tool to extract codes).

General description

5 The electronically controlled fuel and emissions system is linked with many other related engine management systems. It consists mainly of sensors, output actuators and an Electronic Control Module (ECM) or Powertrain Control Module (PCM) (see Section 1). Completing the system are various other components which respond to commands from the ECM/PCM.

6 In many ways, this system can be compared to the central nervous system in the human body. The sensors (nerve endings) constantly gather information and send this data to the ECM/PCM (brain), which processes the data and, if necessary, sends out a command for some type of vehicle change (limbs).

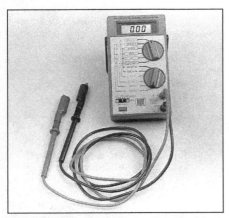

2.1 Digital multimeters can be used for testing all types of circuits; because of their high impedance, they are much more accurate than analog meters for measuring millivolts in low-voltage computer circuits

2.2 Scanners like the Actron Scantool and the AutoXray XP240 are powerful diagnostic aids - programmed with comprehensive diagnostic information, they can tell you just about anything you want to know about your engine management system

Chapter 6 Emissions and engine control systems

2.4 Trouble code tools simplify the task of extracting the trouble codes

2.11a The 12-pin Assembly Line Data Link (ALDL) terminal identification

- A Ground
- B Diagnostic TEST terminal

7   Here's a specific example of how one portion of this system operates: An oxygen sensor, mounted in the exhaust manifold and protruding into the exhaust gas stream, constantly monitors the oxygen content of the exhaust gas as it travels through the exhaust pipe. If the percentage of oxygen in the exhaust gas is incorrect, an electrical signal is sent to the ECM/PCM. The ECM/PCM takes this information, processes it and then sends a command to the fuel injectors, telling it to change the fuel/air mixture. To be effective, all this happens in a fraction of a second, and it goes on continuously while the engine is running. The end result is a fuel/air mixture which is constantly kept at a predetermined ratio, regardless of driving conditions.

Obtaining trouble codes

Refer to illustrations 2.11a and 2.11b

8   One might think that a system which uses exotic electrical sensors and is controlled by an on-board computer would be difficult to diagnose. This is not necessarily the case.

9   The On Board Diagnostic (OBD) system has a built-in self-diagnostic system, which indicates a problem by turning on a "SERVICE ENGINE SOON" light on the instrument panel when a fault has been detected. **Note:** *Since some of the trouble codes do not set the CHECK ENGINE or SERVICE ENGINE SOON light, it is a good idea to access the OBD system and look for any trouble codes that may have been recorded and need tending.*

10   Perhaps more importantly, the ECM/PCM will recognize this fault, in a particular system monitored by one of the various information sensors, and store it in its memory in the form of a trouble code. Although the trouble code cannot reveal the exact cause of the malfunction, it greatly facilitates diagnosis as you or a dealer mechanic can "tap into" the ECM/PCM's memory and be directed to the problem area.

11   To retrieve this information from the ECM on 1993 and earlier models, you must use a short jumper wire to ground a diagnostic terminal. The terminal is part of an electrical connector called the Assembly Line Data Link (ALDL) **(see illustrations)**. The ALDL is located just behind the dashboard, next to the steering column. On some models a small rectangular plate is used to cover the connector and must be pried off to provide access to the terminals. With the electrical connector exposed, push one end of the jumper wire into the diagnostic TEST terminal and the other end into the GROUND terminal. **Note:** *1994 and 1995 models with a 12-pin diagnostic connector do not have a terminal B present in the connector. On these models a scan tool is required to access trouble codes.*

12   On 1994 and later models, a scan tool must be connected to the Assembly Line Data Link (ALDL). The scan tool is a hand held digital computer scanner that interfaces with the on-board computer. The scan tool is a very powerful tool; it not only reads the trouble codes but also displays the actual operating conditions of the sensors and actuators. Scan tools are necessary to accurately diagnose a modern computerized fuel-injected engine. Scan tools are available from auto parts stores and specialty tool companies.

13   It should be noted that the self-diagnosis feature built into this system does not detect all possible faults. If you suspect a problem with the On Board Diagnostic (OBD) system, but the CHECK ENGINE or SERVICE ENGINE SOON light has not come on and no trouble codes have been stored, and performing the checks described in Section 4 doesn't pinpoint a problem, take the vehicle to a dealer service department or other qualified repair shop for diagnosis.

14   Furthermore, when diagnosing an engine performance, fuel economy or exhaust emissions problem (which is not accompanied by a CHECK ENGINE or SERVICE ENGINE SOON light) do not automatically assume the fault lies in this system. Perform all standard troubleshooting procedures, as indicated elsewhere in this manual, before turning to the On Board Diagnostic (OBD) system.

15   Finally, since this is an electronic system, you should have a basic knowledge of automotive electronics before attempting any diagnosis. Damage to the ECM/PCM, Programmable Read Only Memory (PROM) calibration unit or related components can easily occur if care is not exercised.

Clearing trouble codes

16   To clear the trouble codes from the ECM's memory on a 1993 or earlier model, unplug the ECM electrical pigtail at the positive battery cable and wait at least 30 seconds before plugging it back in. If the vehicle you are working on does not have this connector, disconnect the cable from the negative terminal of the battery for at least 30 seconds. **Caution 1:** *To prevent damage to the ECM, the ignition switch must be turned OFF when disconnecting or connecting power to the ECM.* **Caution 2:** *If the vehicle is equipped with a Delco Loc II audio system, make sure you have the correct activation code before disconnecting the battery. See the information at the front of this manual for the radio reactivation procedure.*

2.11b The 16 pin Data Link Connector (DLC) found on models equipped with OBD II

17   To clear the codes from the ECM/PCM memory on 1994 and later models, install the SCAN tool, scroll the menu for the function that describes "CLEARING CODES" and follow the prescribed method for that particular SCAN tool or momentarily remove the PCM/IGN fuse from the fuse box for 30 seconds. Clearing codes may also be accomplished by removing the fusible link (main power fuse) located near the battery positive terminal (see Chapter 12) or by disconnecting the cable from the negative terminal (-) of the battery. **Caution:** *If the vehicle is equipped with a Delco Loc II audio system, make sure you have the correct activation code before disconnecting the battery. See the information at the front of this manual for the radio reactivation procedure.*

18   Disconnecting the power to the ECM/PCM to clear the memory can be an important diagnostic tool, especially on intermittent problems.

19   **Note:** *Not all codes apply to all models.*

2-digit trouble codes

Code	Code definition
12	Diagnostic mode
13	Oxygen sensor or circuit
14	Coolant sensor or circuit/high temperature indicated
15	Coolant sensor or circuit/low temperature indicated
16	System voltage out of range
19	Crankshaft position sensor or circuit
21	Throttle Position Sensor (TPS) or circuit - voltage high
22	Throttle Position Sensor (TPS) or circuit - voltage low
23	Mixture Control (M/C) solenoid or circuit (carbureted models)
23	Manifold Air Temperature (MAT) sensor or circuit (1990 and earlier models)
23	Intake Air Temperature (IAT) sensor circuit (fuel-injected models)
24	Vehicle Speed Sensor (VSS) or circuit
25	Manifold Air Temperature (MAT) sensor or circuit - high temperature indicated (1990 and earlier models)
25	Intake Air Temperature (IAT) sensor or circuit - high temperature indicated (1991 and later models)
26	Quad Driver module circuit
27	Quad Driver module circuit
28	Quad Driver module circuit
29	Quad Driver module circuit
31	Park/Neutral Position (PNP) switch circuit
32	BARO sensor or circuit (carbureted models)
32	EGR circuit (fuel-injected models)
33	Manifold Absolute Pressure (MAP) sensor signal voltage high
33	Mass Air Flow (MAF) sensor or circuit - excessive airflow indicated
34	Manifold Absolute Pressure (MAP) sensor signal voltage low
34	Mass Air Flow (MAF) sensor signal - low airflow indicated
35	Idle Speed Control (ISC) switch or circuit (shorted) (carbureted models)
35	Idle Air Control (IAC) valve or circuit
38	Brake switch circuit
39	Torque Converter Clutch (TCC) circuit
41	No distributor signals to ECM, or faulty ignition module (carbureted models)
41	Cylinder select error - MEM-CAL or ECM problem (fuel-injected models)
41	Cam sensor circuit (3.8L engine)
42	Bypass or Electronic Spark Timing (EST) circuit
43	Low voltage at ECM terminal L (carbureted models)
43	Knock sensor circuit
44	Oxygen sensor or circuit - lean exhaust detected
45	Oxygen sensor or circuit - rich exhaust detected
46	Power steering pressure switch circuit
48	Misfire diagnosis
51	PROM, MEM-CAL or ECM problem
52	CALPAK or ECM problem
53	EGR fault (carbureted models only)
53	System over-voltage (ECM over 17.7 volts)
54	Mixture Control (M/C) solenoid or circuit (carbureted models)
54	Fuel pump circuit (1986 and later models)
55	Oxygen sensor circuit or ECM
55	Fuel lean monitor (2.2L engine)
61	Oxygen sensor signal faulty (possible contaminated sensor)
62	Transaxle gear switch signal circuits
63	Manifold Absolute Pressure (MAP) sensor voltage high (low vacuum detected)
64	Manifold Absolute Pressure (MAP) sensor voltage low (high vacuum detected)
66	Air conditioning pressure sensor or circuit

5-digit trouble codes

Code	Code definition
P0101	Mass Air Flow (MAF) sensor error
P0102	Mass Air Flow (MAF) sensor circuit - low frequency detected
P0103	Mass Air Flow (MAF) sensor - high frequency detected
P0106	Manifold Absolute Pressure (MAP) sensor performance fault
P0107	Manifold Absolute Pressure (MAP) sensor circuit low input
P0108	Manifold Absolute Pressure (MAP) sensor circuit high input
P0112	Intake Air Temperature (IAT) sensor circuit low input
P0113	Intake Air Temperature (IAT) sensor circuit high input
P0117	Engine Coolant Temperature (ECT) sensor circuit low input
P0118	Engine Coolant Temperature (ECT) sensor circuit high input
P0121	Throttle Position Sensor (TPS) range/performance fault
P0122	Throttle Position Sensor (TPS) circuit low input
P0123	Throttle Position Sensor (TPS) circuit high input
P0125	Engine Coolant Temperature (ECT) takes too long to enter closed loop
P0131	Upstream heated O2 sensor circuit low voltage (Bank 1, Sensor 1)
P0132	Upstream heated O2 sensor circuit high voltage (Bank 1, Sensor 1)
P0133	Upstream heated O2 sensor slow response (lazy sensor) (Bank 1, Sensor 1)
P0134	Upstream heated O2 sensor insufficient activity (Bank 1, Sensor 1)
P0135	Upstream heated O2 sensor - heater circuit fault (Bank 1, Sensor 1)
P0137	Downstream heated O2 sensor circuit low voltage (Bank 1, Sensor 2)
P0138	Downstream heated O2 sensor circuit high voltage (Bank 1, Sensor 2)
P0140	Downstream heated O2 sensor insufficient activity (Bank 1, Sensor 2)
P0141	O2 sensor heater circuit fault (Bank 1, Sensor 2)
P0143	HO2S Circuit Low Voltage Bank 1 Sensor 3
P0144	HO2S Circuit High Voltage Bank 1 Sensor 3
P0146	HO2S Circuit Insufficient Activity Bank 1 Sensor 3
P0147	HO2S Heater Performance Bank 1 Sensor 3
P0151	HO2S Circuit Low Voltage Bank 2 Sensor 1
P0152	HO2S Circuit High Voltage Bank 2 Sensor 1
P0153	HO2S Slow Response Bank 2 Sensor 1
P0154	HO2S Circuit Insufficient Activity Bank 2 Sensor 1
P0155	HO2S Heater Performance Bank 2 Sensor 1
P0171	System Adaptive fuel too lean
P0172	System Adaptive fuel too rich
P0191	Injector Pressure sensor system performance
P0192	Injector Pressure sensor circuit low input
P0193	Injector Pressure sensor circuit high input
P0200	Injector circuit control
P0300	Cylinder misfire detected (random)
P0301	Cylinder number 1 misfire detected

Chapter 6 Emissions and engine control systems

5-digit trouble codes

Code	Code definition
P0302	Cylinder number 2 misfire detected
P0303	Cylinder number 3 misfire detected
P0304	Cylinder number 4 misfire detected
P0305	Cylinder number 5 misfire detected
P0306	Cylinder number 6 misfire detected
P0321	Crankshaft Position sensor circuit fault
P0325	Knock sensor circuit fault
P0326	Knock sensor circuit performance
P0327	Knock sensor noise channel - low voltage
P0335	Crankshaft Position sensor circuit
P0336	Crankshaft reference signal circuit
P0337	Crankshaft Position sensor circuit
P0338	Crankshaft Position sensor circuit
P0339	Crankshaft Position (CKP) Sensor Circuit Intermittent
P0340	Camshaft Position sensor circuit
P0341	Camshaft Position sensor circuit
P0342	Camshaft Position sensor circuit, low voltage
P0351	COP ignition coil 1 primary circuit fault
P0352	COP ignition coil 2 primary circuit fault
P0353	COP ignition coil 3 primary circuit fault
P0354	COP ignition coil 4 primary circuit fault
P0400	EGR flow fault
P0401	Exhaust Gas Recirculation (EGR) system flow insufficient
P0402	EGR excessive flow detected
P0410	Secondary Air Injection (AIR) system
P0418	Secondary air injection pump relay system
P0420	Catalyst system efficiency below threshold (Bank 1)
P0421	Catalyst system efficiency below threshold (Bank 1)
P0430	Catalyst system efficiency below threshold (Bank 2)
P0431	Catalyst system efficiency below threshold (Bank 2)
P0440	EVAP system
P0441	EVAP system - no flow during purge cycle
P0442	EVAP system, small leak detected
P0443	EVAP VMV circuit fault
P0446	EVAP canister vent blocked
P0452	EVAP fuel tank pressure sensor low input
P0453	EVAP fuel tank pressure sensor high input
P0461	Fuel level sensor circuit
P0462	Fuel level sensor, low voltage
P0463	Fuel level sensor, high voltage
P0500	VSS circuit error
P0502	VSS circuit low input
P0503	VSS circuit range performance
P0506	IAC system rpm lower than expected
P0507	IAC system rpm higher than expected
P0530	A/C refrigerant pressure sensor circuit
P0560	System voltage out of range
P0562	System voltage low
P0563	System voltage high
P0600	PCM serial data communication link fault
P0601	PCM memory problem
P0602	PCM control module programming error
P0650	Quad driver module circuit
P0703	TCC brake switch input circuit fault
P0705	Transaxle Range sensor circuit malfunction
P0706	Transaxle Range sensor performance
P0712	Transaxle fluid temperature sensor circuit low input
P0713	Transaxle fluid temperature sensor circuit high input
P0740	TCC circuit fault
P0755	Shift solenoid circuit fault
P1106	Manifold Absolute Pressure (MAP) Sensor Circuit Intermittent High Voltage
P1107	Manifold Absolute Pressure (MAP) Sensor Circuit Intermittent Low Voltage
P1111	Intake Air Temperature (IAT) Sensor Circuit Intermittent High Voltage
P1112	Intake Air Temperature (IAT) Sensor Circuit Intermittent Low Voltage
P1114	Engine Coolant Temperature (ECT) Sensor Circuit Intermittent Low Voltage
P1115	Engine Coolant Temperature (ECT) Sensor Circuit Intermittent High Voltage
P1121	Throttle Position (TP) Sensor Circuit Intermittent High Voltage
P1122	Throttle Position (TP) Sensor Circuit Intermittent Low Voltage
P1133	O2 sensor, insufficient switching
P1134	HO2S Transition Time Ratio Bank 1 Sensor 1
P1153	HO2S Insufficient Switching Bank 2 Sensor 1
P1154	HO2S Transition Time Ratio Bank 2 Sensor 1
P1171	Fuel system lean during acceleration
P1336	CKP, system variation not learned
P1345	Crankshaft Position (CKP) or Camshaft Position (CMP) Correlation
P1351	Ignition Coil Control Circuit High Voltage
P1361	Ignition Control (IC) Circuit Low Voltage
P1380	EBCM DTC detected, data unusable
P1381	Misfire detected, no EBCM/PCM serial data
P1406	Exhaust Gas Recirculation (EGR) Position Sensor Performance
P1415	Secondary air injection system malfunction - left bank
P1416	Secondary air injection system malfunction - right bank
P1441	EVAP system flow during non-purge cycle
P1508	Idle Speed Low - Idle Air Control (IAC) System Not Responding
P1509	Idle Speed High - Idle Air Control (IAC) System Not Responding
P1621	PCM memory performance
P1626	Theft deterrent system, fuel-enable circuit
P1631	Theft deterrent system, incorrect password
P1632	Theft deterrent system, fuel disabled

3 Powertrain Control Module (PCM) - check and replacement

Note: *Four-cylinder MFI and V6 TBI models are equipped with a replaceable MEM-CAL in the PCM that must be installed into a new computer (PCM) when exchanged. V6 CMFI and CSEFI models are equipped with an (non-replaceable) EEPROM that must be recalibrated with a special factory SCAN tool (TECH 1) after replacement of the PCM.*

PCM check

1   The PCM on V6 CMFI and CSEFI engines (OBD II) is located in the right corner of the engine compartment near the shock tower. The PCM on four-cylinder MFI and V6 TBI engines (OBD I) is located under the passenger side glovebox.

2   Using the tips of the fingers, tap vigorously on the side of the computer while the engine is running. If the computer is not functioning properly, the engine may stumble or stall and display glitches on the engine data stream obtained using a SCAN tool or other diagnostic equipment.

3   If the PCM fails this test, check the electrical connectors. Each connector is color coded to fit the respective slot in the computer body. If their are no obvious signs of damage, have the unit checked at a dealer service department.

PCM replacement

Refer to illustrations 3.6, 3.7 and 3.10

Caution: *To prevent damage to the PCM, the ignition switch must be turned Off when disconnecting or connecting in the PCM connectors.*

4   The Powertrain Control Module (PCM) on four-cylinder MFI, V6 CMFI and CSEFI engines (OBD II) is located in the right corner of the engine compartment near the shock tower. The PCM on four-cylinder MFI and V6 TBI engines (OBD I) is located under the passenger side glovebox.

5   Disconnect the cable from the negative battery terminal. **Caution:** *On models equipped with a Delco-Loc II audio system,*

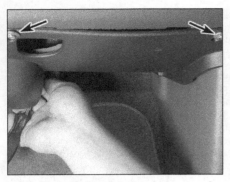

3.6 Remove the glovebox lower panel bolts (arrows) for access to the PCM (OBD-I models)

3.7 Remove the bolts (arrows) from the PCM (1995 V6 OBD II model shown)

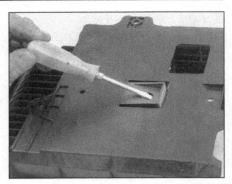

3.10 Lift the clip and separate the PCM cover from the computer

be sure the lockout feature is turned off before disconnecting the battery cable.

6   On four-cylinder MFI and V6 TBI models, remove the lower glovebox panel screws and panel **(see illustration)**. Remove the mounting nuts and lower the PCM.
7   On OBD II models, remove the bolts that retain the PCM to its mounting location in the engine compartment **(see illustration)**.
8   Carefully lift the PCM from the engine compartment without damaging the electrical connectors and wiring harness to the computer.
9   Unplug the electrical connectors from the PCM. Each connector is color coded to fit its respective receptacle in the computer.
10   On V6 CMFI and CSEFI models, pry the tang and lift the plastic protector cover from the PCM **(see illustration)**.
11   Installation is the reverse of removal.

MEM-CAL replacement (four-cylinder MFI and V6 TBI engines)

Refer to illustrations 3.14 and 3.15

12   Four-cylinder MFI and V6 TBI models are equipped with a memory calibration chip called the MEM-CAL. New PCMs are not supplied with a MEM-CAL, therefore it is necessary to remove the original and install it into the new PCM. Use care not to damage the MEM-CAL when working with these delicate electrical components.
13   Remove the PCM from the area under the glovebox (see Step 6).
14   Remove the screws and detach the MEM-CAL cover from the PCM **(see illustration)**.
15   Using two fingers, carefully push both retaining clips back away from the MEM-CAL and lift the unit straight out of the computer (PCM) **(see illustration)**. **Note:** *There are two types of clips used on the MEM-CAL sockets; hollow type tabs or solid type tabs.*
16   Installation is the reverse of removal. Be sure to use the alignment notches in the seat area of the computer when sliding the MEM-CAL back in place. Press only on the ends of the MEM-CAL assembly until it snaps back into place.

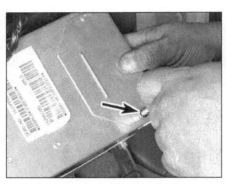

3.14 Remove the MEM-CAL cover screws (arrows) from the PCM (1995 four-cylinder MFI model shown)

EEPROM replacement (V6 CMFI and CSEFI engines)

17   The V6 CMFI and CSEFI OBD II models are equipped with Electrical Erasable Programmable Read Only Memory (EEPROM) chip that is permanently soldered to the computer circuit boards. The EEPROM can be reprogrammed using the GM dealer's TECH 1 SCAN tool. Do not attempt to remove this component from the PCM. Have the EEPROM reprogrammed at a dealer service department.

---

4   Information sensors - check and replacement

Caution 1: *When performing the following tests, use only a high-impedance (10 megohm) digital multi-meter to prevent damage to the PCM.*
Caution 2: *On models equipped with a Delco-Loc II audio system, be sure the lockout feature is turned off before disconnecting the battery cable.*
Note: *The following checking procedures do not include models equipped with the OBD II system. These systems require a special SCAN tool in order to read out the various levels of coded information. Have the vehicle diagnosed by a dealer service department in the event of any fuel injection or emissions component failure. After performing any checking procedure to any of the informa-*

3.15 Press the retaining tabs out to release the MEM-CAL

tion sensors, be sure to clear the PCM of all trouble codes by removing the PCM BAT fuse from the fuse box or disconnecting the cable from the negative terminal of the battery for at least ten seconds.

Engine Coolant Temperature (ECT) sensor

General description

1   The coolant temperature sensor is a thermistor (a resistor which varies the value of its resistance in accordance with temperature changes). The change in the resistance value will directly affect the voltage signal from the coolant thermosensor. As the sensor temperature DECREASES, the resistance will INCREASE. As the sensor temperature INCREASES, the resistance will DECREASE. A failure in the coolant sensor circuit should set a Code 14, 15, P0117, P0118, P1114 or P1115. These codes indicate a failure in the coolant temperature circuit, so the appropriate solution to the problem will be either repair of a wire or replacement of the sensor. The sensor can also be checked with an ohmmeter.

Check

Refer to illustrations 4.2, 4.3, 4.4 and 4.5

2   To check the sensor, check the resistance of the coolant temperature sensor while it is completely cold (50 to 80-degrees F = approximately 5,700 to 2,200 ohms). Next, start the engine and warm it up until it reaches operating temperature. The resistance should

Chapter 6 Emissions and engine control systems

4.2 Location of the engine coolant temperature (ECT) sensor (arrow) on the four-cylinder MFI engine

4.3 Check the REFERENCE voltage from the computer to the ECT sensor

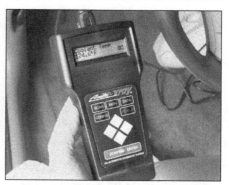

4.4 Plug the SCAN tool connector into the ALDL, switch the SCAN tool into ECT mode and observe the engine temperature through the computer data stream

4.5 To prevent coolant leakage, be sure to wrap the temperature sensor threads with Teflon tape before installation

VOLTAGE RANGE	ALTITUDE
3.8 - 5.5V	Below 1,000
3.6 - 5.3V	1,000 - 2,000
3.5 - 5.1V	2,000 - 3,000
3.3 - 5.0V	3,000 - 4,000
3.2 - 4.8V	4,000 - 5,000
3.0 - 4.6V	5,000 - 6,000
2.9 - 4.5V	6,000 - 7,000
2.8 - 4.3V	7,000 - 8,000
2.6 - 4.2V	8,000 - 9,000
2.5 - 4.0V	9,000 - 10,000 FEET
LOW ALTITUDE = HIGH PRESSURE = HIGH VOLTAGE	

4.9 Typical Manifold Absolute Pressure (MAP) sensor altitude (pressure) vs. voltage values

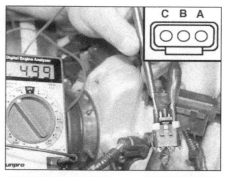

4.10 Check the MAP sensor REFERENCE voltage on the black wire (-) and the gray wire (+). It should be approximately 5.0 volts

be lower (170 to 200-degrees F = approximately 170 to 450 ohms). **Note:** *The coolant temperature sensor on the four-cylinder engine is located on the thermostat housing* **(see illustration)** *and on the V6 engine it is located on the front of the intake manifold. On V6 engines, access to the coolant temperature sensor makes it difficult to position probes of the meter onto the terminals. It may be necessary to remove the sensor and perform the tests in a pan of heated water to simulate the conditions.*

3 Check the reference voltage from the PCM to the ECT with the ignition key ON (engine not running). It should be approximately 5.0 volts **(see illustration)**.

4 Access to the coolant temperature sensor is limited, making it difficult to monitor the resistance changes without removing many components from the engine. If available, install a SCAN tool and switch to the ECT mode and monitor the temperature of a cold engine. The SCAN tool should indicate between 75 to 90-degrees F. Allow the engine to idle for several minutes and observe the coolant temperature increase as the engine warms up **(see illustration)**. The temperature should indicate between 180 to 210-degrees F at normal operating temperature. If there is not a definite change in temperature, remove the coolant temperature sensor and check the resistance in a pan of heated water to simulate warm-up conditions. If the sensor tests are

good, check the wiring harness from the sensor to the computer.

Replacement

Warning: *Wait until the engine is completely cool before beginning this procedure.*

5 Before installing the new sensor, wrap the threads with Teflon sealing tape to prevent leakage and thread corrosion **(see illustration)**.

6 To remove the sensor, release the locking tab, unplug the electrical connector, then carefully unscrew the sensor. **Caution:** *Handle the coolant sensor with care. Damage to this sensor will affect the operation of the entire fuel injection system.*

7 Installation is the reverse of removal. Check the coolant level and add some, if necessary (see Chapter 1).

Manifold Absolute Pressure (MAP) sensor

General description

Refer to illustration 4.9

8 The Manifold Absolute Pressure (MAP) sensor monitors the intake manifold pressure changes resulting from changes in engine load and speed and converts the information into a voltage output. The PCM uses the MAP sensor to control fuel delivery and ignition timing. A failure in the MAP sensor circuit should set a Code 33, 34, P0106, P0107, P0108, P1106 or P1107.

9 The MAP sensor SIGNAL voltage to the PCM varies from below 2 volts at idle (high vacuum) to above 4 volts with the ignition key ON (engine not running) or wide open throttle (WOT) (low vacuum). These values correspond with the altitude and pressure changes the vehicle experiences while driving **(see illustration)**. **Note:** *The MAP sensor is located near the throttle body on the intake manifold.*

Check

Refer to illustrations 4.10, 4.11a, 4.11b and 4.12

10 Check the REFERENCE voltage from the computer to the MAP sensor. Unplug the sensor, turn the ignition On and probe terminal A (gray [+]) and terminal C (black [-]) - the two outside terminals. The voltage should be approximately 5.0 volts **(see illustration)**.

11 Turn the ignition off, plug in the electrical connector and check the SIGNAL voltage. Connect the positive lead of the voltmeter to terminal B (light green [+]) of the sensor electrical connector and the negative lead to terminal C (black) [-]). With the ignition ON (engine not running) the voltage

4.11a Check the MAP sensor SIGNAL voltage on the light green wire (+) and the black wire (-) with the ignition key ON (engine not running) (V6 CMFI engine shown)

4.11b Start the engine and observe the MAP sensor voltage at idle. With the engine idling, vacuum in the intake manifold is high, therefore the voltage value will be low

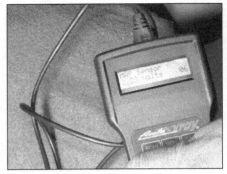
4.12 Connect a SCAN tool to the ALDL and switch to MAP sensor mode. Observe the voltage values as the engine rpm goes from idle to full throttle. The voltage should increase as engine rpm increases (vacuum decreases)

4.13 Removing the MAP sensor mounting bolts (arrows) on a four-cylinder MFI engine

4.15 Check the REFERENCE voltage of the IAT sensor. It should be approximately 5.0 volts

4.16 Check the IAT sensor resistance and compare the value at the existing temperature

reading should be about 4.5 to 5 volts (see illustration). Note: *It will probably be necessary to use pins or straightened-out paper clips to backprobe the terminals in the connector.* Start the engine and let it warm up. The voltage should be approximately 1.5 to 2.0 volts at idle (see illustration). Raise the engine rpm and observe that as vacuum decreases, voltage increases. At wide open throttle the MAP sensor SIGNAL should produce approximately 4.5 volts.

12 An alternate method of diagnosing the MAP sensor is by the use of an electronic SCAN tool. Install the SCAN tool and switch to the MAP mode and monitor the voltage signal with the engine at idle and high rpm (see illustration). The SCAN tool should indicate between 1.2 to 1.6 volts at idle. Raise the engine rpm and observe that as engine rpm increases (decreasing vacuum) the MAP SIGNAL voltage increases. If the MAP sensor voltage readings are incorrect, replace the MAP sensor.

Replacement

Refer to illustration 4.13

13 To replace the sensor, detach the vacuum hose, unplug the electrical connector and remove the mounting screws (see illustration). Installation is the reverse of removal.

Intake Air Temperature (IAT) sensor

General description

14 The Intake Air Temperature (IAT) sensor is located inside the air duct directly downstream of the air filter housing. This sensor acts as a thermistor (a resistor which changes the value of its resistance as the temperature changes). When the intake air is cold, the sensor resistance is high, therefore the PCM will read a high signal voltage. If the intake air is warm, the resistance is low giving the PCM a low voltage reading. The sensor values range from 2,700 ohms at 70-degrees F to 240 ohms at 190-degrees F. The PCM supplies approximately 5.0 volts (REFERENCE voltage) to the IAT sensor. A failure in the IAT sensor circuit should set a Code 23, 25, P0112, P0113, P1111 or P1112.

Check

Refer to illustrations 4.15 and 4.16

15 Measure the REFERENCE voltage. Disconnect the IAT electrical sensor and with the ignition key ON (engine not running), probe terminal A ground (black) and terminal B REF (tan) (see illustration). The meter should read approximately 5.0 volts.

16 Next, measure the resistance across the sensor terminals (see illustration). By measuring its resistance when cold, then warming it up (a hair dryer can be used for this) and taking another measurement. The resistance of the IAT sensor should be HIGH when the temperature is low. Next, start the engine and let it idle. Wait awhile and let the engine reach operating temperature. Turn the ignition key OFF, disconnect the IAT temperature sensor and check the resistance across the terminals. The resistance should be LOW when the air temperature is high. If the sensor does not exhibit this change in resistance, replace the sensor with a new part.

17 An alternate method of diagnosing the IAT sensor is by the use of an electronic SCAN tool. Install the SCAN tool and switch to the IAT mode and monitor the temperature with the engine cold and then completely warmed-up. The SCAN tool temperature ranges should increase while the IAT sensor resistance decreases. If the IAT sensor temperature/resistance readings are incorrect, replace the IAT sensor.

Replacement

Refer to illustration 4.18

18 To remove an IAT sensor, unplug the electrical connector and remove the sensor from the air intake duct. Carefully twist the sensor to release it from the rubber boot (see illustration).

19 Installation is the reverse of removal.

Chapter 6 Emissions and engine control systems

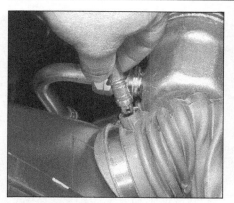

4.18 Remove the IAT sensor from the air intake duct

4.26a Check the O₂ sensor millivolt signal from the single wire O₂ sensor connector (four-cylinder MFI engine shown)

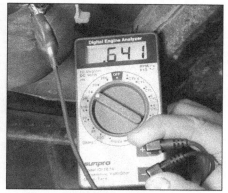

4.26b Backprobe the purple (+) wire on the O₂ connector to monitor the millivolt SIGNAL as the engine goes from cold to warm running conditions (V6 CMFI engine shown)

Oxygen sensor

General description

20 The oxygen sensor, which is located in the exhaust manifold, monitors the oxygen content of the exhaust gas stream. The oxygen content in the exhaust reacts with the oxygen sensor to produce a voltage output which varies from 0.1-volt (high oxygen, lean mixture) to 0.9-volts (low oxygen, rich mixture). The PCM constantly monitors this variable voltage output to determine the ratio of oxygen to fuel in the mixture. The PCM alters the air/fuel mixture ratio by controlling the pulse width (open time) of the fuel injectors. A mixture ratio of 14.7 parts air to 1 part fuel is the ideal mixture ratio for minimizing exhaust emissions, thus allowing the catalytic converter to operate at maximum efficiency. It is this ratio of 14.7 to 1 which the PCM and the oxygen sensor attempt to maintain at all times. Four-cylinder MFI engines are equipped with a single wire O₂ sensor (no heater), V6 TBI models are equipped with a single O₂ sensor (heated) and the V6 CMFI and CSEFI models are equipped with a left bank O₂ sensor (left exhaust manifold) and a post-catalytic converter O₂ sensor. When checking the oxygen sensor system, it will be necessary to test all three O₂ sensors.

21 The oxygen sensor produces no voltage when it is below its normal operating temperature of about 600-degrees F. During this initial period before warm-up, the PCM operates in OPEN LOOP mode.

22 If the engine reaches normal operating temperature and/or has been running for two or more minutes, and the computer detects a malfunction, the PCM will set one or more of the following codes, depending on the type of system your vehicle has. General oxygen sensor malfunction: 13, 44 or 45. Left bank oxygen sensor malfunction: P0131, P0132, P0133 or P0134. Right bank oxygen sensor malfunction: P0137 P0138, P0140 or P0141. Catalyst monitoring oxygen sensor malfunction: P0143, P0144, P0146 or P0147.

23 When there is a problem with the oxygen sensor or its circuit, the PCM operates in the open loop mode - that is, it controls fuel delivery in accordance with a programmed default value instead of feedback information from the oxygen sensor.

24 The proper operation of the oxygen sensor depends on four conditions:

a) *Electrical* - The low voltages generated by the sensor depend upon good, clean connections which should be checked whenever a malfunction of the sensor is suspected or indicated.

b) *Outside air supply* - The sensor is designed to allow air circulation to the internal portion of the sensor. Whenever the sensor is removed and installed or replaced, make sure the air passages are not restricted.

c) *Proper operating temperature* - The PCM will not react to the sensor signal until the sensor reaches approximately 600 degrees-F. This factor must be taken into consideration when evaluating the performance of the sensor.

d) *Unleaded fuel* - The use of unleaded fuel is essential for proper operation of the sensor. Make sure the fuel you are using is of this type.

25 In addition to observing the above conditions, special care must be taken whenever the sensor is serviced.

a) The oxygen sensor has a permanently attached pigtail and electrical connector which should not be removed from the sensor. Damage or removal of the pigtail or electrical connector can adversely affect operation of the sensor.

b) Grease, dirt and other contaminants should be kept away from the electrical connector and the louvered end of the sensor.

c) Do not use cleaning solvents of any kind on the oxygen sensor.

d) Do not drop or roughly handle the sensor.

e) The silicone boot must be installed in the correct position to prevent the boot from being melted and to allow the sensor to operate properly.

Check

Refer to illustrations 4.26a, 4.26b and 4.28

Note: *Access to the oxygen sensor(s) makes it very difficult but not impossible to backprobe the harness electrical connectors for testing purposes. The exhaust manifolds and pipes are extremely hot and will melt stray electrical probes and leads that touch the surface during testing. If possible, use a SCAN tool that plugs into the ALDL (diagnostic link). This tool will access the PCM data stream and indicates the millivolt changes for each individual O₂ sensor.*

26 Check the O₂ sensor millivolt signal. Locate the oxygen sensor electrical connector and carefully backprobe it using a long pin(s) into the appropriate wire terminals:

Four-cylinder MFI OBD I engines:

O₂ sensor harness   purple (+) (signal wire) and body (-) (ground)

V6 TBI OBD I engines:

O₂ sensor harness   purple (+) (signal wire) and black wire (-) (ground)

V6 CMFI and CSEFI engines:

Bank 1 sensor (left)   purple (+) (signal wire) and black wire (-) (ground)

Bank 2 sensor (after catalytic converter)   purple (+) (signal wire) and black wire (-) (ground)

Connect the positive probe of a voltmeter onto the signal wire pin and the negative probe to the ground wire. **Note:** *Consult the wiring diagrams at the end of Chapter 12 for additional information on the oxygen sensor electrical connector wire color designations.* Monitor the SIGNAL voltage (millivolts) as the engine goes from cold to warm **(see illustrations)**.

27 The oxygen sensor will produce a steady voltage signal of approximately 0.1 to 0.2 volts with the engine cold (open loop). After a period of approximately two minutes, the engine will reach operating temperature and the oxygen sensor will start to fluctuate between 0.1 to 0.9 volts (closed loop). If the oxygen sensor fails to reach the closed loop mode or there is a very long period of time until it does switch into closed loop mode, replace the oxygen sensor with a new part.

28 Inspect the oxygen sensor heater **(see**

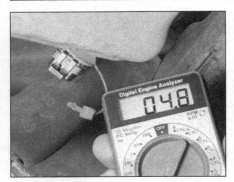

4.28 Check the resistance of the O₂ sensor heater

4.35 Use a special slotted socket to remove the O₂ sensor from the exhaust pipe

4.42 Check the TPS REFERENCE voltage on the gray (+) wire and the black (-) wire. It should be about 5.0 volts

4.43a Check the SIGNAL voltage on the dark blue (+) and black (-) wires at idle . . .

4.43b . . . then at wide open throttle. The voltage should increase smoothly to approximately 4.5 volts

illustration). Disconnect the oxygen sensor electrical connector and working on the O₂ sensor side, connect an ohmmeter between the:

V6 TBI engines - black wire (-) and pink wire (+)

V6 CMFI and CSEFI engines - black wire (-) and pink wire (+)

It should measure approximately 5 to 7 ohms. **Note 1:** *The four-cylinder engine is not equipped with heated O₂ sensors. This can be determined by the single wire.* **Note 2:** *On models equipped with O₂ sensor heaters, the wire colors often change from the harness to the actual O₂ sensor wires according to manufacturer's specifications. Follow the wire colors to the O₂ sensor electrical connector and determine the matching wires and their colors before testing the heater resistance.*

29 Check for proper supply voltage to the heater. Disconnect the oxygen sensor electrical connector and working on the computer side of the harness, measure the voltage between the:

V6 TBI engines - black wire (-) and pink wire (+)

V6 CMFI and CSEFI engines - black wire (-) and pink wire (+)

on the oxygen sensor electrical connector. There should be battery voltage with the ignition key ON (engine not running). If there is no voltage, check the circuit between the main relay, the PCM and the sensor. **Note:** *It is important to remember that supply voltage will only last approximately 2 seconds because the system uses a relay to divert the*

voltage (air conditioning relay), therefore it will be necessary to have an assistant turn the ignition key ON.

30 If the oxygen sensor fails any of these tests, replace it with a new part.

31 Access to the oxygen sensors can be difficult making it impossible to monitor the SIGNAL voltage changes without removing several components to gain access to the O₂ sensor electrical connectors. Install the SCAN tool and switch to the Oxygen Sensor mode and monitor the O₂ sensor crosscounts (varying millivolt signals). The SCAN tool should indicate approximately 100 to 200 millivolts when cold (LEAN condition) and then fluctuate from 300 to 800 millivolts warm (CLOSED LOOP).

Replacement

Refer to illustration 4.35

Note: *Because it is installed in the exhaust manifold or pipe, which contracts when cool, the oxygen sensor may be very difficult to loosen when the engine is cold. Rather than risk damage to the sensor (assuming you are planning to reuse it in another manifold or pipe) or the threads which it screws into, start and run the engine for a minute or two, then shut it off. Be careful not to burn yourself during the following procedure.*

32 Disconnect the cable from the negative terminal of the battery. **Caution:** *On models equipped with a Delco-Loc II audio system, be sure the lockout feature is turned off before disconnecting the battery cable.*

33 Raise the vehicle and place it securely on jackstands.

34 Disconnect the electrical connector from the sensor.

35 Carefully unscrew the sensor from the exhaust manifold **(see illustration)**.

36 Anti-seize compound must be used on the threads of the sensor to facilitate future removal. The threads of new sensors will already be coated with this compound, but if an old sensor is removed and reinstalled, recoat the threads.

37 Install the sensor and tighten it securely.

38 Reconnect the electrical connector of the pigtail lead to the main engine wiring harness.

39 Lower the vehicle, take it on a test drive and check to see that no trouble codes set.

Throttle Position Sensor (TPS)

General description

40 The Throttle Position Sensor (TPS) is located on the end of the throttle shaft on the throttle body. By monitoring the output voltage from the TPS, the PCM can determine fuel delivery based on throttle valve angle (driver demand). A broken or loose TPS can cause intermittent bursts of fuel from the injector and an unstable idle because the PCM thinks the throttle is moving.

41 A problem in any of the TPS circuits will set a Code 21, 22, P0121, P0122, P0123, P1121 or P1122. Once a trouble code is set, the PCM will use an artificial default value for TPS and some vehicle performance will return.

Check

Refer to illustrations 4.42, 4.43a and 4.43b

42 Locate the TPS on the throttle body. Using a voltmeter, check the REFERENCE voltage from the PCM. Install the positive probe (+) onto the gray (reference wire) and negative probe (-) onto the black (ground wire) **(see illustration)**. The meter should read approximately 5.0 volts.

43 Check the TPS signal voltage. With the throttle fully closed, install the positive probe (+) of the voltmeter onto the dark blue wire and the negative probe (-) onto the black wire **(see illustration)**. Gradually open the throttle valve and observe the TPS sensor voltage. Confirm a smooth change in the voltage values as the sensor travels from idle to full throttle. The voltage should increase to approximately 4.5

Chapter 6 Emissions and engine control systems

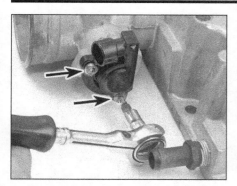

4.46 Remove the Torx bolts (arrows) from the TPS (plenum removed for clarity)

4.53 Check for battery voltage on the air conditioning clutch relay connector

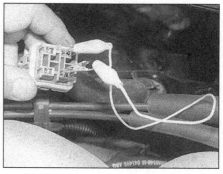

4.54 Apply battery voltage to the air conditioning clutch terminal using a jumper wire, making sure the clutch activates

to 5.0 volts **(see illustration)**. If the readings are incorrect, replace the TPS sensor.

44 An alternate method of diagnosing the TPS sensor is by the use of an electronic SCAN tool. Install the SCAN tool and switch to the TPS mode and monitor the voltage signal with the engine at idle and high rpm. The SCAN tool should indicate between 1.2 to 1.6 volts at idle. Raise the engine rpm and observe that as engine rpm increases (throttle angle) the TPS voltage increases to approximately 4.5 to 5.0 volts. If the TPS sensor voltage readings are incorrect, replace the TPS sensor.

Replacement
Refer to illustration 4.46

45 Disconnect the TPS electrical connector.
46 Remove the Torx drive bolts from the TPS **(see illustration)** and remove the TPS from the throttle body.
47 When installing the TPS, be sure to align the socket locating tangs on the TPS with the throttle shaft in the throttle body.
48 Installation is the reverse of removal.

Neutral Start switch

49 The Neutral Start switch, located on the rear upper part of the automatic transmission, indicates to the PCM when the transaxle is in Park or Neutral. This information is used for Transmission Converter Clutch (TCC), Exhaust Gas Recirculation (EGR) and Idle Air Control (IAC) valve operation. **Caution:** *The vehicle should not be driven with the Neutral Start switch disconnected because idle quality will be adversely affected.*
50 For more information regarding the Neutral Start switch, which is part of the Neutral start and back-up light switch assembly, see Chapter 7B.

Air conditioning control
Air conditioning clutch control

51 During air conditioning operation, the PCM controls the application of the air conditioning compressor clutch. The PCM controls the air conditioning clutch control relay to delay clutch engagement after the air conditioning is turned ON to allow the IAC valve to adjust the idle speed of the engine to compensate for the additional load. The PCM also controls the relay to disengage the clutch on WOT (wide open throttle) to prevent excessively high rpm on the compressor. Be sure to check the air conditioning system as detailed in Chapter 3

before attempting to diagnose the air conditioning clutch or electrical system.

Air conditioning "On" signal
Refer to illustrations 4.53 and 4.54

52 Turning on the air conditioning supplies battery voltage to the air conditioning compressor clutch of PCM electrical connector to increase idle air rate and maintain idle speed. In most cases, if the air conditioning does not function, the problem is probably related to the air conditioning system relays and switches and not the PCM.
53 Remove the air conditioning relay from the relay center and check for battery voltage to the relay **(see illustration)**. Battery voltage should exist with the ignition key ON (engine not running).
54 If battery voltage exists, install a jumper wire into the relay connector and observe that the air conditioning clutch activates **(see illustration)**. If the air conditioning clutch and relay system are working properly, check the air conditioning system pressures (see Chapter 3).

Vehicle Speed Sensor (VSS)
General description
Refer to illustration 4.55

55 The Vehicle Speed Sensor (VSS) is located in the transmission housing at the rear section near the output shaft **(see illustration)**. On 4WD models, the VSS is located in the top-left rear of the transfer case housing. This permanent magnet generator sends

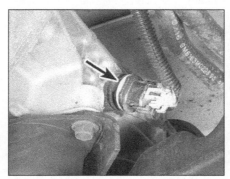

4.55 Location of the VSS on a manual transmission (four-cylinder MFI model shown)

a pulsing voltage signal to the PCM, which the PCM converts to miles per hour. The VSS is part of the Transmission Converter Clutch (TCC) system. A problem in the VSS control circuits will set a Code 24 or P0500.

Check
Refer to illustration 4.56

56 To check the VSS, disconnect the electrical connector in the wiring harness near the sensor. Check the REFERENCE voltage from the PCM by probing the yellow (+) and the purple (-) with a voltmeter **(see illustration)**. There should be approximately 5.0 volts present. If reference voltage is present, have the VSS diagnosed by a dealer service department.

Replacement

57 Detach the sensor retaining screw and bracket, unplug the sensor and remove it.
58 Installation is the reverse of removal.

Camshaft sensor (V6 CMFI and CSEFI engines)

Note: *For additional information concerning the camshaft sensor, refer to Chapter 5.*

General description

59 The camshaft sensor is located in the distributor which is located at the rear of the intake manifold. This component functions as an ignition device in the Enhanced Distributor Ignition system. The camshaft sensor's main function is to detect misfire sequences and set diagnostic trouble codes for the Powertrain

4.56 Check for REFERENCE voltage to the VSS (V6 MFI 4WD model shown)

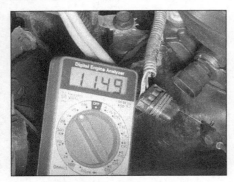

4.63 Check for battery voltage to the crankshaft sensor on the red wire (+)

4.64 Check the crankshaft sensor resistance on the purple (+) wire and the yellow wire (-) (four-cylinder MFI engine shown)

4.67 Remove the crankshaft sensor bolt (arrow) (four-cylinder MFI engine shown)

Control Module. In the event the reference signal is lost or interrupted, the PCM will continue to pulse the fuel injectors using a default value but the synchronizing effect may be slightly off (stumble or misfire) due to the lack of precision. If the cam signal is not received by the PCM, a code 41, P0340 or P0341 will set.

Crankshaft sensor (four-cylinder MFI and V6 CMFI and CSEFI engines)

Note: *For additional information concerning the crankshaft sensor circuit, refer to Chapter 5, illustration 6.13b (four-cylinder engine) and 6.23b (V6 CMFI and CSEFI) and the appropriate wiring diagrams at the end of Chapter 12.*

General description

60   The four-cylinder MFI engine and the V6 CMFI engine are equipped with crankshaft sensors. The crankshaft sensor on the four-cylinder is mounted next to the ignition module on the side of the engine block. The sensor protrudes into the block approximately within 0.050 inch of the crankshaft reluctor. The reluctor is a timing wheel cast into the crankshaft with seven slots machined into the perimeter of the wheel. Each slot is spaced 60-degrees apart and the seventh slot is spaced 10-degrees from one slot to serve as a synchronizing mark for TDC. Based on the pulses from the crankshaft sensor, the computer detects engine speed and crankshaft position which in turn allows it to calculate precise timing for the fuel injection and the ignition system. If a loss of signal is detected by the PCM, a code P0336, P0337, P0338 or P0339 will set.

4.69 A knock sensor mounted to the engine block (typical shown)

61   The crankshaft sensor on the 4.3L CMFI and CSEFI engines is mounted on the engine front cover and perpendicular to the crankshaft target wheel. This sensor reads positions on the crankshaft target wheel from 3 keyed slots positioned 60-degrees apart. The air gap between the target wheel and the sensor is preset at the factory and cannot be adjusted. The sensor detects pulses as the target wheel rotates and creates changes in the magnetic field between the slots and the crankshaft position sensor. These pulses relay information to the computer which allows precise timing controls for the fuel injection and ignition systems.

Check

Refer to illustrations 4.63 and 4.64

62   To check the crankshaft sensor, it is necessary to use a special SCAN tool. The crankshaft sensor and circuit must be monitored while the engine is in motion. However, there is a quick resistance check that can be performed on the crankshaft sensor to indicate a possible defective sensor. It is recommended the sensor be diagnosed by a dealer service department or other qualified repair shop.
63   Locate the crankshaft sensor, disconnect the electrical connector. On V6 CMFI engines, check for battery voltage from the PCM to the sensor **(see illustration)**. **Note:** *On four-cylinder MFI engines, check for battery voltage to the ignition module on the pink wire (see Chapter 5).*
64   Working on the sensor side, check the resistance of the crankshaft sensor **(see illustration)**. It should be between 800 to 1,200 ohms.

   Four-cylinder MFI engines - purple wire (+) and yellow wire (-)

   V6 CMFI engines - blue wire (+) and black wire (-)

Replacement

Refer to illustration 4.67

65   Disconnect the negative terminal from the battery.
66   Disconnect the electrical connector from the crankshaft sensor.
67   Remove bolts from the crankshaft sensor and remove the sensor **(see illustration)**.
68   Installation is the reverse of removal. Tighten the bolts to the torque listed in this Chapter's Specifications.

Knock sensors (V6 CMFI and CSEFI engines)

General description

Refer to illustration 4.69

69   On early model vehicles, a knock sensor is located on each side of the engine block; one sensor designated for each bank for a total of two **(see illustration)**. On later model vehicles, there is only one sensor located at the top of the engine near the distributor. Octane ratings vary the performance of engines and often detonation leads to "spark knock." To control spark knock, the knock control system detects abnormal vibration in the engine. This system is designed to reduce spark knock up to 10-degrees during periods of heavy detonation. This allows the engine to use maximum spark advance to improve driveability. The knock sensor produces an AC output voltage which increases with the severity of the knock. The signal is fed into the PCM and the timing is retarded up to 10 degrees to compensate for the severe detonation. Any problems with the knock sensor circuit will set a code 43.

Check

70   Disconnect the electrical connector for the knock sensor and with the ignition key ON (engine not running), check the REFERENCE voltage from the computer using a voltmeter. It should be approximately 5.0 volts.
71   Disconnect the electrical connector for the knock sensor and using an ohmmeter, check the resistance of the sensor. It should be approximately 8,200 ohms.
72   If the knock sensor resistance is correct but there is no REFERENCE voltage available, have the vehicle checked at a dealer service department or other repair shop equipped with the necessary tools.

Replacement

Note: *On later model vehicles where the sensor is located near the distributor, it may be necessary to remove other components for access to the knock sensor.*

73   Disconnect the cable from the negative terminal of the battery.
74   If necessary, raise the vehicle and support it securely on jackstands.
75   Disconnect the knock sensor electrical harness connector.
76   Unscrew the sensor from the engine block.
77   Installation is the reverse of removal.

Chapter 6 Emissions and engine control systems

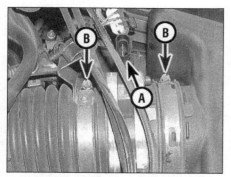

4.79 The electrical connector (A) and hose clamps (B) (typical shown)

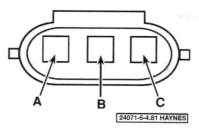

4.81 MAF engine harness connector details

A   MAF sensor signal
B   Ground
C   Ignition 1 voltage

5.6 Apply vacuum directly to the EGR valve and observe the engine rpm decrease and then gradually stall the engine

Mass Airflow (MAF) sensor
General description

Refer to illustration 4.79

78 A Mass Airflow sensor (MAF) is used on V6 models only. The MAF sensor measures the amount of air passing through the sensor body and ultimately entering the engine through the throttle body. The PCM uses this information to control fuel delivery - the more air entering the engine (acceleration), the more fuel required. A failure in the MAF sensor circuit should set a Code 33, 34, P0101, P0102 or P0103.
79 The PCM sends a voltage to the MAF sensor SIGNAL circuit that the sensor uses to create a frequency. The frequency varies between 2000 Hertz (at idle) to about 10,000 Hertz (when the engine is at maximum load). **Note:** *The MAF sensor is part of the air intake duct directly downstream of the air filter housing* **(see illustration).**

Check

Refer to illustration 4.81

80 With the engine at proper operating temperature, disconnect the electrical connector from the MAF sensor **(see illustration 4.79)**.
81 Check terminal B of the harness connector for proper ground **(see illustration)**. There should be no resistance. Resistance of 20 ohms or more can result in a code being set.
82 Check the IGNITION 1 voltage going to the MAF sensor. With the sensor disconnected, turn the ignition ON then probe terminal C (+) and terminal B (-) of the harness connector. A voltage reflecting the ignition feed circuit should be present **(see illustration 4.81)**.
83 Check the REFERENCE voltage from the computer to the MAF sensor. With the sensor disconnected, turn the ignition ON then probe terminal A (+) and terminal B (-) of the harness connector. The voltage should be approximately 5.0 volts; no more than 6 volts and no less than 4 volts **(see illustration 4.81)**.
84 A scan tool, that can read MAF signal parameters, is necessary for any further testing of the MAF sensor and related circuits.

Replacement

85 Disconnect the electrical connector from the intake air temperature (IAT) sensor and the MAF sensor.
86 Loosen the hose clamp securing the air intake duct to the MAF sensor and remove the duct **(see illustration 4.79)**.
87 Loosen the hose clamp retaining the MAF sensor to the air filter cover and remove the sensor.
88 Installation is the reverse of removal.

5 Exhaust Gas Recirculation (EGR) system

Negative Backpressure EGR system (four-cylinder MFI engine - OBD-I)
General description

1 The EGR system meters exhaust gases into the engine induction system through passages cast into the intake manifold. From there the exhaust gases pass into the fuel/air mixture for the purpose of lowering combustion temperatures, thereby reducing the amount of oxides of nitrogen (NOx) formed.
2 The amount of exhaust gas admitted is regulated by a vacuum or backpressure controlled (EGR) valve in response to engine operating conditions. The EGR valve is under the control of the EGR vacuum control solenoid valve, which, in turn, is under the control of the PCM. The vacuum signal to the EGR valve is controlled by varying the duty cycle (on-time) of the solenoid valve. The duty cycle is calculated by the computer using information from the ECT, VSS and IAT sensors. Problems with the EGR control circuit or the EGR valve will set a Code 32, P0401, P0404, P0405, or P1404.
3 Common engine problems associated with the EGR system are rough idling or stalling at idle, rough engine performance during light throttle application and stalling during deceleration.

Check

Refer to illustration 5.6

4 Start the engine, warm it up to normal operating temperature and allow the engine to idle. Manually lift the EGR diaphragm. The engine should stall, or at least the idle should drop considerably. **Note:** *The EGR valve is located on the backside of the cylinder head under the hood cowl.*
5 If the EGR valve appears to be in proper operating condition, carefully check all hoses connected to the valve for breaks, leaks or kinks. Replace or repair the valve/hoses as necessary.
6 With the engine idling at normal operating temperature, disconnect the vacuum hose from the EGR valve and connect a vacuum pump **(see illustration)**. When vacuum is applied, the engine should stumble or die, indicating the vacuum diaphragm is operating properly. **Note:** *Some models use a backpressure-type EGR valve. On models so equipped, backpressure must be created in the exhaust system before the vacuum pump will actuate the valve. To create backpressure, install a large socket into the tailpipe and clamp it to the pipe to prevent it from dropping out during the test. The 1/2 inch hole in the socket will allow the engine to idle but still create backpressure on the EGR valve. Don't restrict the exhaust system any longer than necessary to perform this test.* Replace the EGR valve with a new one if the test does not affect the idle.
7 Check the operation of the EGR control solenoid. Check for vacuum to the solenoid. If vacuum exists, check for battery voltage to the solenoid.
8 Further testing of the EGR system, vacuum solenoid and the PCM will require a SCAN tool to access computer information that directly controls the EGR system. Have the vehicle checked by a dealer service department or other qualified repair facility.

Component replacement
EGR valve

9 Disconnect the vacuum hose at the EGR valve.
10 Remove the nuts or bolts which secure the valve to the intake manifold or adapter.
11 Lift the EGR valve from the engine.
12 Clean the mounting surfaces of the EGR valve. Remove all traces of gasket material.
13 Place the new EGR valve, with a new gasket, on the intake tube or adapter and tighten the attaching nuts or bolts.
14 Connect the vacuum signal hose.

EGR vacuum solenoid

15 Remove the hoses from the EGR vacuum control solenoid, labeling them to ensure proper installation.
16 Remove the solenoid and replace it with the new one.

EGR valve cleaning

17 With the EGR valve removed, inspect the passages for excessive deposits.
18 Place a rag securely in the passage opening to keep debris from entering. Clean the passages by hand, using a drill bit.

5.42 Remove the bolts from the base of the Linear EGR valve

Port EGR system (V6 TBI with manual transmission)

General description

19 The EGR system meters exhaust gases into the engine induction system through passages cast into the intake manifold. From there the exhaust gases pass into the fuel/air mixture for the purpose of lowering combustion temperatures, thereby reducing the amount of oxides of nitrogen (NOx) formed.
20 The amount of exhaust gas admitted is regulated by a vacuum controlled (EGR) valve in response to engine operating conditions. The valve is controlled by a flexible diaphragm which is spring loaded to hold the valve closed. Vacuum applied to the top of the diaphragm overcomes the spring pressure and opens the valve in the exhaust gas port. The EGR valve is under the control of the EGR vacuum regulator valve (EVRV), which, in turn, is under the control of the PCM. The vacuum signal to the EGR valve is controlled by varying the duty cycle (on-time) of the solenoid valve. The duty cycle is calculated by the computer using information from the ECT, VSS and IAT sensors. Problems with the EGR control circuit or the EGR valve will set a Code 32.
21 Common engine problems associated with the EGR system are rough idling or stalling at idle, rough engine performance during light throttle application and stalling during deceleration.

Check

22 Start the engine, warm it up to normal operating temperature and allow the engine to idle. Manually lift the EGR diaphragm. The engine should stall, or at least the idle should drop considerably.
23 If the EGR valve appears to be in proper operating condition, carefully check all hoses connected to the valve for breaks, leaks or kinks. Replace or repair the valve/hoses as necessary.
24 With the engine idling at normal operating temperature, disconnect the vacuum hose from the EGR valve and connect a vacuum pump. When vacuum is applied, the engine should stumble or die, indicating the vacuum diaphragm is operating properly. Replace the EGR valve with a new one if the test does not affect the idle.
25 Check the operation of the EGR control solenoid. Check for vacuum to the solenoid. If vacuum exists, check for battery voltage to the solenoid.
26 Further testing of the EGR system, vacuum solenoid and the PCM will require a SCAN tool to access computer information that directly controls the EGR system. Have the vehicle checked by a dealer service department or other qualified repair facility.

Component replacement

EGR valve
27 Remove the air cleaner (see Chapter 4).
28 Disconnect the vacuum hose at the EGR valve and remove the nuts which secure the valve to the adapter.
29 Lift the EGR valve from the engine.
30 Clean the mounting surfaces of the EGR valve. Remove all traces of gasket material.
31 Place the new EGR valve, with a new gasket, on the intake tube or adapter and tighten the attaching nuts or bolts.
32 Connect the vacuum signal hose.

EGR vacuum solenoid
33 Remove the hoses from the EGR vacuum control solenoid, labeling them to ensure proper installation.
34 Remove the solenoid and replace it with the new one.

EGR valve cleaning
35 With the EGR valve removed, inspect the passages for excessive deposits.
36 It is a good idea to place a rag securely in the passage opening to keep debris from entering. Clean the passages by hand, using a drill bit.

Linear EGR valve system (four-cylinder MFI [OBD-II], V6 TBI automatic transmission and V6 CMFI and CSEFI)

General description
37 The linear EGR valve feeds small amounts of exhaust gas back into the intake manifold and then into the combustion chamber independent of intake manifold vacuum.
38 The valve controls exhaust gas flow from the exhaust to the intake manifold through a single orifice with a PCM controlled pintle. This EGR valve operates similar to the stepper motor type particular to IAC valves that control idle quality.
39 If the EGR system will not allow the PCM to control the position of the EGR valve pintle, it will set a diagnostic trouble code. Have the EGR system checked by a dealer service department or qualified independent repair facility in the event of EGR system failure.

Check
40 Special electronic diagnostic equipment is needed to check this valve and should be left to a dealer service department or other qualified repair facility.

Replacement
Refer to illustration 5.42
41 Disconnect the electrical connector from the EGR valve.
42 Remove the two mounting bolts and remove the EGR valve from the intake manifold (V6) or EGR valve adapter **(see illustration)**.
43 Remove the EGR valve and gasket.
44 Clean the mounting surface of the EGR valve. Remove all traces of gasket material from the intake manifold and from the valve if it is to be reinstalled. Clean both mating surfaces with a cloth dipped in lacquer thinner or acetone.
45 Install a new gasket and the EGR valve and tighten the bolts securely.
46 Connect the electrical connector onto the EGR valve.

6 Evaporative Emissions Control (EVAP) System

Note: *1996 and later models are equipped with an Enhanced Evaporative Emission (EVAP) system. This EVAP system will conduct up to eight different tests using the On Board Diagnostic system to detect leaks, pressure variations or electrical problems within this closed network. Because it is governed by the PCM and OBD II emissions regulations, have the system checked by a dealer service department in the event of malfunction.*

General description
Refer to illustration 6.2

1 This system is designed to trap and store fuel that evaporates from the throttle body and fuel tank which would normally enter the atmosphere and contribute to hydrocarbon (HC) emissions.
2 The system consists of a charcoal-filled canister and lines running to and from the canister. These lines include a vent line from the gas tank, a vent line from the throttle body, an EVAP purge control solenoid (electronic), a charcoal canister and the intake manifold **(see illustration)**. In addition, there is an EVAP control valve in the canister. On some later models, the PCM controls the vacuum to the purge solenoid valve with an electrically operated solenoid. The fuel tank cap is also an integral part of the system. An indication that the system is not operating properly is a strong fuel odor.
3 The vacuum source to the EVAP canister is controlled in three different ways depending upon the engine and fuel system:

Ported vacuum source - this type uses ported vacuum from the throttle body directly to the EVAP control valve

Manifold vacuum with EVAP purge control valve - this type uses a manifold vacuum signal to a separate purge control valve that is mounted on the engine or on the canister itself.

EVAP purge control solenoid - this type uses a computer control solenoid to regulate the vacuum signal to the EGR using a pulse modulated signal

Check
Refer to illustrations 6.8a and 6.8b

4 Maintenance and replacement of the charcoal canister filter is covered in Chapter 1.
5 Check all lines in and out of the canister for kinks, leaks and breaks along their entire lengths. Repair or replace as necessary.
6 Check the gasket in the gas cap for signs of drying, cracking or breaks. Replace the gas cap with a new one if defects are found.
7 On four-cylinder engines, using a hand-held vacuum pump, apply approximately 15 in-Hg to the control vacuum tube located on the pressure control valve. After 10 seconds there should be about 5 in-Hg vacuum remaining.

Note: *Be sure the vacuum pump does not leak internally and there all hoses are sealed tight. If the vacuum leaks down faster and more than the specified amount, replace the valve.*

8 On 1995 V6 CMFI models, the computer operates the purge control solenoid by changing its frequency signal. The EVAP pressure sensor detects abnormal high pressure in the purge lines. Check for battery voltage to the purge control solenoid and the pressure sensor with the ignition key ON (engine not running) **(see illustrations)**. If no battery voltage is present, have the PCM and the EVAP system circuit checked at a dealer service department. If battery voltage exists, connect the solenoid and the pressure sensor and if there is no obvious sounds from the purge control solenoid (buzzing), have it checked by a dealer service department. Remember, the purge control solenoid will not be activated by the computer until fuel tank pressure exceeds 0.7 psi.

Component replacement

9 Replacement of the canister filter is covered in Chapter 1.
10 When replacing any line running to or from the canister, make sure the replacement line is a duplicate of the one you are replacing. These lines are often color coded to denote their particular usage.

Canister purge valve

Four-cylinder models

Note: *The canister purge solenoid is located on the right side of the engine block, adjacent to the ignition coil.*

11 Loosen the right front wheel lug nuts, then raise the front of the vehicle and support it securely on jackstands. Remove the wheel.
12 Disconnect the electrical connector, then detach the hoses from the purge valve.
13 Remove the mounting nut and detach the purge valve from its bracket.
14 Installation is the reverse of removal.

V6 models

Note: *The canister purge valve is located on the right side of the intake manifold, adjacent to the ignition coil.*

15 Disconnect the electrical connector, then detach the hose from the valve.
16 Remove the two mounting nuts and detach the valve from the manifold.
17 Installation is the reverse of removal.

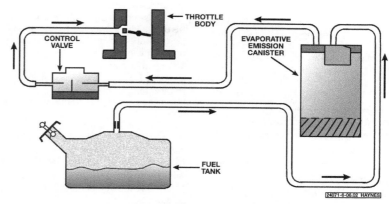

6.2 Details of a typical EVAP system

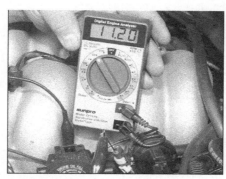

6.8a Remove the purge control solenoid electrical connector and check for battery voltage

6.8b Check for battery voltage on the EVAP pressure sensor

7 Positive Crankcase Ventilation (PCV) system

General description

Refer to illustration 7.2

1 The positive crankcase ventilation system reduces hydrocarbon emissions by circulating fresh air through the crankcase to pick-up blow-by gases, which are then rerouted through the throttle body and burned in the engine.
2 The main components of this system are vacuum hoses and a PCV valve, which regulates the flow of gases according to engine speed and manifold vacuum **(see illustration)**.

Check and component replacement

3 Checking the system and PCV valve replacement are covered in Chapter 1.

8 Transmission Converter Clutch (TCC) system

General information

1 The purpose of the Torque Converter Clutch (TCC) system, equipped in automatic transmissions, is to eliminate the power loss of the torque converter stage when the vehicle is in the cruising mode (usually above 35 mph). This economizes the automatic transmission to the fuel economy of the manual transmission. The lock-up mode is controlled by the PCM through the activation of the TCC apply solenoid which is built into the automatic transmission. When the vehicle reaches a specified speed, the PCM energizes the solenoid and allows the torque converter to lock-up and mechanically couple the engine to the transmission, under which conditions emissions are at their minimum. However, because of other operating condition demands (deceleration, passing, idle, etc.), the transmission must also function in its normal, fluid-coupled mode. When such latter conditions exist, the solenoid de-energizes, returning the torque converter to normal operation. The converter also returns to normal operation whenever the brake pedal is depressed.

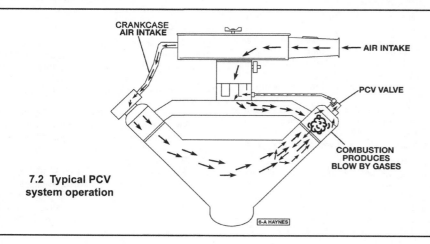

7.2 Typical PCV system operation

Chapter 6 Emissions and engine control systems

Check

2 Due to the requirement of special diagnostic equipment for the testing of this system, and the possible requirement for dismantling of the automatic transmission to replace components of this system, checking and replacing of the components should be handled by a dealer service department or other qualified repair facility.

9 Secondary Air Injection (AIR) system - general description and component replacement

General description

Note: *Some 2000 and later models are equipped with a Secondary Air Injection (AIR) system.*

1 The air injection exhaust emission control system reduces the level of unburned hydrocarbons (HC) and carbon monoxide (CO) in the exhaust gases by injecting outside air into the hot exhaust gases flowing through the exhaust manifolds. When fresh air is mixed with the hot exhaust gases, oxidation is increased, reducing the concentration of hydrocarbons and carbon monoxide and converting them into harmless carbon dioxide and water.

2 The AIR system consists of the electric AIR pump, the vacuum control solenoid, the shut-off valve, the check valves, and the hoses and pipes connecting all of these components.

3 **AIR pump** - The AIR pump draws in filtered, outside air and pumps it into the exhaust manifolds. The pump is turned on and off by a PCM-controlled relay.

4 **AIR vacuum control solenoid** - The AIR vacuum control solenoid controls the AIR shut-off valve. When the PCM turns on the AIR system, it also grounds the vacuum control solenoid, which allows intake manifold vacuum to reach the AIR shut-off valve.

5 **AIR shut-off valve** - The AIR shut-off valve is operated by vacuum. When the AIR system is operating, intake manifold vacuum is applied to the shut-off valve, which opens the valve and allows air from the AIR pump to reach the check valves.

6 **AIR check valves** - The AIR check valves allow air from the AIR pump to flow to the exhaust manifolds but prevent the backflow of exhaust gases into the AIR system.

7 When the coolant temperature is 45-degrees F or more, or when the vehicle is decelerating (high intake vacuum, rich air/fuel mixture, lots of unburned hydrocarbons and carbon monoxide in the exhaust), the PCM energizes the AIR pump relay, at which point the pump begins pumping air into the exhaust system.

8 The PCM may turn off the AIR pump when the system is in closed-loop operation, when one or more Diagnostic Trouble Codes (DTCs) are set by the PCM, when the system is in the power-enrichment mode for too long, when intake manifold vacuum is low or when the rise in intake vacuum is too quick (rapid deceleration).

Component replacement

2.2L engine

AIR check valve

9 Locate the AIR check valve on the reactor pipe assembly on the exhaust manifold.
10 Remove the shut-off valve hose from the AIR check valve.
11 Loosen the pipe locknut, then separate the check valve from the reactor pipe.
12 Installation is the reverse of removal. Tighten all fasteners securely.

AIR shut-off valve

13 Raise the vehicle and support it securely on jackstands.
14 Disconnect the vacuum line from the shut-off valve.
15 Loosen the spring clamp, disconnect the crossover hose from the shut-off valve, then remove the shut-off valve.
16 Installation is the reverse of removal.

AIR pump

17 Raise the vehicle and support it securely on jackstands.
18 Disconnect the cable from the negative battery terminal (see Chapter 5).
19 Loosen the spring clamps, then disconnect the inlet and outlet hoses from the AIR pump.
20 Disconnect the electrical connectors from the AIR pump and from the AIR solenoid valve.
21 Remove the AIR pump mounting bolts, then remove the pump.
22 Installation is the reverse of removal. Tighten all fasteners securely.

AIR solenoid valve

23 Remove the AIR pump (see Steps 17 through 21).
24 Remove the solenoid valve retaining screw from the mounting bracket and remove the solenoid valve.
25 Installation is the reverse of removal.

4.3L V6 engine

AIR check valve/pipe

26 Raise the vehicle and place it securely on jackstands.
27 On 2WD models, remove the AIR shield from below the radiator support.
28 On 4WD models, remove the steering linkage shield from below the crossmember.
29 Working at the exhaust manifold(s), remove the reactor pipe assembly retaining bolts and remove the reactor pipe assembly.
30 Remove the reactor pipe assembly gasket.
31 Remove the spring clamps and separate the hoses from the reactor pipes.
32 Loosen the pipe locknut, then separate the check valve from the reactor pipe.
33 Installation is the reverse of removal. Tighten all fasteners securely.

AIR shut-off valve

34 Raise the vehicle and place it securely on jackstands.
35 On 2WD models, remove the AIR shield from below the radiator support.
36 On 4WD models, remove the steering linkage shield from below the crossmember.
37 Disconnect the vacuum line from the shut-off valve.
38 Loosen the spring clamp, disconnect the inlet and outlet hoses from the shut-off valve, then remove the shut-off valve.
39 Installation is the reverse of removal.

AIR pump

40 Raise the vehicle and place it securely on jackstands.
41 On 2WD models, remove the AIR shield from below the radiator support.
42 On 4WD models, remove the steering linkage shield from below the crossmember.

10.1 Location of the catalytic converter

43 Loosen the spring clamps, then disconnect the AIR check valve/pipe assembly (see Steps 26 through 32).
44 Disconnect the electrical connectors from the AIR pump and from the AIR solenoid valve.
45 Remove the AIR pump mounting bolts, then remove the pump.
46 Installation is the reverse of removal. Tighten all fasteners securely.

AIR solenoid valve

47 Remove the AIR pump (see Steps 40 through 45).
48 Remove the solenoid valve retaining screw from the mounting bracket and remove the solenoid valve.
49 Installation is the reverse of removal.

10 Catalytic converter

General description

Refer to illustration 9.1

1 The catalytic converter is an emission control device added to the exhaust system to reduce pollutants from the exhaust gas stream. These systems are equipped with a single bed monolith catalytic converter. This monolithic converter contains a honeycomb mesh which is also coated with two types of catalysts. One type is the oxidation catalyst while the other type is a three-way catalyst that contains platinum and palladium. The three-way catalyst lowers the levels of oxides of nitrogen (NOx) as well as hydrocarbons (HC) and carbon monoxide (CO) emissions. The oxidation catalyst lowers the levels of hydrocarbons and carbon monoxide **(see illustration)**. If the PCM detects a reduction in the efficiency of the catalytic converter a code P0420 or P0430 will set.

Check

2 The test equipment for a catalytic converter is expensive and highly sophisticated. If you suspect the converter is malfunctioning, take it to a dealer service department or authorized emissions inspection facility for diagnosis and repair.
3 Whenever the vehicle is raised for service of underbody components, check the converter for leaks, corrosion and other damage. If damage is discovered, the converter should be replaced.
4 Because the converter is welded to the exhaust system, converter replacement requires removal of the exhaust pipe assembly (see Chapter 4). Take the vehicle, or the exhaust system, to a dealer service department or a muffler shop.

Chapter 7 Part A
Manual transmission

Contents

	Section		Section
General information	1	Manual transmission overhaul - general information	5
Manual transmission - removal and installation	4	Shift lever - removal and installation	2
Manual transmission lubricant - change	See Chapter 1	Shift lever housing (NV1500/3500) - removal and installation	3
Manual transmission lubricant level - check	See Chapter 1		

Specifications

Torque specifications

Ft-lbs (unless otherwise indicated)

Shift lever housing bolts (NV1500/3500)	
Through 1998	89 in-lbs
1999 and later	180 in-lbs
Transmission-to-clutch housing bolts (Borg-Warner T-5)	55
Transmission-to-engine bolts (New Venture Gear NV1500/3500)	
V6 engine	35
Four cylinder engine	
Through 1998	66
1999 and later	35
Transmission braces	
1994 and 1995	
Lower transmission brace bolts	37
Transmission brace nuts	41
Upper transmission brace bolts	66
1996 and later	
Transmission-to-transfer case brace bolts	35
Transmission brace-to-engine bolts (four-cylinder engine)	37
Transmission brace-to-engine stud (V6 engine)	35

1 General information

Vehicles covered by this manual are equipped with either a five-speed manual or a four-speed automatic transmission. Information on the manual transmission is included in this Part of Chapter 7. Information on the automatic transmission can be found in Part B of this Chapter. You'll also find certain procedures common to both transmissions - such as oil seal replacement - in Part B. Information on the transfer case used on 4WD models is in Part C.

The Borg-Warner T-5 transmission is used on 1994 and 1995 vehicles with a four-cylinder engine; 1996 models with a four-cylinder engine use a New Venture Gear NV1500 transmission. The New Venture Gear 3500 transmission is used on all models with V6 engines.

Depending on the expense involved in having a transmission overhauled, it might be a better idea to consider replacing it with either a new or rebuilt unit. Your local dealer or transmission shop should be able to supply information concerning cost, availability and exchange policy. Regardless of how you decide to remedy a transmission problem, you can still save a lot of money by removing and installing the unit yourself.

7A-2 Chapter 7 Part A Manual transmission

2.1 Back off the shift lever knob locknut, then unscrew and remove the knob

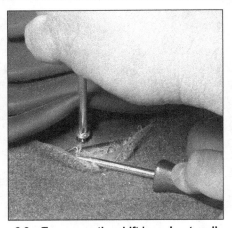

2.2a To remove the shift lever boot, pull up the edge of the boot and remove all boot retaining screws

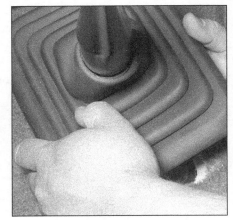

2.2b Slide the boot off the shift lever

2 Shift lever - removal and installation

Refer to illustration 2.1, 2.2a, 2.2b and 2.3

1 Back off the shift lever knob locknut by turning it clockwise **(see illustration)** and remove the shift lever knob and nut from the shift lever.
2 On models with floor console, remove the console (see Chapter 11). Remove the boot retainer screws and the boot **(see illustrations)**.
3 Back off the shift lever nut **(see illustration)**, then unscrew and remove the shift lever.
4 Install the shift lever nut. Do not tighten the nut yet.
5 Install the shift lever, then back the nut up against the lever and tighten it securely.
6 Install the boot and, if equipped, the retainer, install the boot retaining screws and tighten them securely
7 Install the shift lever knob and its locknut. Screw on the knob until it stops, then back it off to align the shift pattern on the knob, and back the nut up against the underside of the knob. Tighten the nut securely.

3 Shift lever housing (NV1500/3500) - removal and installation

1 Place the shift lever in third or fourth gear. **Caution:** *Do not put the transmission in any other gear while the shift housing is removed.*
2 Remove the shift lever (see Section 2).
3 Remove the four housing bolts which attach the shift lever housing to the transmission. Remove the insulator plate. **Caution:** *Do not disassemble the shift lever housing if you're planning to reuse it. The shift lever housing is not serviceable.*
4 Lay the insulator and shift lever housing in position on the transmission case. Install the four housing retaining bolts and tighten

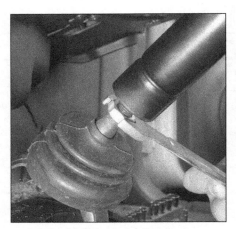

2.3 Back off the locknut at the lower end of the shift lever, then unscrew and remove the shift lever

them to the torque listed in this Chapter's Specifications.
5 Install the shift lever (see Section 2).

4 Manual transmission - removal and installation

Removal

Refer to illustrations 4.10, 4.14a, 4.14b, 4.14c and 4.14d

1 Disconnect the negative battery cable from the battery. **Caution:** *On models equipped with a Delco Loc II anti-theft audio system, be sure the lockout feature is turned off before performing any procedure which requires disconnecting the battery.*
2 On models with a Borg-Warner T5 transmission, shift the transmission into Neutral; on models with a New Venture Gear NV1500 or NV3500, shift the transmission into Third or Fourth. Remove the shift lever (see Section 2).
3 On models with a New Venture Gear NV1500 or NV3500, remove the shift lever housing (see Section 3).

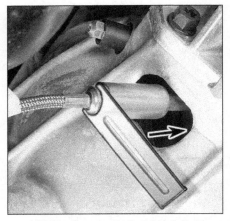

4.10 To disconnect the clutch fluid hydraulic line from the release cylinder on 1996 and later models, use a tool like this to push in on the quick-connect fitting

4 Raise the vehicle and support it securely on jackstands.
5 Disconnect the parking brake cable for clearance and set it aside (see Chapter 9).
6 Remove the driveshaft(s) (see Chapter 8).
7 On 4WD models, remove the skid plate and transfer case (see Chapter 7C), then label and unplug all electrical connectors from the transmission.
8 On 1994 and 1995 models, disconnect the fuel lines - at the manifold on V6 engines or at the top cover on four-cylinder engines (see Chapter 4).
9 Disconnect and remove the exhaust pipe and/or catalytic converter as necessary for clearance (see Chapter 4).
10 On 1994 and 1995 models, remove the clutch release cylinder. On 1996 and later models, disconnect the clutch hydraulic line quick-connect fitting from the clutch (concentric) release cylinder. To disconnect the fitting, you must depress the white plastic sleeve on the quick-connect fitting. This can be done with a special tool **(see illustration)**, but it

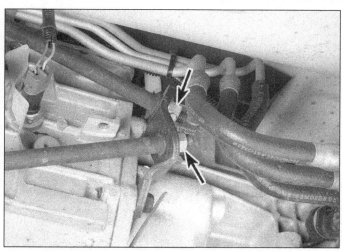

4.14a 1994 and 1995 four-cylinder engines have rear transmission braces like these; remove these two nuts (arrows)

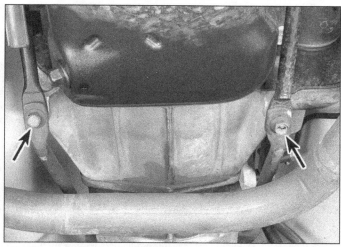

4.14b ... and these bolts (arrows) to remove the lower braces. The front ends of the upper braces are secured by two of the transmission-to-engine bolts

can also be done with two small flat-bladed screwdrivers positioned 180-degrees apart on the sleeve. While depressing the sleeve, pull out the hydraulic line.

11 Support the engine/transmission assembly by placing a floor jack and a block of wood under the engine oil pan.

12 Remove the transmission mount bolts.

13 Remove the hanger supporting the catalytic converter hanger.

14 Unbolt and remove all support braces, including any transfer case braces (see illustrations). The braces for an earlier four-cylinder engine are shown in the photos; other models have similar braces.

15 On 1994 and 1995 models with a V6 engine and on all 1996 and later models, unbolt and remove the cover plate from the lower front of the clutch housing (bellhousing).

16 Place a jack under the transmission and secure the transmission to the jack with safety chains. Raise the transmission slightly with the jack. Remove the crossmember bolts, then remove the crossmember and transmission mount. On 1996 and later models, detach the wiring harness from the crossmember.

17 On 1996 and later models, lower the engine/transmission assembly enough to detach the fuel line retaining clips from the top of the transmission.

18 On 1994 and 1995 models with a four-cylinder engine, remove the transmission-to-bellhousing bolts. On 1994 and 1995 models with a V6 engine, and on all 1996 and later models, remove the transmission-to-engine bolts. It's a good idea to install guide pins in the upper holes before removing the lower bolts (you can make the guide pins by cutting the heads off old bolts).

19 Make a final check that all wires have been disconnected from the transmission, then move the transmission and jack toward the rear of the vehicle until the transmission input shaft is clear of the clutch or clutch housing. Keep the transmission level as you pull it to the rear.

20 Once the input shaft is clear, lower the transmission and remove it from under the vehicle. **Caution:** *Do not depress the clutch pedal while the transmission is out of the vehicle.*

21 Inspect the clutch components (on 1994 and 1995 models with a four-cylinder engine, you'll have to remove the clutch housing first). In most cases, new clutch components should be routinely installed when the transmission is removed (see Chapter 8).

Installation

22 Install the clutch components, if they were removed (see Chapter 8). On 1994 and 1995 models with a four-cylinder engine, install the clutch housing, if it was removed. Tighten the clutch housing bolts to the torque listed in the Chapter 8 Specifications.

23 With the transmission secured to the jack, raise it into position behind the engine or clutch housing, and then carefully slide it forward, engaging the input shaft with the clutch plate hub. Do not use excessive force to install the transmission - if the input shaft won't slide into place, readjust the angle of the transmission so that it's level and/or turn the input shaft so the splines engage properly with the clutch.

24 Once the transmission is flush with the clutch housing or engine, install the transmission-to-engine or transmission-to-clutch housing bolts. Tighten the bolts to the torque

4.14c On 1994 and 1995 four-cylinder engines, detach the front braces from the engine by removing this bolt (arrow) ...

4.14d ... the bolt and cover (right arrow) and the brace bolt (left arrow)

listed in this Chapter's Specifications.

25 Install the crossmember and transmission mount. Tighten all nuts and bolts securely.

26 Remove the jacks supporting the transmission and the engine.

27 Install the various components removed previously. Be sure to tighten all transmission and transfer brace fasteners to the torque listed in this Chapter's Specifications. Refer to Chapter 7, Part C, for installation of the transfer case (if equipped), Chapter 8 for the installation of the driveshaft(s) and Chapter 4 for information regarding the exhaust system components. To connect the hydraulic line quick-connect fitting, simply press it securely into place until you hear a "click" and the line feels secure.

28 Make a final check to verify all wires and hoses have been reconnected and the transmission has been filled with lubricant to the proper level (see Chapter 1). Lower the vehicle.

29 Install the shift lever and boot (see Section 3).

30 Connect the negative battery cable. Road test the vehicle and check for leaks.

5 Manual transmission overhaul - general information

Overhauling a manual transmission is a difficult job for the do-it-yourselfer. It involves the disassembly and reassembly of many small parts. Numerous clearances must be precisely measured and, if necessary, changed with select fit spacers and snap-rings. As a result, if transmission problems arise, it can be removed and installed by a competent do-it-yourselfer, but overhaul should be left to a transmission repair shop. Rebuilt transmissions may be available - check with your dealer parts department and auto parts stores. At any rate, the time and money involved in an overhaul is almost sure to exceed the cost of a rebuilt unit.

Nevertheless, it's not impossible for an inexperienced mechanic to rebuild a transmission if the special tools are available and the job is done in a deliberate step-by-step manner so nothing is overlooked.

The tools necessary for an overhaul include internal and external snap-ring pliers, a bearing puller, a slide hammer, a set of pin punches, a dial indicator and possibly a hydraulic press. In addition, a large, sturdy workbench and a vise or transmission stand will be required.

During disassembly of the transmission, make careful notes of how each piece comes off, where it fits in relation to other pieces and what holds it in place.

Before taking the transmission apart for repair, it will help if you have some idea what area of the transmission is malfunctioning. Certain problems can be closely tied to specific areas in the transmission, which can make component examination and replacement easier. Refer to the *Troubleshooting* Section at the front of this manual for information regarding possible sources of trouble.

Chapter 7 Part B
Automatic transmission

Contents

	Section
Automatic transmission fluid and filter change	See Chapter 1
Automatic transmission fluid level check	See Chapter 1
Automatic transmission - removal and installation	8
Diagnosis - general	2
Extension housing seal - replacement	5
General information	1
Park/neutral switch - check, adjustment and replacement	6
Shift cable (1995 and later models) - removal, installation and adjustment	4
Shift linkage (1994 models) - removal, installation and adjustment	3
Transmission mount - check and replacement	7

Specifications

General

Transmission fluid type .... See Chapter 1

Torque specifications   Ft-lbs

Shift linkage swivel screw (1994 models)	18
Park/Neutral switch (1996 and later models)	
Switch retaining bolts	
1995 models	20
1996 and later models	27
Manual lever nut	
1995 models	20
1996 and later models	21
Transmission-to-engine bolts	
1994 and 1995 models	23
1996 and later models	
Four-cylinder engine	
Through 1998	66
1999 and later models	35
V6 engines	34
Torque converter cover bolts	46
Torque converter-to-flywheel bolts	46
Transmission mount bolts (to transmission)	18
Transmission mount nuts (to crossmember)	38
Crossmember to frame bolts	35

1 General information

The vehicles equipped with an automatic transmission use a four-speed 4L60-E electronically controlled unit. The 4L60-E has a lock-up torque converter, known as a Torque Converter Clutch (or TCC). The clutch provides a direct connection between the engine and the drive wheels for improved efficiency and fuel economy. The 4L60-E is not equipped with a Throttle Valve (TV) cable assembly. The transmission functions normally controlled by the TV cable are controlled electronically. Due to the complexity of the clutches and the hydraulic control system, and because of the special tools and expertise needed to overhaul an automatic transmission, diagnosis and repair of this transmission must be handled by a dealer service department or a transmission repair shop. The procedures in this Chapter are limited to general diagnosis, routine maintenance, adjustment and transmission removal and installation. However, even though the repair work must be done by a transmission specialist, you can save money by removing and installing the transmission yourself. You can also check and adjust the shift linkage or shift cable, replace the extension housing seal and check and replace the neutral start switch. **Caution:** *If a vehicle with an automatic transmission is disabled, do NOT tow it at speeds greater than 30 mph or distances over 50 miles.*

2 Diagnosis - general

Note: *Automatic transmission malfunctions may be caused by five general conditions: poor engine performance, improper adjustments, hydraulic malfunctions, mechanical malfunctions or malfunctions in the computer or its signal network. Diagnosis of these problems should always begin with a check of the easily repaired items: fluid level and condition (see Chapter 1), shift linkage adjustment and throttle linkage adjustment. Next, perform a road test to determine if the problem has been corrected or if more diagnosis is necessary. If the problem persists after the preliminary tests and corrections are completed, additional diagnosis should be done by a dealer service department or transmission repair shop. Refer to the Troubleshooting Section at the front of this manual for information on symptoms of transmission problems.*

Preliminary checks

1 Drive the vehicle to warm the transmission to normal operating temperature.
2 Check the fluid level as described in Chapter 1:

  a) If the fluid level is unusually low, add enough fluid to bring the level within the designated area of the dipstick, then check for external leaks (see below).

7B-2   Chapter 7 Part B   Automatic transmission

b) *If the fluid level is abnormally high, drain off the excess, then check the drained fluid for contamination by coolant. The presence of engine coolant in the automatic transmission fluid indicates that a failure has occurred in the internal radiator walls that separate the coolant from the transmission fluid (see Chapter 3).*

c) *If the fluid is foaming, drain it and refill the transmission, then check for coolant in the fluid or a high fluid level.*

3   Check the engine idle speed. **Note:** *If the engine is malfunctioning, do not proceed with the preliminary checks until it has been repaired and runs normally.*

4   Inspect the shift linkage (see Section 3) or shift cable (see Section 4). Make sure it's properly adjusted and operates smoothly.

Fluid leak diagnosis

5   Most fluid leaks are easy to locate visually. Repair usually consists of replacing a seal or gasket. If a leak is difficult to find, the following procedure may help.

6   Identify the fluid. Make sure it's transmission fluid and not engine oil or brake fluid (automatic transmission fluid is a deep red color).

7   Try to pinpoint the source of the leak. Drive the vehicle several miles, then park it over a large sheet of cardboard. After a minute or two, you should be able to locate the leak by determining the source of the fluid dripping onto the cardboard.

8   Make a careful visual inspection of the suspected component and the area immediately around it. Pay particular attention to gasket mating surfaces. A mirror is often helpful for finding leaks in areas that are hard to see.

9   If the leak still cannot be found, clean the suspected area thoroughly with a degreaser or solvent, then dry it.

10   Drive the vehicle for several miles at normal operating temperature and varying speeds. After driving the vehicle, visually inspect the suspected component again.

11   Once the leak has been located, the cause must be determined before it can be properly repaired. If a gasket is replaced but the sealing flange is bent, the new gasket will not stop the leak. The bent flange must be straightened.

12   Before attempting to repair a leak, check to make sure the following conditions are corrected or they may cause another leak. **Note:** *Some of the following conditions cannot be fixed without highly specialized tools and expertise. Such problems must be referred to a transmission repair shop or a dealer service department.*

Gasket leaks

13   Check the pan periodically. Make sure the bolts are tight, no bolts are missing, the gasket is in good condition and the pan is flat (dents in the pan may indicate damage to the valve body inside).

14   If the pan gasket is leaking, the fluid level or the fluid pressure may be too high, the vent may be plugged, the pan bolts may be too tight, the pan sealing flange may be

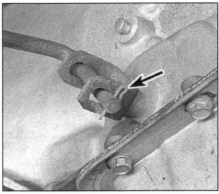

3.2 To remove the transmission control lever from the transmission shift lever, pull out the retaining clip (arrow) and pull the transmission control lever out of the shift lever

warped, the sealing surface of the transmission housing may be damaged, the gasket may be damaged or the transmission casting may be cracked or porous. If sealant instead of gasket material has been used to form a seal between the pan and the transmission housing, it may be the wrong sealant.

Seal leaks

15   If a transmission seal is leaking, the fluid level or pressure may be too high, the vent may be plugged, the seal bore may be damaged, the seal itself may be damaged or improperly installed, the surface of the shaft protruding through the seal may be damaged or a loose bearing may be causing excessive shaft movement.

16   Make sure the dipstick tube seal is in good condition and the tube is properly seated. Periodically check the area around the speedometer gear or sensor for leakage. If transmission fluid is evident, check the O-ring for damage.

Case leaks

17   If the case itself appears to be leaking, the casting is porous and will have to be repaired or replaced.

18   Make sure the oil cooler hose fittings are tight and in good condition.

Fluid comes out vent pipe or fill tube

19   If this condition occurs, the transmission is overfilled, there is coolant in the fluid, the case is porous, the dipstick is incorrect, the vent is plugged or the drain back holes are plugged.

---

3   Shift linkage (1994 models) - removal, installation and adjustment

Removal

Refer to illustration 3.2

1   Apply the parking brake.
2   Remove the retaining clip **(see illustra-**

3.9 Loosen the bolt (arrow) securing the shift rod swivel to the transmission control lever

tion). **Note:** *As you disassemble the shift linkage assembly, note the relationship of all washers, the spring and the insulator to insure proper reassembly.*

3   Disconnect the rod from the column lever.

4   Remove the bolt and washer from the swivel.

5   Remove the rod.

6   Clean the metal parts with solvent and the rubber or nylon parts with soapy water. Wipe all parts with a clean, dry cloth.

Installation

7   Installation is the reverse of removal, but be sure to use new retaining clips. After reassembling the shift linkage, adjust it (see below).

Adjustment

Refer to illustration 3.9

8   Apply the parking brake.
9   Loosen the swivel screw **(see illustration)**.
10   Put the column selector lever in the N (Neutral) position (put the lever into the Neutral gate - don't rely on the indicator to find the neutral position).
11   Put the transmission in Neutral by moving the transmission shift lever to its forward position, then back to the second detent.
12   Hold the shift linkage rod tightly in the swivel and tighten the shift lever nut to the torque listed in this Chapter's Specifications.
13   Put the steering column shift lever in the P (Park) position.
14   Check the adjustment. The shift lever must go into all positions and the engine must start only in the P or N positions. If necessary, readjust the shift linkage and the park/neutral switch (see Section 6) until your adjustment meets both these criteria. **Warning:** *With the shift lever in the Park position, the parking pawl must freely engage the rear (reaction) internal gear lugs or output ring gear lugs to prevent the vehicle from rolling, which could result in personal injury.*
15   If the needle doesn't line up with the numbers/letters on the shift indicator when the shift linkage is adjusted properly, alter the

Chapter 7 Part B Automatic transmission

4.5 Pry the cable end from the manual lever

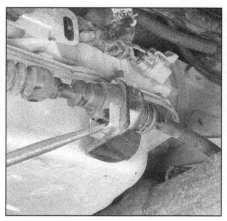

4.7 Carefully pry the horseshoe retainer from the shift cable housing

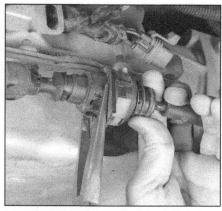

4.8 To disengage the cable from its bracket on the transmission, squeeze the cable retainer with long-nose pliers while pulling the cable toward the back of the vehicle

position of the needle by changing the position of the indicator cable clip on the steering column shroud.
16 Release the parking brake.

4 Shift cable (1995 and later models) - removal, installation and adjustment

Removal

Refer to illustrations 4.5, 4.7, 4.8, 4.10, 4.11a, 4.11b and 4.13

1 Detach the cable from the negative terminal of the battery.
2 Apply the parking brake.
3 Block the rear wheels so the vehicle will not roll in either direction, then shift the transmission into the Neutral position.
4 On models so equipped, remove the transfer case shield (see Chapter 7C).
5 Pry the cable end from the transmission control lever stud **(see illustration)**.
6 Remove the cable from the transmission oil pan clip.
7 Remove the horseshoe retainer from the shift cable housing **(see illustration)**.
8 Squeeze the cable retainer **(see illustration)** while pulling the cable toward the back of the vehicle and disconnect the cable.
9 If the vehicle is equipped with a column shift, proceed by removing the steering column trim filler panel. If the vehicle is equipped with a floor shift, skip to step 12.
10 Pry the shift cable from the shift lever **(see illustration)**.
11 Remove the horseshoe clip that retains the cable to the bracket on the steering column, then squeeze the cable retainer to remove the cable from the bracket **(see illustrations)**.
12 Remove the floor console, if equipped (see Chapter 11). Remove the cowl trim panel and the door sill trim plate, then remove the seat and the front carpet.
13 Remove any cable floor clips or tape, and pry the cable grommet from the floor **(see illustration)**.
14 Remove the cable and grommet from the vehicle.

Installation

15 Installation is the reverse of removal. Make sure you don't damage the rubber boot.

4.10 Pry the shift cable from the shift lever

Adjustment

16 Apply the parking brake.
17 Remove the retaining clip from the rear support bracket.

4.11a Carefully pry the horseshoe retainer from the shift cable housing to disengage the cable from its bracket on the steering column . . .

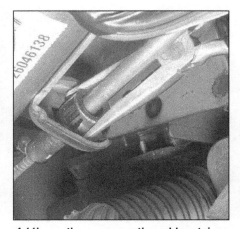

4.11b . . . then squeeze the cable retainer while pulling the cable toward the front of the vehicle

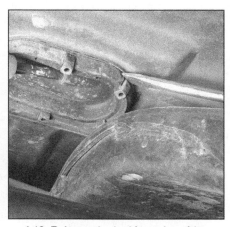

4.13 Release the locking tabs with a screwdriver and push the cable grommet through the floorpan

5.5a Pry out the old extension housing seal with a large screwdriver . . .

5.5b . . . or, if that won't work, with a seal removal tool

5.7 To install a new extension housing seal, tap it into place with a large socket and hammer

18 Separate the body-core adjuster from the range selector pin.
19 Place the transmission shift selector shaft in Neutral.
20 Place the transmission shift lever into the Neutral position.
21 Remove the horseshoe retainer from the shift cable **(see illustration 4.7)**.
22 Remove the cable from the transmission oil pan clip.
23 Install the cable back onto the transmission oil pan clip.
24 Install the horseshoe retainer onto the shift cable.
25 Make sure the transmission shift lever and the shift selector shaft are still in the Neutral position. Reposition if necessary.
26 Place, then snap, the body-core adjustment onto the transmission selector shaft. Ensure the spring loaded body-core adjustment moves freely.
27 Push the locking tab into the body-core until it snaps. If necessary, wiggle the body-core while pushing in the locking tab.
28 Insert the horseshoe retainer into the shift cable.
29 Check the adjustment. The shift lever must go into all positions and the engine must start only in the P or N positions only. If necessary, readjust the shift cable and the park/neutral switch (see Section 6) until your adjustment meets both these criteria.

5 Extension housing seal - replacement

Refer to illustrations 5.5a, 5.5b and 5.7

Note: *This procedure applies to both manual and automatic transmissions.*

1 Oil leaks frequently occur due to wear of the extension housing oil seal and bushing (if equipped), and/or the speedometer drive gear oil seal and O-ring. Replacement of these seals is relatively easy, since the repairs can usually be performed without removing the transmission from the vehicle.
2 The extension housing oil seal is located at the extreme rear of the transmission, where the driveshaft is attached. If leakage at the seal is suspected, raise the vehicle and support it securely on jackstands. If the seal is leaking, transmission lubricant will be built up on the front of the driveshaft and may be dripping from the rear of the transmission.
3 Remove the driveshaft (see Chapter 8).
4 Using a soft-face hammer, carefully tap the dust shield (if equipped) to the rear and remove it from the transmission. Be careful not to distort it. On cast iron case (heavy duty) five-speed manual transmissions, remove the nut and U-joint flange.
5 Using a screwdriver or seal removal tool, carefully pry the oil seal and bushing (if equipped) out of the rear of the transmission **(see illustrations)**. Do not damage the splines on the transmission output shaft.
6 If the oil seal and bushing cannot be removed with a screwdriver or pry bar, a special oil seal removal tool (available at auto parts stores) will be required.
7 Using a large section of pipe or a very large deep socket as a drift, install the new oil seal **(see illustration)**. Drive it into the bore squarely and make sure it's completely seated. Install a new bushing using the same method.
8 Reinstall the U-joint flange and nut (if equipped) and tighten the nut to the torque listed in this Chapter's Specifications. Reinstall the dust shield (if equipped) by carefully tapping it into place.
9 Lubricate the splines of the transmission output shaft and the outside of the driveshaft sleeve yoke with light-weight grease, then install the driveshaft. Be careful not to damage the lip of the new seal.

6 Park/neutral switch - check, replacement and adjustment

Check

Refer to illustrations 6.3 and 6.4

1 The Park/Neutral switch prevents the engine from starting in any gear other than

6.3 Unplug the electrical connector from the Park/Neutral switch (1995 and 1996 models); if you're replacing the switch, remove the manual lever nut (left arrow) and remove the two switch retaining screws (not visible in this photo)

Park or Neutral (On 1995 and later models, it also closes the circuit for the back-up lights when the shift lever is moved to Reverse). If the engine starts in any position other than Park or Neutral, it's either out of adjustment or defective. First, check the switch to make sure that it's operating properly.
2 On 1994 models, remove the steering column insulator panel (see Chapter 11) and unplug the electrical connector from the Park/Neutral switch terminals.
3 On 1995 and later models, raise the vehicle and place it securely on jackstands. Unplug the electrical connector from the Park/Neutral switch **(see illustration)**.
4 Make sure the ignition key is turned to Off, then use an ohmmeter to check the continuity across the two switch terminals on 1994 models, or across the indicated switch terminals on 1995 and 1996 models **(see illustration)**. With the shift lever in Park or Neutral, there should be continuity; with the shift lever in any other position, there should be no continuity (on 1995 and later models,

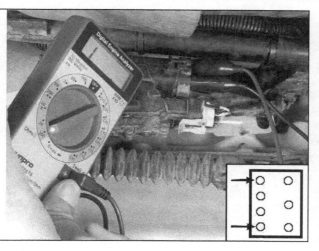

6.4 To check the Park/Neutral switch on 1995 and later models, check continuity between the upper left and lower left terminals with the shift lever in each position; there should be continuity when the shift lever is in Park or Neutral, but not in any other position (if your ohmmeter shows continuity in Reverse, you're touching the terminals for the back-up light circuit)

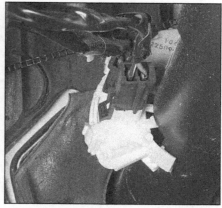

6.7 The Park/Neutral switch is located on the lower steering column mast

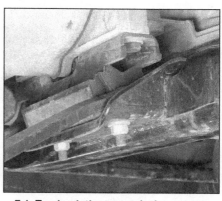

7.1 To check the transmission mount, insert a large screwdriver or prybar between the mount and the crossmember and try to lever the transmission up off its mount; if the transmission moves significantly, replace the mount

there's also continuity when the shift lever is in Reverse, but only at the terminals for the back-up lights). If a 1994 switch doesn't operate as described, replace it (see below). If a 1995 or 1996 switch doesn't operate as described, try adjusting it (see Steps 13 through 24), then retest it. If it still fails to operate properly, replace it (see below).

Replacement and adjustment

1994 models

Refer to illustration 6.7

5 Remove the steering column insulator panel (see Chapter 11) if you haven't already done so.
6 Unplug the electrical connector from the switch, if you haven't already done so.
7 To remove the switch from the steering column, pull it straight out **(see illustration)**.
8 Align the actuator on the switch with the holes in the shift tube.
9 Apply the parking brake and put the shift lever in the Neutral position.
10 Press down on the front of the switch until the tangs snap into the rectangular holes in the steering column jacket.
11 To adjust the switch, move the shift lever to the Park position. The main housing and the housing back should ratchet the switch to its proper position.
12 The remainder of installation is the reverse of removal.

1995 and later models

Note: *You will need a special switch adjustment tool, available at most auto parts stores, to adjust the Park/Neutral switch on these models.*

13 Apply the parking brake and place the shift lever in Neutral.
14 Disconnect the cable from the negative battery terminal.
15 Raise the vehicle and place it securely on jackstands.
16 Remove the nut from the manual lever shaft **(see illustration 6.3)** and remove the manual lever.
17 Unplug the electrical connector from the Park/neutral switch **(see illustration 6.3)**.

18 Remove the switch retaining bolts.
19 Slide the switch off the manual shaft and remove it.
20 Install the special adjustment tool on the switch in accordance with the manufacturer's instructions. Make sure the two detents on the switch (where the manual shaft is inserted through the switch) are aligned with the two lower tabs on the tool. Rotate the tool until its upper locator pin is lined up with the locator on the switch.
21 Before sliding it onto the manual shaft, make sure the end of the shaft isn't burred. If it is, lightly file the edge of the shaft to remove any burrs. Align the switch hub flats with the flats on the manual shaft. Slide the switch onto the manual shaft until the switch mounting bracket contacts the mounting bosses on the transmission.
22 Install the switch mounting bolts and tighten them to the torque listed in this Chapter's Specifications.
23 Remove the special adjustment tool from the switch.
24 The remainder of installation is the reverse of removal. Be sure to tighten the manual lever retaining nut to the torque listed in this Chapter's Specifications.

7 Transmission mount - check and replacement

Check

Refer to illustration 7.1

1 Insert a large screwdriver or pry bar into the space between the transmission and the crossmember and try to pry the transmission up slightly **(see illustration)**.
2 The transmission should not move away from the insulator much. If there is any separation of the rubber, the mount is worn out.

Replacement

3 To replace the mount, remove the nut attaching the insulator to the crossmember and the bolts attaching the insulator to the transmission.

4 Raise the transmission slightly with a jack and remove the insulator.
5 Installation is the reverse of the removal procedure. Be sure to tighten the nuts/bolts securely.

8 Automatic transmission - removal and installation

Removal

Refer to illustrations 8.9, 8.10, 8.13 and 8.14

1 Disconnect the negative cable from the battery. **Caution:** *On models equipped with a Delco Loc II anti-theft audio system, be sure the lockout feature is turned off before performing any procedure which requires disconnecting the battery.*
2 Raise the vehicle and support it securely on jackstands.
3 Drain the transmission fluid (see Chapter 1), then reinstall the pan.
4 Unplug all electrical connectors.
5 On 1994 models, disconnect the shift linkage (see Section 3); on 1995 and later models, disconnect the shift cable (see Section 4).

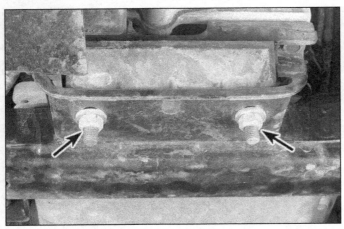

8.9 To detach the transmission mount from the crossmember, remove these nuts (arrows) (the mount shown is for a V6 engine; there's only one nut on the mount for the four-cylinder engine)

8.10 To detach the crossmember from the frame, remove these bolts and nuts

6   Remove the driveshaft (see Chapter 8); if the vehicle is equipped with 4WD, remove the front driveshaft also (see Chapter 8).
7   If the vehicle is equipped with 4WD, remove the transfer case (see Chapter 7C).
8   Relieve the fuel system pressure and disconnect the fuel line fittings (see Chapter 4).
9   Remove the rear transmission mount-to-crossmember nut(s) **(see illustration)**.
10 Remove the crossmember-to-frame bolts **(see illustration)**.
11 Remove any exhaust components which will interfere with transmission removal (see Chapter 4).
12 Remove the starter motor (see Chapter 5).
13 Remove the torque converter cover **(see illustration)**.
14 Mark the torque converter and the driveplate with a scribe or chalk so they can be installed in the same position **(see illustration)**.
15 Remove the driveplate-to-torque converter bolts. Turn the crankshaft (in a clockwise direction only, viewed from the front) for access to each bolt.
16 Support the engine with a jack. Use a block of wood under the oil pan to spread the load.
17 Support the transmission with a jack - preferably a jack made for this purpose. Safety chains will help steady the transmission on the jack.
18 Raise the transmission enough to allow removal of the crossmember.
19 Detach the fuel line and wiring harness brackets from the transmission.
20 Remove the transmission-to-engine bolts.
21 Remove the transmission dipstick tube.
22 Lower the transmission slightly and disconnect and plug the transmission fluid cooler lines.
23 Remove the transfer case shifter (see Chapter 7C) and put it aside.

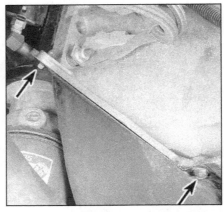

8.13 To remove the torque converter cover, remove these bolts (arrows)

24 Move the transmission to the rear to disengage it from the engine block dowel pins and make sure the torque converter is detached from the driveplate. Secure the torque converter to the transmission so it won't fall out during removal.

Installation

25 Prior to installation, make sure the torque converter hub is securely engaged in the pump.
26 With the transmission secured to the jack, raise it into position. Be sure to keep it level so the torque converter doesn't slide out. Connect the transmission fluid cooler lines.
27 Turn the torque converter until the marks on the converter and driveplate are aligned.
28 Move the transmission forward carefully until the dowel pins engage with the holes in the bellhousing.
29 Install the transmission housing-to-engine bolts. Tighten them securely.
30 Install the driveplate-to-torque converter

8.14 Mark the relationship of the torque converter to the driveplate to ensure proper dynamic balance is maintained

bolts and tighten them to the torque listed in this Chapter's Specifications.
31 Install the crossmember and lower the mount stud into its hole. Tighten the bolts and nuts securely.
32 Remove the jacks supporting the transmission and the engine.
33 Install the dipstick tube.
34 Install the starter motor (see Chapter 5).
35 Connect the shift linkage or cable.
36 Plug in the transmission wire harness connectors.
37 Install the torque converter cover.
38 If the vehicle is equipped with 4WD, install the transfer case (see Chapter 7C).
39 Install the driveshaft(s).
40 Install the shift linkage (see Section 3) or cable (see Section 4).
41 Install any exhaust system components that were removed or disconnected.
42 Lower the vehicle.
43 Fill the transmission with the specified fluid (Chapter 1), run the engine and check for fluid leaks.

Chapter 7 Part C
Transfer case

Contents

	Section		Section
Electronic shift motor (NP 233) - replacement	7	Transfer case control module (NP 233) - replacement	8
General information	1	Transfer case - removal and installation	10
Oil seals - replacement	2	Transfer case switch (NP 233) replacement	9
Selector switch (NP 231) - check and replacement	3	Vacuum actuator - replacement	See Chapter 8
Shift cable - (NP 231) - replacement and adjustment	5	Vacuum switch - check and replacement	6
Shift lever (NP 231) - replacement	4		

Specifications

Torque specifications

	Ft-lbs
Transfer case output shaft yoke nut	110
Shift lever pivot bolt	75
Transmission-to-transfer case bolts	
Automatic	35
Manual	41

1 General information

Four-wheel drive (4WD) models are equipped with a transfer case mounted on the rear of the transmission. Drive is transmitted from the engine, through the transmission and the transfer case to the front and rear axles by driveshafts.

4WD models are equipped with one of the following transfer cases: Manual-shift models are equipped with a four-mode New Process model NP 231(Neutral, 2WD high, 4WD high, 4WD low); electronic-shift models use a three-mode New Process model NP 233 (2WD high, 4WD high, 4WD low); models with full-time all-wheel-drive (AWD) use a Borg-Warner BW 4472 unit.

1999 and later models may have an NVG 236/246-NP8 transfer case, which has an "automatic" 4WD-Hi position, as well as manually-selected 2WD-Hi, 4WD-Hi and 4WD-Lo positions. Some models may have a "one-speed automatic" NVG-136 transfer case, which offers full-time auto 4WD, with no manual mode selections.

We don't recommend trying to rebuild any of these transfer cases at home. They're difficult to overhaul without special tools, and rebuilt units are available for less than it would cost to rebuild your own. However, there are a number of components that you *can* check, adjust and/or replace - and those are the items covered in this Chapter.

2 Oil seals - replacement

Refer to illustrations 2.4 and 2.6
Note: *This procedure applies to both the front output shaft and rear extension housing seals.*

1 Raise the vehicle and support it securely on jackstands.

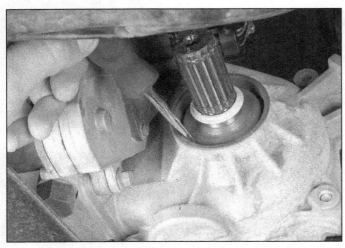

2.4 Using a screwdriver or a seal removal tool, carefully pry out the seal

2.6 Using a large socket, install the new seal

2 If you're replacing the front seal, remove the front driveshaft; if you're replacing the rear seal, remove the rear driveshaft (see Chapter 8).

3 If you're replacing the front seal on a 1994 Blazer or Jimmy model, remove the nut, washers, yoke and shield; if you're replacing the rear seal on a 1995 or 1996 AWD model, remove the nut, washers and flange.

4 Pry out the seal with a screwdriver or a seal removal tool (see illustration). Don't damage the seal bore.

5 Lubricate the new seal lips with petroleum jelly.

6 Drive the seal into place with a large socket (see illustration). The outside diameter of the socket should be slightly smaller than the outside diameter of the seal.

7 The remainder of installation is the reverse of removal.

3 Selector switch (NP 231) - check and replacement

1 Disconnect the cable from the negative battery terminal. **Caution:** *On models equipped with a Delco Loc II anti-theft audio system, be sure the lockout feature is turned off before performing any procedure which requires disconnecting the battery.*

2 If equipped with a manual transmission or a floor-shift automatic transmission, remove the knob from the transmission shift lever (see Chapter 7A or 7B). Remove the knob from the transfer case shift lever by pulling out the horseshoe-shaped clip below the knob with long-nose pliers.

3 Remove the center console (see Chapter 11).

4 Unplug the electrical connector from the switch. The switch is located on the left side of the transfer case shift lever assembly.

5 Remove the two switch retaining screws.

6 Remove the switch.

7 Installation is the reverse of removal. Make sure the switch contact slide is aligned with the actuating pin on the lever. Tighten the switch retaining screws securely.

4 Shift lever (NP 231) - replacement

1 Detach the cable from the negative battery terminal. **Caution:** *On models equipped with a Delco Loc II anti-theft audio system, be sure the lockout feature is turned off before performing any procedure which requires disconnecting the battery.*

2 If equipped with a manual transmission or floor-shift automatic transmission, remove the knob from the transmission shift lever (see Chapter 7A or 7B). Remove the knob from the transfer case shift lever by pulling out the horseshoe-shaped clip directly below the shift knob with long-nose pliers.

3 Remove the center console (see Chapter 11).

4 To disconnect the shift cable from the shift lever, pop it loose with a screwdriver.

5 Remove the horseshoe-shaped cable retainer clip from the shift lever base by pulling it out with long-nose pliers.

6 Remove the shift lever base retaining bolts.

7 Installation is the reverse of removal.

5 Shift cable (NP 231) - replacement and adjustment

1 Raise the vehicle and place it securely on jackstands.

2 Remove the transfer case skid plate.

3 Using long-nose pliers, remove the horseshoe-shaped clip retaining the shift cable to the bracket on the transfer case.

4 Pop the end of the shift cable off the lever on the transfer case with a screwdriver.

5 Slide the control cable out of the bracket.

6 Remove the jackstands and lower the vehicle.

7 Remove the knob from the transfer case shift lever by pulling out the horseshoe-shaped clip below the knob with long-nose pliers.

8 Remove the center console (see Chapter 11).

9 Pop off the cable end from the shift lever.

10 Remove the retainer clip from the shift lever base.

11 Remove the cable retainer clip screws.

12 Remove the cable by pulling it up, through the floorpan, into the passenger compartment.

13 Route the new cable through the floor.

14 Raise the vehicle and place it securely on jackstands.

15 Put the transfer case lever in the Neutral position.

16 Attach the cable to the lever.

17 Slide the cable into its bracket on the transfer case and install the retainer clip.

18 Remove the jackstands and lower the vehicle.

19 Slide the cable into its slot in the shift lever base and install the retainer clip.

20 Place the shift lever in Neutral.

21 Loosen the cable locknut and turn the threaded cable end until it's aligned with the pin on the shift lever. Tighten the locknut and attach the cable end to the shift lever.

22 The remainder of installation is the reverse of removal.

6 Vacuum switch - check and replacement

1 Raise the vehicle and place it securely on jackstands.

2 Detach the vacuum line from the switch.

3 Unscrew and remove the vacuum switch from the transfer case.

4 Coat the threads of the new switch with

Chapter 7 Part C Transfer case

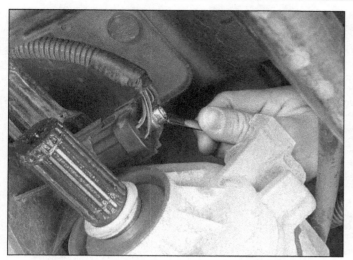

7.4 Before unplugging the electrical connector from the electronic shift motor, remove this lock bolt - on 1999 and later models, this connector is a push-on type with clip-tab

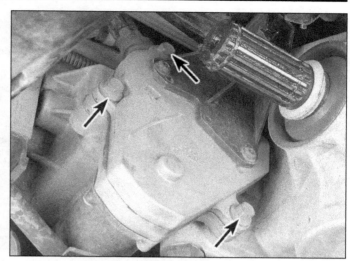

7.7 Remove the electronic shift motor-to-transfer case bolts (arrows)

Teflon tape or thread sealant and install the switch.
5 Attach the vacuum line to the switch.
6 Remove the jackstands and lower the vehicle.

7 Electronic shift motor (NP 233 and NP 236) - replacement

Refer to illustrations 7.4 and 7.7

1 Disconnect the cable from the negative battery cable. **Caution:** *On models equipped with a Delco Loc II anti-theft audio system, be sure the lockout feature is turned off before performing any procedure which requires disconnecting the battery.*
2 Raise the vehicle and place it securely on jackstands.
3 Remove the skid plate.
4 Unplug the electrical connector from the motor **(see illustration)**.
5 Disconnect the front driveshaft (see Chapter 8).
6 Remove the front output shaft yoke, if equipped.
7 Remove the motor-to-transfer case bolts **(see illustration)**.
8 Remove the motor.
9 Installation is the reverse of removal. Be sure to tighten the motor-to-transfer case bolts securely.

8 Transfer case control module (NP 233) - replacement

Refer to illustrations 8.5a and 8.5b

1 The transfer case control module controls the motor on the NP/NV 233 transfer case. On 1994 pick-up models, it's on the back of the right side of the cowl trim panel. On 1995 and 1996 models, it's located in the middle of the dash, behind the instrument cluster bezel. On 1999 and later models, it is located behind the passenger's kick panel.
2 Disconnect the cable from the negative battery terminal. **Caution:** *On models equipped with a Delco Loc II anti-theft audio system, be sure the lockout feature is turned off before performing any procedure which requires disconnecting the battery.*
3 On 1994 pick-up models, remove the cowl side panel. On all other models, remove the instrument cluster bezel (see Chapter 11).
4 On 1994 pick-up models, remove the module retaining screws **(see illustration 8.1)**.
5 On all other models, remove the module retaining screws **(see illustration)**, pull the module down and unplug the electrical connector **(see illustration)**.
6 Installation is the reverse of removal.

8.5a On 1995 and 1996 models, remove the module retaining screws (arrows) . . .

8.5b . . . pull the module down and unplug the electrical connector

9 Transfer case switch (NP 233) - replacement

Refer to illustration 9.2

1 Disconnect the cable from the negative battery terminal. **Caution:** *On models equipped with a Delco Loc II anti-theft audio system, be sure the lockout feature is turned off before performing any procedure which requires disconnecting the battery.*
2 Simply pry the switch out of the dash, unplug the electrical connector **(see illustration)** and remove the switch.
3 Installation is the reverse of removal.

10 Transfer case - removal and installation

Removal

1 Disconnect the cable from the negative battery terminal. **Caution:** *On models equipped with a Delco Loc II anti-theft audio system, be sure the lockout feature is turned off before performing any procedure which requires disconnecting the battery.*
2 On models with a manually-shifted transfer case, put the transfer case in 4-High.
3 Raise the vehicle and support it securely on jackstands.
4 Remove the skid plate (if equipped).
5 Drain the transfer case lubricant (see Chapter 1).
6 Remove the front and rear driveshafts (see Chapter 8).
7 Detach all vacuum and vent lines and unplug all electrical connectors.
8 On manually shifted 1994 through 1996 models, disconnect the shift cable from the transfer case (see Section 5).
9 Remove any support braces.
10 Support the transmission with a jack or jackstand. The transmission should remain supported at all times while the transfer case is out of the vehicle.
11 Support the transfer case with a jack - preferably a special jack made for this purpose. Safety chains will help steady the transfer case on the jack.
12 Remove the transmission-to-transfer case bolts (manual transmission) or the adapter-to-transfer case bolts (automatic transmission). Don't lose the washers. Note the location of the shift cable bracket, if equipped (on manual transmissions, it's attached to the left rear of the transmission; on automatic transmissions, it's attached to the left side of the adapter).
13 Make a final check that all wires and hoses have been disconnected from the transfer case, then move the transfer case and jack toward the rear of the vehicle until the transfer case is clear of the transmission.

9.2 To remove the transfer case switch, pry the bottom edge of the switch loose, pull out the switch and unplug the electrical connector

Keep the transfer case level as this is done. Once the input shaft is clear, lower the transfer case and remove it from under the vehicle.

Installation

14 Installation is the reverse of removal. Be sure to tighten the transmission-to-transfer case bolts to the torque listed in this Chapter's Specifications.

Chapter 8
Clutch and driveline

Contents

	Section
Axle (rear) - removal and installation	19
Axles - description and check	14
Axleshaft (rear) - removal and installation	15
Axleshaft bearing (rear) - replacement	17
Axleshaft oil seal (rear) - replacement	16
Clutch - description and check	2
Clutch components - removal, inspection and installation	5
Clutch hydraulic system - bleeding	8
Clutch master cylinder - removal and installation	3
Clutch release bearing (1994 and 1995 models) - removal, inspection and installation	6
Clutch release cylinder - removal and installation	4
Clutch start switch - check and replacement	9
Front differential carrier (4WD models) - removal and installation	25
Front differential carrier bushing (4WD models) - removal and installation	26
Differential lubricant change	See Chapter 1

	Section
Differential lubricant level check	See Chapter 1
Differential output shaft pilot bearing (4WD models) - replacement	24
Differential pinion seal (rear) - replacement	18
Driveaxle (4WD models) - removal and installation	20
Driveaxle boot replacement and CV joint overhaul (4WD models)	21
Driveaxle CV joint boot check (4WD models)	See Chapter 1
Driveline inspection	11
Driveshaft - removal and installation	12
Driveshaft(s) and universal joints - general information	10
General information	1
Output shaft, tube and tube seal (4WD models) - removal and installation	23
Pilot bearing - inspection and replacement	7
Universal joints - removal, overhaul and installation	13
Vacuum actuator and shift cable (4WD models) - replacement	22

Specifications

General

Clutch fluid type	See Chapter 1
Clutch disc lining thickness (minimum)	1/16-inch

Torque specifications

Ft-lbs (unless otherwise indicated)

Clutch

Clutch housing (bellhousing)-to-engine bolts (1994 and 1995 four-cylinder engine only)	66
Pressure plate-to-flywheel bolts	
1994	30
1995	
Four-cylinder engine	28
V6 engine	29
1996 and later	
Four-cylinder engine	33
V6 engine	29
Release cylinder	
1994 (nuts)	156 in-lbs
1995 (nuts)	
Four-cylinder engine	204 in-lbs
V6 engine	156 in-lbs
1996 (bolts)	216 in-lbs
1997 (bolts)	80 in-lbs
1998 and later (bolts)	71 in-lbs

Driveshaft

U-joint strap-to-pinion flange bolts	
Front (4WD)	55
Rear	
Four-cylinder engine	180 in-lbs
V6 engine	33
Center support bearing bolts	
Through 1998	25
1999 and later	50

Rear axle

Pinion shaft lock bolt	25

Chapter 8 Clutch and driveline

Torque specifications (continued)   Ft-lbs (unless otherwise indicated)

Front driveaxle (4WD models)
Hub nut
   1994 through 1996 180
   1997 and later 103
Driveaxle-to-output shaft flange bolts 60

Output shaft and tube assembly (4WD models)
Tube flange-to-differential carrier bolts 36
Tube-to-support bracket nuts/bolts 55
Shift cable housing bolts 36
Shift cable coupling nut .... 90 in-lbs

1 General information

The Sections in this Chapter deal with the components from the rear of the engine to the rear wheels (except for the transmission and transfer case, which are dealt with in Chapter 7) and forward to the front wheels on four-wheel drive (4WD) models. In this Chapter, the components are grouped into three categories: clutch, driveshaft(s) and axle(s). Separate Sections within this Chapter cover checks and repair procedures for components in each of these three groups.

Since nearly all these procedures involve working under the vehicle, make sure it's safely supported on sturdy jackstands or a hoist where the vehicle can be safely raised and lowered.

2 Clutch - description and check

1 All vehicles with a manual transmission have a single dry plate, diaphragm spring-type clutch. The clutch disc has a splined hub which allows it to slide along the splines of the transmission input shaft. The clutch and pressure plate are held in contact by spring pressure exerted by the diaphragm in the pressure plate.

2 The clutch release system is operated by hydraulic pressure. The hydraulic release system consists of the clutch pedal, a master cylinder and fluid reservoir, the hydraulic line, a release (or slave) cylinder which actuates the clutch release lever and the clutch release (or throwout) bearing. The release cylinder and release bearing are an integral unit on 1996 models; this design is referred to as a "concentric" release cylinder because it's installed on the transmission input shaft with the release bearing.

3 On 1994 and 1995 models, when pressure is applied to the clutch pedal to release the clutch, hydraulic pressure is exerted against the outer end of the clutch release lever. As the lever pivots the shaft fingers push against the release bearing. The bearing pushes against the fingers of the diaphragm spring of the pressure plate assembly, which in turn releases the clutch plate. On 1996 and later models, the operation is very similar, except that hydraulic pressure is exerted directly against the release bearing (there is no release lever).

4 Terminology can be a problem when dis-

3.2 To disconnect the clutch master cylinder pushrod from the clutch pedal, remove the retainer clip (arrow) and push out the clevis pin

cussing the clutch components because common names are in some cases different from those used by the manufacturer. For example, the driven plate is also called the clutch plate or disc, the clutch release bearing is sometimes called a throwout bearing, the release cylinder is sometimes called the slave cylinder.

5 Other than to replace components with obvious damage, some preliminary checks should be performed to diagnose clutch problems.

  a) The first check should be of the fluid level in the clutch master cylinder. If the fluid level is low, add fluid as necessary and inspect the hydraulic system for leaks. If the master cylinder reservoir is dry, bleed the system as described in Section 8 and recheck the clutch operation.

  b) To check "clutch spin-down time," run the engine at normal idle speed with the transmission in Neutral (clutch pedal up - engaged). Disengage the clutch (pedal down), wait several seconds and shift the transmission into Reverse. No grinding noise should be heard. A grinding noise would most likely indicate a bad pressure plate or clutch disc.

  c) To check for complete clutch release, run the engine (with the parking brake applied to prevent vehicle movement) and hold the clutch pedal approximately 1/2-inch from the floor. Shift the transmission between 1st gear and Reverse several times. If the shift is rough, component failure is indicated. On 1994 and 1995 models, check the release cylinder pushrod travel. With the clutch pedal depressed completely, the

3.3a To disconnect the clutch hydraulic fluid line from the clutch master cylinder, tap out this retaining pin with a 7/64-inch punch or drill bit, then pull it out . . .

release cylinder pushrod should extend substantially (you may have to remove an inspection plug to see the pushrod). If it doesn't, check the fluid level in the clutch master cylinder.

  d) Visually inspect the pivot bushing at the top of the clutch pedal to make sure there's no binding or excessive play.

  e) On 1994 and 1995 models, crawl under the vehicle and make sure the clutch release lever is securely attached to the ball stud.

3 Clutch master cylinder - removal and installation

Removal

Refer to illustrations 3.2, 3.3a, 3.3b and 3.4

1 Disconnect the negative cable from the battery. **Caution:** On models equipped with a Delco Loc II anti-theft audio system, be sure the lockout feature is turned off before performing any procedure which requires disconnecting the battery.

2 Working under the dash, remove the lower filler panel (see Chapter 11) and the lower left air conditioning duct (if equipped), then disconnect the pushrod **(see illustration)** from the top of the clutch pedal. It's held in place with a spring clip.

3 To disconnect the clutch fluid hydraulic line, drive out the roll pin holding the hydraulic tube to the master cylinder with a 7/64-inch punch **(see illustration)**. Have rags handy, as some fluid will be lost as the line is removed

Chapter 8 Clutch and driveline   8-3

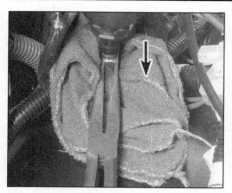

3.3b ... and pull the line straight down

3.4 To remove the master cylinder from the firewall, rotate it clockwise 45-degrees and pull it out

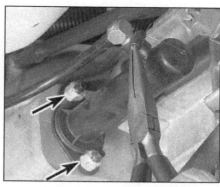

4.3 To detach the release cylinder from the transmission on a 1994 or 1995 model, pull out this cotter pin, disconnect the clutch hydraulic line and remove the two cylinder retaining nuts (arrows)

(see illustration). **Note:** *On 1999 and later models, the fluid line is integral with the master cylinder. Both are replaced as a unit.* **Caution:** *Don't allow brake fluid to come into contact with the paint - it will damage the finish. Also have a plug ready and immediately plug the line to prevent leakage and fluid.*

4 To detach the master cylinder from the firewall, rotate the master cylinder clockwise 45-degrees **(see illustration)**. On later models with integral fluid line, detach the line at the clutch housing using a special tool available to release the quick-connector.

Installation

5 Using a new seal, attach the master cylinder to the firewall by holding it at a 45-degree angle and rotating it counterclockwise. Don't rotate it too far or damage will occur.
6 Use a new roll pin to connect the hydraulic fluid line to the master cylinder. On later models, connect the integral fluid line to the release master cylinder.
7 Working inside the vehicle, connect the pushrod to the clutch pedal.
8 Install the lower left air conditioning duct (if equipped) and the lower filler panel (see Chapter 11).
9 Fill the clutch master cylinder reservoir with the fluid specified in Chapter 1 and bleed the clutch system (see Section 8).

4 Clutch release cylinder - removal and installation

All models

1 Disconnect the retainer clip from the clutch master cylinder pushrod to ensure that, if the clutch pedal is depressed, the release cylinder isn't damaged (see Section 3).
2 Raise the vehicle and support it securely on jackstands.

1994 and 1995 models

Refer to illustration 4.3

3 To disconnect the hydraulic line from the release cylinder, pull out the cotter pin **(see illustration)**, then pull the hydraulic line straight off its fitting.
4 Remove the two release cylinder mounting nuts.
5 Detach the release cylinder.

6 Install the release cylinder on the clutch housing. Make sure the pushrod is seated in the release lever pocket.
7 Connect the hydraulic line to the release cylinder and install a new cotter pin.

1996 and later models

8 Remove the transmission (see Chapter 7A).
9 Remove the two release cylinder retaining bolts.
10 Slide the release cylinder off the transmission input shaft.
11 While the transmission is removed, inspect the clutch disc and pressure plate assembly (see Section 5) and replace as necessary.
12 Clean off the transmission input shaft and coat it with multi-purpose grease. Slide the new release cylinder onto the input shaft, install the retaining bolts and tighten them to the torque listed in this Chapter's Specifications.
13 Install the transmission (see Chapter 7A).

All models

14 Fill the clutch fluid reservoir with the recommended fluid (see Chapter 1).
15 Bleed the clutch hydraulic system (see Section 8).
16 Remove the jackstands and lower the vehicle.

5 Clutch components - removal, inspection and installation

Warning: *Dust produced by clutch wear and deposited on clutch components is hazardous to your health. DO NOT blow it out with compressed air and DO NOT inhale it. DO NOT use gasoline or petroleum-based solvents to remove the dust. Brake system cleaner should be used to flush the dust into a drain pan. After the clutch components are wiped clean with a rag, dispose of the contaminated rags and cleaner in a covered, marked container.*

Removal

Refer to illustrations 5.4, 5.6 and 5.7

1 Access to the clutch components is nor-

mally accomplished by removing the transmission, leaving the engine in the vehicle. However, if the engine is being removed for major overhaul, then check the clutch for wear and replace worn components as necessary. The relatively low cost of the clutch components compared to the time and trouble spent gaining access to them warrants their replacement anytime the engine or transmission is removed, unless they are new or in near-perfect condition. The following procedures are based on the assumption the engine will stay in place.
2 On 1994 and 1995 models, remove the release cylinder (see Section 4). On 1996 and later models, disconnect the clutch hydraulic fluid line from the release cylinder (see the transmission removal and installation procedure in Chapter 7, Part A for details).
3 Remove the transmission from the vehicle (see Chapter 7, Part A). Support the engine while the transmission is out. An engine hoist should be used to support it from above. If you use a jack underneath the engine instead, make sure a piece of wood is positioned between the jack and oil pan to spread the load. **Caution:** *The pick-up for the oil pump is very close to the bottom of the oil pan. If the pan is bent or distorted in any way, engine oil starvation could occur.*
4 On 1994 and 1995 models with four-cylinder engines, remove the bellhousing-to-engine bolts **(see illustration)**, then detach

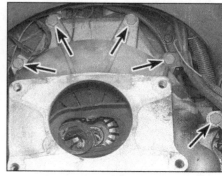

5.4 Locations of the bellhousing-to-engine bolts (1994 and 1995 models)

5.6 Use a clutch alignment tool to hold the clutch plate prior to loosening or to center it before tightening the pressure plate bolts

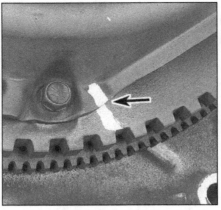

5.7 If you're going to re-use the same pressure plate, mark its relationship to the flywheel (arrow)

5.10 Inspect the surface of the flywheel for cracks, dark-colored areas (signs of overheating) and other obvious defects; resurfacing will correct minor defects (the surface of this flywheel is in fairly good condition, but resurfacing is always a good idea)

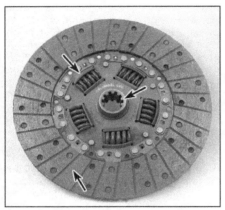

5.12 Inspect the clutch plate lining, springs and splines (arrows) for wear

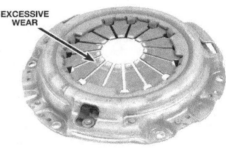
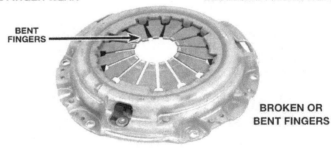

5.14 Replace the pressure plate if excessive wear or damage is noted

the housing. It may have to be gently pried off the alignment dowels with a screwdriver or prybar.

5 The clutch release lever and release bearing can remain attached to the housing for the time being.

6 To support the clutch disc during removal, install an alignment tool through the clutch disc hub **(see illustration)**.

7 Carefully inspect the flywheel and pressure plate for indexing marks. The marks are usually an X, an O or a white letter. If they cannot be found, paint a mark so the pressure plate and the flywheel will be in the same alignment during installation **(see illustration)**.

8 Loosen the pressure plate-to-flywheel bolts in 1/4-turn increments until they can be removed by hand. Work in a criss-cross pattern until all spring pressure is relieved, then hold the pressure plate securely and completely remove the bolts, followed by the pressure plate and clutch disc.

Inspection

Refer to illustrations 5.10, 5.12 and 5.14

9 Ordinarily, when a problem occurs in the clutch, it can be attributed to wear of the clutch driven plate assembly (clutch disc). However, all components should be inspected at this time. **Note:** *If the clutch components are contaminated with oil, there will be shiny, black glazed spots on the clutch disc lining, which will cause the clutch to slip. Replacing clutch components won't completely solve the problem - be sure to check the crankshaft rear oil seal and the transmission input shaft seal for leaks. If it looks like a seal is leaking, be sure to install a new one to avoid the same problem with the new clutch.*

10 Check the flywheel for cracks, heat checking, grooves and other obvious defects **(see illustration)**. If the imperfections are slight, a machine shop can machine the surface flat and smooth, which is highly recommended regardless of the surface appearance. Refer to Chapter 2, Part A, for the flywheel removal and installation procedure.

11 Inspect the pilot bearing (see Section 7).

12 Check the lining on the clutch disc. There should be at least 1/16-inch of lining above the rivet heads. Check for loose rivets, distortion, cracks, broken springs and other obvious damage **(see illustration)**. As mentioned above, ordinarily the clutch disc is routinely replaced, so if you're in doubt about its condition, replace it.

13 The release bearing should also be replaced along with the clutch disc (see Section 6).

14 Check the machined surfaces and the diaphragm spring fingers of the pressure

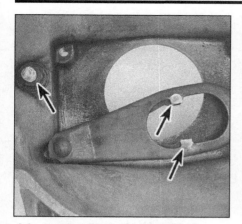

6.6a Lubricate the clutch release lever at the indicated points with multi-purpose grease

6.6b Pack the recess in the release bearing and lightly lubricate the groove where the lever rides with multi-purpose grease

plate **(see illustration)**. If the surface is grooved or otherwise damaged, replace the pressure plate. Also check for obvious damage, distortion, cracks, etc. Light glazing can be removed with medium-grit emery cloth. If a new pressure plate is required, new and factory-rebuilt units are available.

Installation

15   Before installation, clean the flywheel and pressure plate machined surfaces with lacquer thinner or acetone. It's important that no oil or grease is on these surfaces or the lining of the clutch disc. Handle the parts only with clean hands.
16   Position the clutch disc and pressure plate against the flywheel with the clutch held in place with an alignment tool **(see illustration 5.6)**. Make sure it's installed properly (most replacement clutch plates will be marked "flywheel side" or something similar - if it's not marked, install the clutch disc with the damper springs toward the transmission).
17   Tighten the pressure plate-to-flywheel bolts only finger-tight, working around the pressure plate.
18   Center the clutch disc by ensuring the alignment tool extends through the splined hub and into the pilot bearing in the crankshaft. Wiggle the tool up, down or from side-to-side as needed to bottom the tool in the pilot bearing. Tighten the pressure plate-to-flywheel bolts a little at a time, working in a criss-cross pattern, to prevent distorting the cover. After all the bolts are snug, tighten them to the torque listed in this Chapter's Specifications. Remove the alignment tool.
19   Using high-temperature grease, lubricate the inner groove of the release bearing (see Section 6). Also place grease on the release lever contact areas and the transmission input shaft bearing retainer.
20   Install the clutch release bearing (see Section 6).
21   On 1994 and 1995 models with four-cylinder engines, install the bellhousing and tighten the bolts to the torque listed in this Chapter's Specifications.
22   Install the transmission, release cylinder (external type) and all components removed previously. Tighten all fasteners to the recommended torque.

6   Clutch release bearing (1994 and 1995 models) - removal, inspection and installation

Warning: *Dust produced by clutch wear and deposited on clutch components is hazardous to your health. DO NOT blow it out with compressed air and DO NOT inhale it. DO NOT use gasoline or petroleum-based solvents to remove the dust. Brake system cleaner should be used to flush the dust into a drain pan. After the clutch components are wiped clean with a rag, dispose of the contaminated rags and cleaner in a covered, marked container.*

Note: *On 1996 and later models, the release bearing is an integral part of the clutch release cylinder, and the release cylinder/bearing assembly is serviced as a unit. See Section 4 for the removal and installation procedure.*

Removal

1   Remove the release cylinder (see Section 4).
2   On all models, remove the transmission (see Chapter 7, Part A).
3   On 1994 and 1995 models with four-cylinder engines, remove the bellhousing (see Section 5).
4   Detach the clutch release lever from the ball stud, then remove the bearing from the lever.

Inspection

5   Hold the center of the bearing and rotate the outer portion while applying pressure. If the bearing doesn't turn smoothly or if it's noisy, replace it with a new one. Wipe the bearing with a clean rag and inspect it for damage, wear and cracks. Don't immerse the bearing in solvent - it's sealed for life and to do so would ruin it.

Installation

Refer to illustrations 6.6a and 6.6b
6   Lightly lubricate the clutch release lever, the groove for the lever in the release bearing, and the inner groove of the bearing with high-temperature grease **(see illustrations)**.
7   Attach the release bearing to the clutch release lever.
8   Lubricate the clutch release lever ball socket with high-temperature grease and push the lever onto the ball stud until it's securely seated.
9   Apply a light coat of high-temperature grease to the face of the release bearing, where it contacts the pressure plate diaphragm fingers.
10   On 1994 and 1995 models, install the bellhousing and tighten the bolts to the torque listed in this Chapter's Specifications.
11   Prior to installing the transmission, apply a light coat of grease to the sleeve of the transmission front bearing retainer.
12   The remainder of installation is the reverse of the removal procedure. Tighten all bolts to the correct torque.

7   Pilot bearing - inspection and replacement

Refer to illustrations 7.9 and 7.10
1   The clutch pilot bearing is a needle roller type bearing which is pressed into the rear of the crankshaft. It's greased at the factory and does not require additional lubrication. Its primary purpose is to support the front of the transmission input shaft. The pilot bearing should be inspected whenever the clutch components are removed from the engine. Because of its inaccessibility, replace it with a new one if you have any doubt about its condition. **Note:** *If the engine has been removed from the vehicle, disregard the following Steps which don't apply.*
2   Remove the transmission (see Chapter 7A). On 1994 and 1995 four-cylinder engine models, remove the bellhousing (see Section 5).
3   Remove the clutch components (see Section 5).
4   Using a flashlight, inspect the bearing for excessive wear, scoring, dryness, roughness and any other obvious damage. If any of these conditions are noted, replace the bearing.
5   Removal can be accomplished with a special puller available at most auto parts stores, but an alternative method also works very well.
6   Find a solid steel bar which is slightly smaller in diameter than the bearing. Alternatives to a solid bar would be a wood dowel or a socket with a bolt fixed in place to make it solid.
7   Check the bar for fit - it should just slip into the bearing with very little clearance.

7.9 You can remove the pilot bearing by packing the recess behind the bearing with heavy grease and forcing it out hydraulically with a steel rod slightly smaller than the bore in the bearing - when the hammer strikes the rod, the bearing will pop out of the crankshaft

7.10 Tap the bearing into place with a bushing driver or a socket slightly smaller than the outside diameter of the bearing

8 Pack the bearing and the area behind it (in the crankshaft recess) with heavy grease. Pack it tightly to eliminate as much air as possible.
9 Insert the bar into the bearing bore and strike the bar sharply with a hammer, which will force the grease to the back side of the bearing and push it out **(see illustration)**. Remove the bearing and clean all grease from the crankshaft recess.
10 To install the new bearing, lightly lubricate the outside surface with grease, then drive it into the recess with a soft-face hammer **(see illustration)**. Some bearings have an O-ring seal, which must face out.
11 Install the clutch components, transmission and all other components removed previously. Tighten all fasteners to the recommended torque.

8 Clutch hydraulic system - bleeding

1 The hydraulic system should be bled to remove all air whenever any part of the system has been removed or if the fluid level has been allowed to fall so low that air has been drawn into the master cylinder. The procedure is very similar to bleeding a brake system.
2 Fill the master cylinder with new brake fluid conforming to DOT 3 specifications. **Caution:** *Do not re-use any of the fluid coming from the system during the bleeding operation or use fluid which has been inside an open container for an extended period of time.*
3 Raise the vehicle and place it securely on jackstands to gain access to the release cylinder, which is located on the left side of the clutch housing.
4 Remove the dust cap which fits over the bleeder valve and push a length of plastic hose over the valve. Place the other end of the hose into a clear container with about two inches of brake fluid. The hose end must be in the fluid at the bottom of the container.
5 Have an assistant depress the clutch pedal and hold it. Open the bleeder valve on the release cylinder, allowing fluid to flow through the hose. Close the bleeder valve when your assistant signals the clutch pedal is at the bottom of its travel. Once closed, have your assistant release the pedal.
6 Continue this process until all air is evacuated from the system, indicated by a solid stream of fluid being ejected from the bleeder valve each time with no air bubbles in the hose or container. Keep a close watch on the fluid level inside the clutch master cylinder reservoir - if the level drops too low, air will be sucked back into the system and the process will have to be started all over again.
7 Install the dust cap and lower the vehicle. Check carefully for proper operation before placing the vehicle in normal service.

9 Clutch start switch - check and replacement

Check

1 The clutch start switch, which is mounted on a bracket at the top of the clutch pedal, prevents the engine from being started unless the clutch pedal is depressed.
2 To test the switch, verify that the engine will not start unless the clutch pedal is depressed, and that it does start when the pedal is depressed.
3 If the engine starts without depressing the clutch pedal, the switch is probably bad, but check the switch circuit first. Make sure there's voltage to the switch.
4 If there's voltage to the switch, verify that there's no voltage on the other side of the switch unless the clutch pedal is depressed.
5 If there's a short or open in the circuit on either side of the switch, repair it and retest.
6 If the circuit is okay, replace the switch.

Replacement

7 Remove the lower filler panel (see Chapter 11). On 1995 and 1996 models, remove the lower left air conditioning duct.
8 On 1994 and 1995 models, unplug the electrical connector from the switch and remove the switch.
9 On 1996 and later models, remove the plastic retainer tab from the switch, disengage the switch from the pushrod, remove the switch and unplug the electrical connector from the switch.
10 Installation is the reverse of removal.

10 Driveshaft(s) and universal joints - general information

Refer to illustration 10.1
1 A driveshaft is a tube, or a pair of tubes, that transmits power between the transmission (or transfer case on 4WD models) and the differential. Universal joints are located at either end of the driveshaft and in the center on two-piece driveshafts **(see illustration)**.
2 Driveshafts on 2WD models employ a splined yoke at the front, which slips into the extension housing of the transmission. This arrangement allows the driveshaft to slide back-and-forth within the transmission during vehicle operation. An oil seal prevents leakage of fluid at this point and keeps dirt from entering the transmission. If leakage is evident at the front of the driveshaft, replace the oil seal (see Chapter 7, Part B).
3 Driveshafts on 4WD models have either a splined yoke or companion flange at the transfer case end. If two-piece driveshafts are used, a slip joint is usually employed at the front of the rear driveshaft section.
4 Center bearings support the driveline when two-piece driveshafts are used. The center bearing is a ball-type bearing mounted in a rubber cushion attached to a frame crossmember. The bearing is pre-lubricated and sealed at the factory.
5 The driveshaft assembly requires very little service. The universal joints are lubricated for life and must be replaced if problems develop. The driveshaft must be removed from the vehicle for this procedure.
6 Since the driveshaft is a balanced unit, it's important that no undercoating, mud, etc. be allowed to stay on it. When the vehicle is raised for service it's a good idea to clean the driveshaft and inspect it for any obvious damage. Also, make sure the small weights used to originally balance the driveshaft are in place and securely attached. Whenever the driveshaft is removed it must be reinstalled in the same relative position to preserve the balance.
7 Problems with the driveshaft are usually indicated by a noise or vibration while driving the vehicle. A road test should verify if the problem is the driveshaft or another vehicle component. Refer to the *Troubleshooting* Section at the front of this manual. If you suspect trouble, inspect the driveline (see the next Section).

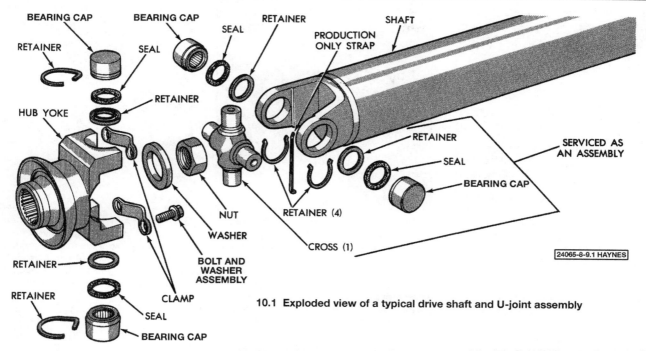

10.1 Exploded view of a typical drive shaft and U-joint assembly

11 Driveline inspection

1 Raise the rear of the vehicle and support it securely on jackstands. Block the front wheels to keep the vehicle from rolling off the stands.

2 Crawl under the vehicle and visually inspect the driveshaft. Look for any dents or cracks in the tubing. If any are found, the driveshaft must be replaced.

3 Check for oil leakage at the front and rear of the driveshaft. Leakage where the driveshaft enters the transmission or transfer case indicates a defective transmission/transfer case seal (see Chapter 7). Leakage where the driveshaft joins the differential indicates a defective pinion seal (see Section 18).

4 While under the vehicle, have an assistant rotate a rear wheel so the driveshaft will rotate. As it does, make sure the universal joints are operating properly without binding, noise or looseness. Listen for any noise from the center bearing (if equipped), indicating it's worn or damaged. Also check the rubber portion of the center bearing for cracking or separation, which will necessitate replacement.

5 The universal joint can also be checked with the driveshaft motionless, by gripping your hands on either side of the joint and attempting to twist the joint. Any movement at all in the joint is a sign of considerable wear. Lifting up on the shaft will also indicate movement in the universal joints.

6 Finally, check the driveshaft mounting bolts at the ends to make sure they're tight.

7 On 4WD models, the above driveshaft checks should be repeated on all driveshafts. In addition, check for grease leakage around the sleeve yoke, indicating failure of the yoke seal.

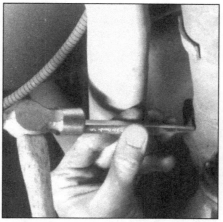

12.3a Use a punch and hammer to mark the relationship of the driveshaft U-joint to the pinion flange . . .

8 Check for leakage where the driveshafts connect to the transfer case and front differential. Leakage indicates worn oil seals.

9 At the same time, check for looseness in the joints of the front driveaxles. Also check for grease or oil leakage from around the driveaxles by inspecting the rubber boots and both ends of each axle. Oil leakage at the differential junction indicates a defective side oil seal. Leakage at the wheel side indicates a defective front hub seal, while leakage at the boots means a damaged rubber boot. For servicing of these components, see the appropriate Sections.

12 Driveshaft - removal and installation

1 Disconnect the negative cable from the

12.3b . . . or put paint marks on the U-joint and the pinion flange

battery. **Caution:** *On models equipped with a Delco Loc II anti-theft audio system, be sure the lockout feature is turned off before performing any procedure which requires disconnecting the battery.*

Rear driveshaft

Removal

Refer to illustrations 12.3a, 12.3b and 12.4

2 Raise the vehicle and support it securely on jackstands. Place the transmission in Neutral with the parking brake off. Block the front wheels to prevent the vehicle from rolling.

3 Using a scribe, a hammer and punch, or paint, make marks on the driveshaft and the differential flange in line with each other **(see illustrations)**. This is to make sure the driveshaft is reinstalled in the same position to preserve the balance.

12.4 Insert a screwdriver through the U-joint to prevent the driveshaft from turning as you break loose the four U-joint strap bolts

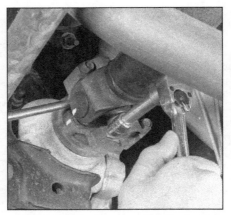

12.15 Before disconnecting the front driveshaft from the differential (or from the transfer case on 1994 models), mark the relationship of the U-joint(s) to the differential pinion flange and transfer output shaft flange, then insert a screwdriver through the U-joint to prevent the driveshaft from turning while you break loose the strap bolts

12.17 To remove the front driveshaft on a 1995 and later 4WD model, remove the strap bolts from the front U-joint, drop down the front end of the driveshaft and slide the rear end off the splined output shaft of the transfer case, as shown here

13.3 Remove the universal joint snap-ring with a small pair of pliers

4   Remove the rear universal joint bolts and straps **(see illustration)**. Turn the driveshaft (or wheels) as necessary to bring the bolts into the most accessible position. If the vehicle is a 4WD and has a companion flange where the driveshaft connects to the transfer case, remove the four bolts, nuts and washers from the flange.
5   On vehicles with a two-piece driveshaft, remove the bolts, nuts and washers from the center support bearing bracket.
6   Tape the bearing caps to the spider to prevent the caps from coming off during removal.
7   Lower the rear of the driveshaft. Slide the front of the driveshaft out of the transmission or transfer case or, if equipped with a companion flange, separate the flange at the transfer case.
8   On vehicles with a splined yoke (instead of a companion flange) at the transmission or transfer case end, wrap a plastic bag over the transmission or transfer case housing and hold it in place with a rubber band. This will prevent loss of fluid and protect against contamination while the driveshaft is out.
9   If you're replacing the center support bearing on a two-piece driveshaft, unscrew the collar, separate the two sections of the driveshaft, then slide the collar and bearing off the front section. Install the new bearing and collar, reattach the two halves of the driveshaft and screw on the collar.

Installation

10   Remove the plastic bag from the transmission or transfer case and wipe the area clean. Inspect the oil seal carefully (see Chapter 7, Part A or B). Slide the front of the driveshaft into the transmission or transfer case or connect the companion flange, installing the fasteners finger-tight.
11   Raise the rear of the driveshaft into position, checking to be sure the marks are in alignment. If not, turn the rear wheels to match the pinion flange and the driveshaft.
12   Remove the tape securing the bearing caps and install the straps and bolts. Tighten all bolts to the torque listed in this Chapter's Specifications. Lower the vehicle and connect the negative battery cable.

Front driveshaft (4WD models)
Removal

Refer to illustrations 12.15 and 12.17
13   Raise the vehicle and place it securely on jackstands.
14   Remove the skid plate (if equipped).
15   Mark the relationship of the driveshaft to the front differential flange and, on 1994 models, the transfer case flange **(see illustration)**.
16   Remove the bolts and straps from the differential flange and, on 1994 models, the transfer case flange.
17   On 1994 models, push the rear half of the driveshaft forward far enough to separate it from the transfer case flange, then lower it and separate it at the differential flange. On 1995 and later models, the rear end of the driveshaft uses a constant velocity (CV) joint splined to the transfer case output shaft. On these models, drop down the front end of the driveshaft first, then slide the driveshaft CV joint off the transfer case splined output shaft **(see illustration)**.

Installation

18   Attach the front end of the driveshaft to the front differential flange and install the straps and bolts finger-tight.
19   Extend or compress the driveshaft as necessary, attach the rear end to the transfer case flange, install the straps and bolts and tighten all bolts to the torque listed in this Chapter's Specifications.
20   Install the skid plate (if equipped).
21   Lower the vehicle and connect the negative battery cable.

13   Universal joints - removal, overhaul and installation

Note: *Always purchase a universal joint service kit for your model vehicle before beginning this procedure. Also, read through the entire procedure before beginning work.*
1   Remove the driveshaft (see Section 11).

Outer snap-ring type

Refer to illustrations 13.3, 13.4, 13.5 and 13.6
2   Place the driveshaft on a bench equipped with a vise.
3   Remove the snap-rings with a small pair of pliers **(see illustration)**.
4   Support the cross (also called a spider or trunnion) on a short piece of pipe or a large socket and use another socket to press out the cross by closing the vise **(see illustration)**.
5   Press the cross through as far as possible, then grip the bearing cup with locking pliers and remove it **(see illustration)**.
6   A universal joint repair kit will contain a

Chapter 8 Clutch and driveline

13.4 To remove the U-joint from the driveshaft, use a vise as a press; the small socket will push the cross and bearing cup into the large socket

13.5 Grip the bearing cup with locking pliers and remove it from the yoke

new trunnion, seals, bearings, cups and snap-rings **(see illustrations)**.
7 Inspect the bearing cup housing in the driveshaft for wear and damage.
8 If the bearing cup housings in the yoke are so worn that the cups are a loose fit in the yokes, replace the driveshaft.
9 Make sure the dust seals are properly located on the trunnion so the cavities face the trunnion.
10 Using a vise, press a bearing cup into the yoke about 1/4-inch.
11 Use multi-purpose grease to hold the needle rollers in place in the cup.
12 Insert the trunnion into the partially installed bearing cup, taking care not to dislodge the needle rollers.
13 Stick the needle bearings into the opposite cup, hold the trunnion in correct alignment and press both cups into place by slowly and carefully closing the jaws of the vise.
14 Use a socket slightly smaller in diameter than the cups to press them into the yoke. Press in one side, install the snap-ring, then press the other side to shift the trunnion

assembly tight against the installed snap-ring and install the other snap-ring.
15 Repeat the operations for the remaining two bearing cups.

Injected-plastic (inner snap-ring) type

16 If the joint has been previously rebuilt, remove the snap-rings (bearing retainers) located on the inner part of each bearing cup.
17 If this is the first time the joint is being rebuilt, there won't be any snap-rings present; the pressing operation will shear the molded plastic retaining material.
18 Press out the bearing cups as described in Steps 3 through 5.
19 Remove the trunnion (cross) and clean all plastic material from the yoke. Use a small punch to remove the plastic from the injection holes.
20 Reassembly is the same as for the outer snap-ring joint described in Steps 9 through 14, except that the snap-rings are on the inner part of each bearing cup.
21 When installing the bearing cup, press it in until the snap-ring can be installed. If difficulty is encountered, strike the yoke sharply

with a hammer. This will spring the yoke ears slightly and allow the snap-ring groove to move into position.

14 Axles - description and check

Description
1 The rear axle assembly is a hypoid, semi-floating type (the centerline of the pinion gear is below the centerline of the ring gear). When the vehicle goes around a corner, the differential allows the outer rear tire to turn more quickly than the inner tire. The axleshafts are splined to the differential side gears, so when the vehicle goes around a corner, the inner tire, which turns more slowly than the outer tire, turns its side gear more slowly than the outer tire turns its side gear. The differential pinion gears roll around the slower side gear, driving the outer side gear - and tire - more quickly. The differential is housed within a casting with a pressed steel cover, known as the "carrier." The steel axle tubes are pressed into and welded to the carrier.
2 On 4WD models, a fully independent front axle assembly is used. This consists of a differential and a pair of driveaxles. Each driveaxle has an inner and outer constant velocity (CV) joint. Because the differential - like the transfer case - is offset to the left, the distance between the differential and the right front wheel is greater than the distance from the differential to the left wheel. In order to use two equal-length driveaxles, an extension axleshaft is employed on the right side to make up the difference.
3 An optional locking limited-slip rear axle is also available. This differential allows for normal operation until one wheel loses traction. A limited-slip unit is similar in design to a conventional differential, except for the addition of a pair of clutch "cones" which slow the rotation of the differential case when one wheel is on a firm surface and the other on a slippery one. The difference in wheel rota-

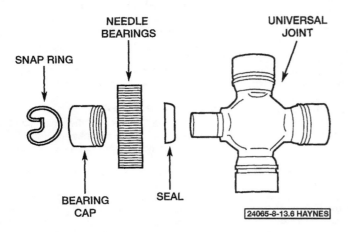

13.6a U-joint kit components (outer snap-ring type)

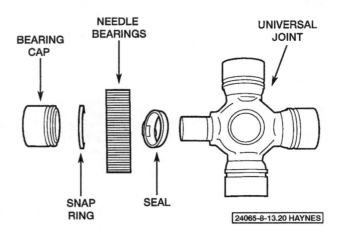

13.6b U-joint repair kit components (injected plastic type)

8-10   Chapter 8   Clutch and driveline

tional speed produced by this condition applies additional force to the pinion gears and through the cone, which is splined to the axleshafts, equalizes the rotation speed of the axleshaft driving the wheel with traction.

Check

4   Often, a suspected "axle" problem lies elsewhere. Do a thorough check of other possible causes before assuming the axle is the problem.
5   The following noises are those commonly associated with axle diagnosis procedures:

a) Road noise is often mistaken for mechanical faults. Driving the vehicle on different surfaces will show whether the road surface is the cause of the noise. Road noise will remain the same if the vehicle is under power or coasting.
b) Tire noise is sometimes mistaken for mechanical problems. Tires which are worn or low on pressure are particularly susceptible to emitting vibrations and noises. Tire noise will remain about the same during varying driving situations, where axle noise will change during coasting, acceleration, etc.
c) Engine and transmission noise can be deceiving because it will travel along the driveline. To isolate engine and transmission noises, make a note of the engine speed at which the noise is most pronounced. Stop the vehicle and place the transmission in Neutral and run the engine to the same speed. If the noise is the same, the axle is not at fault.

6   Because of the special tools needed, overhauling the differential isn't cost effective for a do-it-yourselfer. The procedures included in this Chapter describe axleshaft removal and installation, axleshaft oil seal replacement, axleshaft bearing replacement and removal of the entire unit for repair or replacement. Any further work should be left to a dealer service department or other qualified repair shop.

Identification

7   If the rear axle must be replaced, refer to the identification code and manufacturer's code stamped on the right rear axle tube on the front side of the rear axle tube. This number contains information on the rear axle ratio, differential type, manufacturer and build date information, all of which are necessary to ensure that you get the right axle. On front axles, the ID code is stamped at the top of the carrier.

15 Axleshaft (rear) - removal and installation

Removal

1   Raise the rear of the vehicle, support it securely on jackstands and block the front wheels. Remove the wheel and brake drum.

15.3 Remove the lock bolt and pull out the differential pinion gear shaft (arrow) (standard differential shown, locking differential similar)

15.4 Remove the C-lock from the axleshaft

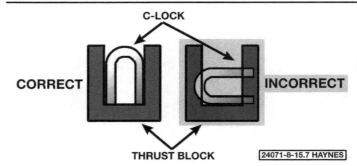

15.7 The C-lock must be positioned as shown before the axleshaft can be removed (locking differential)

On locking differential equipped models, remove both rear wheels and brake drums.
2   Remove the differential cover and allow the lubricant to drain into a container (see Chapter 1).

Conventional differential

Refer to illustrations 15.3 and 15.4

3   Remove the lock bolt and pull out the pinion shaft **(see illustration)**.
4   Have an assistant push in on the outer flanged end of the axleshaft while you remove the C-lock from the groove in the inner end of the shaft **(see illustration)**.

Locking differential

Refer to illustration 15.7

5   Rotate the differential for access, support the pinion shaft so it won't fall into the case, then remove the lock bolt **(see illustration 15.3)**.
6   Withdraw the pinion shaft part way. Rotate the differential until the shaft touches the case, providing enough clearance for access to the C-locks.
7   Use a screwdriver to rotate the C-lock until the open end points in **(see illustration)**.
8   With the C-lock in position so it will pass through the end of the thrust block, push the axleshaft in and remove the C-lock. Repeat the operation for the opposite axleshaft.

All models

9   With the C-lock removed, withdraw the axleshaft, taking care not to damage the oil seal. Some models have a thrust washer in the differential; make sure it doesn't fall out when the axleshaft is removed.

Installation

All models

10   To install, carefully insert the axleshaft into the housing and seat it securely in the differential.

Conventional differential

11   Install the C-lock in the axleshaft groove and pull out on the flange to lock it.
12   Insert the pinion shaft, align the hole in the shaft with the lock bolt hole and install the lock bolt. Tighten the lock bolt to the torque listed in this Chapter's Specifications.

Locking differential

13   Install the C-locks with the pinion shaft still partially withdrawn. Make sure the C-locks are positioned correctly **(see illustration 15.7)**.
14   Withdraw the axleshaft carefully until the C-lock clears the thrust block.
15   After the C-locks are installed, push the pinion shaft into place with the groove lined up with the lock bolt hole, then install the lock bolt. Tighten the lock bolt to the torque listed in this Chapter's Specifications.

Chapter 8 Clutch and driveline

16.2 You can pry out the old seal with the axleshaft

16.3 Install the new axleshaft oil seal with a seal driver or large socket

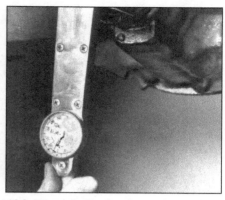

18.3 Use an inch-pound torque wrench to check the torque necessary to rotate the pinion shaft

All models

16 Install the cover and fill the differential with the lubricant specified in Chapter 1.
17 Install the brake drums and wheels and lower the vehicle.

16 Axleshaft oil seal (rear) - replacement

Refer to illustrations 16.2 and 16.3

1 Remove the axleshaft (see Section 15).
2 Pry the oil seal out of the end of the axle housing with a large screwdriver or the inner end of the axleshaft **(see illustration)**.
3 Apply high-temperature grease to the oil seal recess and tap the new seal evenly into place with a hammer and seal installation tool, large socket or piece of pipe so the lips are facing in and the metal face is visible from the end of the axle housing **(see illustration)**. When correctly installed, the face of the oil seal should be flush with the end of the axle housing.

17 Axleshaft bearing (rear) - replacement

1 Remove the axleshaft (see Section 15) and the oil seal (see Section 16).
2 A bearing puller which grips the bearing from behind will be required for this job. These tools, adapted to screw onto slide hammers, are available from tool stores and auto many parts stores.
3 Attach a slide hammer to the puller and extract the bearing from the axle housing.
4 Clean out the bearing recess and drive in the new bearing with a bearing installer or a piece of pipe positioned against the outer bearing race. Make sure the bearing is tapped in to the full depth of the recess and the numbers on the bearing are visible from the outer end of the axle housing.
5 Install a new oil seal (see Section 16), then install the axleshaft.

18 Differential pinion seal (rear) - replacement

Refer to illustrations 18.3, 18.4, 18.6 and 18.10

1 Loosen the rear wheel lug nuts, raise the rear of the vehicle and support it securely on jackstands. Block the front wheels to keep the vehicle from rolling off the stands. Remove the wheels (this will allow you to obtain a more accurate pinion shaft preload reading).
2 Disconnect the driveshaft and fasten it out of the way.
3 On semi-floating axles, use an inch-pound torque wrench to check the torque required to rotate the pinion. Record it for use later **(see illustration)**.
4 Scribe or punch alignment marks on the pinion shaft, nut and flange **(see illustration)**.
5 Count the number of threads visible between the end of the nut and the end of the pinion shaft and record it for use later.
6 A special tool can be used to keep the companion flange from moving while the self-locking pinion nut is loosened. If the special tool isn't available, try using a large pair of pliers **(see illustration)**.
7 Remove the pinion nut.
8 Withdraw the companion flange. It may be necessary to use a two or three-jaw puller engaged behind the flange to draw it out. Do NOT attempt to pry behind the flange or hammer on the end of the pinion shaft.
9 Pry out the old seal and discard it.
10 Lubricate the lips of the new seal with high-temperature grease and tap it evenly into position with a seal installation tool or a large socket. Make sure it enters the housing squarely and is tapped in to its full depth **(see illustration)**.
11 Align the mating marks made before disassembly and install the companion flange. If necessary, tighten the pinion nut to draw the flange into place. Do not try to hammer the flange into position.
12 Apply non-hardening sealant to the ends of the splines visible in the center of the

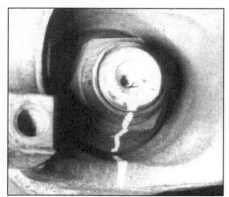

18.4 Mark the relative positions of the pinion, nut and flange (arrows) before removing the nut

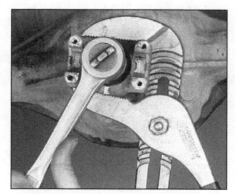

18.6 If you don't have the factory tool or one like it, try holding the flange with a large pair of pliers while you loosen the locknut

18.10 Use a special installation tool with adapter, a large socket (shown) or a piece of pipe to tap the pinion oil seal into place

8-12   Chapter 8 Clutch and driveline

20.2 To remove the driveaxle hub nut, hold the driveaxle by inserting a large screwdriver or prybar between the wheel studs as shown and use a large breaker bar to loosen the nut

20.3 Mark the relationship of the inner CV joint to the flange, then, with the prybar again between the wheel studs, loosen the flange bolts

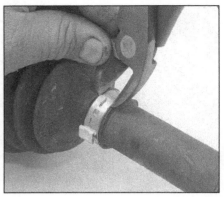

21.3 Cut off the old boot clamps with a pair of diagonal cutters

flange so oil will be sealed in.
13   Install the washer (if equipped) and pinion nut. Tighten the nut carefully until the original number of threads are exposed.
14   Measure the torque required to rotate the pinion and tighten the nut in small increments until it matches the figure recorded in Step 5. In order to compensate for the drag of the new oil seal, the nut should be tightened more until the rotational torque of the pinion slightly exceeds what was recorded earlier, but not by more than 5 in-lbs.
15   Connect the driveshaft, install the wheels and lower the vehicle. Tighten the lug nuts to the torque listed in the Chapter 1 Specifications.

19 Axle (rear) - removal and installation

1   Loosen the rear wheel lug nuts, raise the rear of the vehicle and support it securely on jackstands placed under the frame (not under the axle). Block the front wheels to keep the vehicle from rolling off the stands. Remove the rear wheels.
2   Position a jack under the rear axle differential case.
3   Disconnect the driveshaft from the rear axle companion flange. Fasten the driveshaft out of the way with a piece of wire from the underbody.
4   Disconnect the shock absorbers at the lower mounts, then compress them to get them out of the way. **Note:** *Some later models are equipped with an axle vibration damper on the right side. It looks like a shock absorber and is removed in a similar manner.*
5   If equipped, disconnect the vent hose from the fitting on the axle housing and fasten it out of the way.
6   Disconnect the brake hose from the junction block on the axle housing, then plug the hose to prevent fluid leakage.
7   Remove the brake drums.
8   Disconnect the parking brake cables from the actuating levers and the backing

plate (see Chapter 9).
9   Disconnect the spring U-bolts. Remove the spacers and clamp plates.
10   Lower the jack under the differential, then remove the rear axle assembly from under the vehicle.
11   Installation is the reverse of removal. Lower the vehicle weight onto the wheels before tightening the U-bolt nuts completely.
12   Bleed the brakes (see Chapter 9).

20 Driveaxle (4WD models) - removal and installation

Refer to illustrations 20.2 and 20.3
1   Loosen the wheel lug nuts, raise the vehicle and support it securely on jackstands. Remove the wheel.
2   Place a prybar or large screwdriver between the wheel studs to hold the driveaxle and break the driveaxle hub nut loose with a large breaker bar **(see illustration)**. Remove the nut and washer.
3   Using the same setup, hold the driveaxle and remove the driveaxle flange bolts **(see illustration)**. Be sure to mark the relationship of the inner CV joint to the flange before removing the bolts. On 1999 and later models, the driveaxles are not flanged to the differential, but are retained by snap rings at the inner end of the tri-pot joints. To release the driveaxles, hit the inner part of the tri-pot housing with a hammer and piece of wood until the snap-ring releases.
4   Remove the steering knuckle (see Chapter 10).
5   Remove the driveaxle assembly.
6   Installation is the reverse of removal. Be sure to tighten the driveaxle hub nut and the CV joint flange bolts to the torque listed in this Chapter's Specifications.

21 Driveaxle boot replacement and CV joint overhaul (4WD models)

Note: *If the CV joints exhibit wear, indicating the need for an overhaul (usually due to torn boots), explore all options before beginning*

the job. Complete rebuilt driveaxles may be available on an exchange basis, which eliminates a lot of time and work. Whatever is decided, check on the cost and availability of parts before disassembling the joints.
1   Remove the driveaxle (see Section 20).
2   Place the driveaxle in a vise lined with rags to avoid damage to the shaft.

Inner CV joint

Refer to illustrations 21.3, 21.4, 21.10, 21.11, 21.13a and 21.13b
3   Cut off the boot seal retaining clamps **(see illustration)** and slide the boot towards the center of the driveaxle. Mark the tri-pot housing and driveaxle so they can be reinstalled in the same relative positions, then slide the housing off the spider assembly.
4   Remove the spider assembly from the axle by first removing the inner retaining ring and sliding the spider assembly back to expose the outer retaining ring. Remove the outer retaining ring and slide the joint off the driveaxle **(see illustration)**.
5   Use tape or a cloth wrapped around the three-bearing spider assembly to retain the bearings during removal and installation.
6   Remove the spider assembly from the axle.
7   Slide the boot off the axle.
8   Clean all of the old grease out of the housing and spider assembly. Carefully disassemble each section of the spider assembly, one at a time, and clean the needle bearings with solvent. Inspect the rollers, spider, bearings and housing for scoring, pitting and other signs of abnormal wear. Apply a coat of CV joint grease to the inner bearing surfaces to hold the needle bearings in place when reassembling the spider assembly.
9   Pack the housing with half of the grease furnished with the new boot and place the remainder in the boot.
10   Wrap the driveaxle splines with tape to avoid damaging the boot, then slide the boot onto the axle **(see illustration)**.
11   Install the spider assembly with the recess in the counterbore facing the end of the driveaxle **(see illustration)**.
12   Install the tri-pot housing.
13   Seat the boot in the housing and axle

Chapter 8 Clutch and driveline

21.4 Use snap-ring pliers to remove both the inner and outer retaining rings

21.10 Before installing the CV joint boot, wrap the axle splines with tape to prevent damage to the boot

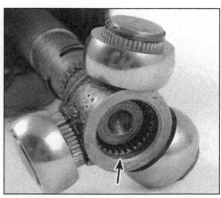

21.11 When installing the spider assembly on the driveaxle, make sure the recess in the counterbore (arrow) is facing the end of the driveaxle

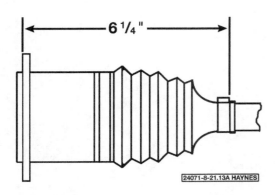

21.13a Before tightening the boot clamp, adjust the tri-pot joint until the dimension is as specified (flanged tri-pot joints)

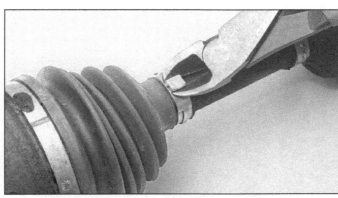

21.13b If you don't have the special factory tool (or an aftermarket version) for tightening the new boot clamps, use a pair of diagonal cutters to crimp the clamp as shown - do this gently so you don't damage the clamp

seal grooves, then adjust the collapsed dimension of the joint **(see illustration)**. Install the retaining clamps **(see illustration)**, then install the driveaxle (see Section 20). On later (non-flanged) tri-pot joints, adjust the position of the small clamp so that when the joint is collapsed together as far as it will go, the distance from the small clamp to the inner end of the tri-pot housing is 9 inches.

Outer CV joint

Refer to Illustrations 21.14, 21.16, 21.19, 21.20, 21.21, 21.22, 21.23a, 21.23b, 21.26 and 21.27

14 Tap lightly around the outer circumference of the seal retaining ring with a hammer and punch to dislodge and remove it. Be very careful not to deform the ring or it won't seal properly **(see illustration)**.
15 Cut off the band retaining the boot to the shaft.
16 Remove the snap-ring and slide the joint assembly off **(see illustration)**.
17 Slide the old boot off the driveaxle.
18 Place marks on the inner race and cage so they both can be installed facing out when reassembling the joint.

21.14 Carefully tap around the circumference of the retaining ring to remove it from the housing

21.16 Use snap-ring pliers to remove the inner retaining ring

8-14　Chapter 8　Clutch and driveline

21.19 Gently tap the inner race with a brass punch to tilt it enough to allow the ball bearings to be removed

21.20 Using a dull screwdriver, carefully pry the balls out of the cage

21.21 Tilt the inner race and cage 90-degrees, then align the windows in the cage with the lands and rotate the inner race up and out of the outer race

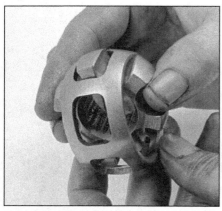

21.22 Align the inner race lands with the cage windows and rotate the inner race out of the cage

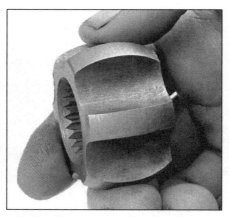

21.23a Check the inner race lands and grooves for pitting and score marks

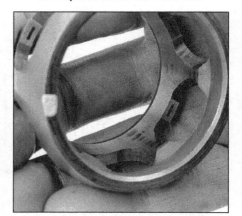

21.23b Check the cage for cracks, pitting and score marks - shiny spots are normal and don't affect operation

19 Press down on the inner race far enough to allow a ball bearing to be removed. If it's difficult to tilt, tap the inner race with a brass punch and hammer **(see illustration)**.
20 Pry the balls out of the cage, one at a time, with a dull screwdriver or wooden tool **(see illustration)**.
21 With all of the balls removed from the cage and the cage/inner race assembly tilted 90-degrees, align the cage windows with the outer race lands and remove the assembly from the outer race **(see illustration)**.
22 Remove the inner race from the cage by turning the inner race 90-degrees in the cage, aligning the inner lands with the cage windows and rotating the inner race out of the cage **(see illustration)**.
23 Clean the components with solvent to remove all traces of grease. Inspect the cage and races for pitting, score marks, cracks and other signs of wear and damage. Shiny, polished spots are normal and won't adversely affect CV joint operation **(see illustrations)**.
24 Install the inner race in the cage by reversing the technique described in Step 22.
25 Install the inner race and cage assembly in the outer race by reversing the procedure in Step 21. The marks that were previously

21.26 Align the cage windows and the inner and outer race grooves, then tilt the cage and inner race to insert the balls

applied to the inner race and cage must both be visible after the assembly is installed in the outer race.
26 Press the balls into the cage windows **(see illustration)**.
27 Pack the CV joint assembly with grease through the inner splined hole. Force the

21.27 Apply grease through the splined hole, then insert a wooden dowel (about 15/16-inch diameter) through the splined hole and push down - the dowel will force the grease into the joint

grease into the bearing by inserting a wooden dowel through the splined hole and pushing it to the bottom of the joint. Repeat this procedure until the bearing is completely packed **(see illustration)**.

Chapter 8 Clutch and driveline

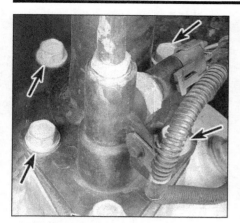

22.6a To detach the shift cable housing from the differential carrier, unplug the electrical connector from the 4WD indicator switch and remove the three lower bolts (arrows); to detach the output shaft tube assembly, remove the two upper bolts (arrows)

22.6b To disengage the shift cable from the shift fork shaft, you'll need to pull out the shift cable housing about 3/4-inch to an inch

23.10 To detach the output shaft tube from its support bracket, remove these two nuts (arrows)

28 Install the boot on the driveaxle as described in Step 10. Apply a liberal amount of grease to the inside of the axle boot.
29 Position the CV joint assembly on the driveaxle, aligning the splines. Using a soft-face hammer, drive the CV joint onto the driveaxle until the retaining ring is seated in the groove.
30 Seat the inner end of the boot in the seal groove and install the retaining clamp.
31 Install the seal retainer securely by tapping evenly around the outer circumference with a hammer and punch.
32 Install the driveaxle as described in Section 22.

22 Vacuum actuator and shift cable (4WD models) - replacement

Shift cable

Removal

Refer to illustrations 22.6a and 22.6b

1 Disconnect the negative cable from the battery. **Caution:** *On models equipped with a Delco Loc II anti-theft audio system, be sure the lockout feature is turned off before performing any procedure which requires disconnecting the battery.*
2 The vacuum actuator is located either on top of the right fenderwell or on the front of the right fenderwell, underneath the battery tray.
3 If the actuator is located below the battery, remove the battery (see Chapter 5).
4 To disconnect the shift cable from the vacuum actuator, disengage the retainer clip from the cable by pushing the straight part of the clip sideways and pushing in on the vacuum actuator diaphragm to release the grooved cable end. Squeeze the two locking fingers of the cable ferrule and pull the cable out of the bracket.

5 Raise the vehicle and support it securely on jackstands.
6 Remove the cable housing bolts **(see illustration)**. Pull out the housing about 3/4-inch **(see illustration)**.
7 To disconnect the shift cable from the shift fork shaft, disengage the retainer clip from the cable by pushing the straight part of the clip sideways and pulling the cable end from the shaft.
8 To detach the cable from the shift cable housing, unscrew the coupling nut. **Note:** *Do not unscrew the cable coupling nut unless the cable is being replaced.*
9 Remove the shift cable assembly.

Installation

10 Guide the end of the new cable through the hole in the shift cable housing, and push the cable end into the shift fork shaft. It should snap into place. Now carefully thread the coupling nut into the shift cable housing by hand to avoid cross-threading it. Tighten the coupling nut to the torque listed in this Chapter's Specifications.
11 Install the shift cable housing on the differential carrier, install the shift cable housing bolts and tighten the bolts to the torque listed in this Chapter's Specifications.
12 Route the cable to the actuator. Make sure it has no kinks or bends.
13 Lower the vehicle.
14 Push the end of the cable through the hole in the actuator bracket, spread the retainer clip, engage the cable end with the actuator and release the retainer clip. Grasp the cable ferrule and push it toward the actuator until the locking fingers on the ferrule snap into their locked position.

Vacuum actuator

15 Detach the vacuum hose from the actuator.
16 Disconnect the shift cable from the actuator (see Step 4).
17 Remove the actuator retaining bolts.
18 Remove the actuator.
19 Installation is the reverse of removal.

20 Connect the shift cable to the actuator (see Step 14).
21 Attach the vacuum hose to the actuator.

23 Output shaft, tube and tube seal (4WD models) - removal and installation

Refer to illustration 23.10

1 If the vacuum actuator is located under the battery tray, remove the battery and tray (see Chapter 5).
2 Disconnect the shift cable from the vacuum actuator (see Section 22).
3 Unlock the steering column so that the linkage is free to move.
4 Loosen the wheel lug nuts, raise the vehicle, place it securely on jackstands and remove the right front wheel.
5 Remove the front axle skidplate, if equipped.
6 Remove the right steering knuckle (see Chapter 10).
7 Remove the right driveaxle assembly (see Section 20).
8 Unplug the 4WD indicator electrical connector from the switch.
9 Remove the three lower shift cable housing bolts **(see illustration 22.6a)**.
10 Remove the retaining nuts, washers and bolts that attach the output shaft tube to its support bracket **(see illustration)**.
11 Remove the two upper shift cable housing bolts.
12 Carefully remove the output shaft tube. Make sure you don't allow the thrust washers, carrier connector, sleeve or output shaft to fall out of the tube. If any of these parts hit the floor, they could be damaged.
13 Remove the output shaft from the tube by striking the inside of the flange with a soft-faced hammer while holding the tube.
14 Using a large screwdriver, pry the output shaft tube seal from the tube.
15 Remove the tube bearing using a bearing removal tool.
16 Drive the shift cable housing seal out of

the tube flange with a drift punch or socket.
17   Install the new shift cable housing seal using an appropriate size socket.
18   Install the tube bearing (a special tool is machined to provide the proper bearing recess).
19   Install the output shaft tube seal using a socket equivalent to the diameter of the seal outer flange. The flange of the seal must be flush with the tube outer surface when installed.
20   Before installing the tube assembly, make sure that the sleeve, thrust washers and connectors are in place in the differential.
21   Apply Loctite 514 or equivalent to the tube-to-differential carrier mating surface.
22   Apply some grease to the thrust washer and install the washer, making sure the tab on the washer is aligned with the notch in the face of the output shaft tube.
23   Carefully install the output shaft tube assembly and install the first mounting bolt at the one o'clock position - just start it, don't tighten it.
24   Pull the assembly down and install the cable (see Section 22) and switch housing, and the remaining four switch housing bolts. Tighten the switch housing bolts to the torque listed in this Chapter's Specifications.
25   Install the two tube retaining bolts, washers and nuts and tighten the nuts to the torque listed in this Chapter's Specifications.
26   Install the shift cable by pushing it into the shift fork hole (the cable will automatically snap into place).
27   Plug in the electrical connector for the 4WD indicator switch.
28   Install the driveaxle assembly (see Section 20).
29   Install the steering knuckle (see Chapter 10).
30   Install the skid plate (if equipped) and drivebelt shield.
31   Install the wheel, remove the jackstands and lower the vehicle.
32   Push the shift cable into place in the vacuum actuator (see Section 22).

24   Differential output shaft pilot bearing (4WD models) - replacement

1   Remove the tube and output shaft assembly (see Section 23).
2   Remove the pilot bearing from the output shaft using a special tool available at most auto parts stores.
3   Install the replacement bearing using a special tool.

25   Front differential carrier (4WD models) - removal and installation

1   Disconnect the cable from the negative battery cable. **Caution:** *On models equipped with a Delco Loc II anti-theft audio system, be sure the lockout feature is turned off before performing any procedure which requires disconnecting the battery.*
2   Unlock the steering column so the steering linkage is free to move.
3   If the actuator is underneath the battery, remove the battery and the battery tray (see Chapter 5).
4   Disconnect the shift cable from the vacuum actuator (see Section 22).
5   Loosen the wheel lug nuts, raise the vehicle and place it securely on jackstands. Remove the wheels.
6   Remove the front axle skid plate, if equipped.
7   Remove the front driveaxles (see Section 20) and the front driveshaft (see Section 12).
8   Drain the lubricant from the differential (see Chapter 1).
9   Detach the vent hose from the carrier.
10   Unplug the electrical connector from the 4WD indicator switch, and remove the shift cable housing and output shaft tube (see Sections 22 and 23).
11   Disconnect the relay rod from the idler arm and Pitman arm (see Chapter 10).
12   Remove the two carrier-to-frame mounting bolts. Use an 18mm combination wrench inserted through the frame to prevent the upper nut from turning.
13   Lift up and rotate the carrier for clearance and remove it from the vehicle.
14   Installation is the reverse of removal.

26   Front differential carrier bushing (4WD models) - replacement

1   Remove the differential carrier (see Section 25).
2   Take the carrier assembly to an automotive machine shop and have the old bushings pressed out and new bushings pressed in.
3   Install the differential carrier (see Section 25).

Chapter 9 Brakes

Contents

	Section		Section
Anti-lock brake system - general information	2	Drum brake shoes - replacement	6
Brake check	See Chapter 1	Front wheel bearing check, repack and adjustment (2WD models only)	See Chapter 1
Brake disc - inspection, removal and installation	5	General information	1
Brake fluid level check	See Chapter 1	Master cylinder - removal, overhaul and installation	8
Brake hoses and lines - check and replacement	9	Parking brake - adjustment	11
Brake hydraulic system - bleeding	14	Parking brake cables - replacement	12
Brake light switch - check and replacement	16	Parking brake shoes - removal and installation	15
Combination valve (non-ABS models) - removal and installation	10	Power brake booster - removal and installation	13
Disc brake caliper - removal, overhaul and installation	4	Wheel cylinder - removal, overhaul and installation	7
Disc brake pads - replacement	3		

Specifications

General
Brake fluid type .... See Chapter 1

Disc brakes
Brake pad minimum thickness .... See Chapter 1
Disc lateral runout limit
    Front .... 0.003 inch
    Rear .... 0.004 inch
Disc minimum (discard) thickness .... Cast into disc
Front (ABS) wheel sensor resistance
    -40 to 40 degrees F .... 920 to 1440 ohms
    41 to 110 degrees F .... 1125 to 1700 ohms
    111 to 200 degrees F .... 1305 to 2000 ohms
    201 to 302 degrees F .... 1530 to 2310 ohms

Drum brakes
Minimum brake lining thickness .... See Chapter 1
Maximum drum diameter .... Cast into drum

Torque specifications
Ft-lbs (unless otherwise indicated)

Caliper mounting bolts
    Front caliper guide pins
        Single-piston front caliper 37
        Dual-piston front caliper 85
    Rear caliper guide pins 23
    Rear caliper mounting plate bolts 52
Front brake hose-to-caliper banjo bolt 32
Rear brake hose-to-caliper banjo bolt
    Through 1998 20
    1999 and later 40
Master cylinder mounting nuts
    1994 25
    1995 on 27
Vacuum booster mounting nuts 27
Wheel cylinder mounting bolts .... 115 in-lbs
Wheel lug nuts .... See Chapter 1

1 General information

General

All vehicles covered by this manual are equipped with hydraulically operated front and rear brake systems. All front brake systems are disc type; rear brake systems are drum brakes on some models and disc type on others. A quick visual inspection will tell you what type of brakes you have at the rear of your vehicle. All brakes are self-adjusting.

The hydraulic system has separate circuits for the front and rear brakes. If one circuit fails, the other circuit will remain functional and a warning indicator will light up on the dashboard when a substantial amount of brake fluid is lost, showing that a failure has occurred.

Master cylinder

The master cylinder is located under the hood, mounted to the power brake booster, and can be identified by the large fluid reservoir on top. The master cylinder has separate primary and secondary piston assemblies for the front and rear circuits.

Power brake booster

The power brake booster uses engine manifold vacuum to provide assistance to the brakes. It is mounted on the firewall in the engine compartment, directly behind the master cylinder.

Anti-lock Brake System (ABS)

A Rear Wheel Anti-Lock brake system (RWAL) is used on these vehicles to improve directional stability and control during hard braking. Some models are equipped with a Four-Wheel Anti-Lock brake system (4WAL) that prevents wheel skid at all four wheels.

Parking brake

The parking brake system is mechanically operated and cable-actuated. It's applied by a foot-operated pedal adjacent to the left kick panel and released by a pull-handle under the left end of the dash. When the parking brake pedal is applied on models with rear drum brakes, each rear cable pulls on a parking brake lever attached to the rear shoe of each brake assembly. When the parking brake pedal is applied on models with rear discs, each cable activates a lever that expands a horseshoe-shaped one-piece rear parking brake shoe against the inside of the rear disc.

Precautions

There are some general precautions and warnings related to the brake system:

a) *Use only brake fluid conforming to DOT 3 specifications.*
b) *The brake pads and linings contain fibers which are hazardous to your health if inhaled. Whenever you work on brake system components, clean all parts with brake system cleaner or denatured alcohol. Do not allow the fine dust to become airborne.*
c) *Safety should be paramount whenever any servicing of the brake components is performed. Do not use parts or fasteners which are not in perfect condition, and be sure all clearances and torque specifications are adhered to. If you are at all unsure about a certain procedure, seek professional advice. Upon completion of any brake system work, test the brakes carefully in a controlled area before driving the vehicle in traffic. If a problem is suspected in the brake system, don't drive the vehicle until it's fixed.*

2 Anti-lock brake system - general information

Description

The Anti-lock brake system is designed to maintain vehicle maneuverability, directional stability and optimum deceleration under severe braking conditions on most road surfaces. It does so by monitoring the rotational speed of the wheels and controlling the brake line pressure during braking. This prevents the wheels from locking up prematurely.

Two types of systems are used: Rear Wheel Anti-Lock (RWAL) and Four Wheel Anti-Lock (4WAL). RWAL only controls lockup on the rear wheels, whereas 4WAL prevents lockup on all four wheels.

Actuator assembly

The actuator assembly includes the

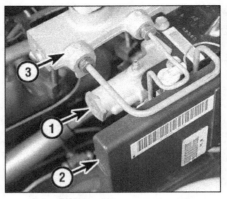

2.4a RWAL system components

1 *Isolation/dump valve*
2 *Control module*
3 *Master cylinder*

master cylinder and a control valve which consists of a dump valve and an isolation valve. The valve operates by changing the brake fluid pressure in response to signals from the control unit.

Control unit

Refer to illustrations 2.4a and 2.4b

The control unit for the anti-lock brakes is called the Electro-Hydraulic Control Unit (EHCU) on 4WAL systems and Control Module on RWAL systems. The unit is mounted in the engine compartment below (4WAL) or adjacent to (RWAL) the master cylinder and is the "brain" for the system **(see illustrations)**. The function of the control unit is to accept and process information received from the speed sensor(s) and brake light switch to control the hydraulic line pressure, avoiding wheel lockup. The control unit also constantly monitors the system, even under normal driving conditions, to find faults within the system.

If a problem develops within the system, the BRAKE (RWAL system) or ANTI-LOCK (4WAL system) warning light will glow on the dashboard. A diagnostic code will also be stored, which, when retrieved by a service technician, will indicate the problem area or component.

Speed sensor

On 4WAL systems, each wheel has a speed sensor. On RWAL systems, a rear wheel speed sensor is located in the transmission extension housing on 2WD models and in the transfer case on 4WD models. The speed sensor(s) sends a signal to the control unit indicating wheel rotational speed.

Brake light switch

The brake light switch signals the control unit when the driver steps on the brake pedal. Without this signal the anti-lock system won't activate.

Checks

Although a special electronic tester is necessary to properly diagnose the system, the home mechanic can perform a few preliminary checks before taking the vehicle to a dealer service department or other repair

2.4b 4WAL system components

shop which is equipped with this tester:

a) *Check the fuses.*
b) *Check the electrical connectors at the EBCM and the hydraulic modulator/motor pack.*
c) *Follow the wiring harness to the speed sensors and brake light switch and make sure all connections are secure and the wiring isn't damaged.*
d) *Make sure the brake lines, calipers and wheel cylinders are in good condition.*
e) *Check the resistance of the front wheel speed sensors and compare your readings with the values listed in this Chapter's Specifications (see Chapter 6 for the Vehicle Speed Sensor check, which monitors the speed of the rear wheels). If any sensor is out of range, replace it (the sensors are integral with the hub/wheel bearing assemblies).*

If the above preliminary checks don't rectify the problem, the vehicle should be diagnosed by a dealer service department or other qualified repair shop.

Scan tool diagnostics

At the time of writing, most generic scan tools will not access ABS trouble codes. If your scan tool will access the codes, however, follow the instructions on the scan tool display to check for any stored ABS trouble codes. The following chart lists the codes that may be encountered. Once repairs are made, use the scan tool to delete the trouble code(s) - these codes can't be erased by any other method.

3.3 To make room for the new pads, use a C-clamp to depress the piston into its bore before removing the caliper

ABS trouble codes

Code	Code definition
C0221	Right front wheel speed sensor - open circuit condition *
C0222	Right front wheel speed sensor - no signal detected
C0223	Right front wheel speed sensor - erratic signal detected
C0225	Left front wheel speed sensor - open circuit condition *
C0226	Left front wheel speed sensor - no signal detected
C0227	Left front wheel speed sensor - erratic signal detected
C0229	Front wheel speed signals - dropped out (both wheels simultaneously)
C0235	Rear wheel speed sensor - open circuit condition *
C0236	Rear wheel speed sensor - no signal detected
C0237	Rear wheel speed sensor - erratic signal detected
C0238	Wheel/vehicle speed mismatch **
C0241	EBCM control valve circuit (usually caused by an open or shorted solenoid coil in the EHCU)
C0242	See C0241
C0243	See C0241
C0244	See C0241
C0245	See C0241
C0246	See C0241
C0247	See C0241
C0248	See C0241
C0251	See C0241
C0252	See C0241
C0253	See C0241
C0254	See C0241
C0265	EBCM relay circuit - open
C0266	EBCM relay circuit - closed
C0267	Pump motor circuit - open or shorted circuit
C0268	Pump motor circuit - open or shorted circuit
C0269	Excessive dump/isolation time
C0271	EBCM malfunction
C0272	EBCM malfunction
C0273	EBCM malfunction
C0274	Excessive dump/isolation time
C0281	Brake light switch or circuit problem
C0284	EBCM malfunction

* Keep in mind that the sensor itself may not be the problem. Unplug the electrical connector and check the resistance of the sensor first; if the resistance is within specifications, carefully inspect the wiring harness leading to the sensor.

** This code sets when the wheel speed differs from the vehicle speed by more than 10-percent. If this code is stored, the first thing to do is check the tire sizes to make sure they match the size listed on the tire placard (on the end of the driver's door). The tire size is programmed into the EBCM - if smaller or larger diameter tires are installed, the EBCM must be re-calibrated with a scan tool to accurately determine the rotational speed of the wheels.

---

3 Disc brake pads - replacement

Refer to illustrations 3.3, 3.4a through 3.4l and 3.4m through 3.4s

Warning: *Disc brake pads must be replaced on both wheels at the same time - never replace the pads on only one wheel. Also, brake system dust is hazardous to your health. DO NOT blow it out with compressed air and DO NOT inhale it. An approved filtering mask should be worn when working on the brakes. DO NOT use gasoline or solvents to remove the dust. Use brake system cleaner only!*

1 Loosen the front wheel lug nuts, raise the front of the vehicle and support it securely on jackstands. Apply the parking brake. Remove the wheels.

2 Remove about two-thirds of the fluid from the master cylinder reservoir and discard it; as the pistons are pushed in for clearance to allow the pads to be removed, the fluid will be forced back into the reservoir. Position a drain pan under the brake assembly and clean the caliper and surrounding area with brake system cleaner.

3 Push the piston back into the bore with a C-clamp to provide room for the new brake pads **(see illustration)**. As the piston is depressed to the bottom of the caliper bore, the fluid in the master cylinder will rise. Make sure it doesn't overflow. If necessary, siphon

Chapter 9 Brakes

3.4a Wash down the disc and brake pads with brake cleaner to remove brake dust; DO NOT blow off brake dust with compressed air

3.4b Remove the two caliper mounting bolts, check them for thread damage and replace as necessary

3.4c Remove the brake line retaining clip (mounted to the frame rail at the upper left) and separate the brake line from the retainer. Pull off the caliper and remove the inside pad from the caliper

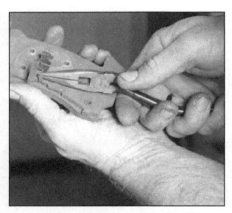

3.4d Remove the anti-rattle clip with pliers

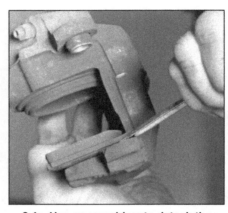

3.4e Use a screwdriver to detach the outer pad from the retainer

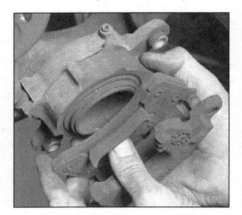

3.4f Separate the outer pad from the caliper

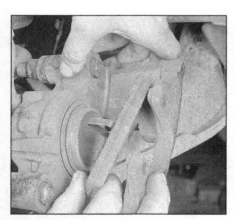

3.4g After cleaning the caliper, install the anti-rattle clip to the new pad and install the inner pad

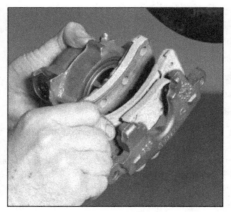

3.4h Attach the outer pad to the caliper

3.4i Apply multi-purpose grease to the ends of the caliper mounting bracket surfaces

off some of the fluid.

4   Work on one brake assembly at a time. If you're replacing the front brake pads, follow the accompanying photos, **beginning with illustration 3.4a; if you're replacing the rear pads, start at illustration 3.4m**. Be sure to stay in order and read the caption under each illustration.

Note: *Some models have two-piston front calipers. The basic pad replacement procedure is the same as for the single-piston calipers shown, except that anti-rattle springs are not used.*

5   While the pads are removed, inspect the caliper for brake fluid leaks and ruptures in the piston boot. Overhaul or replace the caliper as necessary (see Section 4). Also inspect the brake disc carefully (see Section 5). If machining is necessary, follow the information in that Section to remove the disc.

6   Before installing the caliper mounting bolts, clean them and check them for corro-

Chapter 9 Brakes   9-5

3.4j Install the caliper over the disc onto the mounting bracket; don't get any grease on the pad linings

3.4k Install the mounting bolts and tighten them to the torque listed in this Chapter's Specifications, then reattach the brake hose to the retaining clip on the frame

3.4l After refilling the master cylinder and pumping the brake pedal to seat the pads, use adjustable pliers to bend the upper ears of the outer pad until the ears are flush with the caliper housing, with no radial clearance

3.4m Before beginning, wash the disc and caliper assembly with brake cleaner to protect yourself from brake dust, which is hazardous to your health

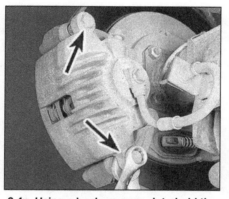

3.4n Using a back-up wrench to hold the caliper guide pins, remove the caliper bolts (arrows)

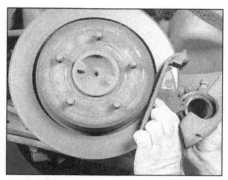

3.4o Remove the caliper assembly, and hang it out of the way with a coat hanger or a piece of wire

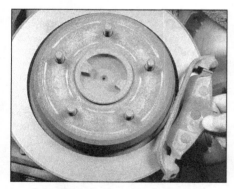

3.4p Remove the outer brake pad

3.4q Remove the inner pad

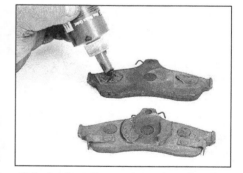

3.4r Apply anti-squeal compound to the backing plates of the new pads

sion and damage. If they're significantly corroded or damaged, replace them. Be sure to tighten the caliper mounting bolts to the torque listed in this Chapter's Specifications.

7 Install the brake pads on the opposite wheel, then install the wheels and lower the vehicle. Tighten the lug nuts to the torque listed in the Chapter 1 Specifications.

8 Add brake fluid to the reservoir until it's full (see Chapter 1). Pump the brakes several times to seat the pads against the disc, then check the fluid level again.

3.4s Install the pads and the caliper, then tighten the caliper bolts to the torque listed in this Chapter's Specifications

4.1a Remove the brake fluid hose banjo bolt (arrow)

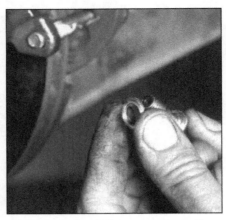

4.1b Always install new copper sealing washers when attaching the brake hose to the caliper

4.4 Compressed air is used to move the piston out of the caliper bore (use a wooden block as a cushion)

9 Check the operation of the brakes before driving the vehicle in traffic. Try to avoid heavy brake applications until the brakes have been applied lightly several times to seat the pads.

4 Disc brake caliper - removal, overhaul and installation

Refer to illustrations 4.1a, 4.1b, 4.4, 4.6a, 4.6b, 4.8, 4.9a, 4.9b and 4.10

Note 1: *If an overhaul is indicated (usually because of fluid leaks, a stuck piston or broken bleeder screw) explore all options before beginning this procedure. New and factory rebuilt calipers are available on an exchange basis, which makes this job quite easy. If it's decided to rebuild the calipers, make sure rebuild kits are available before proceeding. Always rebuild or replace the calipers in pairs - never rebuild just one of them.* **Note 2:** *Some later models have dual-piston front calipers. Overhaul of these calipers is virtually the same as for the single-piston calipers illustrated here, except that there are two pistons, seals and dust boots.*

1 Mark the position of the brake hose-to-caliper banjo fitting relative to the caliper, then remove the banjo bolt **(see illustration)**. **Note:** *If you're removing the caliper for access to other components, leave the hose connected.* Discard the two copper sealing washers on each side of the banjo fitting **(see illustration)**; use new ones when you reattach the brake hose to the caliper. Wrap a plastic bag around the end of the hose to prevent fluid loss and contamination. **Note:** *If the piston was stuck when you tried to depress it with a C-clamp, leave the brake line connected and, with the caliper removed and a rag positioned to catch the piston, use the brake pedal to pump the piston out of the caliper.*

2 Remove the caliper mounting bolts **(see illustration 4.3b)**.

3 Clean the caliper with brake system cleaner. DO NOT use kerosene or petroleum-based solvents.

4 Place several shop towels or a block of wood in the center of the caliper, then force the piston out of the bore by directing compressed air into the inlet opening **(see illustration)**. Only low air pressure should be required; a small tire pump may be adequate. **Warning:** *Keep your hand away from the piston as this is done!*

5 With the piston removed, inspect the surfaces of the piston for nicks and burrs and loss of plating. If surface defects are present, the piston must be replaced. Check the caliper bore in a similar way. Light polishing with crocus cloth is permissible to remove light corrosion and stains.

6 If the components are in good condition, pry out the piston seal and dust boot **(see illustrations)** and discard them. When you obtain an overhaul kit, it will contain all the replaceable items. Do NOT reuse the seal or the boot.

7 Clean the piston and cylinder bore with brake system cleaner, clean brake fluid or denatured alcohol only. **Warning:** *DO NOT, under any circumstances, clean brake system parts with gasoline or petroleum-based solvents.*

8 Lubricate the new piston seal with clean brake fluid and position the seal in the cylinder groove using your fingers only **(see illustration)**.

4.6a Pry the dust boot out of the caliper bore

4.6b Remove the piston seal from its groove in the caliper bore

9 Install the new dust boot in the groove in the end of the piston **(see illustration)**. Dip the piston in clean brake fluid, place it in position, "square" it with the bore **(see illustration)** and depress it to the bottom of the bore.

10 Seat the boot in the caliper counterbore using a boot installation tool or a blunt punch **(see illustration)**.

11 Connect the brake hose to the caliper, using new copper washers. Align the union

4.8 Position the seal in its groove in the caliper bore

Chapter 9 Brakes

4.9a Install a new dust boot in the piston groove with the folds toward the open end of the piston

4.9b Square the piston in the bore

4.10 Use a hammer and driver to seat the boot in the caliper housing counterbore

fitting with the marks made before disassembly to ensure the hose is positioned correctly.
12 Install the caliper and tighten the mounting bolts and the brake hose fitting bolt to the torque listed in this Chapter's Specifications, then bleed the front brake circuit (see Section 14).

5 Brake disc - inspection, removal and installation

Inspection

Refer to illustrations 5.3, 5.4a and 5.4b

1 Loosen the wheel lug nuts, raise the front of the vehicle and support it securely on jackstands. Apply the parking brake. Remove the wheel.
2 Remove the brake caliper (see Section 4). Visually inspect the disc surface for score marks and other damage. Light scratches and shallow grooves are normal after use and won't affect brake operation. Deep grooves - over 0.015-inch (0.38 mm) deep - require disc removal and refinishing by an automotive machine shop. Be sure to check both sides of the disc. If you're inspecting a rear disc, reinstall the wheel lug nuts - flat side toward the disc - to hold the disc in place during the following inspection. It may be necessary to install washers under the lug nuts in order for them to apply pressure to the disc.
3 To check disc runout, place a dial indicator at a point about 1/2-inch from the outer edge of the disc **(see illustration)**. On 4WD models, install two lug nuts, with the flat sides facing in, and tighten them securely to hold the disc in place. Set the indicator to zero and turn the disc. The indicator reading should not exceed 0.003-inch. If it does, the disc should be refinished by an automotive machine shop. **Note:** *The manufacturer recommends resurfacing the brake discs only in the event of pedal pulsations or hard (overheated) spots on the disc. If you elect not to have the discs resurfaced, deglaze them with sandpaper or emery cloth.*

5.3 Measure the brake disc runout with a dial indicator

4 The disc must not be machined to a thickness less than the specified minimum thickness, which is cast into the disc **(see illustration)**. The disc thickness can be checked with a micrometer **(see illustration)**.

Removal and installation

5 On 2WD models, the disc is an integral part of the front hub. Hub removal and installation is done as part of the front wheel bearing maintenance procedure (see Chapter 1). On 4WD models, the disc can be pulled off the hub after the wheel and caliper have been removed (see Section 4).

6 Drum brake shoes - replacement

Refer to illustrations 6.4a through 6.4v
Warning: *Brake shoes must be replaced on both wheels at the same time - never replace the shoes on only one wheel. Also, brake system dust is hazardous to your health. DO NOT blow it out with compressed air and DO NOT inhale it. DO NOT use gasoline or solvent to remove the dust. Use brake system cleaner or denatured alcohol only.*

5.4a The minimum (discard) thickness of the brake disc is cast into the disc itself

5.4b Measure the brake disc thickness at several points with a micrometer

1 Loosen the wheel lug nuts, raise the rear of the vehicle and support it securely on jackstands. Block the front wheels to keep the vehicle from rolling off the stands. Remove the rear wheels.
2 Remove the brake drum. If the drum is stuck because of corrosion between the axle flange and the wheel studs or brake drum,

6.4a Wash down the brake assembly with brake cleaner; DO NOT blow it out with compressed air!

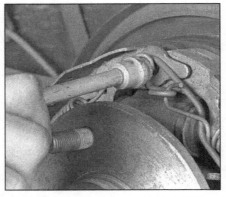

6.4b Remove the shoe return springs - the spring tool shown here is available at most auto parts stores and makes this job much easier and safer

6.4c Pull the bottom of the actuator lever toward the secondary brake shoe, compressing the lever return spring - the actuator link can now be removed from the top of the lever

6.4d Pry the actuator lever spring out with a large screwdriver

6.4e Slide the parking brake strut out from between the axle flange and primary shoe

6.4f Remove the hold-down springs and pins - the hold-down spring tool shown here is available at most auto parts stores

6.4g Remove the actuator lever and pivot - be careful not to let the pivot fall out of the lever

6.4h Spread the top of the shoes apart and slide the assembly around the axle

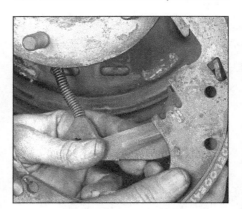

6.4i Unhook the parking brake lever from the secondary shoe - on some models, you'll have to pry off an E-clip with a small screwdriver before the shoe pin will slide out of the lever - you may have to drive the pin out of the old shoe with a hammer and punch and transfer it to the new shoe

spray penetrating oil around the flange and studs and allow it to soak in. Tap around the studs with a hammer and flange to break the drum loose, then tap around the back edge of the drum to remove it.

3   If the drum is locked onto the shoes because of excessive drum wear, knock out the plug in the access hole in the backing plate, insert a screwdriver through the hole, push the actuator lever off the star adjuster (see illustration 6.4v) and, using another screwdriver or a brake adjuster tool, back off the adjuster wheel to retract the shoes. **Note:** *If the plug is already knocked out, make sure it's not lodged somewhere inside the brake assembly.* Be sure to install a plug (available at most auto parts stores) in the hole when you're done to prevent water from entering the brake assembly.

4   Clean the brake assembly with brake system cleaner before beginning work **(see illustration)**. Follow **illustrations 6.4b** through **6.4v** for the inspection and replacement of the brake shoes. Be sure to stay in order and read the caption under each illustration.

Chapter 9 Brakes

6.4j Spread the bottom of the shoes apart and remove the adjusting screw assembly

6.4k Clean the adjusting screw with solvent. Dry it off and lubricate the threads and end with multi-purpose grease, then reinstall the adjusting screw assembly between the new brake shoes

6.4l Lubricate the shoe contact points on the backing plate with high-temperature brake grease

6.4m Insert the parking brake lever into the opening in the secondary brake shoe or insert the pin in the shoe through the lever and install a new E-clip

6.4n Spread the shoes apart and slide them into position on the backing plate

6.4o Install the hold-down pin and spring through the backing plate and primary shoe

6.4p Insert the lever pin into the actuator lever, place the lever over the secondary shoe hold-down pin and install the hold-down spring

6.4q Guide the parking brake strut behind the axle flange and engage the rear end of it in the slot on the parking brake lever - spread the shoes enough to allow the other end to seat against the primary shoe

6.4r Place the shoe guide over the anchor pin

5   Clean the brake drum and check it for score marks, deep grooves, hard spots (which will appear as small discolored areas) and cracks. If the drum is worn, scored or out-of-round, it can be resurfaced by an automotive machine shop. **Note:** *Professionals recommend resurfacing the drums whenever a brake job is done. Resurfacing will eliminate the possibility of out-of-round drums. If the drums are worn so much they can't be resurfaced without exceeding the maximum allowable diameter (stamped into the drum)* **(see illustration)***, new ones will be required. At the very least, if you elect not to have the drums resurfaced, remove the glaz-*ing from the surface with medium-grit emery cloth using a swirling motion.

6   Repeat this procedure for the other rear brake assembly.

7   Install the brake drums. Pump the brake several times, then turn the adjuster star wheels using a screwdriver inserted through the hole in the backing plate until the shoes slightly drag on the drums as the drums are turned. Now, back off the adjuster until the

Chapter 9 Brakes

6.4s Hook the lower end of the actuator link to the actuator lever, then loop the top end over the anchor pin

6.4t Install the lever return spring over the tab on the actuator lever, then push the spring up onto the brake shoe

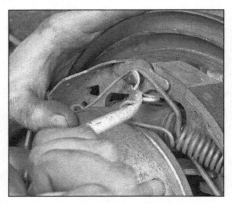

6.4u Install the primary and secondary shoe return springs

shoes don't drag on the drums.

8 Install the rear wheels, install the lug nuts, lower the vehicle and tighten the wheel lug nuts to the torque listed in the Chapter 1 Specifications.

9 Check the brake pedal position. If the brake pedal goes too close to the floor, further adjustment of the brakes is required. Back the vehicle up, making repeated stops, to actuate the self-adjusters, which work only when the vehicle is in reverse. Test the brakes for proper operation before driving in traffic.

7 Wheel cylinder - removal, overhaul and installation

Note: *If an overhaul is indicated (usually because of fluid leakage or sticky operation) explore all options before beginning the job. New wheel cylinders are available, which makes this job quite easy. If you decide to rebuild the wheel cylinder, make sure a rebuild kit is available before proceeding. Never overhaul only one wheel cylinder. Always rebuild both of them at the same time.*

Removal

Refer to illustration 7.2

1 Remove the brake drum and brake shoes (see Section 6).
2 Unscrew the brake line fitting from the rear of the wheel cylinder **(see illustration)**. If available, use a flare-nut wrench to avoid rounding off the corners on the fitting. Don't pull the metal line out of the wheel cylinder - it could bend, making installation difficult.
3 Remove the two bolts securing the wheel cylinder to the brake backing plate.
4 Remove the wheel cylinder.
5 Plug the end of the brake line to prevent the loss of brake fluid and the entry of dirt.

Overhaul

Refer to illustration 7.6

6 To disassemble the wheel cylinder, remove the rubber boot from each end of the

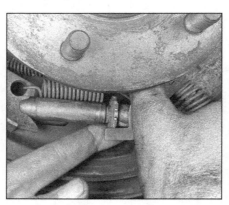

6.4v Pull out on the actuator lever to disengage it from the adjusting screw wheel, turn the wheel to adjust the shoes in or out as necessary - the brake drum should slide over the shoes and turn with a very slight amount of drag (at which point you'll back-off the adjuster until they don't drag)

cylinder, push out the two pistons and cups, and remove the cup return spring and expander **(see illustration)**. Discard the rubber parts and use new ones from the rebuild kit when reassembling the wheel cylinder.

7 Inspect the pistons for scoring and scuff marks. If present, the pistons should be replaced with new ones.

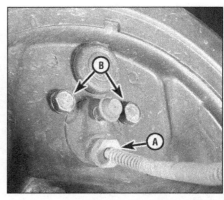

7.2 To remove the wheel cylinder, loosen the brake line threaded fitting (A) with a flare-nut wrench, disconnect the line and remove the wheel cylinder mounting bolts (B)

8 Examine the inside of the cylinder bore for score marks and corrosion. If these conditions exist, the cylinder can be honed slightly to restore it, but replacement is recommended.

9 If the cylinder is in good condition, clean it with brake system cleaner or denatured alcohol. **Warning:** *DO NOT, under any circumstances, use gasoline or petroleum-based solvents to clean brake parts!*

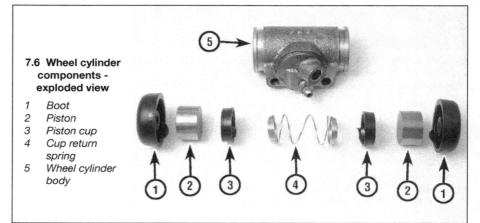

7.6 Wheel cylinder components - exploded view

1 Boot
2 Piston
3 Piston cup
4 Cup return spring
5 Wheel cylinder body

Chapter 9 Brakes

8.2 To detach the master cylinder assembly from the power brake booster, unscrew the brake line threaded fittings with a flare-nut wrench, then remove the two master cylinder mounting nuts (arrow)

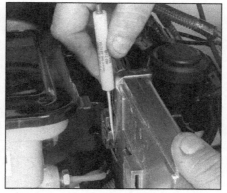

8.3 On models with Rear Wheel Anti-Lock (RWAL) brakes, unclip and remove the RWAL module from its mounting bracket

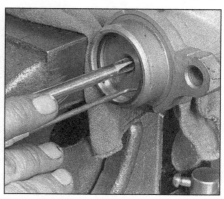

8.7 Push the primary piston in with a Phillips screwdriver and remove the lock-ring

8.8 Pull the primary piston and spring assembly out of the bore

8.9 To remove the secondary piston, tap the cylinder against a block of wood

10   Remove the bleeder screw and make sure the hole is clean.
11   Lubricate the cylinder bore with clean brake fluid, then insert one of the new rubber cups into the bore. Make sure the lip on the rubber cup faces in.
12   Place the expander spring in the opposite end of the bore and push it in until it contacts the rear of the rubber cup.
13   Install the remaining cup in the cylinder bore.
14   Attach the rubber boots to the pistons, then install the pistons and boots.
15   The wheel cylinder is now ready for installation.

Installation

16   Installation is the reverse of removal. Attach the brake line to the wheel cylinder before installing the mounting bolts and tighten the line fitting after the wheel cylinder mountings bolts have been tightened. If available, use a flare-nut wrench to tighten the line fitting.
17   Bleed the brakes (see Section 14). Don't drive the vehicle in traffic until the operation of the brakes has been thoroughly tested.

8   Master cylinder - removal, overhaul and installation

Removal

Refer to illustrations 8.2 and 8.3
Note: *Before deciding to overhaul the master cylinder, check on the availability and cost of a new or factory rebuilt unit and also the availability of a rebuild kit.*

1   Place rags under the brake line fittings and prepare caps or plastic bags to cover the ends of the lines once they're disconnected. **Caution:** *Brake fluid will damage paint. Cover all painted surfaces and avoid spilling fluid during this procedure.*
2   Loosen the tube nuts at the ends of the brake lines where they enter the master cylinder. To prevent rounding off of the flats on these nuts, a flare-nut wrench, which wraps around the nut, should be used **(see illustration)**. Pull the brake lines away from the master cylinder slightly and plug the ends to prevent contamination.
3   On models with a Rear-Wheel Anti-Lock (RWAL) brake system, remove the RWAL module **(see illustration)**. Carefully set the module aside; it's not necessary to unplug the electrical connectors.
4   Remove the two master cylinder mounting nuts and set aside the bracket for the RWAL module and isolation/dump valve; make sure you don't kink the hydraulic lines. Remove the master cylinder from the vehicle.
5   Remove the reservoir cover and diaphragm, then discard any fluid remaining in the reservoir.

Overhaul and installation

Refer to illustrations 8.7, 8.8, 8.9, 8.10, 8.14, 8.15a, 8.15b, 8.15c, 8.15d and 8.16
6   Mount the master cylinder in a padded vise.
7   Remove the primary piston lock-ring by depressing the piston and prying the ring out with a screwdriver **(see illustration)**.
8   Remove the primary piston assembly from the cylinder bore **(see illustration)**.
9   Remove the secondary piston assembly

8.10 Pry the reservoir out of the grommets

from the cylinder bore. It may be necessary to remove the master cylinder from the vise and invert it, carefully tapping it against a block of wood to expel the piston **(see illustration)**.
10   Pry the reservoir off the body with a screwdriver **(see illustration)**. Remove the grommets. **Caution:** *Later model master cylinders have spring-pins securing the reservoir to the master cylinder. Do not try to pry off the reservoir until you check for the presence of these pins. Pins can be removed with a 1/8-inch drift punch.*
11   Do not attempt to remove the quick

9-12   Chapter 9   Brakes

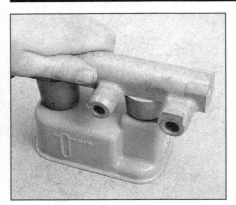

8.14 Lay the reservoir face down on a work bench and push the master cylinder straight down over the reservoir fittings with a rocking motion

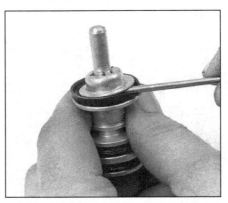

8.15a Pry the secondary piston spring seat off with a small screwdriver, then remove the seal

8.15b Remove the secondary seals from the piston (some pistons have only one seal)

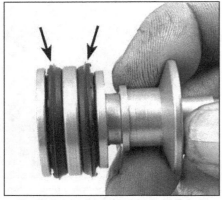

8.15c Install the secondary seals with the lips facing away from each other (on the single-seal design, the seal lip should face away from the center of the piston)

8.15d Install a new primary seal on the secondary piston with the seal lip facing in the direction shown

8.16 Install a new spring seat over the end of the secondary piston and push it into place with a socket

take-up valve from the master cylinder body - it's not serviceable.

12 Inspect the cylinder bore for damage. If any damage is found, replace the master cylinder body with a new one, as abrasives cannot be used on the bore.

13 Lubricate the new reservoir grommets with brake fluid and press them into the master cylinder body. Make sure they're properly seated.

14 Lay the reservoir on a hard surface and press the master cylinder body onto the reservoir, using a rocking motion (see illustration). If it was secured by spring-pins, install them now.

15 Remove the old seals from the secondary piston assembly and install the new secondary seals with the lips facing away from each other (see illustrations). The lip on the primary seal must face in (see illustration).

16 Attach the spring seat to the secondary piston assembly (see illustration).

17 Lubricate the cylinder bore with clean brake fluid and install the spring and secondary piston assembly.

18 Install the primary piston assembly in the cylinder bore, depress it and install the lock-ring.

19 Inspect the reservoir cover and diaphragm for cracks and deformation. Replace any damaged parts with new ones and attach the diaphragm to the cover.

20 Whenever the master cylinder is removed, the entire hydraulic system must be bled. The time required to bleed the system can be reduced if the master cylinder is filled with fluid and bench bled (refer to Steps 21 through 23) before the master cylinder is installed on the vehicle.

21 Fill the reservoirs with brake fluid. The master cylinder should be supported so the brake fluid won't spill during the bench bleeding procedure.

22 Hold your fingers tightly over the holes where the brake lines normally connect to the master cylinder to prevent air from being drawn back into the master cylinder.

23 Stroke the piston several times to ensure all air has been expelled. A large Phillips screwdriver can be used to push on the piston assembly. Wait several seconds each time for brake fluid to be drawn from the reservoir into the piston bore, then depress the piston again, removing your finger as brake fluid is expelled. Be sure to put your fingers back over the holes each time before releasing the piston. When the bleeding procedure is complete, temporarily install plugs in the holes.

24 Carefully install the master cylinder by reversing the removal steps.

25 Bleed the brake system (see Section 14).

9   Brake hoses and lines - check and replacement

1 About every six months, with the vehicle raised and placed securely on jackstands, the flexible hoses which connect the steel brake lines with the front and rear brake assemblies should be inspected for cracks, chafing of the outer cover, leaks, blisters and other damage. These are important and vulnerable parts of the brake system and inspection should be complete. A light and mirror will be needed for a thorough check. If a hose exhibits any of the above defects, replace it with a new one.

Flexible hoses

Refer to illustrations 9.3 and 9.4

2 Clean all dirt away from the ends of the hose.

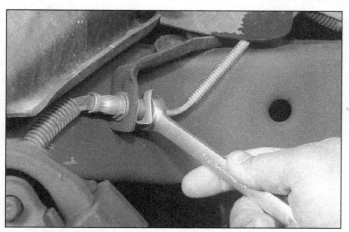

9.3 Unscrew the brake line threaded fitting with a flare-nut wrench so you don't round off the corners of the nut

9.4 Pull off the retainer clip with a pair of pliers

3 Disconnect the brake line from the hose fitting **(see illustration)**. Be careful not to bend the frame bracket or line. If necessary, soak the connections with penetrating oil.

4 Remove the U-clip from the female fitting at the bracket **(see illustration)** and remove the hose from the bracket.

5 Disconnect the hose from the caliper, discarding the copper washers on either side of the fitting.

6 Using new copper washers, attach the new brake hose to the caliper.

7 Pass the female fitting through the frame or frame bracket. With the least amount of twist in the hose, install the fitting in this position. **Note:** *The weight of the vehicle must be on the suspension, so the vehicle should not be raised while positioning the hose.*

8 Install the U-clip in the female fitting at the frame bracket.

9 Attach the brake line to the hose fitting using a back-up wrench on the fitting.

10 Carefully check to make sure the suspension or steering components don't make contact with the hose. Have an assistant push down on the vehicle and also turn the steering wheel lock-to-lock during inspection.

11 Bleed the brake system as described in Section 14.

Metal brake lines

12 When replacing brake lines, be sure to use the correct parts. Don't use copper tubing for any brake system components. Purchase steel brake lines from a dealer parts department or auto parts store.

13 Prefabricated brake line, with the tube ends already flared and fittings installed, is available at auto parts stores and dealer parts departments. These lines are also bent to the proper shapes.

14 When installing the new line make sure it's well supported in the brackets and has plenty of clearance between moving or hot components.

15 After installation, check the master cylinder fluid level and add fluid as necessary. Bleed the brake system as outlined in the next Section and test the brakes carefully before placing the vehicle into normal operation.

10 Combination valve (non-ABS models) - replacement

Refer to illustration 10.3

1 Place rags under the brake line fittings and prepare caps or plastic bags to cover the ends of the lines once they're disconnected. **Caution:** *Brake fluid will damage paint. Cover all painted surfaces and avoid spilling fluid during this procedure.*

2 On RWAL models, remove the RWAL module **(see illustration 8.3)**.

3 Loosen the tube nuts at the ends of the brake lines where they enter the combination valve. To prevent rounding off of the flats on these nuts, a flare-nut wrench, which wraps around the nut, should be used **(see illustration)**. Pull the brake lines away from the combination valve slightly and plug the ends to prevent contamination.

4 Unplug the electrical connectors for the low pressure warning switch and, if used, for the anti-lock pressure valve.

5 Remove the combination valve mounting nuts (they're the mounting nuts for the master cylinder) and remove the combination valve and bracket as a single assembly.

6 Separate the combination valve from its bracket.

7 Installation is the reverse of removal. Be sure to tighten the combination valve retaining nuts and the brake line fittings to the torque listed in this Chapter's Specifications.

8 Bleed the brake hydraulic system (see Section 14).

11 Parking brake - adjustment

Refer to illustrations 11.4a, 11.4b and 11.4c

1 The parking brake is pedal operated and is normally self-adjusting through the automatic adjusters in the rear brake drums. However, supplementary adjustment may be needed in the event of cable stretch, wear in the linkage or after installation of new components. As a rule of thumb, when you push down the parking brake pedal, it should travel at least nine clicks, but not more than 13 clicks. If the pedal travels less than nine clicks, the cable's too tight; if it travels more than 13, the cable's too loose.

2 Raise the rear of the vehicle until the wheels are clear of the ground and support it securely on jackstands. Release the parking brake pedal by pulling on the release lever.

3 Apply the parking brake four "clicks."

4 The adjuster is located under the vehicle, near the center on 1994 models and next to the frame under the driver's side of the vehicle on 1995 and later models. To adjust the rear cables on 1994 models, hold the flat portion of the cable end with a wrench or locking pliers and turn the adjuster nut until you feel a slight drag when the rear wheels

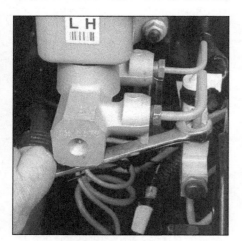

10.3 To prevent rounding off of the flats on the tube fitting nuts, use a flare-nut wrench, which wraps around the nut

Chapter 9 Brakes

11.4a To adjust the parking brake cables on 1994 models, hold the cable and turn the nut with a wrench

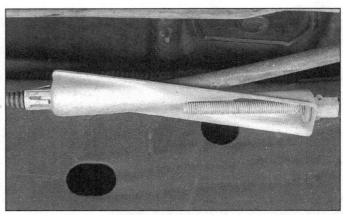

11.4b The parking brake cable equalizer is located next to the outside of the frame, under the driver's side, on 1995 and later models

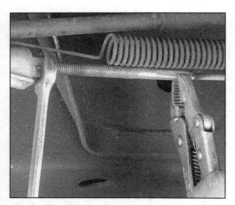

11.4c To adjust the parking brake cables on 1995 and later models, hold the cable and turn the hex on the adjuster with a wrench

12.3 To detach the parking brake pedal assembly from the vehicle, remove these three bolts (arrows)

12.4 Disengage the cable end plug from its hole in the ratcheting cam

are turned **(see illustration)**. To adjust the rear cables on 1995 and later models, hold the flat portion of the threaded cable end with a wrench or locking pliers and turn the adjusting nut at the rear of the equalizer **(see illustrations)** until you feel a slight drag when the rear wheels are turned. **Note:** *If the threads appear rusty, apply penetrating oil before attempting adjustment.*

5   Release the parking brake pedal and make sure there's no longer any drag when the wheels are turned.

6   Tighten the locknut and lower the vehicle to the ground.

12 Parking brake cables - replacement

1   Release the parking brake. If you're removing a rear cable, loosen the rear wheel lug nuts for that side. Raise the vehicle and place it securely on jackstands. Remove the wheel. **Caution:** *On models with plastic-coated cables, do not use lubricant on the cables. It can deteriorate the plastic.* **Note:** *Where multi-fingered ferrules must be squeezed to release a cable, you can use a*

small hose clamp to evenly squeeze the fingers until they clear their bracket.

Front cable

Refer to illustrations 12.3, 12.4, 12.5 and 12.6

2   Referring to Section 11, if necessary, back off the adjusting nut to take tension off the cable.

3   Detach the parking brake pedal assem-

bly from the kick panel **(see illustration)**.

4   Pull the plug on the end of the cable out of its hole in the ratcheting cam **(see illustration)**.

5   Squeeze the tangs on the parking brake cable ferrule and slip the cable through its bracket **(see illustration)**.

6   Working underneath the vehicle again, disengage the front cable from the connector **(see illustration)** located forward of the equalizer.

12.5 Squeeze the fingers on the cable ferrule and slide the cable through its bracket

12.6 The rear end of the front cable and the front end of the left rear cable are attached at this connector, which is located just forward of the equalizer assembly

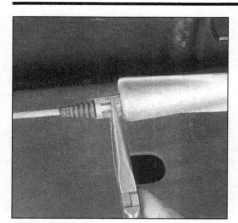

12.10 To disengage the left rear cable from the equalizer, squeeze the fingers on this ferrule and slide the cable through

12.12 To disengage the right rear cable from this bracket, squeeze the fingers on the ferrule and slide the cable through the bracket

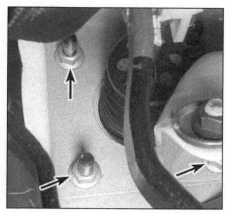

13.6 To detach the power brake booster from the firewall, remove these four nuts (arrows) (upper right nut not visible in this photo)

7   Installation is the reverse of removal. Adjust the cable (see Section 11) when you're done.

Rear cables

Left rear cable

Refer to illustration 12.10

8   Referring to Section 11, if necessary, loosen the cable adjusting nut to put slack in the cables.
9   Disengage the front end of the cable from the connector **(see illustration 12.6)**.
10 Squeeze the fingers on the cable ferrule at the forward end of the equalizer **(see illustration)** and slide the cable out of the equalizer.

Right rear cable

Refer to illustration 12.12

11 Referring to Section 11, if necessary, unscrew the adjuster nut all the way and disconnect the cable from the equalizer.
12 Squeeze the fingers on the cable ferrule **(see illustration)** and slide the cable through the bracket.
13 Follow the cable back to the right rear wheel. On top of the differential, there's a cable clip. Detach the cable from this clip.

Either rear cable (with drum brakes)

14 Remove the rear brake drum and brake shoe assembly (see Section 6), disconnect the parking brake cable from the parking brake lever, squeeze the fingers on the cable ferrule in the backing plate and slide the cable through the backing plate.
15 Installation is the reverse of removal. Adjust the cable (see Section 11) when you're done.

Either rear cable (with disc brakes)

16 Disconnect the park brake cable from the park brake lever.
17 Installation is the reverse of removal. Adjust the cable (see Section 11) when you're done.

13   Power brake booster - removal and installation

Refer to illustration 13.6

1   The power brake booster unit requires no special maintenance apart from periodic inspection of the vacuum hose and the case.
2   Disassembly of the power unit requires special tools and is not ordinarily performed by the home mechanic. If a problem develops, it's recommended that a new or factory rebuilt unit be installed.
3   In the engine compartment, remove the nuts attaching the master cylinder to the booster and carefully pull the master cylinder, along with the RWAL control module and isolation/dump valve, forward until it clears the mounting studs. Be careful not to bend or kink the brake lines.
4   Disconnect the vacuum hose where it attaches to the power brake booster.
5   In the passenger compartment, remove the clip and disconnect the power brake pushrod from the top of the brake pedal.
6   Remove the nuts attaching the booster to the firewall **(see illustration)**.
7   Carefully lift the booster unit away from the firewall and out of the engine compartment.
8   To install the booster, place it into position and tighten the retaining nuts. Connect the brake pedal.
9   Install the master cylinder and vacuum hose.
10 Carefully test the operation of the brakes before placing the vehicle in normal operation.

14   Brake hydraulic system - bleeding

Warning: *Wear eye protection when bleeding the brake system. If the fluid comes in contact with your eyes, immediately rinse them with water and seek medical attention.*

Note: *Bleeding the brake system is necessary to remove any air that's trapped in the system when it's opened during removal and installation of a hose, line, caliper, wheel cylinder or master cylinder.*

1   It will probably be necessary to bleed the system at all four brakes if air has entered the system due to low fluid level, or if the brake lines have been disconnected at the master cylinder.
2   If a brake line was disconnected only at a wheel, then only that caliper or wheel cylinder must be bled.
3   If a brake line is disconnected at a fitting located between the master cylinder and any of the brakes, that part of the system served by the disconnected line must be bled.
4   Remove any residual vacuum (or hydraulic pressure) from the brake power booster by applying the brake several times with the engine off.
5   Remove the master cylinder reservoir cover and fill the reservoir with brake fluid. Reinstall the cover. **Note:** *Check the fluid level often during the bleeding operation and add fluid as necessary to prevent the fluid level from falling low enough to allow air bubbles into the master cylinder.*
6   Have an assistant on hand, as well as a supply of new brake fluid, an empty clear plastic container, a length of 3/16-inch plastic, rubber or vinyl tubing to fit over the bleeder valve and a wrench to open and close the bleeder valve.
7   Beginning at the right rear wheel, loosen the bleeder screw slightly, then tighten it to a point where it's snug but can still be loosened quickly and easily.
8   Place one end of the tubing over the bleeder screw fitting and submerge the other end in brake fluid in the container.
9   Have the assistant pump the brakes a few times to get pressure in the system, then hold the pedal firmly depressed.
10 While the pedal is held depressed, open the bleeder screw just enough to allow a flow of fluid to leave the valve. Watch for air bubbles to exit the submerged end of the tube.

15.3 To remove a rear parking brake shoe assembly, lift it straight up

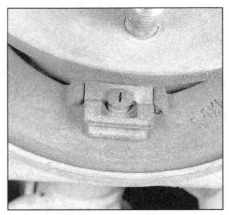

15.5a When installing the new shoe assembly, make sure it's fully seated into this retaining clip . . .

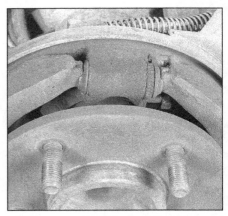

15.5b . . . and both ends of the shoe are properly engaged with the parking brake lever housing

When the fluid flow slows after a couple of seconds, tighten the screw and have your assistant release the pedal.

11 Repeat Steps 9 and 10 until no more air is seen leaving the tube, then tighten the bleeder screw and proceed to the left rear wheel, the right front wheel and the left front wheel, in that order, and perform the same procedure. Be sure to check the fluid in the master cylinder reservoir frequently.

12 Never use old brake fluid. It contains moisture which will deteriorate the brake system components.

13 Refill the master cylinder with fluid at the end of the operation.

14 Check the operation of the brakes. The pedal should feel solid when depressed, with no sponginess. If necessary, repeat the entire process. **Warning:** *Do not operate the vehicle if you are in doubt about the effectiveness of the brake system. It is possible for air to become trapped in the anti-lock brake system valve assembly, so, if the pedal continues to feel spongy after repeated bleedings or the BRAKE or ANTI-LOCK light stays on, have the vehicle towed to a dealer service department or other qualified shop to be bled with the aid of a scan tool.*

15 Parking brake shoes - removal and installation

Refer to illustrations 15.3, 15.5a and 15.5b

Note: *The following procedure applies to models with rear disc brakes, which use a small parking brake shoe assembly and drum inside the rear disc. The parking brake system on models with rear drum brakes uses the rear shoes as the parking brakes (see Section 6).*

1 Loosen the rear wheel lug nuts, raise the vehicle and place it securely on jackstands. Remove the rear wheels.

2 Remove the rear brake calipers (see Section 4) and discs (see Section 5).

3 Slide the parking brake shoe assembly straight down **(see illustration)**.

4 Lubricate the shoe guide pads with high temperature grease.

5 Install the new shoe assembly **(see illustrations)**.

6 Repeat steps 3 through 5 for the other side.

7 Install the brake discs and calipers.

8 Install the rear wheels, remove the jackstands and lower the vehicle. Tighten the wheel lug nuts to the torque listed in the Chapter 1 Specifications.

16 Brake light switch - check and replacement

Check

Refer to illustration 16.1

1 The brake light switch **(see illustration)** is located on the brake pedal bracket. You'll need to remove the trim panel beneath the steering column to get to the switch and connector (see Chapter 11).

2 With the brake pedal in the fully released position, the switch opens the brake light circuit. When the brake pedal is depressed, the switch closes the circuit and sends current to the brake lights.

3 If the brake lights are inoperative, check the fuse and the bulbs (see Chapter 12).

4 If the fuse and bulbs are okay, verify that voltage is available at the switch (see Wiring Diagrams at the end of Chapter 12).

5 If there's no voltage to the switch, use a test light to find the short or open between the battery and the switch (see Wiring Diagrams at the end of Chapter 12). If there is voltage to the switch, close the switch (depress the brake pedal) and verify that there's voltage on the ground side of the switch.

6 If there's no voltage on the ground side of the switch with the brake pedal depressed, look for a short or open in the ground circuit (see Wiring Diagrams at the end of Chapter 12) and repair it. If there is no short or open on the ground side of the switch, replace the switch (see below).

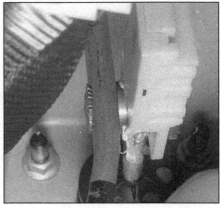

16.1 The brake light switch is located at the top of the brake pedal assembly

Replacement

7 Remove the trim panel below the steering column (see Chapter 11).

8 Unplug the electrical connector from the switch.

9 Remove the retainer from the brake pedal pin.

10 Unsnap the switch from the brake pedal pushrod.

11 Installation is the reverse of removal.

12 Adjust the switch when you're done.

Adjustment

13 Depress the brake pedal, insert the switch into its bracket and push it in until it's fully seated.

14 Slowly pull the brake pedal to the rear until you no longer hear any "clicking" sounds. The switch should now be adjusted.

15 You can check your work with an ohmmeter or continuity tester by verifying that the switch contacts are open at one inch or less of brake pedal travel, and closed thereafter.

16 Installation is otherwise the reverse of removal.

Chapter 10
Suspension and steering systems

Contents

	Section
Balljoints - check and replacement	4
Coil spring (2WD models) - removal and installation	6
Front end alignment - general information	22
Front wheel bearing check, repack and adjustment (2WD models only)	See Chapter 1
General information	1
Hub and bearing assembly (4WD models) - replacement	9
Leaf spring/shackle - removal and installation	13
Lower control arm - removal and installation	8
Power steering fluid level check	See Chapter 1
Power steering pump - removal and installation	19
Power steering system - bleeding	20
Shock absorber (front) - removal and installation	2
Shock absorber (rear) - removal and installation	12

	Section
Stabilizer bar (front) - removal and installation	3
Stabilizer bar (rear) - removal and installation	14
Steering gear - removal and installation	18
Steering knuckle (2WD models) - removal and installation	5
Steering knuckle (4WD models) - removal and installation	10
Steering linkage - inspection, removal and installation	17
Steering wheel - removal and installation	16
Suspension and steering check	See Chapter 1
Tire and tire pressure checks	See Chapter 1
Tire rotation	See Chapter 1
Torsion bar (4WD models) - removal and installation	11
Track bar (ZR2 models) - removal and installation	15
Upper control arm - removal and installation	7
Wheels and tires - general information	21

Specifications

Torque specifications
Ft-lbs (unless otherwise indicated)

Front suspension
2WD models
Balljoints
  Lower balljoint-to-knuckle nut
    1994 70
    1995 on 79
  Upper balljoint-to-knuckle nut
    1994 70
    1995 on 61
Lower control arm-to-frame bolts
  1994 70
  1995 on
    Front bolt 85
    Rear bolt 72
Shock absorber
  Lower bolt
    1994 and 1995 20
    1996 on 22
  Upper nut
    1994 and 1995 .... 132 in-lbs
    1996 on .... 106 in-lbs
Stabilizer bar
  Clamp-to-frame bolts
    1994 24
    1995 on 26
  Link nut .... 150 in-lbs
Upper control arm
  Control arm-to-frame nuts
    1994 and 1995 65
    1996 on 55
  Control arm shaft nuts 85

4WD models
Balljoints
  Lower balljoint 79
  Upper balljoint 61
Hub and bearing assembly bolts 77
Lower control arm 81
Shock absorber upper and lower nuts 54

Torque specifications (continued)

Ft-lbs (unless otherwise indicated)

Front suspension
4WD models
Stabilizer bar
    Clamp-to-frame bolts
        1994 and 1995 30
        1996 on 48
    Link nut-to-lower control arm
        1994 and 1995 .... 123 in-lbs
        1996 on .... 132 in-lbs
Tie-rod-to-steering knuckle nut
    1994 35
    1995 on 39
Torsion bar
    Lower link nut 37
    Upper link nut 48
Upper control arm 84

Rear suspension
Leaf spring
    Leaf spring-to-front bracket nuts 89
    Shackle-to-frame nuts 89
    Shackle-to-spring nuts 89
    U-bolt nuts 73
Shock absorber
    Lower nut
        1994 and 1995 73
        1996 on
            Two-door and truck 62
            Four-door 74
    Upper nut 22
Stabilizer bar
    Clamp U-bolt nuts 44
Stabilizer bar link
    Upper link nut
        Two-door and truck 25
        Four-door 50
    Lower link nut
        Two-door and truck 50
        Four-door 42
Track bar (ZR2 models)
    Axle bracket nuts
        Through 1998 84
        1999 and later 45
    Frame bracket nut 44

Steering
Idler arm
    Idler arm-to-frame
        1994 and 1995 60
        1996 on 79
    Idler arm-to-relay rod
        2WD 35
        4WD 60
Intermediate shaft-to-steering shaft pinch bolt 26
Pitman arm
    Pitman arm-to-steering gear nut 185
    Pitman arm-to-relay rod
        2WD 35
        4WD 60
Power steering hose clamps .... 17 in-lbs
Power steering line fittings 21
Steering damper
    Steering damper-to-front crossmember
        1994 and 1995 26
        1996 on 44
    Steering damper-to-relay rod
        1994 and 1995 45
        1996 on 46
Steering gear-to-frame bolts 55

Chapter 10 Suspension and steering systems

Steering wheel nut
   1994 17
   1995 on 30
Tie-rods
   Tie-rod-to-relay rod 35
   Adjuster tube clamp bolts
      1994 and 1995 .... 168 in-lbs
      1996 on .... 192 in-lbs

1 General information

Refer to illustrations 1.1a, 1.1b and 1.3

The front suspension **(see illustrations)** is fully independent. Each wheel is connected to the frame by a steering knuckle, upper and lower balljoints and upper and lower control arms. Coil springs and shock absorbers are used on 2WD models; 4WD models use shocks and torsion bars. The coil springs are mounted between the spring pockets on the frame and the lower control arms. The shocks are attached to the lower control arms by bolts and nuts; the upper end of each shock is attached to a bracket on the frame. A stabilizer bar, bolted to the frame and to the two lower control arms, reduces vehicle roll during cornering.

The steering linkage consists of a Pitman arm, idler arm, relay rod, two adjustable tie-rod assemblies (each consisting of an inner tie-rod, adjuster tube and outer tie-rod) and, on 4WD and ZR2 models, a steering damper. When the steering wheel is turned, the gear rotates the Pitman arm which forces the relay rod to one side. The tie-rods, which are connected to the relay rod by ball studs, transfer steering force to the wheels. The tie-rods are adjustable and are used for toe-in adjustments. The relay rod is supported by the Pitman arm and idler arm. The idler arm pivots on a support attached to a frame rail. The steering damper, if equipped, is attached to a bracket on the frame and to the relay rod.

The rear suspension **(see illustration)** consists of a pair of multi-leaf springs and two shock absorbers. The rear axle assembly is attached to the leaf springs by U-bolts. The front ends of the springs are attached to the frame at the front hangers, through rubber bushings. The rear ends of the springs are attached to the frame by shackles which allow the springs to alter their length when the vehicle is in operation. Some models also use a stabilizer bar, bolted to the frame and the axle, to reduce vehicle roll during cornering. ZR2 models are equipped with a track bar bolted to the left end of the rear axle and, at its opposite end, to a bracket on the right frame rail.

Frequently, when working on the suspension or steering system components, you may come across fasteners which seem impossible to loosen. These fasteners on the underside of the vehicle are continually subjected to water, road grime, mud, etc., and can become rusted or "frozen," making them extremely difficult to remove. In order to unscrew these stubborn fasteners without damaging them (or other components), be sure to use lots of penetrating oil and allow it to soak in for a while. Using a wire brush to clean exposed threads will also ease removal of the nut or bolt and prevent damage to the threads. Sometimes a sharp blow with a hammer and punch is effective in breaking the bond between a nut and bolt threads, but care must be taken to prevent the punch from slipping off the fastener and ruining the threads. Heating the stuck fastener and surrounding area with a torch sometimes helps too, but isn't recommended because of the obvious dangers associated with fire. Long breaker bars and extension, or "cheater," pipes will increase leverage, but never use an extension pipe on a ratchet - the ratcheting mechanism could be damaged. Sometimes, turning the nut or bolt in the tightening (clockwise) direction first will help to break it loose. Fasteners that require drastic measures to unscrew should always be replaced with new ones.

1.1a Front suspension and steering components (2WD models)

1	Stabilizer bar	6	Idler arm	10	Coil springs
2	Stabilizer bar mounting clamps	7	Inner tie-rods	11	Shock absorber
3	Stabilizer bar-to-control arm links	8	Tie-rod adjuster tubes	12	Lower control arm
4	Pitman arm	9	Outer tie-rods	13	Lower control arm balljoint
5	Relay rod				

1.1b Front suspension and steering components (4WD models)

1. Stabilizer bar
2. Stabilizer bar clamps
3. Outer tie-rod
4. Stabilizer bar links
5. Torsion bars
6. Lower control arms
7. Lower control arm balljoints

1.3 Rear suspension components

1. Shock absorbers
2. Lower shock absorber mount
3. Multi-leaf spring assemblies
4. Leaf spring anchor plates

Chapter 10 Suspension and steering systems

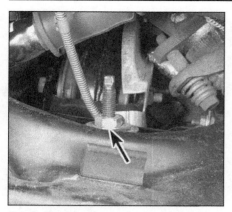

2.1 To detach the upper end of the shock absorber from the frame bracket, remove this nut (arrow), the retainer (large metal washer) and the grommet (large rubber washer)

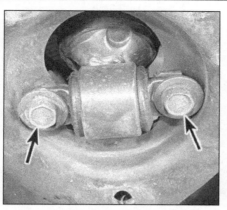

2.4 To detach the lower end of the shock absorber from the lower control arm on a 2WD model, remove these two bolts (arrows)

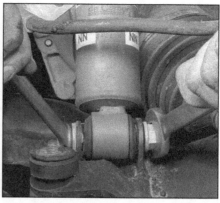

2.8 To detach the lower end of the shock absorber from the lower control arm on 4WD models, remove this nut and bolt

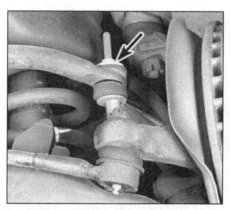

3.2a To disconnect the stabilizer bar from the link bolt on 2WD models, remove this nut (arrow); be sure to keep all the retainers (washers), grommets and spacers in order when you pull out the link bolt

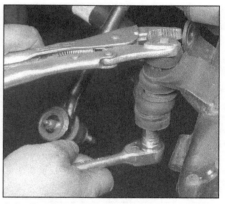

3.2b To disconnect the stabilizer bar from the lower control arm on 4WD models, hold the link bolt nut with a pair of locking pliers and break the link bolt loose

Since most of the procedures that are dealt with in this Chapter involve jacking up the vehicle and working underneath it, a good pair of jackstands will be needed. A hydraulic floor jack is the preferred type of jack to lift the vehicle, and it can also be used to support certain components during various operations. **Warning:** *Never, under any circumstances, rely on a jack to support the vehicle while working on it. Also, whenever any of the suspension or steering fasteners are loosened or removed they must be inspected and, if necessary, replaced with new ones of the same part number or of original equipment quality and design. Torque specifications must be followed for proper reassembly and component retention. Never attempt to heat or straighten suspension or steering components. Instead, replace bent or damaged parts with new ones.*

2 Shock absorber (front) - removal and installation

2WD models

Refer to illustrations 2.1 and 2.4

1 Using a backup wrench on the stem, remove the upper shock mounting nut **(see illustration)**.
2 Remove the retainer (metal washer) and grommet (rubber washer).
3 Raise the vehicle and support it securely on jackstands.
4 Working from underneath the vehicle, remove the two bolts which attach the lower end of the shock absorber to the lower control arm **(see illustration)** and pull the shock out from below.
5 Remove the lower grommet and retainer from the stem.
6 Installation is the reverse of removal. Be sure to tighten the upper mounting nut and the lower mounting bolts to the torque listed in this Chapter's Specifications.

4WD models

Refer to illustration 2.8

7 Raise the vehicle and support it securely on jackstands.
8 Remove the lower shock mounting nut, washer and bolt **(see illustration)**. Note the direction in which the bolt points.
9 Collapse the shock absorber.
10 Remove the upper shock mounting nut, washer and bolt. Again, note the direction in which the bolt points.
11 Remove the shock absorber.
12 Installation is the reverse of removal. Be sure to install the bolts so that they're pointing in the same direction as they were prior to removal. Tighten all fasteners to the torque listed in this Chapter's Specifications.

3 Stabilizer bar (front) - removal and installation

Refer to illustrations 3.2a, 3.2b and 3.3

1 Raise the vehicle and support it securely on jackstands.
2 Remove the nuts from the link bolts and remove the link bolts **(see illustrations)**. **Note:** *Be sure to keep the parts for the left and right sides separate.*
3 Remove the stabilizer bar bracket bolts **(see illustration)**.
4 Remove the stabilizer bar.

3.3 To separate the stabilizer bar from the frame, remove the bushing clamp bolts (arrows) from both frame rails (left clamp shown, right clamp identical)

4.5 On 2WD models, it's easy to check the lower control arm balljoint for wear: The shoulder (arrow) for the grease fitting protrudes 0.050-inch from the surface of the cover. As the balljoint wears, this shoulder moves up into the balljoint assembly; if the shoulder is flush with the cover, or is up inside the hole, replace the balljoint

5   Remove the rubber bushings.
6   Inspect all parts for wear and damage.
7   When you install the rubber bushings on the stabilizer bar, be sure to position them so the slits face toward the front of the vehicle.
8   Installation is otherwise the reverse of removal. Be sure to tighten all fasteners to the torque listed in this Chapter's Specifications.

4 Balljoints - check and replacement

Check

1   Inspect the control arm balljoints for looseness anytime either of them is separated from the steering knuckle. See if you can turn the ballstud in its socket with your fingers. If the balljoint is loose, or if the ballstud can be turned, replace the balljoint. You can also check the balljoints with the suspension assembled as follows.
2   Loosen the wheel lug nuts, raise the front of the vehicle and support it securely on jackstands. Remove the wheel.
3   Place jack stands under the lower control arms. Position the stands as close to each balljoint as possible. Make sure the vehicle is stable. It should not rock on the stands.

Lower balljoints

Refer to illustration 4.5

4   Wipe each balljoint clean and inspect the seal for cuts and tears. If the seal is damaged, replace the balljoint.
5   The lower balljoints on 2WD models employ a visual wear indicator **(see illustration)** that allows easy diagnosis. The shoulder at the base of the grease nipple protrudes about 0.050-inch from the lower surface of

the balljoint's lower cover when the balljoint is new. As the balljoint wears, this shoulder slowly moves up into the balljoint. If the shoulder is flush with or inside the surface of the lower cover, replace the lower balljoint.
6   To check the lower balljoint on 4WD models, position a dial indicator against the wheel rim and insert a prybar between the lower control arm and the steering knuckle. As you lever the prybar, the needle should not deflect more than 0.125-inch. If it does, replace the lower balljoint.

Upper balljoints

7   Position a dial indicator against the wheel rim, grasp the top and bottom of the tire and "rock" the tire, alternately pushing the top and pulling the bottom, and vice versa. The dial indicator should indicate no more than 0.125-inch deflection. If the indicated reading exceeds this figure, replace the upper balljoint.

Replacement

2WD models (upper balljoint)

Refer to illustration 4.11

8   Raise the vehicle, support it securely on jackstands and remove the front wheel, if you haven't already done so. Remove the brake caliper and hang it out of the way with a piece of wire (see Chapter 9). Disconnect the ABS wheel speed sensor and remove the wiring from the upper control arm (see Section 9).
9   Remove the cotter pin from the balljoint and back off the nut two turns.
10 Place a jack or jackstand under the lower control arm and slightly compress the spring. **Note:** *The jack or jackstand must remain under the control arm during removal and installation of the balljoint to hold the spring and control arm in position.*
11 Separate the balljoint from the steering knuckle using a special tool or a balljoint separator to press the balljoint out of the steering knuckle **(see illustration)**. **Note:** *The use of a "picklefork" type balljoint separator may tear the balljoint boot.*
12 To remove the upper balljoint from the upper control arm, drill out the four rivets from the top of the balljoint, remove the balljoint and clean the control arm. Install the new upper balljoint against the mating surface of the control arm and secure it with the nuts and bolts (the nuts and bolts are normally supplied with the new balljoint). Tighten the nuts and bolts to the torque specified in the balljoint replacement kit instructions.

2WD models (lower balljoint)

13 Remove the lower control arm (see Section 8) and take it to an automotive machine shop to have the balljoint pressed out and a new one pressed in.

2WD models (upper or lower balljoint)

14 Inspect the tapered holes in the steering knuckle, removing any accumulated dirt. If out-of-roundness, deformation or other dam-

4.11 On 2WD models, the best way to separate the upper and/or lower balljoint studs from the steering knuckle is with a special tool; however, you can fabricate a similar tool from a large bolt, nut, washer and socket, as shown

age is noted, the knuckle must be replaced with a new one (see Section 5).
15 Reconnect the balljoints to the steering knuckle and tighten the nuts to the torque listed in this Chapter's Specifications.
16 If the cotter key does not line up with the opening in the castellated nut, tighten (never loosen) the nut just enough to allow installation of the cotter pin.
17 Install the grease fittings and lubricate the new balljoints (Chapter 1).
18 Install the wheels and lower the vehicle.
19 The front end alignment should be checked by a dealer service department or alignment shop.

4WD models

20 The procedure for replacing an upper balljoint on a 4WD model is similar to that described above for 2WD models, except that a picklefork-type balljoint separator is usually necessary for separating the balljoint stud from the steering knuckle. The procedure for replacing a lower balljoint follows.
21 Raise the vehicle and support it securely on jackstands.
22 Remove the wheel.
23 Remove the front splash shield bolts. Pivot the shield out of the way to gain access to the tie-rod end.
24 Disconnect the tie-rod from the relay rod (see Section 17).
25 Remove the driveaxle (see Chapter 8).
26 Center punch the bottom of each rivet.
27 Using a 1/8-inch drill bit, drill a guide hole 1/2-inch deep into the rivet heads.
28 Using a 5/16-inch drill bit, drill the rivet heads off. Drill a hole two-thirds the length of the rivet shank, using the same drill.
29 Knock the rivets out with a hammer and punch.
30 Remove the balljoint stud cotter pin.
31 Support the lower control arm with floor jack.
32 Loosen the balljoint nut a couple of turns.

Chapter 10 Suspension and steering systems

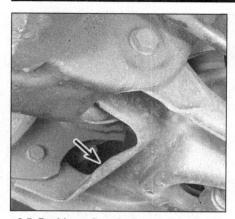

6.5 Position a floor jack under the inner edge of the control arm (arrow)

33 Back off the torsion bar adjusting arm bolt to reduce the tension on the bar (see Section 11). Don't remove the bolt completely.
34 Separate the knuckle from the balljoint with a two-jaw puller or a "picklefork."
35 Remove the nut and separate the balljoint from the lower control arm.
36 Install the new balljoint to the control arm. Install the nuts and bolts and tighten them to the torque specified in the balljoint replacement kit.
37 Raise the lower control arm with a floor jack and insert the balljoint stud into the steering knuckle. Lower the jack until the balljoint stud seats in the hole, then install the nut and tighten it to the torque listed in this Chapter's Specifications. Install a new cotter pin and bend the ends of the pin against the nut.
38 Load the torsion bar (see Section 11).
39 Install the driveaxle assembly (see Chapter 8).
40 Connect the inner tie-rod end to the relay rod, install the tie-rod end nut and tighten it to the torque listed in this Chapter's Specifications.
41 Install the front splash shield and tighten the splash shield bolts securely.
42 Install the wheel and lug nuts.
43 Lower the vehicle.
44 Tighten the wheel lug nuts to the torque listed in the Chapter 1 Specifications.

5 Steering knuckle (2WD models) - removal and installation

1 Raise the front of the vehicle and support it securely on jackstands. Apply the parking brake.
2 Support the lower control arm with a jack so the coil spring is compressed to its normal ride height. **Warning:** *The jack must remain in this position throughout the entire procedure.*
3 Remove the wheel. Disconnect the ABS wheel speed sensor (see Section 9).
4 Remove the brake caliper (see Chapter 9) and the brake disc/hub assembly (see Chapter 1, *Front wheel bearing check, repack and adjustment*).
5 Remove the disc splash shield, if equipped.
6 Disconnect the tie-rod end from the knuckle (see Section 17).
7 Disconnect the balljoints from the steering knuckle (see Section 4).
8 Remove the steering knuckle.
9 Installation is the reverse of removal. Adjust the front wheel bearings (see Chapter 1) and have the front wheel alignment checked by a dealer service department or alignment shop.

6 Coil spring (2WD models) - removal and installation

Removal
Refer to illustration 6.5
1 Loosen the wheel lug nuts, raise the vehicle and support it securely on jackstands.
2 Remove the shock absorber (see Section 2).
3 Disconnect the stabilizer bar from the lower control arm (see Section 3).
4 Loop a length of chain up through the control arm and coil spring and bolt the chain together. Make sure there's enough slack in the chain so it won't inhibit spring extension when the control arm is lowered.
5 Position a floor jack under the inner edge of the lower control arm **(see illustration)**. **Warning:** *Failure to support the lower control arm could result in severe injury.*
6 Raise the jack slightly to relieve spring pressure from the control arm pivot bolts. Remove the rear bolt first, then the front. If they're difficult to remove, tap them out with a hammer and a long, narrow punch.
7 Slowly lower the jack until the coil spring is fully extended.
8 Check that all tension has been removed from the spring, then unbolt the safety chain and maneuver the coil spring out. Do not apply downward pressure on the control arm, as it may damage the balljoint. If the upper insulator isn't on the top of the spring, reach up into the spring pocket and retrieve it.

Installation
9 Place the insulator on top of the coil spring (the top of the spring is flat on the end, with a gripper notch near the end of the spring coil).
10 Install the top of the spring into the spring pocket and the bottom in the lower control arm. The piece of tape on the coil spring must be towards the bottom. The end of the lower coil must cover all of one drain hole, while the other drain hole must remain unobstructed.
11 Raise the lower control arm until the bolt holes are aligned, then install the pivot bolts. The bolts must be inserted from the front. Install the nuts, but don't tighten them completely at this time.
12 Install the shock absorber.
13 Connect the stabilizer bar to the lower control arm.
14 Install the wheel and lug nuts. Lower the vehicle and tighten the lug nuts to the torque specified in Chapter 1.
15 Reach under the vehicle and tighten the pivot bolt nuts to the torque listed in this Chapter's Specifications.

7 Upper control arm - removal and installation

Removal
Refer to illustration 7.4 and 7.6
Note: *This procedure applies to both 2WD and 4WD models.*
1 Loosen the wheel lug nuts, raise the front of the vehicle and support it securely on jackstands. Apply the parking brake. Remove the wheel.
2 Position a floor jack, with a wood block on the jack head (to act as a cushion), under the lower control arm in the area between the spring seat and the balljoint. Raise the jack slightly to take the spring pressure off the upper control arm. **Warning:** *The jack must remain in this position throughout the entire procedure.*
3 Remove the air cleaner extension, if necessary.
4 Disconnect the brake hose bracket from the upper control arm **(see illustration)**. Disconnect the ABS wheel speed sensor and remove the wiring from the upper control arm (see Section 9).
5 Disconnect the upper balljoint from the steering knuckle (see Section 4).
6 On 2WD models, note the location of number of shims on the upper control arm pivot shaft, then remove the pivot shaft nuts, bolts and shims. On 4WD models, mark the positions of the adjuster cams to the frame so they can be returned to the same settings

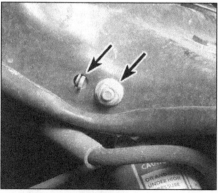

7.4 Before removing the upper control arm, make sure you detach the brake hose bracket (left arrow points to the bracket tang protruding through arm) by removing this nut (right arrow)

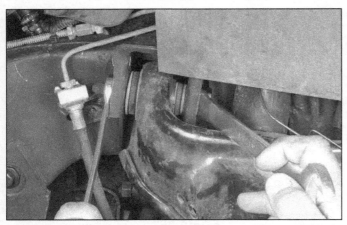

7.6 When removing the upper control arm on a 4WD model, make sure you mark the relationship of the eccentric cams to the frame bracket before removing the front and rear pivot nuts and bolts

8.11a The nut for the 4WD lower control arm's front pivot bolt is located inside this hole in the lower crossmember

when the arm is installed. Remove the upper control arm pivot bolts and nuts **(see illustration)**. Remove the control arm.

7   Inspect the pivot bolt bushings for wear. Replace them if necessary. **Note:** *The bushings on some models are welded in place and can't be removed. If they're worn, you'll have to replace the control arm. On other models, a hydraulic press may be required to accomplish removal and installation of the bushings, in which case you'll have to take the control arm to a dealer service department or other repair shop to have this done for you.*

Installation

8   Position the arm in the frame brackets and install the bolts and nuts. They must be installed with their heads towards the inside, facing each other. Don't tighten the nuts completely at this time. If an alignment kit has been installed, line up the previously applied matchmarks.

9   Attach the balljoint to the steering knuckle (see Section 4).

10   Connect the brake hose bracket to the upper control arm.

11   Install the wheel and lug nuts, then lower the vehicle. Tighten the lug nuts to the torque listed in the Chapter 1 Specifications.

12   Tighten the pivot bolt nuts to the torque listed in this Chapter's Specifications.

13   Install the air cleaner extension, if it was removed.

---

8   Lower control arm - removal and installation

2WD models

Removal

1   Loosen the wheel lug nuts, raise the vehicle and support it securely on jackstands. Remove the wheel.

2   Remove the coil spring (see Section 6).

3   Disconnect the lower balljoint (see Section 4).

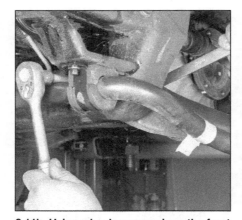

8.11b Using a backup wrench on the front pivot bolt, break the nut loose with a deep socket

4   Take the control arm to a dealer or properly equipped shop to have the balljoint and bushings replaced, if necessary.

Installation

5   Installation is the reverse of removal. Don't tighten the pivot shaft nuts or bolts until the vehicle is at normal ride height, then tighten all fasteners to the torque listed in this Chapter's Specifications.

4WD models

Removal

Refer to illustrations 8.11a, 8.11b and 8.11c

6   Loosen the wheel lug nuts, raise the vehicle and support it securely on jackstands. Remove the wheel.

7   Remove the front splash shield, if equipped, for access to the front end components.

8   Disconnect the shock absorber from the lower control arm (see Section 2).

9   Disconnect the stabilizer bar from the lower control arm (Section 3).

10   Remove the torsion bar (see Section 11).

11   Remove the lower control arm pivot

8.11c Using a backup socket on the rear pivot bolt, break the nut loose with a wrench

bolts, nuts and washers **(see illustrations)**. Remove the lower control arm. Detach the lower balljoint from the steering knuckle (see Section 4).

12   Check the bushings for damage or wear. Some models have welded-in bushings that can't be replaced. If this is the case, replace the control arm. Other models have replaceable bushings, but a press and special adapters are required to remove and install them. Take the control arm to a dealer service department or other repair shop to have the bushings replaced for you.

Installation

13   Raise the control arm into position and install the control arm pivot bolts, washers and nuts, but don't tighten them completely at this time. Make sure the threaded portions of both bolts are facing forward, i.e. the nuts should be at the front.

14   Carefully raise the lower control arm with the floor jack until the balljoint stud can be inserted into the hole in the steering knuckle. Install the balljoint stud nut, tighten it to the torque listed in this Chapter's Specifi-

Chapter 10 Suspension and steering systems

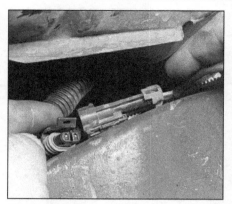

9.4 Locate the electrical connector for the wheel speed sensor on top of the frame rail and unplug it to protect the lead from damage when the hub and bearing assembly is removed from the steering knuckle

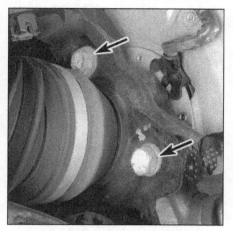

9.5a To detach the hub and bearing assembly from the steering knuckle, remove these bolts (arrows) . . .

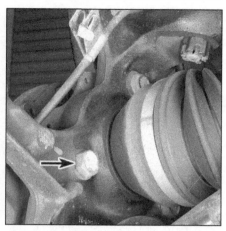

9.5b . . . and this bolt (arrow) (upper bolt is the same upper bolt shown in the previous photo)

cations, then install a new cotter pin.
15   Install the torsion bar (see Section 11).
16   Connect the stabilizer bar to the lower control arm (see Section 3).
17   Install the shock absorber (see Section 2).
18   Install the wheel and lug nuts. Lower the vehicle and tighten the lug nuts to the torque listed in the Chapter 1 Specifications.
19   Tighten the lower control arm pivot bolts to the torque listed in this Chapter's Specifications.
20   Install the splash shield, if equipped.
21   Measure the vehicle's ride height on each side, from equal points on the frame to the ground. If the side that has been worked on is higher or lower than the other side, turn the torsion bar adjusting screw accordingly until the vehicle sits level. This may take a few tries, and it's important to roll the vehicle back and forth and jounce the front end between adjustments, to settle the suspension and get an accurate reading.

9   Hub and bearing assembly (4WD models) - replacement

Refer to illustrations 9.4, 9.5a, 9.5b and 9.6
Note: *The hub and bearing assembly is a sealed unit and isn't serviceable. If it's defective, it must be replaced.*
1   Put the vehicle in gear, apply the parking brake and break loose the driveaxle/hub nut with a socket and large breaker bar.
2   Loosen the wheel lug nuts, raise the vehicle and support it securely on jackstands. Remove the wheel. Remove the driveaxle/hub nut.
3   Unbolt the brake caliper and hang it out of the way with a piece of wire (see Chapter 9). Remove the brake disc.
4   On models with four-wheel ABS, trace the electrical lead for the wheel speed sensor along the upper control arm, then forward along the frame rail and unplug the electrical connector **(see illustration)**. (When the hub and bearing assembly is removed, the brake disc shield is no longer attached and might fall off the steering knuckle; the wheel speed sensor is attached to this shield, so unplugging the connector protects the lead in the event the shield falls.)
5   Remove the hub assembly-to-steering knuckle bolts **(see illustrations)**. Remove the brake disc shield and wheel speed sensor assembly and set it aside.
6   Tap the hub assembly from side-to-side to break it loose from the steering knuckle. Pull the hub assembly off the end of the driveaxle **(see illustration)**. Wrap the end of the driveaxle with a rag to prevent damaging it. If the hub is stuck on the splines on the end of the driveaxle, use a puller to free it.
7   Installation is the reverse of the removal procedure. Be sure to lubricate the driveaxle splines with multi-purpose grease, and tighten all of the fasteners to the torque listed in this Chapter's Specifications.

10   Steering knuckle (4WD models) - removal and installation

Removal
1   Put the vehicle in gear, apply the parking brake and break loose the driveaxle/hub nut with a socket and large breaker bar.
2   Loosen the wheel lug nuts, raise the front of the vehicle and support it securely on jackstands. Apply the parking brake. Remove the wheel. Remove the driveaxle/hub nut.
3   Unbolt the brake caliper, hang it out of the way with a piece of wire and remove the brake disc (see Chapter 9).
4   Disconnect the tie-rod end from the steering knuckle (see Section 17).
5   Support the lower control arm with a floor jack and detach the steering knuckle from the lower balljoint (see Section 4).
Warning: *The jack must remain in this position throughout the entire procedure; the lower control arm is "loaded" by the torsion bar.*
6   Support the steering knuckle and separate it from the upper balljoint (see Section 4).
7   Using a puller, push the driveaxle out of the hub while withdrawing the steering knuckle and hub assembly.
8   If you're planning to replace the hub and bearing assembly or the knuckle, remove the hub and bearing assembly from the knuckle (see Section 9).
9   Inspect the seal on the rear side of the knuckle. If it's damaged or shows signs of deterioration, pry it out with a large screwdriver or prybar. Install a new seal by driving it in with a socket that has an outside diameter slightly smaller than the seal.

Installation
10   Installation is the reverse of the removal procedure. Be sure to lubricate the driveaxle splines with multi-purpose grease and tighten all of the fasteners to the torque listed in this Chapter's Specifications.

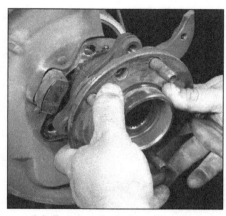

9.6 Remove the hub and bearing assembly from the steering knuckle, then remove the brake disc shield and set it aside

Chapter 10 Suspension and steering systems

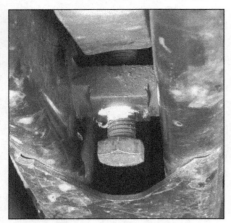

11.2 To ensure proper adjustment of the torsion bar upon reassembly, count the number of threads showing on the torsion bar adjuster bolt and mark the relationship of the bolt to the torsion bar adjuster nut as insurance

11.3 Install a small puller as shown, with the fingers hooked around the flange running along each side of the crossmember; make sure the puller bolt is centered on the dimple in the torsion bar adjuster arm; tighten the puller bolt until all tension is removed from the adjuster nut (see illustration).

11.4 With tension removed from the adjuster nut, remove the adjuster nut

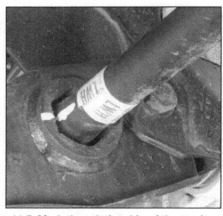

11.5 Mark the relationship of the torsion bar to the lower control arm as shown

11 Torsion bar (4WD models) - removal and installation

Refer to illustrations 11.2, 11.3, 11.4, 11.5, 11.6a, 11.6b, 11.7, 11.8 and 11.9

1 Loosen the front wheel lugs nuts, raise the vehicle and place it securely on jackstands. Remove the wheel.
2 Count the number of threads showing on the torsion bar adjuster bolt and mark the relationship of the bolt to the torsion bar adjuster nut **(see illustration)**.
3 In the torsion bar adjuster arm, there's a small dimple. Install a small puller with its bolt centered on this dimple **(see illustration)**.
4 Turn the puller bolt until all tension is removed from the torsion bar adjuster arm, then remove the torsion bar adjuster nut **(see illustration)**. Remove the puller.
5 Mark the relationship of the forward end of the torsion bar to the lower control arm

6 Push the torsion bar forward, through the lower control arm, until the rear end of the bar clears the crossmember **(see illustration)** and remove the torsion bar adjuster arm **(see illustration)**.
7 Pull the torsion bar down and to the rear as far as it will go. If the front end of the bar hangs up in the lower control arm, drive it out of the control arm with a brass drift **(see illustration)**.
8 Installation is the reverse of removal. Be sure to clean out the hexagonal hole in the lower control arm and lube it with multi-purpose grease **(see illustration)** before inserting the torsion bar into the arm. Also apply some grease to the hex ends of the torsion bar, to the top of the adjuster arm and to the adjuster bolt. Make sure that the alignment marks you made on the torsion bar and the control arm line up. And make sure that the

torsion bar adjuster bolt is tightened until the same number of threads are showing and the marks you made on the adjuster bolt and nut are lined up.
9 Measure the torsion arm-to-crossmember clearance after installation **(see illustra-**

11.6a Slide the torsion bar forward through the lower control arm far enough to pull the rear end of the bar out of the crossmember . . .

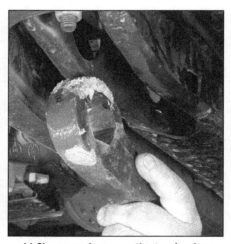

11.6b . . . and remove the torsion bar adjuster arm. Hold your hand under the arm as you slide out the torsion bar to prevent the arm from falling

11.7 If the torsion bar hangs up in the lower control arm, knock it out with a brass drift

Chapter 10 Suspension and steering systems

11.8 Clean up any corrosion inside the hex hole in the control arm and lube it with multi-purpose grease before installing the torsion bar

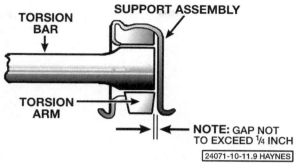

11.9 The clearance measurement indicated here should be checked after torsion bar installation; it should not exceed 1/4-inch

tion) and adjust the torsion bar fore and aft as necessary to bring it within this clearance.
10   Install the wheel, remove the jackstands and lower the vehicle.
11   Tighten the wheel lug nuts to the torque listed in the Chapter 1 Specifications.
12   Measure the vehicle's ride height on each side, from equal points on the frame to the ground. If the side that has been worked on is higher or lower than the other side, turn the torsion bar adjusting screw accordingly until the vehicle sits level. This may take a few tries, and it's important to roll the vehicle back and forth and jounce the front end between adjustments, to settle the suspension and get an accurate reading.

12 Shock absorber (rear) - removal and installation

Refer to illustrations 12.2 and 12.3

1   Raise the rear of the vehicle and support securely on jackstands. Block the front wheels so the vehicle doesn't roll off the stands.
2   Remove the shock absorber upper mounting bolts from the frame **(see illustration)**.
3   Remove the lower mounting nut, washer and bolt from the anchor plate bracket **(see illustration)**.
4   Remove the shock absorber.
5   Installation is the reverse of removal. Make sure you install the nuts and bolts facing in the proper direction. Tighten all fasteners to the torque listed in this Chapter's Specifications.

13 Leaf spring/shackle - removal and installation

Refer to illustrations 13.2, 13.5 and 13.7

1   Raise the rear of the vehicle and support

12.2 To detach the upper end of the shock absorber from the frame, remove these bolts (arrows)

it securely on jackstands. Block the front wheels to keep the vehicle from rolling off the stands. Support the axle with a floor jack and raise it slightly to relieve the tension on the leaf springs.
2   Remove the four U-bolt nuts and washers **(see illustration)**.
3   Remove the anchor plate.
4   Remove the U-bolts and spacer.
5   Remove the nut from the front spring-

13.2 To detach the anchor plate, remove these four nuts from the U-bolts

12.3 To detach the lower end of the shock absorber from the anchor plate, remove this nut, washer and bolt (arrows)

mount bolt **(see illustration)**.
6   Raise the axle slightly off the springs with the floor jack. Be sure the axle is stable. The axle must remain supported by the jack during the entire time the spring is removed from the vehicle. If this will be an extended period of time, it would be a good idea to support the axle with jackstands at this point.
7   While an assistant supports the rear of the spring, remove the shackle-to-frame

13.5 Remove the nut from the front leaf spring mounting bolt (arrow), but don't remove the bolt until the rear of the leaf spring has been detached and the leaf spring is supported

Chapter 10 Suspension and steering systems

13.7 To detach the rear of the leaf spring, have an assistant support the spring, then remove this nut (arrow) and withdraw the shackle bolt

14.2 To detach the stabilizer bar from the lower end of the link, remove this nut and bolt

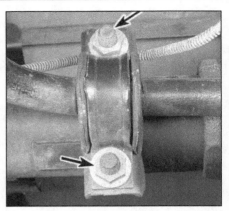

14.3 To detach the stabilizer bar from the rear axle, remove these nuts (arrows) from both pairs of U-bolts

bracket nut, washers and bolt **(see illustration)**. Lower the rear of the spring to the ground. **Note:** *On some pick-ups and four-door Blazers and Jimmys, it will be necessary to remove the fuel tank for access to the bolt.*
8 While having an assistant support the front of the spring, remove the front spring-mount bolt, then remove the spring assembly from the vehicle.
9 Installation is the reverse of removal. Gradually tighten the U-bolt nuts in a criss-cross pattern. Then tighten all the fasteners to the torque listed in this Chapter's Specifications.
10 If the bushings at the ends of the spring are worn or deteriorated, an automotive machine shop or dealer service department can press the old ones out and press new ones in.

14 Stabilizer bar (rear) - removal and installation

Refer to illustrations 14.2 and 14.3

1 Raise the rear of the vehicle and support it securely on jackstands. Block the front wheels to keep the vehicle from rolling off the stands.
2 Remove the lower nuts, washers and bolts from the stabilizer bar-to-frame bracket links **(see illustration)**.
3 Remove the nuts from the U-bolts and remove the stabilizer bar clamps **(see illustration)**.
4 Remove the stabilizer bar assembly.
5 Inspect the stabilizer bar and link bushings for cracks, tears and other deterioration. Replace as necessary.
6 Installation is the reverse of removal. Be sure to tighten all fasteners to the torque listed in this Chapter's Specifications.

15 Track bar (ZR2 models) - removal and installation

1 Raise the rear of the vehicle and support it securely on jackstands. Block the front

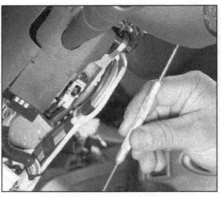

16.4a To detach the airbag module from the steering wheel on 1995 models, rotate the wheel 90-degrees, insert a hard and thin object - such as a scribe - and push in firmly against each of the four ball-lock fasteners to unlock the airbag module from the wheel

wheels to keep the vehicle from rolling off the stands.
2 Support the rear axle at its normal ride height position with a floor jack.
3 Remove the nut, washer and bolt and detach the upper end of the track bar from the right frame rail bracket.
4 Remove the nut, washer and bolt and detach the lower end of the track bar from the left axle bracket.
5 Remove the track bar.
6 Installation is the reverse of removal. Be sure to tighten both nuts to the torque listed in this Chapter's Specifications.

16 Steering wheel - removal and installation

Refer to illustrations 16.4a, 16.4b, 16.5, 16.7 and 16.8

Warning: *Anytime you are working in the vicinity of airbag wiring or components, disable the SIR (airbag) system (see Chapter 12).*
1 Disconnect both cables from the battery (see Chapter 5). **Caution:** *On models*

16.4b Unplug the yellow electrical connector from the back of the airbag module; the arrows point at the four ball-lock type fasteners used on 1995 models

equipped with a Delco Loc II anti-theft audio system, be sure the lockout feature is turned off before performing any procedure which requires disconnecting the battery.
2 On 1994 models, simply pry off the horn cap.
3 On airbag-equipped models, refer to Chapter 12 and disable the airbag system.
4 On 1995 models, turn the steering wheel 90-degrees to gain access to the holes in the backside (the side facing the dash) of the steering wheel. Insert a thin hard tool such as a ballpoint pen or a scribe into the hole for each of the four spring-loaded fasteners **(see illustrations)**. Push in firmly against the head of each fastener with the tool and pull on the airbag module simultaneously with the other hand.
5 On 1996 and later models, turn the steering wheel 90-degrees to gain access to the holes in the backside (the side facing the dash) of the steering wheel. Insert a screwdriver into the hole for each of the four spring clips **(see illustration)** and push the spring aside to release the pin. There are four pins and four springs.
6 Push in and twist the horn contact plunger to release and remove it, then remove the screws securing the horn contact plate. On airbag-equipped models, unplug the yellow electrical connector from the module and set

Chapter 10 Suspension and steering systems

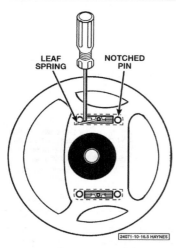

16.5 To detach the airbag module from the steering wheel on 1996 and later models, rotate the wheel 90-degrees, insert a screwdriver into each of the four holes in the backside of the steering wheel and pry each leaf spring aside to release it from its notched pin (there are four of them)

the module aside, with the airbag side of the module facing UP. **Warning:** *DO NOT lay the module down with the airbag side facing down. If it goes off, you could be injured!*

7   Remove the steering wheel retaining nut and mark the position of the steering wheel to the shaft, if marks don't already exist or don't line up **(see illustration)**. **Note:** *On later models with optional control switches mounted in the steering wheel, disconnect the electrical connectors before removing the wheel.*
8   Use a puller to detach the steering wheel from the shaft **(see illustration)**. Don't hammer on the shaft to dislodge the wheel.
9   To install the wheel, align the mark on the steering wheel hub with the mark on the shaft and slide the wheel onto the shaft. Install the nut and tighten it to the torque listed in this Chapter's Specifications.
10   Installation is otherwise the reverse of removal.

16.8 Remove the steering wheel with a steering wheel puller - do not hammer on the steering shaft

16.7 Mark the relationship of the steering wheel to the steering shaft before removing the wheel

17   Steering linkage - inspection, removal and installation

Inspection

1   The steering linkage **(see illustrations 1.1a and 1.1b)** connects the steering gear to the front wheels and keeps the wheels in proper relation to each other. The linkage consists of the Pitman arm, the idler arm, the relay rod, two adjustable tie-rods and a steering damper. The Pitman arm, which is fastened to the steering gear shaft, moves the relay rod back-and-forth. The relay rod is supported on the other end by a frame-mounted idler arm. The back-and-forth motion of the relay rod is transmitted to the steering knuckles through a pair of tie-rod assemblies. Each tie-rod is made up of an inner and outer tie-rod end, a threaded adjuster tube and two clamps.
2   Set the wheels in the straight-ahead position and lock the steering wheel.
3   Raise one side of the vehicle until the tire is approximately 1-inch off the ground.
4   Mount a dial indicator with the needle resting on the outside edge of the wheel. Grasp the front and rear of the tire and, using light pressure, wiggle the wheel back-and-forth and note the dial indicator reading. The gauge reading should be less than 0.108-inch. If the play in the steering system is more than specified, inspect each steering linkage pivot point and ball stud for looseness and replace parts, if necessary.

17.8 Remove the cotter pin, then loosen - but don't yet remove - the castellated nut on the tie-rod end ballstud

5   Raise the vehicle and support it on jackstands. Push up, then pull down on the relay rod end of the idler arm, exerting a force of approximately 25 pounds each way. Measure the total distance the end of the arm travels. If the play is greater than 1/4-inch, replace the idler arm.
6   Check for torn ball stud boots, frozen joints and bent or damaged linkage components.

Removal and installation

Tie-rod

Refer to illustrations 17.8, 17.9, 17.11 and 17.15

Note: *This procedure covers replacing the tie-rod ends as well as the entire tie-rod. If you'll only be replacing a tie-rod end, ignore the Steps that don't apply.*

7   Loosen the wheel lug nuts, raise the vehicle and support it securely on jackstands. Apply the parking brake. Remove the wheel.
8   Remove the cotter pin and loosen, but do not remove, the castellated nut(s) from the ball stud(s) **(see illustration)**. If only the outer tie-rod end will be replaced, only loosen the outer nut. If only the inner tie-rod end will be replaced, only loosen the inner nut. If the entire tie-rod will be replaced, loosen both nuts.
9   If the outer tie-rod end or the entire tie-rod will be replaced, use a two-jaw puller to separate the tie-rod end from the steering knuckle **(see illustration)**. Remove the castellated nut and pull the tie-rod end from the knuckle.

17.9 Install a small puller on the tie-rod end ballstud and separate the ballstud from the steering knuckle - leaving the nut in place will prevent the parts from separating violently

Chapter 10 Suspension and steering systems

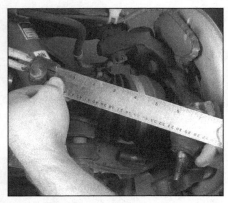

17.11 Measure the distance from the end of the adjuster tube to the center of the ballstud and record your measurement before loosening the adjuster tube clamp and unscrewing the tie-rod end

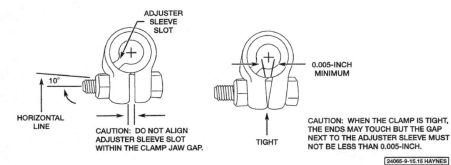

17.15 Note these guidelines when installing the tie-rod ends

10 If the inner tie-rod end or the entire tie-rod will be replaced, separate the inner tie-rod end from the relay rod (see Steps 8 and 9).
11 If the inner or outer tie-rod end must be replaced, measure the distance from the end of the adjuster tube to the center of the ball stud and record it **(see illustration)**. Loosen the adjuster tube clamp bolts and unscrew the tie-rod end.
12 Lubricate the threaded portion of the tie-rod end with chassis grease. Screw the new tie-rod end into the adjuster tube and adjust the distance from the tube to the ball stud to the previously measured dimension. The number of threads showing on the inner and outer tie-rod ends should be equal within three threads. Don't tighten the clamp yet.
13 Connect the disconnected ball stud nut(s). Tighten the nut(s) to the torque listed in this Chapter's Specifications and install a new cotter pin. If the ball stud spins when attempting to tighten the nut, force it into the tapered hole with a large pair of pliers. If necessary, tighten the nut slightly to align a slot in the nut with the hole in the ball stud.
14 Insert the inner tie-rod end ball stud into the relay rod until it's seated. Install the nut and tighten it to the torque listed in this Chapter's Specifications.
15 Tighten the clamp nuts. The center of the bolt should be nearly horizontal and the adjuster tube slot must not line up with the gap in the clamps **(see illustration)**.
16 Install the wheel and lug nuts, lower the vehicle and tighten the lug nuts to the torque listed in the Chapter 1 Specifications. Drive the vehicle to an alignment shop to have the front end alignment checked and, if necessary, adjusted.

Idler arm

17 Raise the vehicle and support it securely on jackstands. Apply the parking brake.
18 Loosen but do not remove the idler arm-to-relay rod nut.
19 Separate the idler arm from the relay rod with a two jaw puller **(see illustration 17.9)**. Remove the nut.

20 Remove the idler arm-to-frame bolts.
21 To install the idler arm, position it on the frame and install the bolts, tightening them to the torque listed in this Chapter's Specifications.
22 Insert the idler arm ball stud into the relay rod and install the nut. Tighten the nut to the specified torque. If the ball stud spins when attempting to tighten the nut, force it into the tapered hole with a large pair of pliers.

Relay rod

23 Raise the vehicle and support it securely on jackstands. Apply the parking brake.
24 Separate the two inner tie-rod ends from the relay rod (see Step 10).
25 Separate the relay rod from the Pitman arm (see Steps 29 and 30).
26 Separate the relay rod from the idler arm (see Steps 18 and 19).
27 Installation is the reverse of the removal procedure. If the ball studs spin when attempting to tighten the nuts, force them into the tapered holes with a large pair of pliers. Be sure to tighten all of the nuts to the torque listed in this Chapter's Specifications.

Pitman arm

28 Raise the vehicle and support it securely on jackstands.
29 Remove the relay rod nut from the Pitman arm ball stud. Discard the nut - don't reuse it.
30 Using a puller, separate the relay rod from the Pitman arm ball stud.
31 Remove the steering gear (see Section 18).
32 Remove the Pitman arm nut and washer. Mark the Pitman arm and the steering gear shaft to ensure proper alignment at reassembly time.
33 Remove the Pitman arm with a Pitman arm puller or a two-jaw puller.
34 Inspect the ball stud threads for damage. Inspect the ball stud seals for excessive wear. Clean the threads on the ball stud.
35 Installation is the reverse of removal. Make sure the marks you made on the Pitman arm and Pitman shaft are aligned. **Note:** *If a clamp type Pitman arm is used, spread the arm just enough, with a wedge, to slip the arm onto the Pitman shaft. Don't spread the arm more than necessary to slip it over the shaft with hand pressure. Do not hammer the arm onto the shaft or you may damage the steering gear.*

Steering damper

36 Inspect the steering damper for fluid leakage. A slight film of fluid near the shaft seal is normal, but if there's excessive fluid present and it's obviously coming from the steering damper, replace the damper.
37 Inspect the steering damper bushing for excessive wear. If it's in bad shape, replace the damper.
38 To test the damper itself, disconnect it from the frame or axle end (see next step). Using as much travel as possible, extend and compress the damper. The resistance should be smooth and constant for each stroke. If any binding or unusual noises are present, replace the damper.
39 Remove the damper ballstud-to-relay rod cotter pin, then remove the nut. Separate the damper from the relay rod, using the technique shown in **illustration 17.9**.
40 Remove the steering damper mounting bolt and nut, then remove the damper.
41 Installation is the reverse of removal. Tighten all the fasteners securely.

18 Steering gear - removal and installation

Removal

Refer to illustrations 18.2, 18.3a, 18.3b, 18.5, 18.6 and 18.7

Warning 1: *If equipped with an airbag, disable the airbag system before working in the vicinity of the steering wheel, instrument panel or any airbag system component. Failure to do so could cause accidental deployment of the airbag resulting in personal injury (see Chapter 12).*
Warning 2: *On models equipped with an airbag, DO NOT allow the steering column shaft to rotate with the steering gear removed or damage to the airbag system could occur. As a method of preventing the shaft from turning, wrap the seat belt around the rim of the steering wheel and buckle the belt in place.*

Chapter 10 Suspension and steering systems

18.2 To disconnect the power steering fluid high-pressure line, unscrew this threaded fitting (lower arrow) with a flare-nut wrench; to disconnect the return line, loosen this hose clamp (upper arrow)

18.3a The intermediate shaft U-joint and pinch bolt are protected by a plastic shield; to remove the shield, pop the two halves apart with a screwdriver, then pull it out

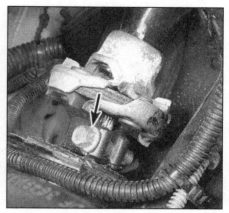

18.3b To detach the intermediate shaft from the steering gear shaft, remove this U-joint pinch bolt (arrow)

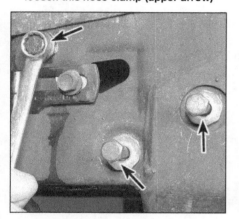

18.5 To remove the steering gear, remove these three bolts (arrows)

18.6 With the steering gear in a bench vise, use a prybar as shown to hold the Pitman arm and break loose the Pitman arm nut (don't remove the nut, just loosen it); be sure to mark the relationship of the Pitman arm to the steering gear shaft before pulling it off

18.7 Remove the Pitman arm with a Pitman arm puller. This specialized tool is available at most auto parts stores

1 Raise the front of the vehicle and support it securely on jackstands. Apply the parking brake.
2 Place a drain pan under the steering gear. Disconnect the power steering hose fittings (see illustration) and cap the ends to prevent excessive fluid loss and contamination.
3 Insert a screwdriver between the two halves of the U-joint plastic shield (see illustration) and pry them apart, then pull out the shield. Mark the relationship of the intermediate shaft lower universal joint to the steering gear input shaft. Remove the intermediate shaft lower pinch bolt (see illustration).
4 Disconnect the Pitman arm from the relay rod (see Section 17).
5 Support the steering gear and remove the mounting bolts (see illustration). Lower the unit, separate the intermediate shaft from the steering gear input shaft and remove the steering gear from the vehicle. **Warning:** *On models equipped with an airbag, DO NOT allow the steering column shaft to rotate with the steering gear removed or damage to the airbag system could occur. As a method of preventing the shaft from turning, wrap the*

seat belt around the rim of the steering wheel and buckle the belt in place.
6 Mark the relationship of the Pitman arm to the shaft so it can be installed in the same position. Remove the Pitman arm nut and washer (see illustration).
7 Remove the Pitman arm from the shaft with a special puller (see illustration).

Installation

8 Slide the Pitman arm onto the shaft. Make sure the marks are aligned. Install the washer and nut and tighten the nut to the torque listed in this Chapter's Specifications.
9 Raise the steering gear and Pitman arm into position and connect the intermediate shaft, aligning the marks.
10 Install the mounting bolts and washers and tighten them to the torque listed in this Chapter's Specifications.
11 Install the lower intermediate shaft pinch bolt and tighten it to the torque listed in this

Chapter's Specifications. Install the plastic shield.
12 Connect the power steering hose fittings to the steering gear and fill the power steering pump reservoir with the recommended fluid (see Chapter 1).
13 Lower the vehicle and bleed the steering system (see Section 20).

19 Power steering pump - removal and installation

Removal

Refer to illustration 19.3

1 Disconnect the cable from the negative terminal of the battery. **Caution:** *On models equipped with a Delco Loc II anti-theft audio system, be sure the lockout feature is turned off before performing any procedure which requires disconnecting the battery.*
2 Remove the upper fan shroud (see Chapter 3). Remove the serpentine drivebelt (see Chapter 1).
3 Using a special power steering pump pulley remover, remove the pulley from the

19.3 You'll need a special puller to remove the power steering pump pulley

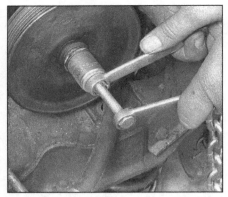

19.8 To install the pulley on the shaft, use a long bolt with the same thread pitch as the internal threads of the power steering pump shaft, a nut, washer and socket that's the same diameter as the pulley hub

pump (see illustration).

4 Position a drain pan under the power steering pump. Disconnect the pressure and return hoses from the backside of the pump. Plug the hoses to prevent contaminants from entering. **Note:** *On 1999 and later model 2.2L engines, the air conditioning compressor must be removed for access, but do not disconnect the hoses (see Chapter 3).*

5 Remove the pump mounting fasteners and lift the pump from the vehicle, taking care not to spill fluid on the painted surfaces. **Note:** *On 1999 and later 4.3L V6 engines, the filler tube bracket must be unbolted from the air conditioning compressor. Remove the filler tube before removing the pump.*

Installation

Refer to illustration 19.8

6 Position the pump in the mounting bracket and install the bolts and nut. Tighten the fasteners securely.
7 Connect the hoses to the pump. Tighten the fittings securely.
8 Press the pulley onto the shaft using a special pulley installer tool. An alternative tool can be fabricated from a long bolt, nut, washer and a socket of the same diameter as the pulley hub (see illustration). Push the pulley onto the shaft until the front of the hub is flush with the shaft, but no further.
9 Install the drivebelt.
10 Fill the power steering reservoir with the recommended fluid (see Chapter 1) and bleed the system following the procedure described in the next Section.

20 Power steering system - bleeding

1 Following any operation in which the power steering fluid lines have been disconnected, the power steering system must be bled to remove all air and obtain proper steering performance.
2 With the front wheels in the straight ahead position, check the power steering fluid level and, if low, add fluid until it reaches the Cold mark on the dipstick.
3 Start the engine and allow it to run at fast idle. Recheck the fluid level and add more if necessary to reach the Cold mark on the dipstick.
4 Bleed the system by turning the wheels from side-to-side, without hitting the stops. This will work the air out of the system. Keep the reservoir full of fluid as this is done.
5 When the air is worked out of the system, return the wheels to the straight ahead position and leave the vehicle running for several more minutes before shutting it off.
6 Road test the vehicle to be sure the steering system is functioning normally and noise free.
7 Recheck the fluid level to be sure it's up to the Hot mark on the dipstick while the engine is at normal operating temperature. Add fluid if necessary (see Chapter 1).

21 Wheels and tires - general information

All vehicles covered by this manual are equipped with metric-size fiberglass or steel belted radial tires. Use of other size or type of tires may affect the ride and handling of the vehicle. Don't mix different types of tires, such as radials and bias belted, on the same vehicle as handling may be seriously affected. It's recommended that tires be replaced in pairs on the same axle, but if only one tire is being replaced, be sure it's the same size, structure and tread design as the other.

Because tire pressure has a substantial effect on handling and wear, the pressure on all tires should be checked at least once a month or before any extended trips (see Chapter 1).

Wheels must be replaced if they're bent, dented, leak air, have elongated bolt holes, are heavily rusted, out of vertical symmetry or if the lug nuts won't stay tight. Wheel repairs that use welding or peening are not recommended.

Tire and wheel balance is important to the overall handling, braking and performance of the vehicle. Unbalanced wheels can adversely affect handling and ride characteristics as well as tire life. Whenever a tire is installed on a wheel, the tire and wheel should be balanced by a shop with the proper equipment.

22 Front end alignment - general information

Refer to illustration 22.1

A front end alignment refers to the adjustments made to the front wheels so they're in proper angular relationship to the suspension and the ground. Front wheels that are out of proper alignment not only affect steering control, but also increase tire wear.

Getting the proper front wheel alignment is a very exacting process, one in which complicated and expensive machines are necessary to perform the job properly. Because of this, you should have a technician with the proper equipment perform these tasks. We will, however, use this space to give you a basic idea of what is involved with front end alignment so you can better understand the process and deal intelligently with the shop that does the work.

Toe-in is the turning in of the front wheels. The purpose of a toe specification is to ensure parallel rolling of the front wheels. In a vehicle with zero toe-in, the distance between the front edges of the wheels will be the same as the distance between the rear edges of the wheels. The actual amount of toe-in is normally only a fraction of an inch. Toe-in adjustment is controlled by the tie-rod end position on the inner tie-rod. Incorrect toe-in will cause the tires to wear improperly by making them scrub against the road surface. Toe-in can be adjusted by turning the adjusting sleeves on the tie-rods equal amounts in the same direction.

Camber is the tilting of the front wheels from the vertical when viewed from the front of the vehicle. When the wheels tilt out at the top, the camber is said to be positive (+). When the wheels tilt in at the top the camber is negative (-). The amount of tilt is measured in degrees from the vertical and this measurement is called the camber angle. This angle affects the amount of tire tread which contacts the road and compensates for changes in the suspension geometry when the vehicle is cornering or traveling over an undulating surface. Camber is adjusted by placing an equal number of shims between the upper control arm pivot shaft and the frame at each bolt location (2WD models). On 4WD models, camber is adjusted by rotating cam-shaped adjusters at the front and rear of the upper control arm.

Caster is the tilting of the top of the front steering axis from the vertical. A tilt toward the rear is positive caster and a tilt toward the front is negative caster. Caster is adjusted by placing unequal numbers of shims between the upper control arm pivot shaft and the frame at each bolt location (2WD models). On 4WD models, caster is adjusted by rotating cam-shaped adjusters at the front and rear of the upper control arm.

Chapter 11 Body

Contents

	Section		Section
Body - maintenance	2	Hinges and locks - maintenance	5
Body repair - major damage	7	Hood - removal, installation and adjustment	9
Body repair - minor damage	6	Hood release latch and cable - removal and installation	10
Bumpers - removal and installation	12	Instrument cluster bezel - removal and installation	25
Console - removal and installation	24	Outside mirror - removal and installation	14
Cowl vent grille - removal and installation	27	Radiator grille - removal and installation	11
Dashboard trim panels - removal and installation	26	Seats - removal and installation	28
Door - removal and installation	15	Steering column cover - removal and installation	23
Door latch, lock cylinder and handles - removal and installation	20	Support struts - replacement	19
Door trim panel - removal and installation	13	Tailgate - removal and installation	17
Door window glass - removal and installation	21	Tailgate glass - removal, installation and adjustment	18
Door window regulator - removal and installation	22	Upholstery and carpets - maintenance	3
Front fender - removal and installation	16	Vinyl trim - maintenance	4
General information	1	Windshield and fixed glass - replacement	8

1 General information

Warning: *1995 and later models covered by this manual are equipped with an airbag. Impact sensors for the airbag system are located in the instrument panel, at the front and underneath the vehicle. The airbag can accidentally deploy if these sensors are disturbed, so be extremely careful when working in these areas. Airbag system components are also located in the steering wheel and the steering column, so be extremely careful when working in these areas and don't disturb any airbag system components or wiring. You could be injured if an airbag accidentally deploys, and the airbag might not deploy correctly in a collision if any of the components or wiring in the system have been disturbed.*
Caution: *On models equipped with a Delco Loc II anti-theft audio system, be sure the lockout feature is turned off before performing any procedure which requires disconnecting the battery.*

The vehicles covered by this manual are built with a body-on-frame construction.

Certain components are particularly vulnerable to accident damage and can be unbolted and repaired or replaced. Among these parts are the doors, seats, tailgate, tailgate glass, bumpers and fenders and door glass.

Only general body maintenance practices and body panel repair procedures within the scope of the do-it-yourselfer are included in this Chapter.

2 Body - maintenance

1   The condition of the body is very important, because the value of the vehicle is dependent on it. It's much more difficult to repair a neglected or damaged body than it is to repair mechanical components. The hidden areas of the body, such as the fender wells and the engine compartment, are equally important, although they obviously don't require as frequent attention as the rest of the body.
2   Once a year, or every 12,000 miles, it's a good idea to have the underside of the body steam cleaned. All traces of dirt and oil will be removed and the underside can then be inspected carefully for damaged brake lines, frayed electrical wiring, damaged cables and other problems.
3   At the same time, clean the engine and the engine compartment with a water soluble degreaser.
4   The body should be washed as needed. Wet the vehicle thoroughly to soften the dirt, then wash it down with a soft sponge and plenty of clean soapy water. If the surplus dirt isn't washed off very carefully, it will in time wear down the paint.
5   Spots of tar or asphalt coating thrown up from the road should be removed with a cloth soaked in solvent.
6   Once every six months, wax the body thoroughly.

3 Upholstery and carpets - maintenance

1   Every three months remove the floormats and clean the interior of the vehicle (more frequently if necessary). Use a stiff whisk broom to brush the carpeting and loosen dirt and dust, then vacuum the upholstery and carpets thoroughly, especially along seams and crevices.
2   Dirt and stains can be removed from carpeting with basic household or automotive carpet shampoos available in spray cans. Follow the directions and vacuum again, then use a stiff brush to bring back the "nap" of the carpet.
3   Most interiors have cloth or vinyl upholstery, either of which can be cleaned and maintained with a number of material-specific cleaners or shampoos available in auto supply stores. Follow the directions on the product for usage, and always spot-test any upholstery cleaner on an inconspicuous area (bottom edge of a back seat cushion) to ensure that it doesn't cause a color shift in the material.
4   After cleaning, vinyl upholstery should be treated with a protectant. **Note:** *Make sure the protectant container indicates the product can be used on seats - some products may make a seat too slippery.* **Caution:** *Do not use protectant on vinyl-covered steering wheels.*
5   Leather upholstery requires special care. It should be cleaned regularly with saddlesoap or leather cleaner. Never use alcohol, gasoline, nail polish remover or thinner to clean leather upholstery.
6   After cleaning, regularly treat leather upholstery with a leather conditioner, rubbed in with a soft cotton cloth. Never use car wax on leather upholstery.
7   In areas where the interior of the vehicle is subject to bright sunlight, cover leather seating areas of the seats with a sheet if the vehicle is to be left out for any length of time.

4 Vinyl trim - maintenance

Vinyl trim should not be cleaned with detergents, caustic soaps or petroleum-based cleaners. Plain soap and water or a mild vinyl cleaner is best for stains. Test a small area for color fastness. Bubbles under the vinyl can be eliminated by piercing them with a pin and then working the air out.

5 Hinges and locks - maintenance

Every 3000 miles or three months, the door, hood and trunk lid hinges should be lubricated with a few drops of oil. The door striker plates should also be given a thin coat of white lithium-based grease to reduce wear and ensure free movement.

These photos illustrate a method of repairing simple dents. They are intended to supplement *Body repair - minor damage* in this Chapter and should not be used as the sole instructions for body repair on these vehicles.

1 If you can't access the backside of the body panel to hammer out the dent, pull it out with a slide-hammer-type dent puller. In the deepest portion of the dent or along the crease line, drill or punch hole(s) at least one inch apart . . .

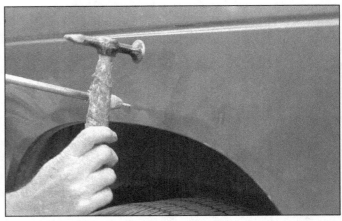

2 . . . then screw the slide-hammer into the hole and operate it. Tap with a hammer near the edge of the dent to help 'pop' the metal back to its original shape. When you're finished, the dent area should be close to its original contour and about 1/8-inch below the surface of the surrounding metal

3 Using coarse-grit sandpaper, remove the paint down to the bare metal. Hand sanding works fine, but the disc sander shown here makes the job faster. Use finer (about 320-grit) sandpaper to feather-edge the paint at least one inch around the dent area

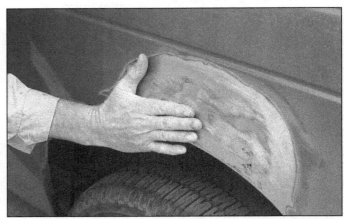

4 When the paint is removed, touch will probably be more helpful than sight for telling if the metal is straight. Hammer down the high spots or raise the low spots as necessary. Clean the repair area with wax/silicone remover

5 Following label instructions, mix up a batch of plastic filler and hardener. The ratio of filler to hardener is critical, and, if you mix it incorrectly, it will either not cure properly or cure too quickly (you won't have time to file and sand it into shape)

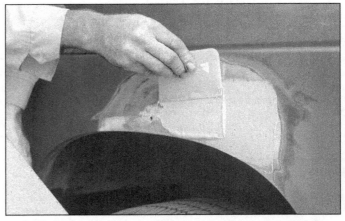

6 Working quickly so the filler doesn't harden, use a plastic applicator to press the body filler firmly into the metal, assuring it bonds completely. Work the filler until it matches the original contour and is slightly above the surrounding metal

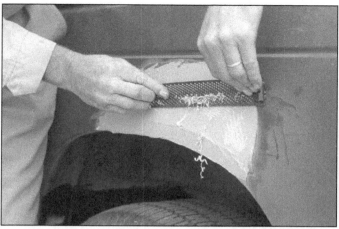

7 Let the filler harden until you can just dent it with your fingernail. Use a body file or Surform tool (shown here) to rough-shape the filler

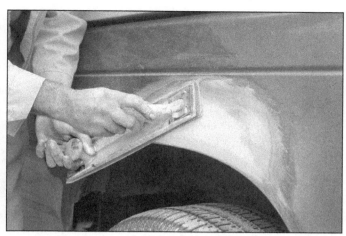

8 Use coarse-grit sandpaper and a sanding board or block to work the filler down until it's smooth and even. Work down to finer grits of sandpaper - always using a board or block - ending up with 360 or 400 grit

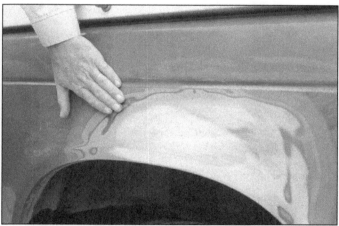

9 You shouldn't be able to feel any ridge at the transition from the filler to the bare metal or from the bare metal to the old paint. As soon as the repair is flat and uniform, remove the dust and mask off the adjacent panels or trim pieces

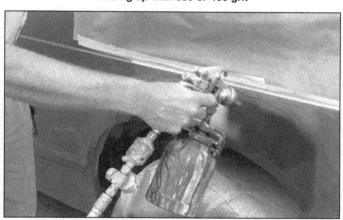

10 Apply several layers of primer to the area. Don't spray the primer on too heavy, so it sags or runs, and make sure each coat is dry before you spray on the next one. A professional-type spray gun is being used here, but aerosol spray primer is available inexpensively from auto parts stores

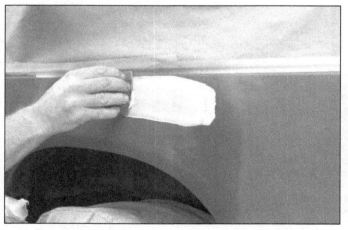

11 The primer will help reveal imperfections or scratches. Fill these with glazing compound. Follow the label instructions and sand it with 360 or 400-grit sandpaper until it's smooth. Repeat the glazing, sanding and respraying until the primer reveals a perfectly smooth surface

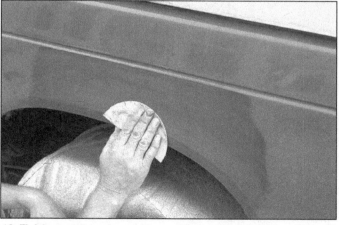

12 Finish sand the primer with very fine sandpaper (400 or 600-grit) to remove the primer overspray. Clean the area with water and allow it to dry. Use a tack rag to remove any dust, then apply the finish coat. Don't attempt to rub out or wax the repair area until the paint has dried completely (at least two weeks)

6 Body repair - minor damage

See color photo sequence

Repair of scratches

1   If the scratch is superficial and does not penetrate to the metal of the body, repair is very simple. Lightly rub the scratched area with a fine rubbing compound to remove loose paint and built up wax. Rinse the area with clean water.

2   Apply touch-up paint to the scratch, using a small brush. Continue to apply thin layers of paint until the surface of the paint in the scratch is level with the surrounding paint. Allow the new paint at least two weeks to harden, then blend it into the surrounding paint by rubbing with a very fine rubbing compound. Finally, apply a coat of wax to the scratch area.

3   If the scratch has penetrated the paint and exposed the metal of the body, causing the metal to rust, a different repair technique is required. Remove all loose rust from the bottom of the scratch with a pocket knife, then apply rust inhibiting paint to prevent the formation of rust in the future. Using a rubber or nylon applicator, coat the scratched area with glaze-type filler. If required, the filler can be mixed with thinner to provide a very thin paste, which is ideal for filling narrow scratches. Before the glaze filler in the scratch hardens, wrap a piece of smooth cotton cloth around the tip of a finger. Dip the cloth in thinner and then quickly wipe it along the surface of the scratch. This will ensure that the surface of the filler is slightly hollow. The scratch can now be painted over as described earlier in this Section.

Repair of dents

See photo sequence

4   When repairing dents, the first job is to pull the dent out until the affected area is as close as possible to its original shape. There is no point in trying to restore the original shape completely as the metal in the damaged area will have stretched on impact and cannot be restored to its original contours. It is better to bring the level of the dent up to a point which is about 1/8-inch below the level of the surrounding metal. In cases where the dent is very shallow, it is not worth trying to pull it out at all.

5   If the back side of the dent is accessible, it can be hammered out gently from behind using a soft-face hammer. While doing this, hold a block of wood firmly against the opposite side of the metal to absorb the hammer blows and prevent the metal from being stretched.

6   If the dent is in a section of the body which has double layers, or some other factor makes it inaccessible from behind, a different technique is required. Drill several small holes through the metal inside the damaged area, particularly in the deeper sections. Screw long, self-tapping screws into the holes just enough for them to get a good grip in the metal. Now the dent can be pulled out by pulling on the protruding heads of the screws with locking pliers.

7   The next stage of repair is the removal of paint from the damaged area and from an inch or so of the surrounding metal. This is easily done with a wire brush or sanding disk in a drill motor, although it can be done just as effectively by hand with sandpaper. To complete the preparation for filling, score the surface of the bare metal with a screwdriver or the tang of a file or drill small holes in the affected area. This will provide a good grip for the filler material. To complete the repair, see the Section on *filling and painting*.

Repair of rust holes or gashes

8   Remove all paint from the affected area and from an inch or so of the surrounding metal using a sanding disk or wire brush mounted in a drill motor. If these are not available, a few sheets of sandpaper will do the job just as effectively.

9   With the paint removed, you will be able to determine the severity of the corrosion and decide whether to replace the whole panel, if possible, or repair the affected area. New body panels are not as expensive as most people think and it is often quicker to install a new panel than to repair large areas of rust.

10   Remove all trim pieces from the affected area except those which will act as a guide to the original shape of the damaged body, such as headlight shells, etc. Using metal snips or a hacksaw blade, remove all loose metal and any other metal that is badly affected by rust. Hammer the edges of the hole on the inside to create a slight depression for the filler material.

11   Wire brush the affected area to remove the powdery rust from the surface of the metal. If the back of the rusted area is accessible, treat it with rust inhibiting paint.

12   Before filling is done, block the hole in some way. This can be done with sheet metal riveted or screwed into place, or by stuffing the hole with wire mesh.

13   Once the hole is blocked off, the affected area can be filled and painted. See the following subsection on *filling and painting*.

Filling and painting

14   Many types of body fillers are available, but generally speaking, body repair kits which contain filler paste and a tube of resin hardener are best for this type of repair work. A wide, flexible plastic or nylon applicator will be necessary for imparting a smooth and contoured finish to the surface of the filler material. Mix up a small amount of filler on a clean piece of wood or cardboard (use the hardener sparingly). Follow the manufacturer's instructions on the package, otherwise the filler will set incorrectly.

15   Using the applicator, apply the filler paste to the prepared area. Draw the applicator across the surface of the filler to achieve the desired contour and to level the filler surface. As soon as a contour that approximates the original one is achieved, stop working the paste. If you continue, the paste will begin to stick to the applicator. Continue to add thin layers of paste at 20-minute intervals until the level of the filler is just above the surrounding metal.

16   Once the filler has hardened, the excess can be removed with a body file. From then on, progressively finer grades of sandpaper should be used, starting with a 180-grit paper and finishing with 600-grit wet-or-dry paper. Always wrap the sandpaper around a flat rubber or wooden block, otherwise the surface of the filler will not be completely flat. During the sanding of the filler surface, the wet-or-dry paper should be periodically rinsed in water. This will ensure that a very smooth finish is produced in the final stage.

17   At this point, the repair area should be surrounded by a ring of bare metal, which in turn should be encircled by the finely feathered edge of good paint. Rinse the repair area with clean water until all of the dust produced by the sanding operation is gone.

18   Spray the entire area with a light coat of primer. This will reveal any imperfections in the surface of the filler. Repair the imperfections with fresh filler paste or glaze filler and once more smooth the surface with sandpaper. Repeat this spray-and-repair procedure until you are satisfied that the surface of the filler and the feathered edge of the paint are perfect. Rinse the area with clean water and allow it to dry completely.

19   The repair area is now ready for painting. Spray painting must be carried out in a warm, dry, windless and dust free atmosphere. These conditions can be created if you have access to a large indoor work area, but if you are forced to work in the open, you will have to pick the day very carefully. If you are working indoors, dousing the floor in the work area with water will help settle the dust which would otherwise be in the air. If the repair area is confined to one body panel, mask off the surrounding panels. This will help minimize the effects of a slight mismatch in paint color. Trim pieces such as chrome strips, door handles, etc., will also need to be masked off or removed. Use masking tape and several thickness of newspaper for the masking operations.

20   Before spraying, shake the paint can thoroughly, then spray a test area until the spray painting technique is mastered. Cover the repair area with a thick coat of primer. The thickness should be built up using several thin layers of primer rather than one thick one. Using 600-grit wet-or-dry sandpaper, rub down the surface of the primer until it is very smooth. While doing this, the work area should be thoroughly rinsed with water and the wet-or-dry sandpaper periodically rinsed as well. Allow the primer to dry before spraying additional coats.

21   Spray on the top coat, again building up the thickness by using several thin layers of paint. Begin spraying in the center of the repair area and then, using a circular motion,

Chapter 11 Body

11-5

9.4 Use a wrench and socket to remove the hood hinge bolt and nut (arrow)

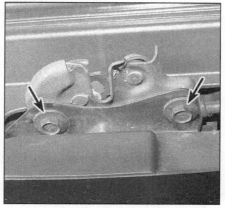

9.6 Loosen the bolts (arrows) and move the hood latch to adjust the hood closed position

9.7 Screw the hood bumpers in or out to adjust the hood flush with the fenders

work out until the whole repair area and about two inches of the surrounding original paint is covered. Remove all masking material 10 to 15 minutes after spraying on the final coat of paint. Allow the new paint at least two weeks to harden, then use a very fine rubbing compound to blend the edges of the new paint into the existing paint. Finally, apply a coat of wax.

7 Body repair - major damage

1   Major damage must be repaired by an auto body/frame repair shop with the necessary welding and hydraulic straightening equipment.
2   If the damage has been serious, it is vital that the structure be checked for proper alignment or the vehicle's handling characteristics may be adversely affected. Other problems, such as excessive tire wear and wear in the driveline and steering may occur.
3   Due to the fact that all of the major body components (hood, fenders, etc.) are separate and replaceable units, any seriously damaged components should be replaced rather than repaired. Sometimes these components can be found in a wrecking yard that specializes in used vehicle components, often at considerable savings over the cost of new parts.

8 Windshield and fixed glass - replacement

Replacement of the windshield and fixed glass requires the use of special fast setting adhesive/caulk materials. These operations should be left to a dealer or a shop specializing in glass work.

9 Hood - removal, installation and adjustment

Caution: *On models equipped with a Delco Loc II anti-theft audio system, be sure the lockout feature is turned off before performing any procedure which requires disconnecting the battery.*
Note: *The hood is somewhat awkward to remove and install - at least two people should perform this procedure.*

Removal and installation

Refer to illustration 9.4

1   Disconnect the negative cable from the battery.
2   Open the hood and place rags or covers over the windshield and fenders to protect them during the removal procedure. Disconnect the underhood light electrical connector and remove the windshield washer hoses from the nozzles.
3   Remove the cowl vent grille end pieces (see Section 27).
4   Remove the hood bolts and nuts and lift off the hood **(see illustration)**.
5   Installation is the reverse of removal.

Adjustment

Refer to illustrations 9.6 and 9.7

6   If necessary after installation, the entire hood latch assembly can be adjusted up-and-down as well as from side-to-side on the upper radiator support so the hood closes securely and is flush with the fenders. To do this, scribe a line around the hood latch mounting bolts to provide a reference point. Then loosen the bolts and reposition the latch assembly as necessary **(see illustration)**. Following adjustment, retighten the mounting bolts.
7   Finally, adjust the hood bumpers on the radiator support so the hood, when closed, is flush with the fenders **(see illustration)**.
8   The hood latch assembly, as well as the hinges, should be periodically lubricated with white lithium-based grease to prevent sticking and wear.

10 Hood release latch and cable - removal and installation

Warning: *1995 and later models are equipped with an airbag. Always turn the steering wheel to the straight ahead position, place the ignition switch in Lock and remove the key, then remove the airbag fuse and unplug the yellow Connector Position Assurance (CPA) connector at the base of the steering column before working in the vicinity of the impact sensors, steering column or instrument panel to avoid the possibility of accidental deployment of the airbag, which could cause personal injury (see Chapter 12).*
Caution: *On models equipped with a Delco Loc II anti-theft audio system, be sure the lockout feature is turned off before performing any procedure which requires disconnecting the battery.*

Removal

1   Detach the clips and remove the plastic air deflector cover.
2   Remove the radiator grille (see Section 11).
3   Detach the cable from the latch assembly and the clips at the front of the vehicle and in the engine compartment. To remove the latch assembly, remove the mounting bolts **(see illustration 9.6)**.
4   Attach a piece of string or thin wire to the end of the cable and unclip all cable retaining clips.
5   Working in the passenger compartment, detach the left side cowl panel, remove two bolts and detach the hood release lever.
6   Remove the cable and grommet rearward, pulling them into the passenger compartment.

Installation

7   Ensure that the new cable has a grommet attached and then connect the string or wire to the new cable and pull it forward into the engine compartment.
8   Connect the cable and secure it with the retaining clips.
9   Ensure that the grommet is in place and install the hood release lever and any other components that were remove.

11.3a Remove the bolts along the top of the grille with a socket and extension tool

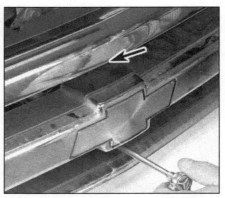

11.3b Remove the bolts inside the grille

12.3 Remove the bumper bolts (arrow)

11 Radiator grille - removal and installation

Refer to illustrations 11.3a and 11.3b

Warning: *These models have airbags. Always turn the steering wheel to the straight ahead position, place the ignition switch in Lock, remove the airbag fuse and unplug the yellow Connector Position Assurance (CPA) connectors at the base of the steering column before working in the vicinity of the impact sensors, steering column or instrument panel to avoid the possibility of accidental deployment of the airbag, which could cause personal injury (see Chapter 12).* **Caution:** *On models equipped with a Delco Loc II anti-theft audio system, be sure the lockout feature is turned off before performing any procedure which requires disconnecting the battery.*

1998 and earlier models

1   Disconnect the negative cable from the battery.
2   Open the hood and unplug the side marker light connectors (some models).
3   Remove the grille mounting bolts **(see illustrations)**.
4   Grasp the grille securely and detach the clips along the lower edge and lift it off.
5   Installation is the reverse of removal.

1999 and later models

6   Later model grilles are attached to the radiator support with several types of clips, depending on model. On 1999 models, all grilles are similarly mounted, except the Bravada, which has a wider grille and more clips. On 2000 and later models, only the GMC grille is like previous years, while Bravada and Chevrolet both have wider grilles. **Note:** *On 2000 and later Diamond models, remove the grille guard before removing the grille.*
7   To remove the grille, pull forward slowly and carefully until all the clips release from the radiator support.
8   On 2000 and later Chevrolet models with headlight washer system, remove the hose from the grille.
9   Remove the fog lamps and turn signals on models with wider grilles.

12.4 Remove the bumper brace-to-frame bolts (arrow)

10   Installation is the reverse of the removal procedure.

12 Bumpers - removal and installation

Warning: *These models have airbags. Always turn the steering wheel to the straight ahead position, place the ignition switch in Lock, remove the airbag fuse and unplug the yellow Connector Position Assurance (CPA) connectors at the base of the steering column before working in the vicinity of the impact sensors, steering column or instrument panel to avoid the possibility of accidental deployment of the airbag, which could cause personal injury (see Chapter 12).* **Caution:** *On models equipped with a Delco Loc II anti-theft audio system, be sure the lockout feature is turned off before performing any procedure which requires disconnecting the battery.*

Front

Refer to illustrations 12.3 and 12.4

1   Remove the radiator grille (see Section 11). **Note:** *On later models, the front bumper is a structural beam bolted to the chassis, and covered by a plastic fascia. Fascia removal varies with the model, but involves removal of bolts and releasing clips. Fasteners are found in the license plate opening, the bottom edge of the fascia, and the ends of the bumpers. In most models, the bumper can be*

12.8 Rear bumper to frame bolt locations (arrows)

removed from the vehicle with the fascia still attached, as long as any fasteners connecting the fascia to the front fenders are removed.
2   Remove the bumper beam-to-chassis bolts and detach the bumper **(see illustration)**. On 2000 and later Bravada, Envoy and Xtreme models, disconnect the fog lights before removing the bumper/fascia.
3   Remove the bumper-to-chassis bolts **(see illustration)**.
4   Remove the bumper brace-to-frame bolts/nuts and detach the bumper **(see illustration)**.
5   Detach the right and left side parking lights from the housings in the bumper.
6   Installation is the reverse of removal.

Rear

Refer to illustration 12.8

7   Remove the license plate light housing from the bumper and disconnect the lamp lead(s) (see Chapter 12).
8   Remove the bumper-to-frame and brace-to-bumper bolts/nuts and detach the bumper **(see illustration)**.
9   Installation is the reverse of removal.

13 Door trim panel - removal and installation

Caution: *On models equipped with a Delco Loc II anti-theft audio system, be sure the*

Chapter 11 Body

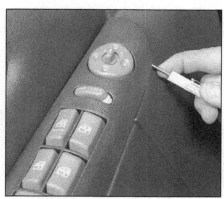

13.1a On models equipped with power windows, mirrors and door locks, gently pry off the armrest control panel . . .

13.1b . . . and disconnect the electrical connectors from the back side

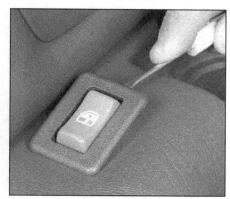

13.1c To remove an individual switch, pry it out . . .

13.1d . . . and disconnect the electrical connectors from the rear side

13.3a Remove the door handle cover plastic fastener by unscrewing the center . . .

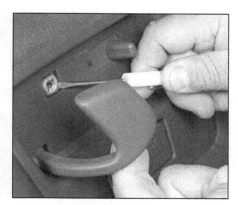

13.3b . . . then pry out the fastener with a small screwdriver

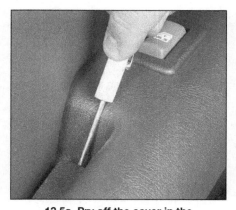

13.5a Pry off the cover in the door pull opening . . .

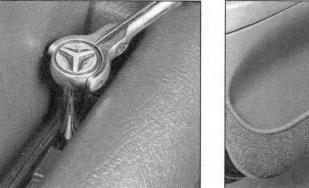

13.5b . . . and remove the bolts

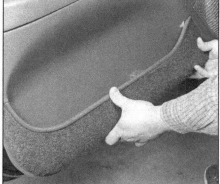

13.5c Pull up on the door trim panel to detach it, then rotate it back and out of the door

lockout feature is turned off before performing any procedure which requires disconnecting the battery.

Removal

Refer to illustrations 13.1a, 13.1b, 13.1c, 13.1d, 13.3a, 13.3b, 13.5a, 13.5b, 13.5c and 13.6

1 On models equipped with power components (power windows and door locks), disconnect the negative cable at the battery. Pry out the control switch assembly and disconnect the electrical connectors **(see illustrations)**.

2 On manual window equipped models, remove the window regulator handle by pressing in on the bearing plate and door trim panel and, with a piece of hooked wire, pull off the spring clip. A special tool is available for this purpose but its use is not essential. With the clip removed, remove the handle and the bearing plate.

3 Remove the inner door handle cover **(see illustrations)**.

4 Remove the power door lock switch (if equipped).

5 Remove any remaining door trim panel retaining bolts/screws, then carefully pull the trim panel up and away from the door **(see illustrations)**. Do not pull straight out from the door or you might break the trim panel retaining hooks. Unplug any wire harness connectors and remove the panel. **Note:** *On 2000 and later extended-cab pickups, there is an optional third door, behind the driver's door,*

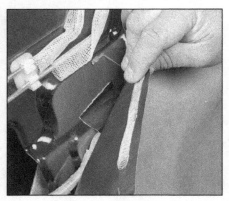

13.6 For access to the inner door components, peel back the water shield - if you're careful, it can be reused

14.3 Remove the outside mirror retaining nuts and on power models, unplug the electrical connector

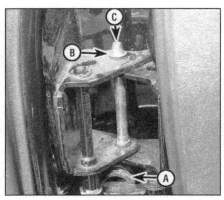

15.5 Before beginning work, remove the spring (A) with the special compressor tool, then remove the clip (B) before driving out the hinge pin (C)

for access to the rear of the cab. The panel for this door is retained by clips, but you must first remove the bottom anchor bolt for the seat belt, then remove the upper trim cover at the shoulder harness mounting point.

6   For access to the inner door, carefully peel back the plastic water shield **(see illustration)**.

Installation

7   Prior to installation of the door trim panel, be sure to reinstall any clips in the panel which may have come out when you removed the panel.

8   Plug in the wire harness connectors for the power door lock switch and the power window switch, if equipped, and place the panel in position in the door. Press the door panel into place and push down until the hooks are seated. Install the trim panel retaining screws. Install the power door lock switch assembly, if equipped. Install the manual regulator crank handle or power window switch assembly.

14 Outside mirror - removal and installation

Refer to illustration 14.3
Caution: *On models equipped with a Delco Loc II anti-theft audio system, be sure the lockout feature is turned off before performing any procedure which requires disconnecting the battery.*

1   Disconnect the negative cable from the battery.
2   Remove the door trim panel (see Section 13)
3   Remove the mirror-to-door retaining nuts and, on power mirrors, unplug the electrical connector **(see illustration)**.
4   Lift off the mirror assembly.
5   Installation is the reverse of removal.

15 Door - removal and installation

Refer to illustration 15.5
Caution: *On models equipped with a Delco Loc II anti-theft audio system, be sure the lockout feature is turned off before performing any procedure which requires disconnecting the battery.*

1   On models equipped with power components (power windows and door locks), disconnect the negative cable from the battery.
2   Remove the door trim panel (see Section 13). On doors with power components, unplug all electrical connections from the door (it is a good idea to label all connections to aid the reassembly process) and remove the electrical harness from the door.
3   Open the door all the way and support it on jacks or blocks covered with cloth or pads to prevent damaging the paint.
4   Use a spring compressor tool, available at most auto parts stores, to compress and remove the hinge spring. **Warning:** *The door hinge spring may fly off the compressor tool during removal. To prevent personal injury, cover the spring with a heavy rag before removing it.*
5   Have an assistant hold the door, remove the hinge pin retainers and drive out the pins then carefully lift off the door **(see illustration)**.
6   If the door does not close properly after installation, the door latch striker can be adjusted both up-and-down and sideways to provide positive engagement with the latch mechanism. This is done by loosening the door striker bolt and moving the striker as necessary.
7   Install the door by reversing the removal procedure.

16 Front fender - removal and installation

Refer to illustrations 16.7 and 16.8
Warning: *1995 and later models are equipped with an airbag. Always turn the steering wheel to the straight ahead position, place the ignition switch in Lock and remove the key, then remove the airbag fuse and unplug the yellow Connector Position Assurance (CPA) connector at the base of the steering column before working in the vicinity of the impact sensors, steering column or*

16.7 Remove the two bolts (arrows) at the rear upper edge of the front fender

instrument panel to avoid the possibility of accidental deployment of the airbag, which could cause personal injury (see Chapter 12).
Caution: *On models equipped with a Delco Loc II anti-theft audio system, be sure the lockout feature is turned off before performing any procedure which requires disconnecting the battery.*

1   Raise the vehicle, support it securely on jackstands and remove the front wheel.
2   Remove the radiator grille (see Section 11).
3   Disconnect all lighting assembly wiring harness connectors and other components that would interfere with fender removal.
4   Remove the front bumper (see Section 12).
5   Remove the hood (see Section 9).
6   Remove the bolts and detach the wheelhouse cover. Remove the cowl vent grille end pieces (see Section 27).
7   Remove the bolts securing the rear edge of the fender to the body. One bolt is accessible at the rear of the wheelhouse (near the bottom of the fender), one inside the door opening and two at the top-rear **(see illustration)**.
8   Remove the bolts securing the top of the fender to the body (not all models).
9   Remove the fender-to-radiator support mounting bolts **(see illustration)**. Depending on the year and model, the front of the fender

Chapter 11 Body

16.9 The front edge of the fender is retained to the radiator support by a bolt (arrow)

17.5 Remove the torque rod bolts (arrow)

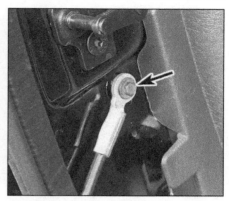

17.6 Remove the bolt and detach the cable (arrow)

17.8 Remove the clip and drive out the hinge pin

18.6 Remove the Torx-head tailgate glass bolts (arrows)

may be bolted to the radiator support (at the top and bottom) and the end of the bumper fascia.
10 Detach the fender. It is a good idea to have an assistant support the fender while it's being moved away from the vehicle to prevent damage to the surrounding body panels.
11 Installation is the reverse of removal. Tighten all fasteners securely.

17 Tailgate - removal and installation

Refer to illustrations 17.5, 17.6 and 17.8

Pick-up models

1 Open the tailgate.
2 Detach the tailgate support cables.
3 Raise the tailgate to a 45-degree angle and with the help of an assistant, lift the left end from the hinge, followed by the right end and detach the tailgate from the vehicle.
4 Installation is the reverse of removal.

Blazer, Jimmy and Bravada models

5 With the tailgate open, remove the trim panels around the tailgate opening, then disconnect the electrical connectors. Working underneath the rear of the vehicle, remove the torque rod bracket bolt **(see illustration)**.
6 Remove the bolts and detach the support cables **(see illustration)**.
7 Detach the support struts (see Section 19).
8 With an assistant supporting the weight of the tailgate, use a small screwdriver to remove the clips, then drive out the retaining pins and remove the tailgate **(see illustration)**.
9 Installation is the reverse of removal.

18 Tailgate glass - removal, installation and adjustment

Refer to illustration 18.6
Caution: *On models equipped with a Delco Loc II anti-theft audio system, be sure the lockout feature is turned off before performing any procedure which requires disconnecting the battery.*
Note: *The tailgate glass is heavy and somewhat awkward to remove and install - at least two people should perform this procedure.*
1 Disconnect the negative cable from the battery.
2 Open the tailgate glass and cover the upper body area around the opening with pads or blankets to protect the painted surfaces when the tailgate glass is removed.
3 Unplug all electrical connectors. On models with high-mounted stoplight attached to the glass, remove the stoplamp mounting screws and pull the light out far enough to disconnect the electrical connector.
4 While an assistant supports the tailgate glass, detach the support struts (see Section 19).
5 Draw around the hinge plates with a marking pen.
6 Remove the Torx-head hinge bolts and detach the tailgate glass from the vehicle **(see illustration)**.
7 Installation is the reverse of removal.
8 If the tailgate glass needs to be adjusted, loosen the bolts and slightly reposition the latch, repeating the procedure as necessary.

19 Support struts - replacement

Refer to illustrations 19.2, 19.3, 19.5a and 19.5b

Tailgate glass

1 Open the tailgate glass and support it securely.
2 On the upper end of the tailgate glass strut, use a small screwdriver to detach the

Chapter 11 Body

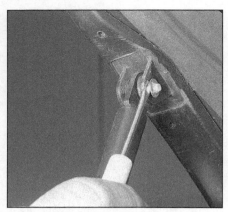

19.2 Pry off the clip and detach the upper end of the strut

19.3 Use a small screwdriver to pry off the clip, then detach the end of the strut from the ballstud

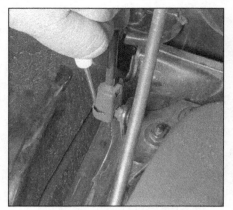

19.5a Pry the retaining clips off with a small screwdriver and detach the lower end . . .

19.5b . . . then the upper end of the strut

20.3 Rotate the plastic retainer off and detach the rod, then drill out the rivets (arrows) and remove the handle - on installation, install new rivets with a pop-riveting tool

20.10 To remove the latch assembly, unscrew the retaining screws at the rear edge of the door

clip and slide the retaining pin out **(see illustration)**.
3 Detach the lower end of the strut by prying off the retaining clip with a small screwdriver, then pry the lower end off the ballstud **(see illustration)**.
4 Installation is the reverse of removal.

Tailgate

5 Open the tailgate and support it securely. Pry the retaining clips off the upper and lower ends of both struts. Pry or pull sharply to detach the ends from the ballstuds **(see illustrations)**.
6 Installation is the reverse of removal.

20 Door latch, lock cylinder and handles - removal and installation

Caution: *On models equipped with a Delco Loc II anti-theft audio system, be sure the lockout feature is turned off before performing any procedure which requires disconnecting the battery.*

1 On models equipped with power components (power windows and door locks), disconnect the cable from the negative battery terminal.

Inside handle

Refer to illustration 20.3

2 With the window glass in the full up position, remove the door trim panel and water shield (see Section 13).
3 Disconnect the lock rod from the handle, drill out the rivets securing the handle to the door frame and remove the handle **(see illustration)**.
4 Installation is the reverse of the removal procedure. **Note:** *If you don't have a rivet tool, install screws, washers and locknuts to retain the handle.*

Outside handle and lock cylinder

5 Remove the door trim panel and water shield (see Section 13).
6 Disengage the handle and lock cylinder from the lock rods.
7 Remove the handle-to-door nuts and pull the handle free. Detach the lock cylinder from the handle.
8 Installation is the reverse of the removal procedure.

Latch

Refer to illustration 20.10

9 From inside the door, disconnect the locking knob rod, inside handle rod and inside handle lock rod from the latch assembly.
10 Remove the latch assembly retaining screws and pull the assembly free **(see illustration)**.
11 Installation is the reverse of the removal procedure.

21 Door window glass - removal and installation

Warning: *Safety glasses and gloves should be worn when performing this procedure.*
Caution: *On models equipped with a Delco Loc II anti-theft audio system, be sure the lockout feature is turned off before performing any procedure which requires disconnecting the battery.*

1 Lower the glass fully in the door. On some models, the glass should be lowered to halfway, so that the regulator rollers can be accessed through the holes in the door. On models equipped with power windows and locks, disconnect the cable from the negative battery terminal.

Chapter 11 Body

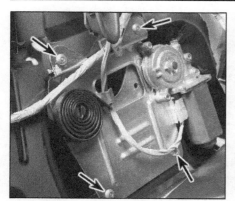

22.4 Drill out the regulator rivets (arrows) - power window regulator shown

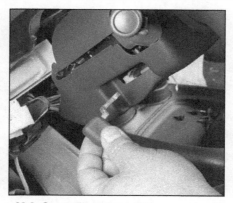

23.3 Grasp the column tilt lever and pull out sharply to remove it

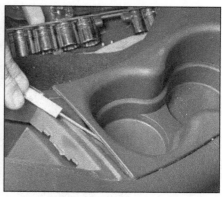

24.2a Use a small screwdriver to pry up on the cup holder to detach it from the console

2   Remove the door trim panel and water shield (see Section 13).

Front door

3   Remove the two bolts and detach the front glass run channel from the door.
4   Pry out the window weatherstrip in the top of the door opening.
5   Detach the glass from the regulator by bending the sash channel tab back, then raising the glass and tilting it forward and out of the rear regulator roller.
6   Tilt the glass up toward the rear, detach the sash roller and lift the glass out.
7   Installation is the reverse of removal.

Rear door

8   Remove the bolts and detach the front run channel and stationary vent glass assemblies from the door.
9   Detach the glass from the regulator by pushing it toward the rear and rotating it out of the regulator rollers.
10   Lift the glass out of the door.
11   Installation is the reverse of removal.

22   Door window regulator - removal and installation

Warning: *Do not remove the electric motor from the regulator assembly. The lift arm is under tension from the counterbalance spring and could cause personal injury if allowed to retract.*
Caution: *On models equipped with a Delco Loc II anti-theft audio system, be sure the lockout feature is turned off before performing any procedure which requires disconnecting the battery.*
1   On models equipped with power components (power windows and door locks), disconnect the cable from the negative battery terminal.

Removal

Refer to illustration 22.4
2   With the window glass in the full up position, remove the door trim panel and water shield (see Section 13).
3   Raise the window glass in the "Up" position and secure to the door frame with duct tape. Do not proceed until you're sure the weight of the window is supported.
4   Punch out the center pins of the rivets that secure the window regulator to the door frame and drill the rivets out with a 3/16-inch drill bit **(see illustration)**.
5   On power window equipped models, unplug the electrical connector.
6   Disengage the regulator from the window glass sash and lift it from the door.

Installation

7   Place the regulator in position in the door and engage to the guide channel assemblies.
8   Secure the regulator to the door using 3/16-inch rivets and a rivet tool.
9   Plug in the electrical connector (if equipped).
10   Install the door trim panel. Connect the negative battery cable.

23   Steering column cover - removal and installation

Refer to illustration 23.3
Warning: *1995 and later models are equipped with an airbag. Always turn the steering wheel to the straight ahead position, place the ignition switch in Lock and remove the key, then remove the airbag fuse and unplug the yellow Connector Position Assurance (CPA) connector at the base of the steering column before working in the vicinity of the impact sensors, steering column or instrument panel to avoid the possibility of accidental deployment of the airbag, which could cause personal injury (see Chapter 12). The yellow wires and connectors routed through the instrument panel are for this system. Do not use electrical test equipment on these yellow wires or tamper with them in any way while working under the instrument panel.*
Caution: *On models equipped with a Delco Loc II anti-theft audio system, be sure the lockout feature is turned off before performing any procedure which requires disconnecting the battery.*
1   Disconnect the negative battery cable.
2   Remove the column cover screws.
3   Grasp the steering column tilt lever securely and detach it by pulling straight out **(see illustration)**.
4   Remove the screws from the bottom cover, then tilt the lower cover back toward the dash to disengage the cover from its clips. The upper cover has one retaining screw. On airbag-equipped models, it will be necessary to remove the steering wheel to provide sufficient clearance for removal of both covers (see Chapter 10).
5   Installation is the reverse of removal.

24   Console - removal and installation

Refer to illustrations 24.2a, 24.2b, 24.2c and 24.3
Warning: *1995 and later models are equipped with an airbag. Always turn the steering wheel to the straight ahead position, place the ignition switch in Lock and remove the key, then remove the airbag fuse and unplug the yellow Connector Position Assurance (CPA) connector at the base of the steering column before working in the vicinity of the impact sensors, steering column or instrument panel to avoid the possibility of accidental deployment of the airbag, which could cause personal injury (see Chapter 12). The yellow wires and connectors routed through the instrument panel are for this system. Do not use electrical test equipment on these yellow wires or tamper with them in any way while working under the instrument panel.*
Caution: *On models equipped with a Delco Loc II anti-theft audio system, be sure the lockout feature is turned off before performing any procedure which requires disconnecting the battery.*
1   Remove the shift knob or lever. On models with an automatic transmission, pry out the clip and pull off the knob. On models with a manual transmission, simply remove the bolts and detach the shifter knob and lever assembly.
2   Pry out the cup holder for access. Remove the retaining screw and detach the console cover, then remove the nuts **(see**

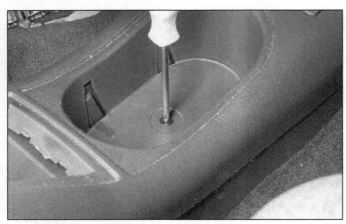

24.2b Remove the screw, the detach the console cover

24.2c Remove the console center retaining nut

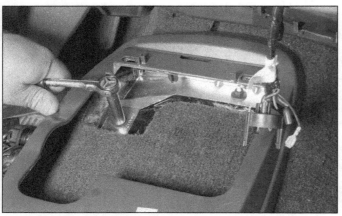

24.3 Remove the nuts retaining the front of the console

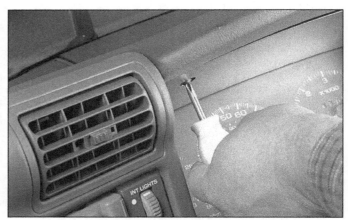

25.1a Remove the screws from the left side of the cluster bezel . . .

illustrations). Open the storage compartment, pull out the compartment liner, and remove the nuts securing the rear of the console.

3 Remove the front console retaining nuts **(see illustration)**.

4 Detach the clips, unplug any electrical connectors and lift the console up and out of the vehicle. **Note:** *Some later models have a short front console, which is secured to the floor with clips instead of nuts. Carefully pull up the console to remove it.*

5 Installation is the reverse of removal.

25 Instrument cluster bezel - removal and installation

Refer to illustrations 25.1a, 25.1b and 25.3

Warning: *1995 and later models are equipped with an airbag. Always turn the steering wheel to the straight ahead position, place the ignition switch in Lock and remove the key, then remove the airbag fuse and unplug the yellow Connector Position Assurance (CPA) connector at the base of the steering column before working in the vicinity of the impact sensors, steering column or instrument panel to avoid the possibility of accidental deployment of the airbag, which could cause personal injury (see Chapter 12). The yellow wires and connectors*

25.1b . . . and the right side of the cluster bezel

routed through the instrument panel are for this system. Do not use electrical test equipment on these yellow wires or tamper with them in any way while working under the instrument panel.

Caution: *On models equipped with a Delco Loc II anti-theft audio system, be sure the lockout feature is turned off before performing any procedure which requires disconnecting the battery.*

1 Remove the two bezel-to-instrument

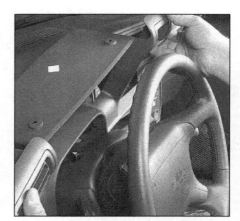

25.3 Pull the instrument cluster bezel straight out to detach the clips

panel retaining screws **(see illustrations)**. On 1999 and later models, remove the driver's knee bolster (see Section 26) for access to the two screws at the bottom of the cluster bezel.

2 Pull the bezel off the instrument panel and unplug the electrical connectors.

3 Grasp the bezel securely and pull back sharply to detach the clips from the instrument panel **(see illustration)**. **Note:** *Later models with three mounting screws do not have clips.*

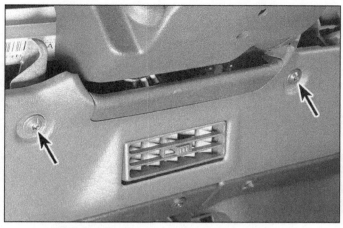

26.2 Remove the knee bolster screws (arrows)

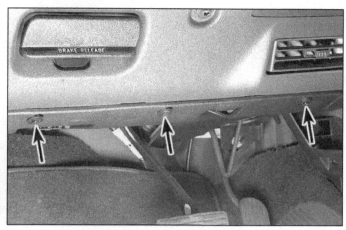

26.7a Left side sound insulator screws (arrows)

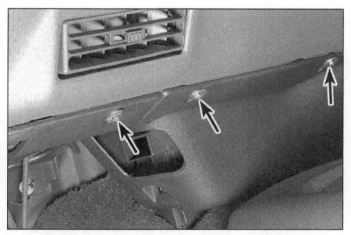

26.7b Left side and center sound insulator screws (arrows)

26.9 Remove the screws (arrows) that attach the glove box to the instrument panel

4 Unplug the electrical connectors and remove the bezel.
5 Installation is the reverse of removal.

26 Dashboard trim panels - removal and installation

Warning: *1995 and later models are equipped with an airbag. Always turn the steering wheel to the straight ahead position, place the ignition switch in Lock and remove the key, then remove the airbag fuse and unplug the yellow Connector Position Assurance (CPA) connector at the base of the steering column before working in the vicinity of the impact sensors, steering column or instrument panel to avoid the possibility of accidental deployment of the airbag, which could cause personal injury (see Chapter 12). The yellow wires and connectors routed through the instrument panel are for this system. Do not use electrical test equipment on these yellow wires or tamper with them in any way while working under the instrument panel.*
Caution: *On models equipped with a Delco Loc II anti-theft audio system, be sure the lockout feature is turned off before performing any procedure which requires disconnecting the battery.*

Knee bolster
Refer to illustration 26.2
1 On earlier models, remove the instrument cluster bezel (see Section 25).
2 Remove the screws at the upper edge of the knee bolster **(see illustration)**. On 1999 and later models, there are four screws, two at the top and two at the bottom.
3 Remove the left side sound insulator panel (see below).
4 If equipped with a tilt-column, tilt the steering column down and place the automatic transmission lever in Low.
5 Remove the remaining screws, lower the knee bolster from the instrument panel, disconnect the parking brake release cable. Unplug any electrical connectors and remove the bolster.
6 Installation is the reverse of the removal procedure.

Sound insulator panels
Refer to illustrations 26.7a and 26.7b
7 Remove the screws, detach the sound insulator panel and lower it from the instrument panel **(see illustrations)**.
8 Installation is the reverse of the removal procedure.

Glove box
Refer to illustration 26.9
9 Remove the screws, slide the glove box out and lower it from the instrument panel **(see illustration)**. On 1999 and later models, push upward on the spring inside the left end of the glove compartment to release the door, then remove the four door-to-dash screws.
10 Installation is the reverse of the removal procedure.

27 Cowl vent grille - removal and installation

Refer to illustrations 27.2, 27.3, 27.4a and 27.4b
1 Mark the position of the windshield wiper blade on the windshield with a wax marking pen.
2 Remove the cowl vent grille end pieces

27.2 Remove the bolts and lift the vent grille end pieces off

27.3 Flip up the plastic cover, remove the nut and detach the wiper arm

27.4a Loosen the plastic retainer screws . . .

27.4b . . . then pry the retainers out with a screwdriver

(see illustration).

3   Pry the wiper arm covers up, remove the nuts and detach the wipers **(see illustration)**.
4   Remove the plastic cowl retainers and detach the cowl grille from the vehicle **(see illustrations)**.
5   Installation is the reverse of removal. Make sure to align the wiper blades with the marks made during removal.

28 Seats - removal and installation

Caution: *On models equipped with a Delco Loc II anti-theft audio system, be sure the lockout feature is turned off before performing any procedure which requires disconnecting the battery.*

1   On power seat models, disconnect the negative cable from the battery.

2   Disconnect any electrical connectors.
3   Remove the seat adjuster trim covers.
4   Remove the securing nuts and remove the seat. On SUV rear seats (split-bench), the passenger side seat section should be removed first. Tilt the seat-back forward to access the rear mounting bolts.
5   Installation is the reverse of the removal procedure. Tighten the nuts securely.

Chapter 12
Chassis electrical system

Contents

	Section		Section
Airbag - general information	28	Instrument cluster - removal and installation	21
Antenna - removal and installation	17	Power door lock system - description and check	25
Bulb replacement	15	Power mirror control system - description and check	23
Circuit breakers - general information	5	Power seats - description and check	27
Composite headlight housing - removal and installation	14	Power window system - description and check	26
Cruise control system - description and check	24	Radio and speakers - general information, removal and installation	16
Electrical troubleshooting - general information	2	Rear window defogger - check and repair	20
Fuses - general information	3	Rear window defogger switch - removal and installation	19
Fusible links - general information	4	Rear wiper switch - removal and installation	11
General information	1	Relays - general information	6
Headlight bulb - replacement	12	Turn signal switch assembly - removal and installation	8
Headlight switch - removal and installation	10	Turn signal/hazard flasher - check and replacement	7
Headlights - adjustment	13	Wiper motor - removal and installation	18
Horn - check and replacement	22	Wiring diagrams - general information	29
Ignition key lock cylinder - replacement	9		

Specifications

Bulbs
Front
  Headlight
    Composite
      Low beam
        1994 .... H9006
        1995 on 9006HB4
      High beam
        1994 .... H9005
        1995 on .... 9006HB3 or 9005HB3
    Sealed beam
      1994 .... 2E1H
      1995 on 2E1
  Side marker .... 194 or 194NA
  Parking/turn signal 3157NA
Interior
  Instrument cluster illumination lights PC168
  Instrument cluster indicator lights PC74
  Dome light 2112
  Courtesy lights 1003
Rear
  Back-up
    1994 3057
    1995 on 3156
  Side marker 194
  License 194
  Tail/turn/stop 3057
Center high-mounted stoplight 577

Torque specifications
Steering column support bracket fasteners .... 22 ft-lbs

1 General information

Warning: *1995 and later models are equipped with an airbag. Always turn the steering wheel to the straight ahead position, place the ignition switch in Lock and remove the key, then remove the airbag fuse and unplug the yellow Connector Position Assurance (CPA) connector at the base of the steering column before working in the vicinity of the impact sensors, steering column or instrument panel to avoid the possibility of accidental deployment of the airbag, which could cause personal injury (see Section 28). The yellow wires and connectors routed through the instrument panel are for this system. Do not use electrical test equipment on these yellow wires or tamper with them in any way while working under the instrument panel.*
Caution: *On models equipped with a Delco Loc II anti-theft audio system, be sure the lockout feature is turned off before performing any procedure which requires disconnecting the battery.*

Chapter 12 Chassis electrical system

The electrical system is a 12-volt, negative ground type. Power for the lights and all electrical accessories is supplied by a lead/acid-type battery which is charged by the alternator.

This Chapter covers repair and service procedures for the various electrical components not associated with the engine. Information on the battery, alternator, ignition system and starter motor can be found in Chapter 5. It should be noted that when portions of the electrical system are serviced, the negative battery cable should be disconnected from the battery to prevent electrical shorts and/or fires.

2 Electrical troubleshooting - general information

A typical electrical circuit consists of an electrical component, any switches, relays, motors, fuses, fusible links or circuit breakers related to that component and the wiring and electrical connectors that link the component to both the battery and the chassis. To help you pinpoint an electrical circuit problem, wiring diagrams are included at the end of this book.

Before tackling any troublesome electrical circuit, first study the appropriate wiring diagrams to get a complete understanding of what makes up that individual circuit. Trouble spots, for instance, can often be narrowed down by noting if other components related to the circuit are operating properly. If several components or circuits fail at one time, chances are the problem is in a fuse or ground connection, because several circuits are often routed through the same fuse and ground connections.

Electrical problems usually stem from simple causes, such as loose or corroded connections, a blown fuse, a melted fusible link or a bad relay. Visually inspect the condition of all fuses, wires and connections in a problem circuit before troubleshooting it.

If testing instruments are going to be utilized, use the diagrams to plan ahead of time where you will make the necessary connections in order to accurately pinpoint the trouble spot.

The basic tools needed for electrical troubleshooting include a circuit tester or voltmeter (a 12-volt bulb with a set of test leads can also be used), a continuity tester, which includes a bulb, battery and set of test leads, and a jumper wire, preferably with a circuit breaker incorporated, which can be used to bypass electrical components. Before attempting to locate a problem with test instruments, use the wiring diagram(s) to decide where to make the connections.

Voltage checks

Voltage checks should be performed if a circuit is not functioning properly. Connect one lead of a circuit tester to either the negative battery terminal or a known good ground. Connect the other lead to an electrical connector in the circuit being tested, preferably nearest to the battery or fuse. If the bulb of the tester lights, voltage is present, which means that the part of the circuit between the electrical connector and the battery is problem free. Continue checking the rest of the circuit in the same fashion. When you reach a point at which no voltage is present, the problem lies between that point and the last test point with voltage. Most of the time the problem can be traced to a loose connection. **Note:** *Keep in mind that some circuits receive voltage only when the ignition key is in the Accessory or Run position.*

Finding a short

One method of finding shorts in a circuit is to remove the fuse and connect a test light or voltmeter in its place to the fuse terminals. There should be no voltage present in the circuit. Move the wiring harness from side-to-side while watching the test light. If the bulb goes on, there is a short to ground somewhere in that area, probably where the insulation has rubbed through. The same test can be performed on each component in the circuit, even a switch.

Ground check

Perform a ground test to check whether a component is properly grounded. Disconnect the battery and connect one lead of a self-powered test light, known as a continuity tester, to a known good ground. Connect the other lead to the wire or ground connection being tested. If the bulb goes on, the ground is good. If the bulb does not go on, the ground is not good.

Continuity check

A continuity check is done to determine if there are any breaks in a circuit - if it is passing electricity properly. With the circuit off (no power in the circuit), a self-powered continuity tester can be used to check the circuit. Connect the test leads to both ends of the circuit (or to the "power" end and a good ground), and if the test light comes on the circuit is passing current properly. If the light doesn't come on, there is a break somewhere in the circuit. The same procedure can be used to test a switch, by connecting the continuity tester to the switch terminals. With the switch turned On, the test light should come on.

Finding an open circuit

When diagnosing for possible open circuits, it is often difficult to locate them by sight because oxidation or terminal misalignment are hidden by the electrical connectors. Merely wiggling an electrical connector on a sensor or in the wiring harness may correct the open circuit condition. Remember this when an open circuit is indicated when troubleshooting a circuit. Intermittent problems may also be caused by oxidized or loose connections.

Electrical troubleshooting is simple if you keep in mind that all electrical circuits are basically electricity running from the battery, through the wires, switches, relays, fuses and fusible links to each electrical component (light bulb, motor, etc.) and to ground, from which it is passed back to the battery. Any electrical problem is an interruption in the flow of electricity to and from the battery.

3 Fuses - general information

Refer to illustrations 3.1 and 3.3
Caution: *On models equipped with a Delco Loc II anti-theft audio system, be sure the lockout feature is turned off before performing any procedure which requires disconnecting the battery.*

1 The electrical circuits of the vehicle are protected by a combination of fuses, circuit breakers and fusible links. The fuse block is located under a cover on the left end of the instrument panel **(see illustration)**.
2 Each of the fuses is designed to protect a specific circuit, and the various circuits are identified on the fuse panel itself.
3 Miniaturized fuses are employed in the fuse block. These compact fuses, with blade terminal design, allow fingertip removal and replacement. If an electrical component fails, always check the fuse first. The best way to check the fuses is with a test light. Check for power at the exposed terminal tips of each fuse. If power is present on one side of the fuse but not the other, the fuse is blown. A blown fuse can also be confirmed by visually inspecting it **(see illustration)**.
4 Be sure to replace blown fuses with the correct type. Fuses of different ratings are physically interchangeable, but only fuses of the proper rating should be used. Replacing a fuse with one of a higher or lower value than specified is not recommended. Each electrical circuit needs a specific amount of protection. The amperage value of each fuse is molded into the fuse body.
5 If the replacement fuse fails immediately, don't replace it again until the cause of the problem is isolated and corrected. In most cases, the cause will be a short circuit in the wiring caused by a broken or deteriorated wire.

3.1 The passenger compartment fuse block, located a the left end of the instrument panel, is accessible after removing the cover

Chapter 12 Chassis electrical system

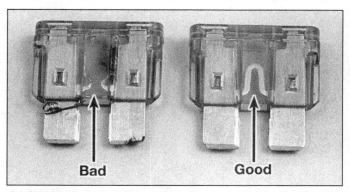

3.3 When a fuse blows, the element between the terminals burns - the fuse on the left is blown the fuse on the right is good

6.2 These relays are for the air conditioning system

4 Fusible links - general information

Caution: *On models equipped with a Delco Loc II anti-theft audio system, be sure the lockout feature is turned off before performing any procedure which requires disconnecting the battery.*

Some circuits are protected by fusible links. Fusible links are used in circuits which are not ordinarily fused, such as the ignition circuit.

Although the fusible links appear to be a heavier gauge than the wire they are protecting, the appearance is due to the thick insulation. Fusible links are several wire gauges smaller than the wire they are designed to protect.

Burned fusible links must be replaced with the same size link. The procedure is as follows:

a) Disconnect the negative cable from the battery.
b) Disconnect the fusible link from the wiring harness.
c) Cut the damaged fusible link out of the wiring just behind the electrical connector.
d) Strip the insulation back approximately 1/2-inch.
e) Position the electrical connector on the new fusible link and crimp it into place.
f) Use rosin core solder at each end of the new link to obtain a good solder joint.
g) Use plenty of electrical tape around the soldered joint. No wires should be exposed.
h) Connect the battery ground cable. Test the circuit for proper operation.

5 Circuit breakers - general information

Circuit breakers protect components such as power windows, power door locks and headlights.

On some models the circuit breaker resets itself automatically, so an electrical overload in a circuit breaker protected system will cause the circuit to fail momentarily, then come back on. If the circuit doesn't come back on, check it immediately. Once the condition is corrected, the circuit breaker will resume its normal function. Some circuit breakers must be reset manually.

For a basic check, pull the circuit breaker up out of its socket on the fuse or relay panel, but just far enough to probe with a voltmeter. The breaker should still contact the sockets.

With the voltmeter negative lead on a good chassis ground, touch each prong of the circuit breaker with the positive meter probe. There should be battery voltage at each end. If there is battery voltage at only one end, the circuit breaker must be replaced.

6 Relays - general information

Refer to illustration 6.2

Several electrical accessories in the vehicle use relays to transmit the electrical signal to the component. If the relay is defective, that component will not operate properly.

The various relays are grouped together in several locations. Some relays are grouped together in the engine compartment **(see illustration)**.

If a faulty relay is suspected, it can be removed and tested by a dealer service department or a repair shop. Defective relays must be replaced as a unit.

7 Turn signal/hazard flasher - check and replacement

Refer to illustration 7.1

Warning: *1995 and later models are equipped with an airbag. Always turn the steering wheel to the straight ahead position, place the ignition switch in Lock and remove the key, then remove the airbag fuse and unplug the yellow Connector Position Assurance (CPA) connector at the base of the steering column before working in the vicinity of the impact sensors, steering column or instrument panel to avoid the possibility of accidental deployment of the airbag, which could cause personal injury (see Section 28). The yellow wires and connectors routed through the instrument panel are for this system. Do not use electrical test equipment on these yellow wires or tamper with them in any way while working under the instrument panel.*

Caution: *On models equipped with a Delco Loc II anti-theft audio system, be sure the lockout feature is turned off before performing any procedure which requires disconnecting the battery.*

Turn signal flasher

1 The turn signal flasher, a small canister-shaped unit located under the dash behind the knee bolster, near the cigarette lighter **(see illustration)**. On the 1999 and later models, the turn signal/hazard flasher is located behind the glove compartment. To access it, remove the glove box (see Chapter 11). The flasher is a black box, next to the large wiring disconnect, and is retained by a plastic strap.

2 When the flasher unit is functioning properly, an audible click can be heard during its operation. If the turn signals fail on one side or the other and the flasher unit does not make its characteristic clicking sound, a faulty turn signal bulb is indicated.

3 If both turn signals fail to blink, the problem may be due to a blown fuse, a faulty flasher unit, a broken switch or a loose or open connection. If a quick check of the fuse box indicates that the turn signal fuse has blown, check the wiring for a short before

7.1 Location of the turn signal and hazard flasher units (earlier models)

8.3 Use a special tool (available at most auto parts stores) to compress the lock plate for access to the retaining ring

8.4 Remove the hazard warning knob screw (arrow) and the knob

8.5 Remove the cancel cam assembly

installing a new fuse.
4   To replace the flasher, remove the under-dash panel or knee bolster (see Chapter 11) and unplug the flasher.
5   Make sure that the replacement unit is identical to the original. Compare the old one to the new one before installing it.
6   Installation is the reverse of removal.

Hazard flasher

7   The hazard flasher, a small canister-shaped unit, located in the instrument panel behind the knee bolster, flashes all four turn signals simultaneously when activated. On the 1999 and later models, the turn signal/hazard flasher is located behind the glove compartment. To access it, remove the glove box (see Chapter 11). The flasher is a black box, next to the large wiring disconnect, and is retained by a plastic strap.
8   The hazard flasher is checked in a fashion similar to the turn signal flasher (see Steps 2 and 3).
9   To replace the hazard flasher, remove the under-dash panel or knee bolster for access (see Chapter 11), then unplug the flasher **(see illustration 7.1)**.
10 Make sure the replacement unit is identical to the one it replaces. Compare the old one to the new one before installing it.
11 Installation is the reverse of removal.

8   Turn signal switch assembly - removal and installation

Warning: *1995 and later models are equipped with an airbag. Always turn the steering wheel to the straight ahead position, place the ignition switch in Lock and remove the key, then remove the airbag fuse and unplug the yellow Connector Position Assurance (CPA) connector at the base of the steering column before working in the vicinity of the impact sensors, steering column or instrument panel to avoid the possibility of accidental deployment of the airbag, which could cause personal injury (see Section 28). The yellow wires and connectors routed through the instrument panel are for*

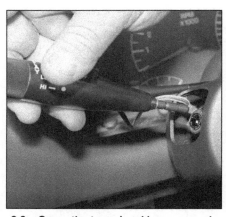

8.6a Grasp the turn signal lever securely and pull it straight out to detach it

this system. Do not use electrical test equipment on these yellow wires or tamper with them in any way while working under the instrument panel.
Caution: *On models equipped with a Delco Loc II anti-theft audio system, be sure the lockout feature is turned off before performing any procedure which requires disconnecting the battery.*

1994 models

Removal

Refer to illustrations 8.3, 8.4, 8.5, 8.6a and 8.6b
1   Detach the cable from the negative battery terminal.
2   Remove the steering wheel (see Chapter 10).
3   Use a lock plate removal tool to depress the lock plate for access to the retaining ring **(see illustration)**. Use a small screwdriver or pick to pry the retaining ring out of the groove in the steering column and remove the lock plate.
4   Use a Phillips head screwdriver to remove the hazard warning knob **(see illustration)**.
5   Remove the turn signal cancel cam **(see illustration)**.
6   Pull the multi-function lever straight out to detach the lever, then remove the turn signal switch mounting screws **(see illustrations)**.

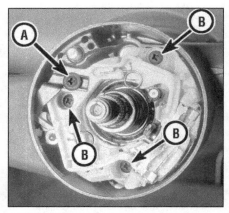

8.6b Remove the turn signal lever arm screw (A), followed by the switch screws (B)

7   Remove the knee bolster panel below the steering column.
8   Locate the turn signal switch electrical connector and unplug it. Detach the wires to the switch assembly from the back of the electrical connector.
9   Loosen or, if necessary, fully unbolt the steering column support bracket.
10 Pull the wiring harness and electrical connector up through the steering column and remove the switch assembly.

Installation

11 Feed the turn signal switch connector and wiring harness down through the column. Use a section of mechanics wire to pull it through, if necessary. Plug in the connector and replace the wiring protector.
12 Install the steering column support bracket and tighten the fasteners to the torque listed in this Chapter's Specifications.
13 Seat the turn signal switch on the column and install the turn signal switch mounting screws and lever arm. Install the hazard knob and multi-function lever. Press the multi-function lever straight in until it snaps in place.
14 Install the cancel cam and the lock plate. Depress the lock plate and install the retaining ring. The remainder of installation is the reverse of removal.

Chapter 12 Chassis electrical system

8.17 Remove the turn signal retaining screw (arrow)

1995 through 1998 models
Removal
Refer to illustrations 8.17 and 8.19

15   Disconnect the negative battery cable. Remove the steering wheel (see Chapter 10).
16   Remove the steering column cover (see Chapter 11).
17   Remove the Torx-head bolt, unplug the electrical connector and lift the switch off **(see illustration)**.

Installation
18   Installation is the reverse of removal. If necessary, center the airbag coil as follows (it will only become off-centered if the spring lock is depressed and the hub is rotated with the coil off the column).
  a) Turn the coil over and depress the spring lock.
  b) Rotate the hub clockwise until it stops.
  c) Rotate the hub in the opposite direction 2-1/2 turns and release the spring lock.

19   Install the wave washer and the airbag coil **(see illustration)**. Pull the slack out of the airbag coil lower wiring harness to keep it tight, or it may be cut when the steering wheel is turned.

1999 and later models
20   The turn signal switch function is incorporated into the multi-function switch, along

9.5 Lift off the buzzer switch and clip with needle-nose pliers

8.19 When properly installed, the airbag coil will be centered with the marks aligned (circle) and the tab fitted between the projections on the top of the steering column (arrow)

with the wiper/washer and headlight dimming controls.
21   Disconnect the negative battery cable. Remove the steering column covers and the driver's knee bolster (see Chapter 11).
22   Remove the two Torx screws securing the multi-function switch to the top of the steering column.
23   Clip the wire-ties securing the multi-function switch harness to the other steering column wires.
24   At the bottom of the steering column, remove the bolt securing the large bulkhead electrical connector to the firewall.
25   Separate the two multi-function switch connectors from the large bulkhead connector.
26   The multi-function switch and its harness must be replaced as a unit
27   Installation is the reverse of removal

9   Ignition key lock cylinder - replacement

Warning: *1995 and later models are equipped with an airbag. Always turn the steering wheel to the straight ahead position, place the ignition switch in Lock and remove the key, then*

9.7 The lock cylinder is held in place by a Torx-head screw (arrow)

remove the airbag fuse and unplug the yellow Connector Position Assurance (CPA) connector at the base of the steering column before working in the vicinity of the impact sensors, steering column or instrument panel to avoid the possibility of accidental deployment of the airbag, which could cause personal injury (see Section 28). The yellow wires and connectors routed through the instrument panel are for this system. Do not use electrical test equipment on these yellow wires or tamper with them in any way while working under the instrument panel.
Caution: *On models equipped with a Delco Loc II anti-theft audio system, be sure the lockout feature is turned off before performing any procedure which requires disconnecting the battery.*

1   Detach the cable from the negative battery terminal.

1994 models
Refer to illustrations 9.5 and 9.7
2   Remove the steering wheel (see Chapter 10).
3   Remove the turn signal switch assembly (see Section 8).
4   Make sure the key is removed from the lock cylinder.
5   Using needle-nose pliers, remove the buzzer switch **(see illustration)**.
6   Insert the key and place the lock cylinder in the Lock position.
7   Remove the lock cylinder retaining screw **(see illustration)**.
8   Remove the lock cylinder.
9   Installation is the reverse of removal.

1995 and later models
Refer to illustration 9.12
10   Remove the steering wheel (see Chapter 10).
11   Remove the steering column covers (see Chapter 11).
12   With the ignition switch in the Start position, insert a small Allen wrench into the hole in the top of the casting, press the release pin and rotate the key to the Run position and pull the lock cylinder out **(see illustration)**.
13   Installation is the reverse of removal.

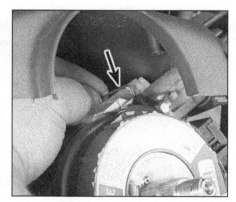

9.12 Use a small Allen wrench (arrow) to press the lock cylinder release pin

12-6　Chapter 12　Chassis electrical system

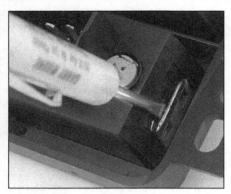

10.4 Release the clips by inserting a small screwdriver and lift the switch out of the housing

12.3 On sealed beam headlights, the bezel is held in place by screws (arrows) on each side of the housing

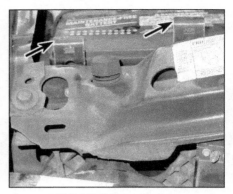

12.9a Pull up on the two headlight retainers . . .

12.9b . . . and lower the headlight assembly for access to the bulbs

10 Headlight switch - removal and installation

Refer to illustration 10.4

Warning: *1995 and later models are equipped with an airbag. Always turn the steering wheel to the straight ahead position, place the ignition switch in Lock and remove the key, then remove the airbag fuse and unplug the yellow Connector Position Assurance (CPA) connector at the base of the steering column before working in the vicinity of the impact sensors, steering column or instrument panel to avoid the possibility of accidental deployment of the airbag, which could cause personal injury (see Section 28). The yellow wires and connectors routed through the instrument panel are for this system. Do not use electrical test equipment on these yellow wires or tamper with them in any way while working under the instrument panel.*
Caution: *On models equipped with a Delco Loc II anti-theft audio system, be sure the lockout feature is turned off before performing any procedure which requires disconnecting the battery.*

1　Detach the cable from the negative battery terminal.
2　Remove the trim panel or instrument cluster bezel and detach the electrical connector (see Chapter 11).
3　Remove the screws and lift the switch housing out of the trim panel or cluster bezel.
4　Use a small screwdriver to release the clips and withdraw the switch from the opening **(see illustration)**. On some models screws are used to attach the switch. Remove the screws.
5　Place the switch in position, then press in until the clips engage. If screws are used, install and tighten them.
6　Install the trim panel or instrument cluster bezel.

11 Rear wiper switch - removal and installation

Warning: *1995 and later models are equipped with an airbag. Always turn the steering wheel to the straight ahead position, place the ignition switch in Lock and remove the key, then remove the airbag fuse and unplug the yellow Connector Position Assurance (CPA) connector at the base of the steering column before working in the vicinity of the impact sensors, steering column or instrument panel to avoid the possibility of accidental deployment of the airbag, which could cause personal injury (see Section 28). The yellow wires and connectors routed through the instrument panel are for this system. Do not use electrical test equipment on these yellow wires or tamper with them in any way while working under the instrument panel.*
Caution: *On models equipped with a Delco Loc II anti-theft audio system, be sure the lockout feature is turned off before performing any procedure which requires disconnecting the battery.*

1　Detach the cable from the negative battery terminal.
2　Remove the trim panel or instrument cluster bezel and detach the electrical connector (see Chapter 11).
3　Remove the screws and lift the switch housing out of the trim panel or cluster bezel.
4　Use a small screwdriver to release the clips and withdraw the switch from the opening **(see illustration 10.4)**.
5　Place the switch in position, then press in until the clips engage.
6　Install the trim panel or instrument cluster bezel.

12 Headlight bulb - replacement

Refer to illustrations 12.3, 12.9a, 12.9b and 12.10

1　Disconnect the negative cable from the battery. **Caution:** *On models equipped with a Delco Loc II anti-theft audio system, be sure the lockout feature is turned off before performing any procedure which requires disconnecting the battery.*

Sealed beam-type

2　Remove the radiator grille (see Chapter 11).
3　Remove the retainer screws, taking care not to disturb the adjusting screws **(see illustration)**.
4　Remove the retainer and pull the headlight out enough to allow the connector to be unplugged.
5　Remove the headlight.
6　To install the headlight, plug the connector in, place the headlight in position and install the retainer and screws. Tighten the screws securely.
7　Install the radiator grille.

Composite headlight

8　Open the hood. **Warning:** *Halogen bulbs on composite headlights are gas-filled and under pressure, and may shatter if the surface is scratched or the bulb is dropped. Wear eye protection and handle the bulbs carefully, grasping only the base whenever possible. Don't touch the surface of the bulb with your fingers because oil from your skin could cause it to overheat and fail prematurely. If you do touch the bulb surface, clean it with rubbing alcohol.* **Note 1:** *On 1999 and later models, pull up on the two black headlight housing retainers, then pull the headlight housing out to the front to access the bulb in back.* **Note 2:** *Some 1999 and later Envoy and Diamond models are equipped with optional HID (Hid Intensity Discharge) headlights. It is suggested that headlight replacement on these models be performed at your local dealership.*
9　Remove the two headlight retainers **(see illustration)** by pulling up and out toward the engine compartment. This will allow the headlight assembly to drop forward and down, exposing the wiring and bulb sockets **(see illustration)**.

Chapter 12 Chassis electrical system

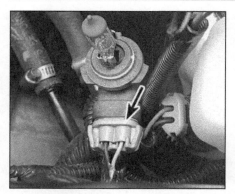

12.10 Withdraw the bulb holder and unplug the electrical connector (arrow)

10   Reach behind the headlight assembly, grasp the bulb holder, turn it counterclockwise, then pull out the bulb assembly and unplug it from the electrical connector **(see illustration)**.
11   Plug the electrical connector into the new bulb assembly, insert the assembly into its receptacle and rotate in a clockwise direction to lock it into place.

13   Headlights - adjustment

Refer to illustrations 13.1a, 13.1b, 13.1c and 13.3
Note: *The headlights must be aimed correctly. If adjusted incorrectly they could blind the driver of an oncoming vehicle and cause a serious accident or seriously reduce your ability to see the road. The headlights should be checked for proper aim every 12 months and any time a new headlight is installed or front end body work is performed. It should be emphasized that the following procedure*

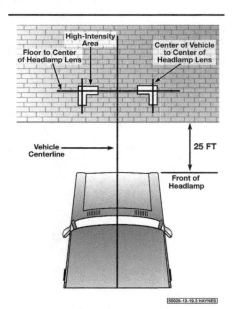

13.3 Headlight aiming details

13.1a The sealed beam headlight vertical adjustment screw is located at the top of the headlight and the horizontal screw is on the side of the headlight (arrows) - a Torx-head tool will be required for making headlight adjustments

is only an interim step which will provide temporary adjustment until the headlights can be adjusted by a properly equipped shop.
1   Sealed beam headlights have two spring loaded adjusting screws, one on the top controlling up-and-down movement and one on the side controlling left-and-right movement **(see illustration)**. Composite headlights have four Torx-head screw adjusters to control vertical and horizontal movement **(see illustration)**.
2   There are several methods of adjusting the headlights. The simplest method requires a blank wall 25-feet in front of the vehicle and a level floor.
3   Position masking tape vertically on the wall in reference to the vehicle centerline and the centerlines of both headlights **(see illustration)**.
4   Position a horizontal tape line in reference to the centerline of all the headlights.
Note: *It may be easier to position the tape on the wall with the vehicle parked only a few inches away.*
5   Adjustment should be made with the vehicle sitting level, the gas tank half-full and no unusually heavy load in the vehicle.
6   Starting with the low beam adjustment, position the high intensity zone so it's two inches below the horizontal line and two inches to the right of the headlight vertical line. Adjustment is made by turning the top adjusting screw clockwise to raise the beam and counterclockwise to lower the beam. The adjusting screw on the side should be used in the same manner to move the beam left or right.
7   With the high beams on, the high intensity zone should be vertically centered with the exact center just below the horizontal line. **Note:** *It may not be possible to position the headlight aim exactly for both high and low beams. If a compromise must be made, keep in mind that the low beams are the most used and have the greatest effect on driver safety.*
8   Have the headlights adjusted by a

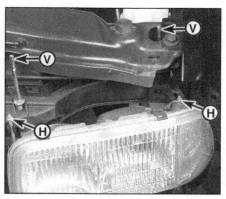

13.1b 1997 and earlier composite headlights have two vertical (V) and two horizontal (H) adjusters

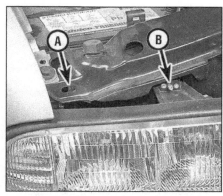

13.1c 1998 and later composite headlights have one vertical (A) and one horizontal (B) adjuster

dealer service department or service station at the earliest opportunity.

14   Composite headlight housing - removal and installation

Refer to illustrations 14.4a, 14.4b and 14.4c
Warning: *1995 and later models have an airbag. Always turn the steering wheel to the straight ahead position, place the ignition switch in Lock and remove the key, then remove the airbag fuse and unplug the yellow Connector Position Assurance (CPA) connector at the base of the steering column before working in the vicinity of the impact sensors, steering column or instrument panel to avoid the possibility of accidental deployment of the airbag, which could cause personal injury (see Section 28). The yellow wires and connectors routed through the instrument panel are for this system. Do not use electrical test equipment on these yellow wires or tamper with them in any way while working under the instrument panel.* **Caution:** *On models equipped with a Delco Loc II anti-theft audio system, be sure the lockout feature is turned off before performing any procedure which requires disconnecting the battery.*

12-8 Chapter 12 Chassis electrical system

14.4a Remove the bolts (arrow) at each side of the housing . . .

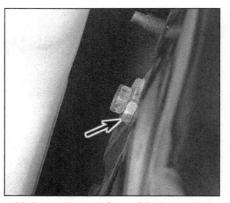

14.4b . . . the bolt (arrow) in the engine compartment, then . . .

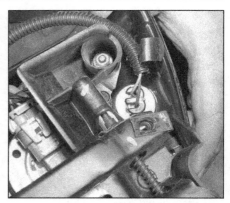

14.4c . . . rotate the housing forward and unplug the connectors

1 Open the hood.
2 Remove the radiator grille (see Chapter 11).
3 If you're removing a right-side headlight, remove the battery (see Chapter 5). To remove the left-side headlight, remove the air cleaner assembly (see Chapter 4). **Note:** *On 1999 and later models, pull up on the two black headlight housing retainers, then pull the headlight housing out to the front to access the bulb in back.*
4 Remove the retaining bolts and pull housing out and unplug the electrical connectors **(see illustrations)**.
5 Installation is the reverse of removal. After you're done, check the headlight adjustment (see Section 13).

15 Bulb replacement

Note: *On some 1999 and later models with the park-turn lights and fog lights in the grille, the grille and front bumper may have to be removed to access the bulbs.*
Caution: *On models equipped with a Delco Loc II anti-theft audio system, be sure the lockout feature is turned off before performing any procedure which requires disconnecting the battery.*

15.4 Remove the Torx-head screws and detach the lens for access to the side marker light bulb

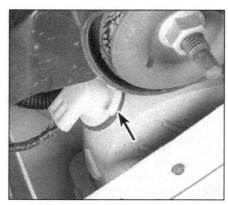

15.1 The parking light bulb holder (arrow) is accessible from behind the bumper

Front
Parking
Refer to illustrations 15.1 and 15.2
1 Reach up under the front bumper and remove the bulb holder by turning it counterclockwise **(see illustration)**.
2 Depress the clips and remove the bulb **(see illustration)**.
3 Installation is the reverse of removal.

Side marker
Refer to illustrations 15.4 and 15.7
4 On sealed beam headlights, remove the

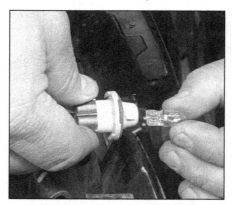

15.7 Rotate the side marker counterclockwise to remove it - the bulb pulls straight out

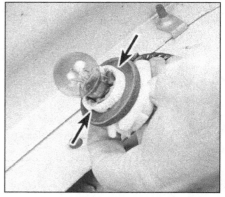

15.2 Rotate the holder counterclockwise to remove it, then squeeze the clips (arrows) and remove the bulb

screws, detach the lens and remove the bulb **(see illustration)**.
5 Installation is the reverse of removal.
6 On composite headlight models, remove the headlight housing (see Section 14).
7 Rotate the holder counterclockwise and remove it, then pull the bulb straight out **(see illustration)**.
8 Installation is the reverse of removal.

Interior
Dome light
9 Squeeze the lens and lift it off (pickup) or

15.12 Detach the lens for access to the cargo light bulb

Chapter 12 Chassis electrical system

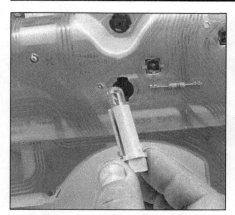

15.16 Rotate the holder counterclockwise and detach it from the cluster

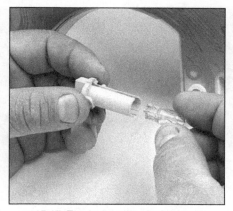

15.17 The bulb pulls straight out of the holder

15.23 Remove the screws and rotate the tail light out for access

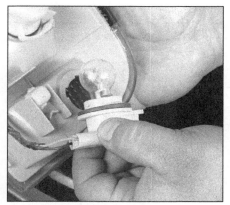

15.24 Rotate the holders from the housing - squeeze the plastic tabs to remove the bulbs

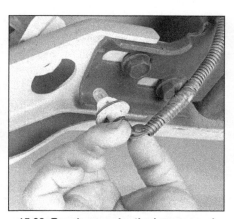

15.26 Reach up under the bumper and detach the license plate light bulb holder - remove the bulb by pulling straight out

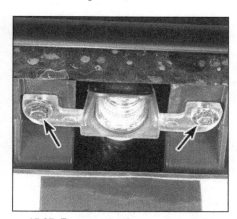

15.27 Remove the license plate light housing bolts (arrows)

pry it off (sport utility).
10 Detach the bulb by using a small screwdriver to pry back one of the contacts.
11 Installation is the reverse of removal.

Cargo lamp

Refer to illustration 15.12

12 Use a small screwdriver to detach the lens, then lower it from the housing **(see illustration)**.
13 Detach the bulb by using a small screwdriver to pry back one of the contacts. Don't pry on the glass.
14 Installation is the reverse of removal.

Instrument panel light

Refer to illustrations 15.16 and 15.17

15 To gain access to the instrument panel lights, the instrument cluster will have to be removed first (see Section 21).
16 Rotate the bulb holder counterclockwise and pull it out of the cluster **(see illustration)**.
17 To remove the bulb from the holder, simply pull it straight out **(see illustration)**.
18 Insert a new bulb into the holder and install the assembly into the instrument cluster.
19 Install the instrument cluster (see Section 21).

Rear

Center high-mounted brake light

Note: *The center high-mounted brake light on four-door models uses special bulbs that must be replaced by a dealer.*

20 Remove the screws and detach the light housing.
21 Twist the bulb holder counterclockwise to remove it, then pull the bulb straight out of the holder.
22 Installation is the reverse of removal.

Tail/back-up lights

Refer to illustrations 15.23 and 15.24

23 Open the tailgate or liftgate and remove the screws, then rotate the housing out for access to the bulb holders **(see illustration)**.
24 Turn the bulb holder counterclockwise and withdraw it from the housing, then remove the bulb by pulling it straight out **(see illustration)**.
25 Installation is the reverse of removal.

License plate light

Refer to illustrations 15.26, 15.27 and 15.28

Pick-up

26 Reach up behind the rear bumper, rotate the bulb holder counterclockwise, then

15.28 Detach the housing, then pull the bulb out

withdraw the holder from the housing and pull the bulb out **(see illustration)**. On some models, the license light housing is held to the bumper with clips, and, on models with step-bumpers the housing is removed by pulling the housing upwards while pulling out at the bottom.

Blazer, Jimmy, Bravada, Hombre

27 Remove the bolts and detach the bulb lens assembly **(see illustration)**.
28 Pull the bulb straight out **(see illustration)**.
29 Installation is the reverse of removal.

12-10   Chapter 12   Chassis electrical system

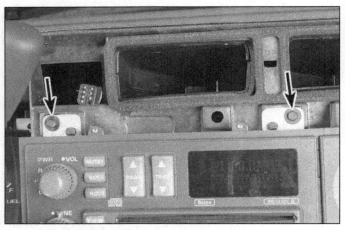

16.3a The radio is secured by two screws at the top . . .

16.3b . . . and two at the bottom of the center compartment (arrows) (the center compartment can be detached from the radio after removal)

16 Radio and speakers - general information, removal and installation

Warning: *1995 and later models are equipped with an airbag. Always turn the steering wheel to the straight ahead position, place the ignition switch in Lock and remove the key, then remove the airbag fuse and unplug the yellow Connector Position Assurance (CPA) connector at the base of the steering column before working in the vicinity of the impact sensors, steering column or instrument panel to avoid the possibility of accidental deployment of the airbag, which could cause personal injury (see Section 28). The yellow wires and connectors routed through the instrument panel are for this system. Do not use electrical test equipment on these yellow wires or tamper with them in any way while working under the instrument panel.*
Caution: *On models equipped with a Delco Loc II anti-theft audio system, be sure the lockout feature is turned off before performing any procedure which requires disconnecting the battery.*

Radio

Refer to illustrations 16.3a and 16.3b
1 Detach the cable from the negative terminal of the battery.
2 Remove the instrument cluster bezel (Chapter 11).
3 Remove the screws **(see illustrations)** and pull the radio, then disconnect the electrical connection and antenna lead.
4 Remove the radio from the dash.
5 Installation is the reverse of removal.

Speakers

Refer to illustration 16.7
6 On door mounted speakers, remove the door trim panel (Chapter 11).
7 Drill out the rivets, detach the speaker and unplug the electrical connector **(see illustration)**. A small hand-operated rivet gun will be necessary when installing the new speaker.
8 Installation is the reverse of removal.
9 On front speakers, remove the speaker grille, then remove the screws and detach the speaker. On rear speakers, remove the trim panel for access. Remove the nuts and remove the speaker.

17 Antenna - removal and installation

Refer to illustrations 17.1, 17.2 and 17.3
1 Use a small open-end wrench to unscrew the antenna mast **(see illustration)**.
2 Unplug the antenna cable **(see illustration)**.
3 Remove the cap nut and lift off the upper adapter and gasket, then remove the screws detach the antenna assembly from the fender **(see illustration)**.
4 Installation is the reverse of removal.

18 Wiper motor - removal and installation

1 Detach the cable from the negative terminal of the battery. **Caution:** *On models equipped with a Delco Loc II anti-theft audio system, be sure the lockout feature is turned off before performing any procedure which requires disconnecting the battery.*

16.7 Drill out the rivets (arrows), detach the speaker and unplug the electrical connector

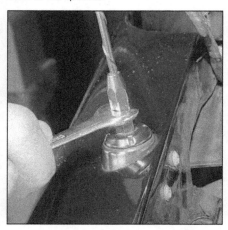

17.1 Use a small wrench to unscrew the antenna mast

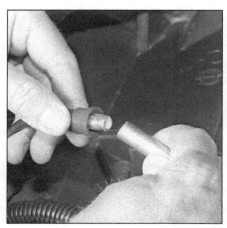

17.2 Unplug the antenna connector

Chapter 12 Chassis electrical system

12-11

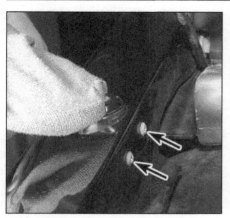

17.3 Using a rag to protect the surface, unscrew the nut with pliers, then remove the screws (arrows)

18.3a Remove the nut (arrow) and detach the wiper linkage

18.3b Remove the bolts and unplug the electrical connector (arrows), then lift the windshield wiper motor assembly out of the engine compartment

18.5 Pry out the cover, then remove the nut, wiper arm, nuts and washers from the motor pivot

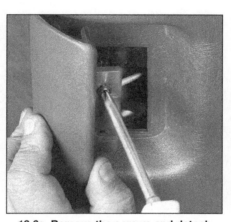

18.6a Remove the screws and detach the handle

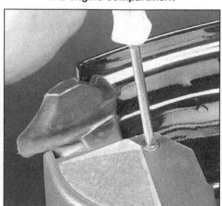

18.6b Remove the screws from the ends and . . .

Windshield wiper

Refer to illustrations 18.3a and 18.3b

2   Remove the cowl vent grille (Chapter 11).

3   Remove the wiper linkage stud nut, detach the wiper link and unplug the electrical connector, then remove the motor assembly-to-firewall bolts, detach the motor and lift it from the engine compartment (see illustrations).

4   Installation is the reverse of removal.

Rear wiper

Refer to illustrations 18.5, 18.6a, 18.6b, 18.6c, 18.6d and 18.7

5   Detach the wiper arm and remove the nut, bezel and gasket (see illustration).

6   Open the tailgate and remove the trim panel and cover (see illustrations).

7   Remove the retaining bolts, unplug the electrical connector, then lift the wiper motor from the vehicle.

8   Installation is the reverse of removal.

18.6c . . . top of the trim panel

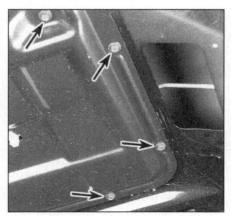

18.6d Remove the bolts (arrows) and detach the tailgate cover

18.7 Remove the bolts and unplug the electrical connectors (arrows), then lift the motor out

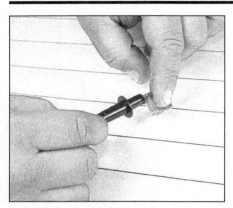

20.4 When measuring the voltage at the rear window defogger grid, wrap a piece of aluminum foil around the positive probe of the voltmeter and press the foil against the wire with your finger

20.5 To determine if a heating element has broken, check the voltage at the center of each element - if the voltage is 6-volts, the element is unbroken - if the voltage is 12-volts, the element is broken between the center and the ground side - if there is no voltage, the element is broken between the center and the positive side

20.7 To find the break, place the voltmeter negative lead against the defogger ground terminal, place the voltmeter positive lead with the foil strip against the heating element at the positive terminal end and slide it toward the negative terminal end - the point at which the voltmeter reading changes abruptly is the point at which the element is broken

19 Rear window defogger switch - removal and installation

Warning: *1995 and later models are equipped with an airbag. Always turn the steering wheel to the straight ahead position, place the ignition switch in Lock and remove the key, then remove the airbag fuse and unplug the yellow Connector Position Assurance (CPA) connector at the base of the steering column before working in the vicinity of the impact sensors, steering column or instrument panel to avoid the possibility of accidental deployment of the airbag, which could cause personal injury (see Section 28). The yellow wires and connectors routed through the instrument panel are for this system. Do not use electrical test equipment on these yellow wires or tamper with them in any way while working under the instrument panel.*
Caution: *On models equipped with a Delco Loc II anti-theft audio system, be sure the lockout feature is turned off before performing any procedure which requires disconnecting the battery.*

1997 and earlier models

1 Detach the cable from the negative battery terminal.
2 Remove the trim panel or instrument cluster bezel and detach the electrical connector (see Chapter 11).
3 Remove the screws and lift the switch housing out of the trim panel or cluster bezel.
4 Use a small screwdriver to release the clips and withdraw the switch from the opening **(see illustration 10.4)**.
5 Place the switch in position, then press in until the clips engage.
6 Install the trim panel or instrument cluster bezel.

1998 and later models

7 On 1998 and later models, the defogger switch is incorporated into the HVAC control assembly. To replace the control assembly, remove the instrument cluster trim (see chapter 11).
8 Press the tabs securing the control assembly and pull the assembly out of the carrier. Disconnect the electrical and vacuum lines from the control assembly.
9 Installation is the reverse of removal.

20 Rear window defogger - check and repair

1 The rear window defogger consists of a number of horizontal elements baked onto the glass surface.
2 Small breaks in the element can be repaired without removing the rear window.

Check

Refer to illustrations 20.4, 20.5 and 20.7
3 Turn the ignition switch and defogger system switches to the ON position. Using a voltmeter, place the positive probe against the defogger grid positive terminal and the negative lead against the ground terminal. If the battery voltage is not indicated, check the fuse, defogger switch and related wiring.
4 When measuring voltage during the next two tests, wrap a piece of aluminum foil around the tip of the voltmeter positive probe and press the foil against the heating element with your finger **(see illustration)**.
5 Check the voltage at the center of each heating element **(see illustration)**. If the voltage is 6-volts, the element is okay (there is no break). If the voltage is 12-volts, the element is broken between the center of the element and the ground side. If the voltage is 0-volts the element is broken between the center of the element and positive side.
6 If none of the elements are broken, connect the negative lead to a good body ground. The reading should stay the same, if it doesn't the ground connection is bad.
7 To find the break, place the voltmeter negative lead against the defogger ground terminal. Place the voltmeter positive lead with the foil strip against the heating element at the positive terminal end and slide it toward the negative terminal end. The point at which the voltmeter deflects from several volts to zero is the point at which the heating element is broken **(see illustration)**.

Repair

Refer to illustration 20.13
8 Repair the break in the element using a repair kit specifically recommended for this purpose, such as Dupont paste No. 4817 (or equivalent). Included in this kit is plastic conductive epoxy.
9 Prior to repairing a break, turn off the system and allow it to cool off for a few minutes.
10 Lightly buff the element area with fine steel wool, then clean it thoroughly with rubbing alcohol.
11 Use masking tape to mask off the area being repaired.
12 Thoroughly mix the epoxy, following the

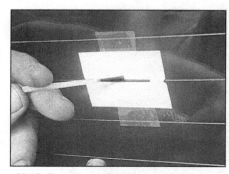

20.13 To use a defogger repair kit, apply masking tape to the inside of the window at the damaged area, then brush on the special conductive coating

21.3a Remove the bolts - there is one at each corner of the cluster

21.3b Detach the clips and rotate the cluster out

instructions provided with the repair kit.
13 Apply the epoxy material to the slit in the masking tape, overlapping the undamaged area about 3/4-inch on either end **(see illustration)**.
14 Allow the repair to cure for 24 hours before removing the tape and using the system.

21 Instrument cluster - removal and installation

Refer to illustrations 21.3a and 21.3b
Warning: *1995 and later models are equipped with an airbag. Always turn the steering wheel to the straight ahead position, place the ignition switch in Lock and remove the key, then remove the airbag fuse and unplug the yellow Connector Position Assurance (CPA) connector at the base of the steering column before working in the vicinity of the impact sensors, steering column or instrument panel to avoid the possibility of accidental deployment of the airbag, which could cause personal injury (see Section 28). The yellow wires and connectors routed through the instrument panel are for this system. Do not use electrical test equipment on these yellow wires or tamper with them in any way while working under the instrument panel.*

1 Detach the cable from the negative battery terminal. **Caution:** *On models equipped with a Delco Loc II anti-theft audio system, be sure the lockout feature is turned off before performing any procedure which requires disconnecting the battery.*
2 Remove instrument cluster bezel (Chapter 11).
3 Remove the four bolts, detach the instrument cluster from the electrical wiring harness connector and lift it from the dash **(see illustrations)**.
4 Installation is the reverse of removal.

22 Horn - check and replacement

Caution: *On models equipped with a Delco Loc II anti-theft audio system, be sure the*

lockout feature is turned off before performing any procedure which requires disconnecting the battery.
1 The horn assembly is located inside the right fender, next to the battery. On 1999 and later models, there are two horns, one in each fender. To access the left horn, remove the windshield tank first.
2 Unplug the electrical connector from the horn.
3 To test the horn, connect battery voltage to the two terminals with a pair of jumper wires. If the horn doesn't sound, check the current draw with an ammeter - it should be around five amps at battery voltage.
4 If the amperage is over 20 amps, replace the horn.
5 If the reading is around 18 amps, the contact points are not open. Turn the adjustment screw counterclockwise a quarter turn at a time to open the points and lower the amperage until the proper level is reached.
6 If the horn doesn't sound, it could mean the problem lies in the switch, relay, the wiring between the components or open points.
7 Unplug the electrical connector, remove the bracket bolt and lift the horn from the engine compartment.
8 Installation is the reverse of removal.

23 Power mirror control system - description and check

1 Electric rear view mirrors use two motors to move the glass; one for up-and-down adjustments and one for left-to-right adjustments.
2 The control switch has a selector portion which sends voltage to the left or right side mirror. With the ignition ON but the engine OFF, roll down the windows and operate the mirror control switch through all functions (left-right and up-down) for both the left and right side mirrors.
3 Listen carefully for the sound of the electric motors running in the mirrors.
4 If the motors can be heard but the mirror glass doesn't move, there's probably a problem with the drive mechanism inside the mirror. Remove and disassemble the mirror to locate the problem.
5 If the mirrors don't operate and no sound comes from the mirrors, check the fuse (see Section 3).
6 If the fuse is OK, remove the mirror control switch from its mounting without disconnecting the wires attached to it. Turn the ignition ON and check for voltage at the switch. There should be voltage at one terminal. If there's no voltage at the switch, check for an open or short in the wiring between the fuse panel and the switch.
7 If there's voltage at the switch, disconnect it. Check the switch for continuity in all its operating positions. If the switch does not have continuity, replace it.
8 Re-connect the switch. Locate the wire going from the switch to ground. Leaving the switch connected, connect a jumper wire between this wire and ground. If the mirror works normally with this wire in place, repair the faulty ground connection.
9 If the mirror still doesn't work, remove the cover and check the wires at the mirror for voltage with a test light. Check with ignition ON and the mirror selector switch on the appropriate side. Operate the mirror switch in all its positions. There should be voltage at one of the switch-to-mirror wires in each switch position (except the neutral "off" position).
10 If there's not voltage in each switch position, check the wiring between the mirror and control switch for opens and shorts.
11 If there's voltage, remove the mirror and test it off the vehicle with jumper wires. Replace the mirror if it fails this test (see Chapter 11).

24 Cruise control system - description and check

1 The cruise control system maintains vehicle speed with a vacuum actuated servo or an electronically operated cable and module located in the engine compartment, which is connected to the throttle linkage by a cable. The system consists of the electronic control module, brake switch, control switches, a relay, the vehicle speed sensor and associated wiring. Listed below are some general procedures that may be used to locate common cruise control problems.
2 Locate and check the fuse (see Section 3).
3 Have an assistant operate the brake lights while you check their operation (voltage from the brake light switch deactivates the cruise control).
4 If the brake lights don't come on or don't shut off, correct the problem and retest the cruise control.
5 Inspect the cable linkage between the cruise control actuator and the throttle linkage. The cruise control module is located on the firewall on the left (drivers) side of the vehicle.

6 Visually inspect the wires connected to the cruise control actuator and check for damage and broken wires.
7 The vehicle speed sensor is located on the transmission extension housing on 2WD models and on transfer case on 4WD models. Refer to Chapter 6, Section 4 to check the sensor for proper operation.
8 Test drive the vehicle to determine if the cruise control is now working. If it isn't, take it to a dealer service department or an automotive electrical specialist for further diagnosis and repair.

25 Power door lock system - description and check

1 Power door lock systems are operated by bi-directional solenoids located in the doors. The lock switches have two operating positions: Lock and Unlock. These switches activate a relay which in turn connects voltage to the door lock solenoids. Depending on which way the relay is activated, it reverses polarity, allowing the two sides of the circuit to be used alternately as the feed (positive) and ground side.
2 Always check the circuit protection first. Some vehicles use a combination of circuit breakers and fuses.
3 Operate the door lock switches in both directions (Lock and Unlock) with the engine off. Listen for the faint click of the relay operating.
4 If there's no click, check for voltage at the switches. If no voltage is present, check the wiring between the fuse panel and the switches for shorts and opens.
5 If voltage is present but no click is heard, test the switch for continuity. Replace it if there's not continuity in both switch positions.
6 If the switch has continuity but the relay doesn't click, check the wiring between the switch and relay for continuity. Repair the wiring if there's no continuity.
7 If the relay is receiving voltage from the switch but is not sending voltage to the solenoids, check for a bad ground at the relay case. If the relay case is grounding properly, replace the relay.
8 If all but one lock solenoid operates, remove the trim panel from the affected door (see Chapter 11) and check for voltage at the solenoid while the lock switch is operated. One of the wires should have voltage in the Lock position; the other should have voltage in the unlock position.
9 If the inoperative solenoid is receiving voltage, replace the solenoid.
10 If the inoperative solenoid isn't receiving voltage, check for an open or short in the wire between the lock solenoid and the relay.
Note: *It's common for wires to break in the portion of the harness between the body and door (opening and closing the door fatigues and eventually breaks the wires).*

26 Power window system - description and check

1 The power window system consists of the control switches, the motors, glass mechanisms (regulators), and associated wiring.
2 Power windows are wired so they can be lowered and raised from the master control switch by the driver or by remote switches located at the individual windows. Each window has a separate motor which is reversible. The position of the control switch determines the polarity and therefore the direction of operation. The system is equipped with a relay that controls current flow to the motors.
3 The power window system operates when the ignition switch is ON. In addition, these models have a window lockout switch at the master control switch which, when activated, disables the switches at the rear windows and, sometimes, the switch at the passenger's window also. Always check these items before troubleshooting a window problem.
4 These procedures are general in nature, so if you can't find the problem using them, take the vehicle to a dealer service department or other qualified repair shop.
5 If the power windows don't work at all, check the fuse or circuit breaker.
6 If only the rear windows are inoperative, or if the windows only operate from the master control switch, check the rear window lockout switch for continuity in the unlocked position. Replace it if it doesn't have continuity.
7 Check the wiring between the switches and fuse panel for continuity. Repair the wiring, if necessary.
8 If only one window is inoperative from the master control switch, try the other control switch at the window. **Note:** *This doesn't apply to the drivers door window.*
9 If the same window works from one switch, but not the other, check the switch for continuity.
10 If the switch tests OK, check for a short or open in the wiring between the affected switch and the window motor.
11 If one window is inoperative from both switches, remove the trim panel from the affected door and check for voltage at the motor while the switch is operated.
12 If voltage is reaching the motor, disconnect the glass from the regulator (see Chapter 11). Move the window up and down by hand while checking for binding and damage. Also check for binding and damage to the regulator. If the regulator is not damaged and the window moves up and down smoothly, replace the motor (see Chapter 11). If there's binding or damage, lubricate, repair or replace parts, as necessary.
13 If voltage isn't reaching the motor, check the wiring in the circuit for continuity between the switches and motors. Check that the relay is grounded properly and receiving voltage from the switches. Also check that the relay sends voltage to the motor when the switch is turned on. If it doesn't, replace the relay.
14 Test the windows after you are done to confirm proper repairs.

27 Power seats - description and check

1 Power seats allow you to adjust the position of the seat with little effort. The optional power seats on these models adjust forward and backward, up and down and tilt forward and backward.
2 The power seat system consists of a motor, a switch on the seat and a relay and fuse in the engine compartment fuse block.
3 Look under the seat for any objects which may be preventing the seat from moving.
4 If the seat won't work at all, check the fuse.
5 With the engine off to reduce the noise level, operate the seat controls in all directions and listen for sound coming from the seat motor(s).
6 If the motor runs or clicks but the seat doesn't move, the integral the seat drive mechanism is damaged and the motor assembly must be replaced.
7 If the motor doesn't work or make noise, check for voltage at the motor while an assistant operates the switch.
8 If the motor is getting voltage but doesn't run, test it off the vehicle with jumper wires. If it still doesn't work, replace it.
9 If the motor isn't getting voltage, check for voltage at the switch. If there's no voltage at the switch, check the wiring between the fuse panel and the switch. If there's voltage at the switch, obtain the wiring diagrams for the vehicle and check the switch for continuity in all its operating positions. Replace the switch if there's no continuity.
10 If the switch is OK, check for a short or open in the wiring between the switch and motor. If there's a relay between the switch and motor, check that it's grounded properly and there's voltage to the relay. Also check that there's voltage going from the relay to the motor when the when the switch is operated. If there's not, and the relay is grounded properly, replace the relay.
11 Test the completed repairs.

28 Airbag - general information

Caution: *On models equipped with a Delco Loc II anti-theft audio system, be sure the lockout feature is turned off before performing any procedure which requires disconnecting the battery.*

Description

1 These models are equipped with a Supplemental Inflatable Restraint (SIR) system, more commonly called an airbag. The SIR system is designed to protect the driver and passenger from serious injury in the event of a head-on or frontal collision.

Chapter 12 Chassis electrical system

12-15

28.3a There are two crash sensors like this one (arrow) at the front, plus . . .

28.3b . . . an arming sensor (arrow) mounted under the vehicle

28.9 Remove the airbag fuse (refer to the fuse block cover to make sure it's the right one)

2 The SIR system consists of an airbag in the center of the steering wheel and the passenger's side of the dash, three crash sensors mounted at the front and under the vehicle and the Diagnostic Energy Reserve Module (DERM) behind the passenger's side kick panel.

Sensors

Refer to illustrations 28.3a and 28.3b

3 The system has three sensors: two crash sensors at the front of the vehicle behind the bumper and an arming sensor located under the left side of the vehicle **(see illustrations)**.
4 The front crash sensors are basically pressure sensitive switches that complete an electrical circuit during an impact of sufficient G force. The arming sensor is calibrated to send voltage to the airbag module when it senses a dramatic drop in vehicle speed readying it for deployment. The electrical signal from the crash sensors is sent to the DERM, that then completes the circuit and inflates the airbag.

Diagnostic/Energy Reserve Module

5 The diagnostic module (DERM) contains an on-board microprocessor which monitors the operation of the system. It checks this system every time the vehicle is started, causing the AIRBAG light to go on, then off seven times, if the system is operating properly. If there is a fault in the system, the light will go on and stay on and the airbag control module will store fault codes indicating the nature of the fault. If the AIRBAG light does go on and stay on, the vehicle should be taken to your dealer immediately for service.

Operation

6 For the airbag to deploy, an impact of sufficient G force must occur within 30-degrees of the vehicle centerline. When this condition occurs, the circuit to the airbag inflator is closed and the airbag inflates. If the battery is destroyed by the impact, or is too low to power the inflator, a back-up power unit inside the DERM supplies current to the airbag.

Self-diagnosis system

7 A self-diagnosis circuit in the module displays a light when the ignition switch is turned to the On position. If the system is operating normally, the light should go out after seven flashes. If the light doesn't come on, or doesn't go out after seven flashes, or if it comes on while you're driving the vehicle, there's a malfunction in the SIR system. Have it inspected and repaired as soon as possible. Do not attempt to troubleshoot or service the SIR system yourself. Even a small mistake could cause the SIR system to malfunction when you need it.

Servicing components near the SIR system

8 Nevertheless, there are times when you need to remove the steering wheel, radio or service other components on or near the dashboard. At these times, you'll be working around components and wire harnesses for the SIR system. SIR wires are easy to identify: They're all covered with bright yellow conduit. Do not unplug the connectors for these wires. And do not use electrical test equipment on yellow wires. *ALWAYS DISABLE THE SIR SYSTEM BEFORE WORKING NEAR THE SIR SYSTEM COMPONENTS OR RELATED WIRING.*

Disabling the SIR system

Refer to illustrations 28.9 and 28.11
Warning: *Anytime you are working in the vicinity of airbag wiring or components, DISABLE THE SIR SYSTEM.*
9 Turn the steering wheel to the straight ahead position, place the ignition switch in Lock, remove the key, then remove the airbag fuse from the fuse block **(see illustration)**.
10 Unplug the yellow Connector Position Assurance (CPA) connector at the base of the steering column as described in the following steps.
11 Remove the knee bolster and sound insulator panel (see Chapter 11) below the instrument panel and unplug the yellow Connector Position Assurance (CPA) steering column harness connector **(see illustration)**.

Enabling the SIR system

12 After you've disabled the airbag and performed the necessary service, plug in the steering column (driver's side) and passenger side Connector Position Assurance (CPA) connectors. Reinstall the knee bolster and the glove box.
13 Install the airbag fuse.

29 Wiring diagrams - general information

Since it isn't possible to include all wiring diagrams for every year covered by this manual, the following diagrams are those that are typical and most commonly needed.
Prior to troubleshooting any circuit, check the fuse and circuit breakers (if equipped) to make sure they're in good condition. Make sure the battery is properly charged and check the cable connections (see Chapter 1).
When checking a circuit, make sure that all electrical connectors are clean, with no broken or loose terminals. When unplugging an electrical connector, do not pull on the wires. Pull only on the connector housings themselves.

28.11 The airbag Connector Position Assurance (CPA) connector is found at the base of the steering column

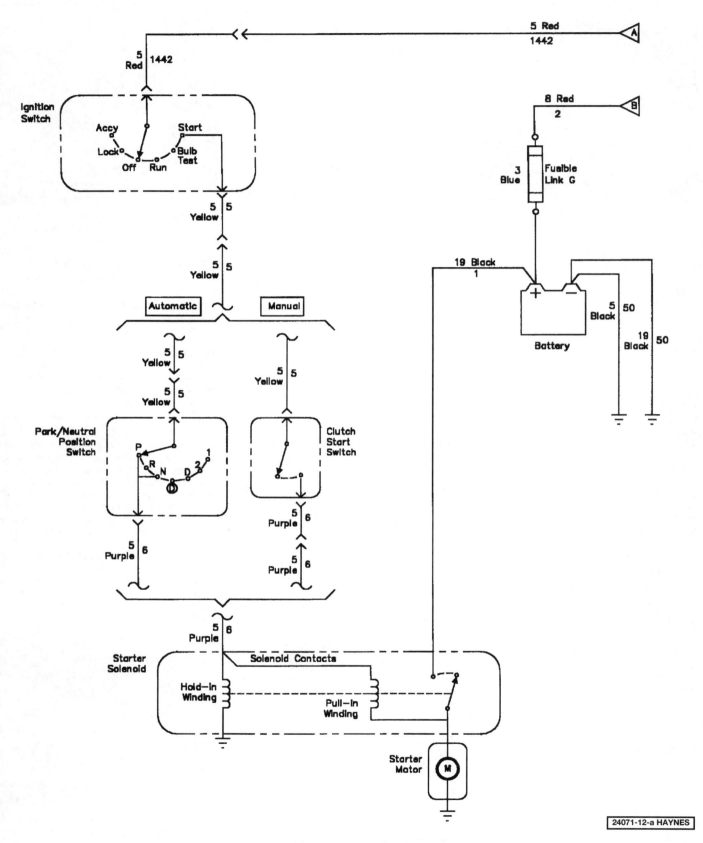

Typical starting and charging system wiring diagram (1996 and earlier shown) (1 of 2)

Chapter 12 Chassis electrical system 12-17

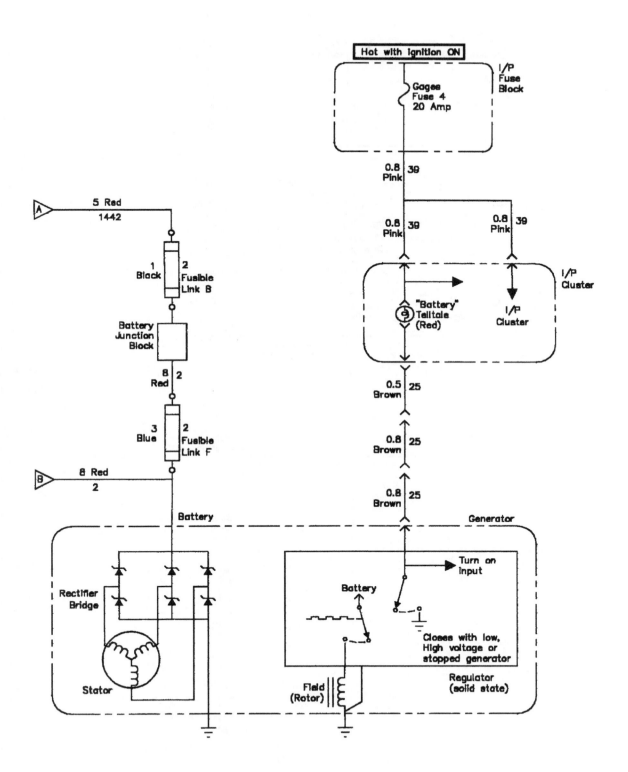

Typical starting and charging system wiring diagram (1996 and earlier shown) (2 of 2)

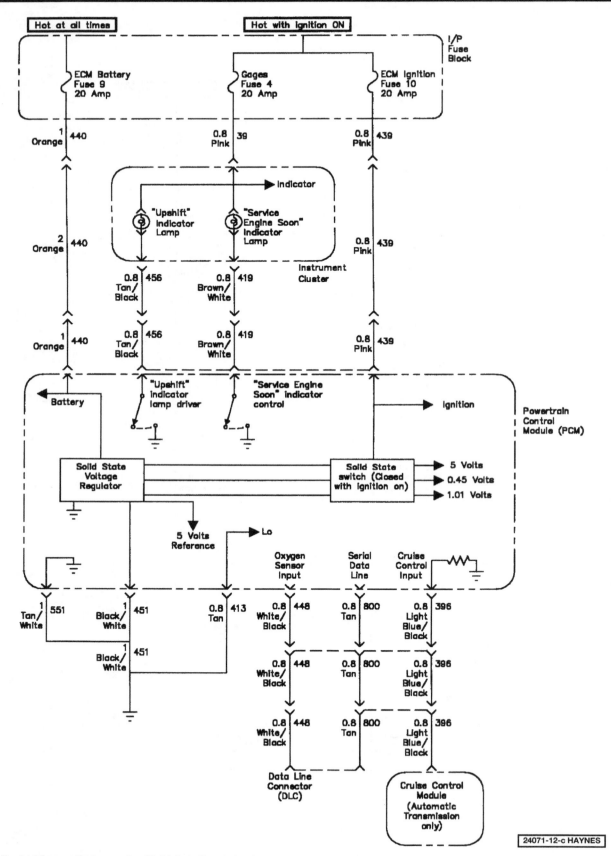

Typical four-cylinder engine Multi-Port Fuel Injection system wiring diagram (1995 and earlier shown) (1 of 3)

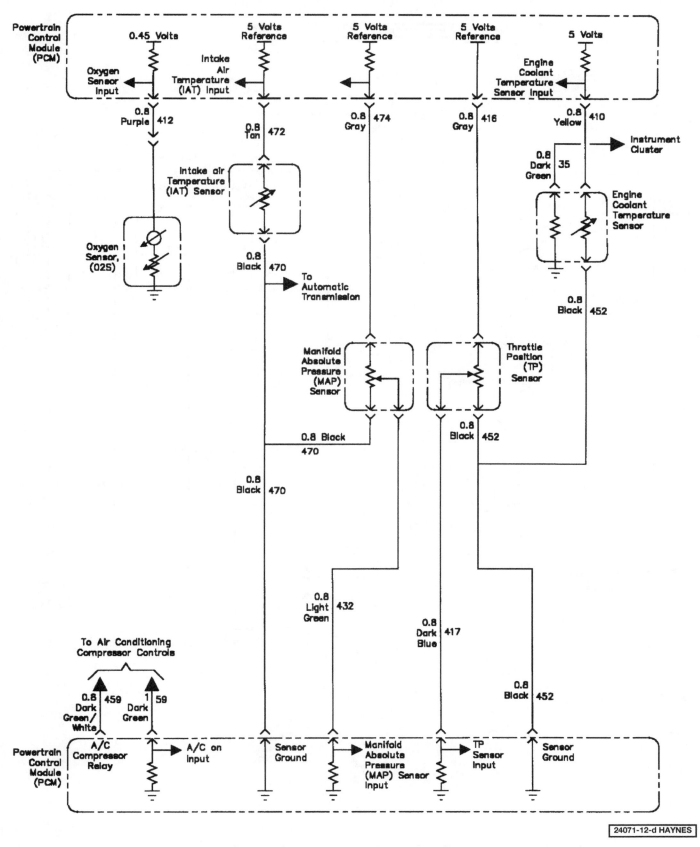

Typical four-cylinder engine Multi-Port Fuel Injection system wiring diagram (1995 and earlier shown) (2 of 3)

Chapter 12 Chassis electrical system

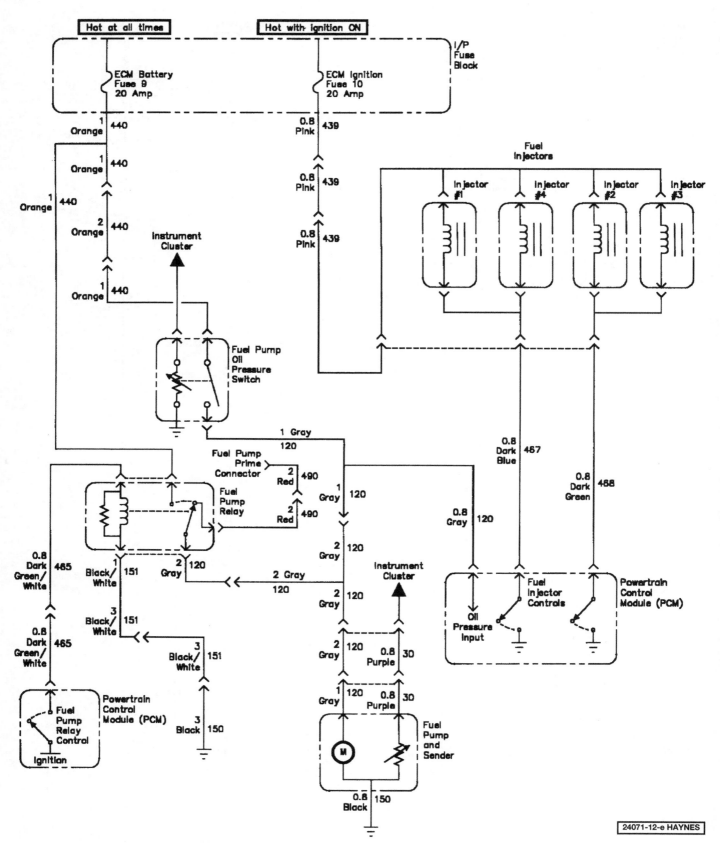

Typical four-cylinder engine Multi-Port Fuel Injection system wiring diagram (1995 and earlier shown) (3 of 3)

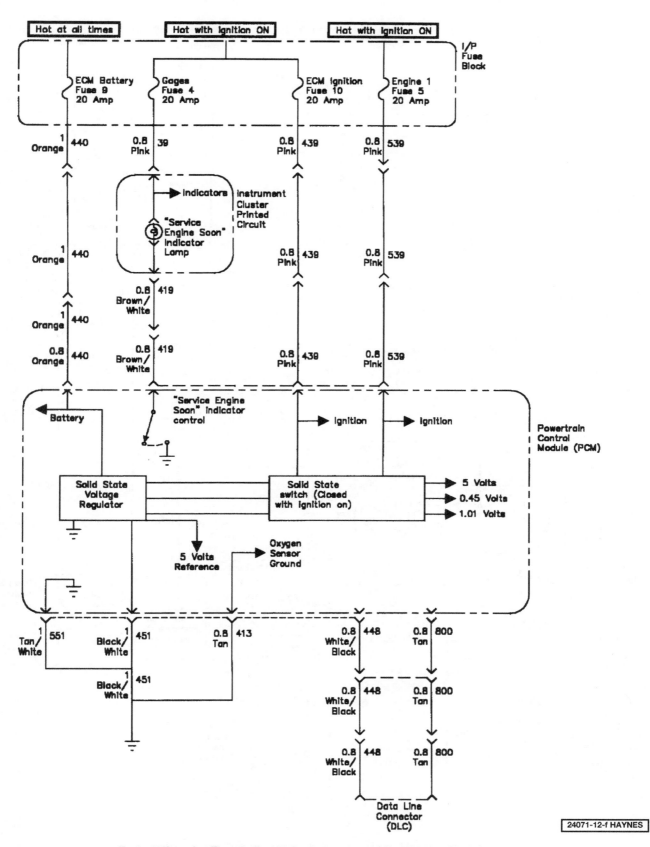

Typical V6 engine Throttle Body Injection system wiring diagram (1 of 3)

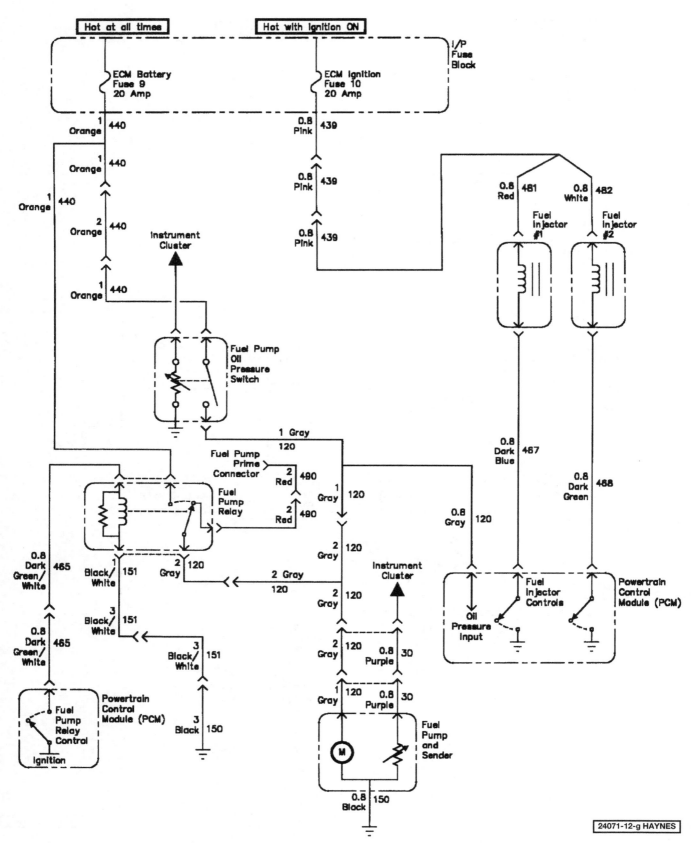

Typical V6 engine Throttle Body Injection system wiring diagram (2 of 3)

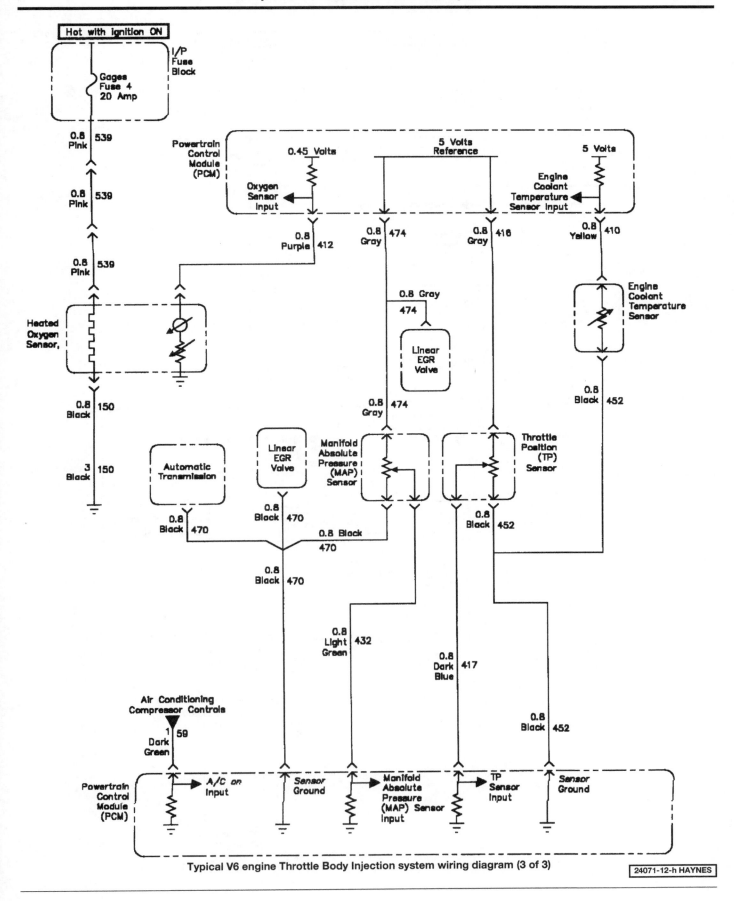

Typical V6 engine Throttle Body Injection system wiring diagram (3 of 3)

12-24　Chapter 12　Chassis electrical system

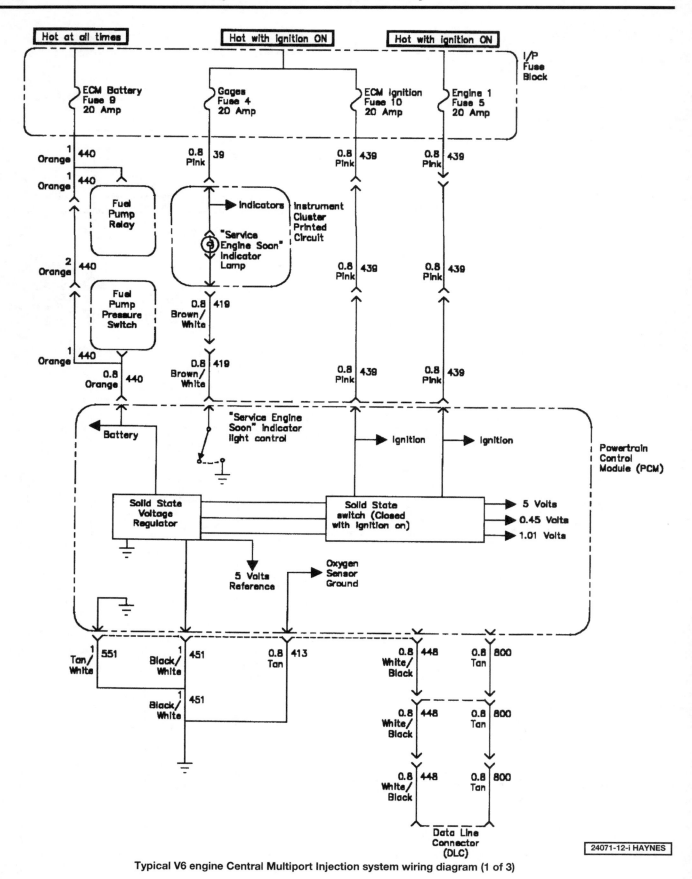

Typical V6 engine Central Multiport Injection system wiring diagram (1 of 3)

Chapter 12 Chassis electrical system 12-25

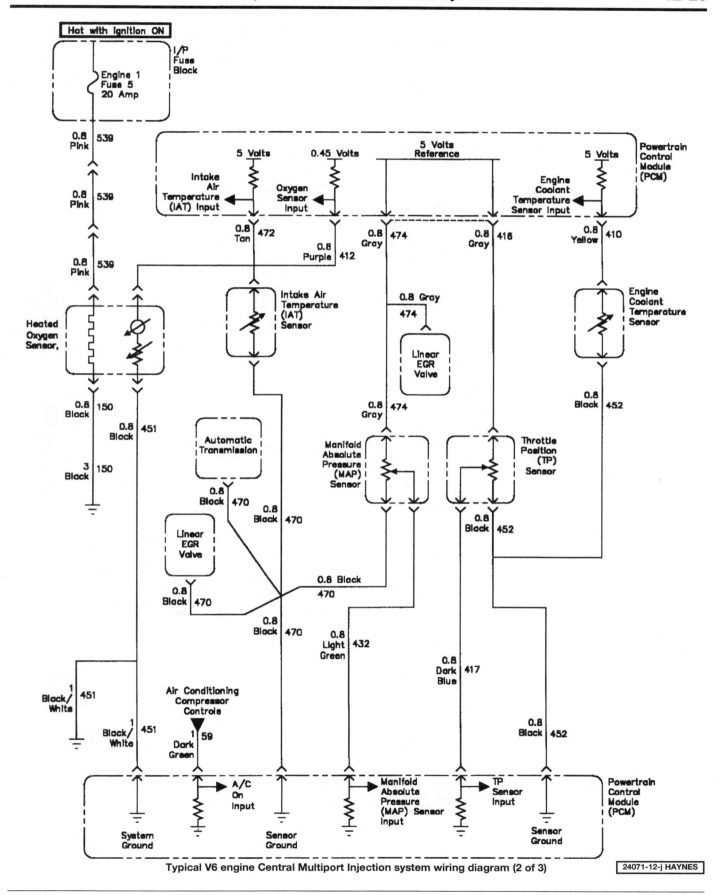

Typical V6 engine Central Multiport Injection system wiring diagram (2 of 3)

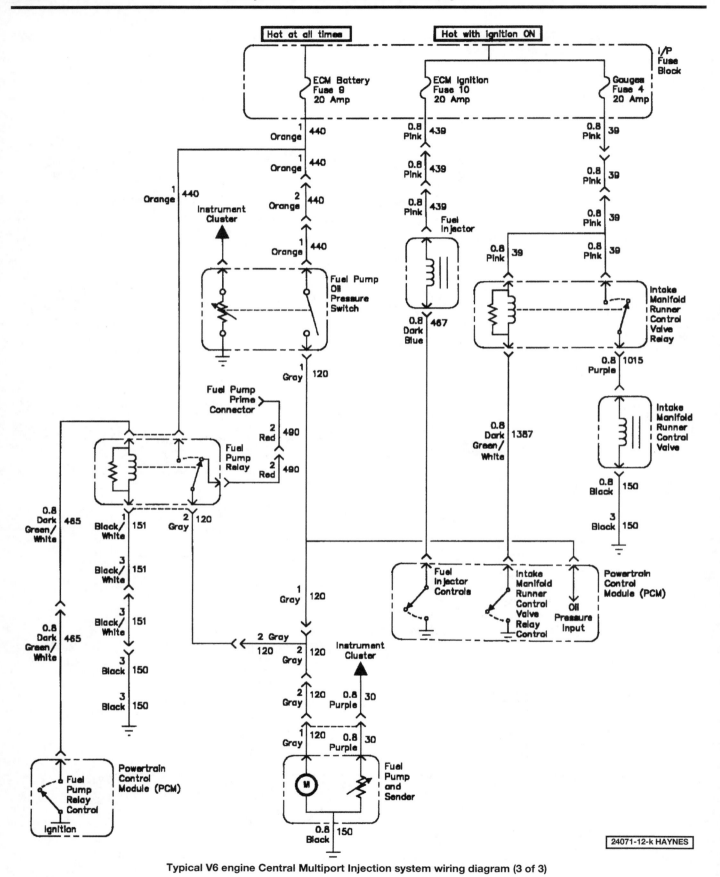

Typical V6 engine Central Multiport Injection system wiring diagram (3 of 3)

Chapter 12 Chassis electrical system

12-27

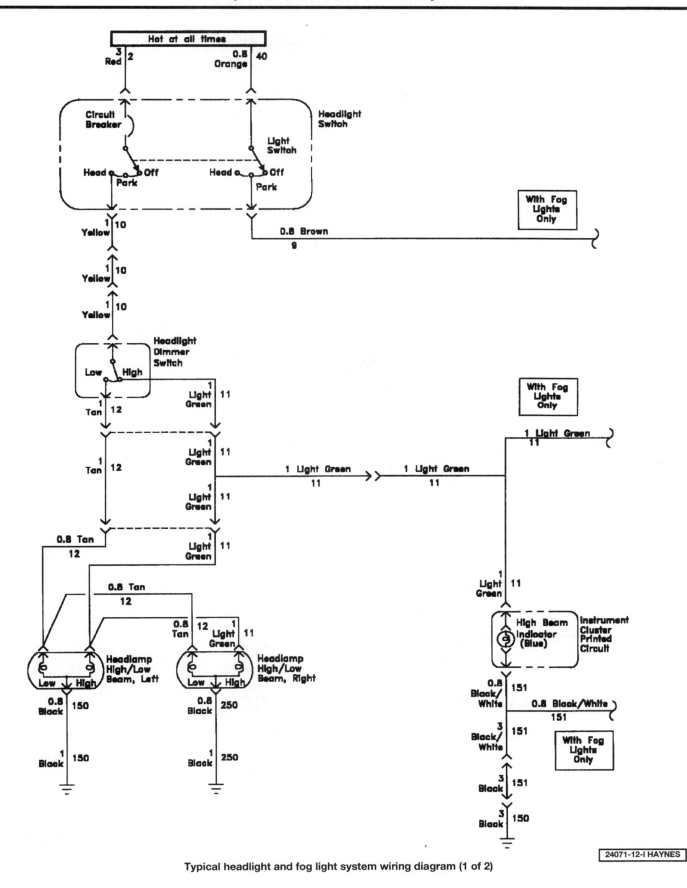

Typical headlight and fog light system wiring diagram (1 of 2)

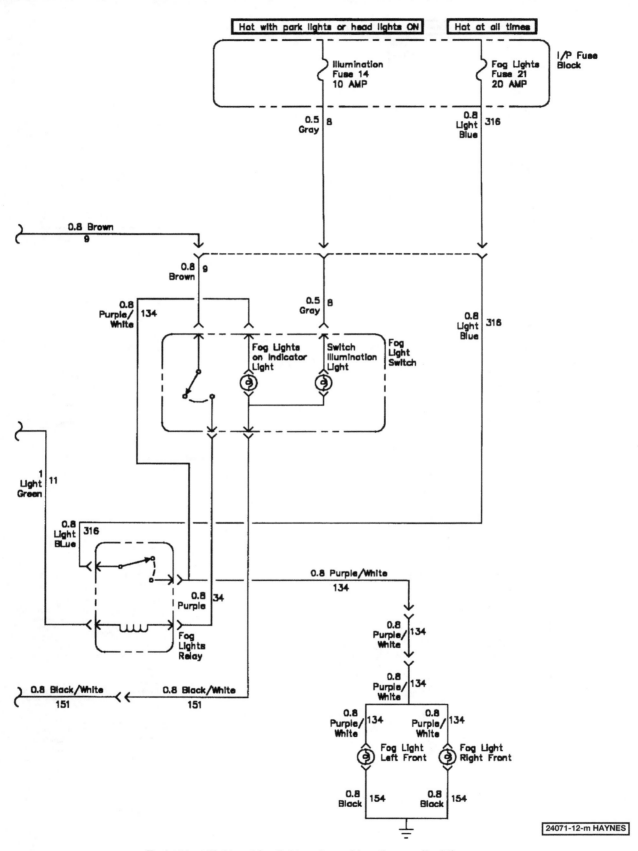

Typical headlight and fog light system wiring diagram (2 of 2)

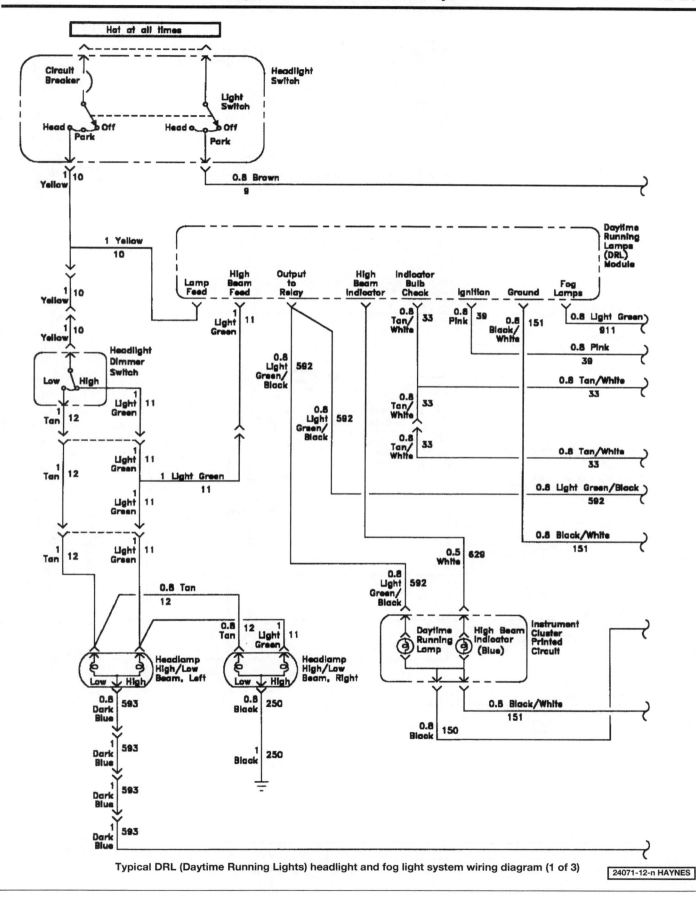

Typical DRL (Daytime Running Lights) headlight and fog light system wiring diagram (1 of 3)

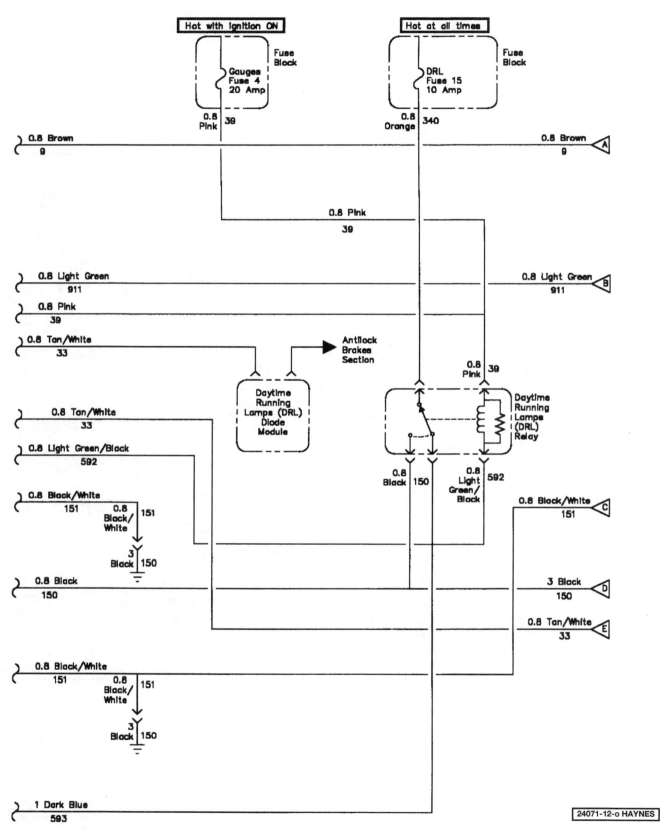

Typical DRL (Daytime Running Lights) headlight and fog light system wiring diagram (2 of 3)

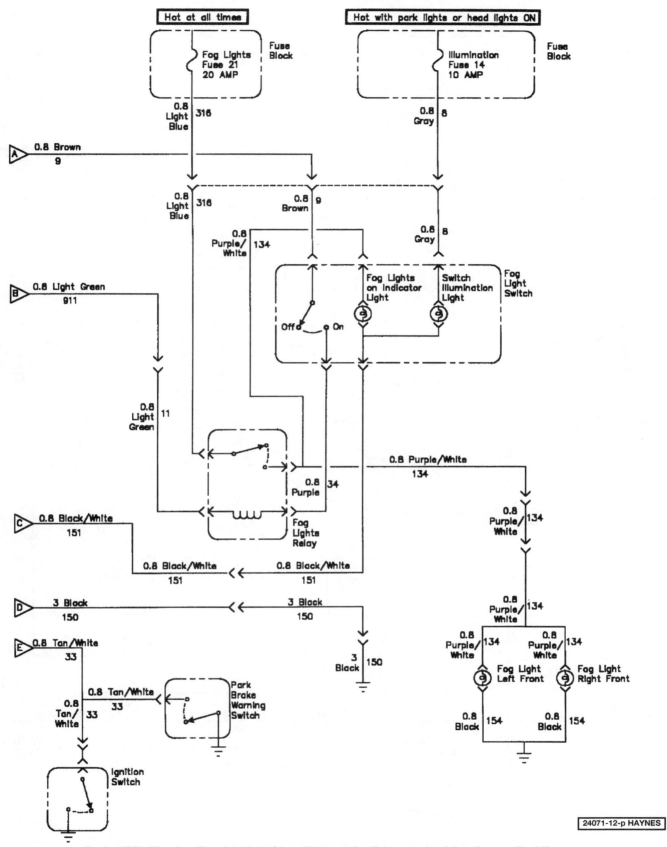

Typical DRL (Daytime Running Lights) headlight and fog light system wiring diagram (3 of 3)

12-32　　Chapter 12　Chassis electrical system

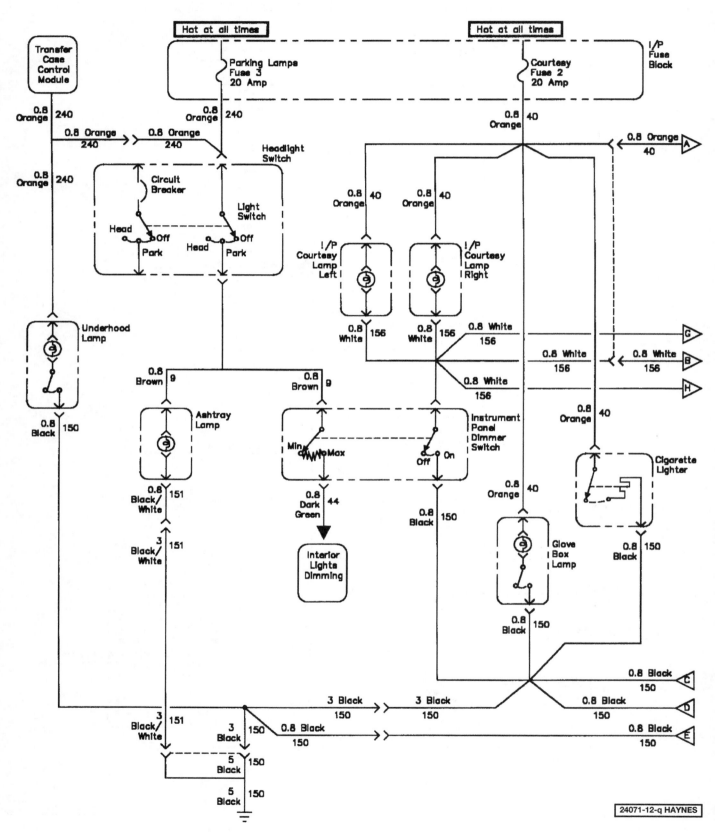

Typical interior lighting system wiring diagram (1 of 3)

Chapter 12 Chassis electrical system

12-33

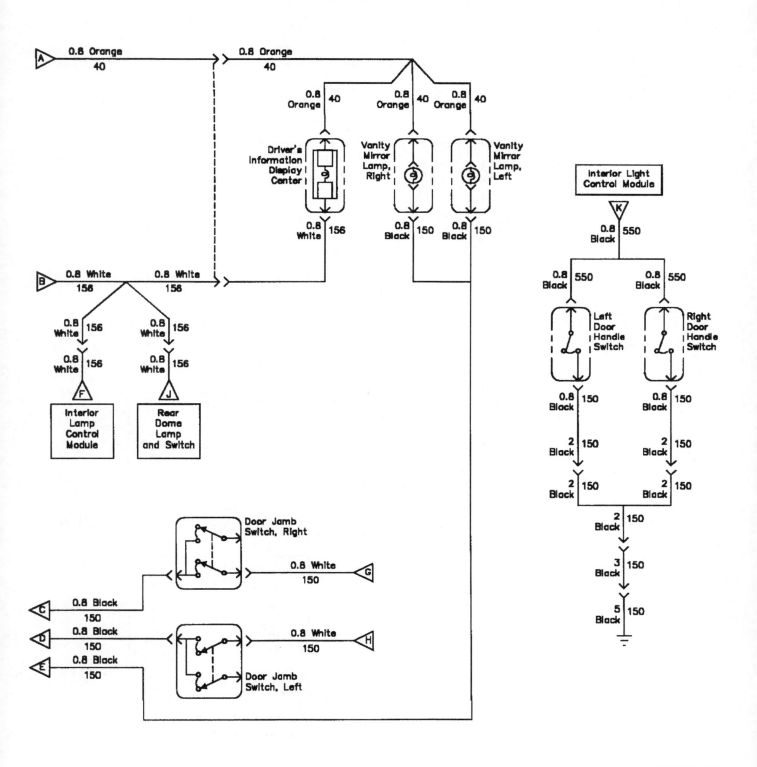

Typical interior lighting system wiring diagram (2 of 3)

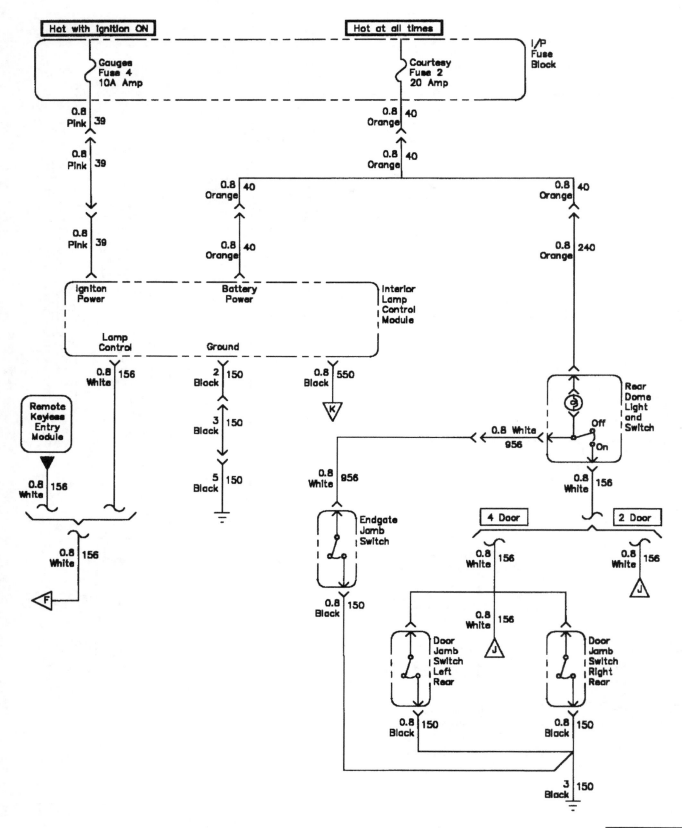

Typical interior lighting system wiring diagram (3 of 3)

Chapter 12 Chassis electrical system

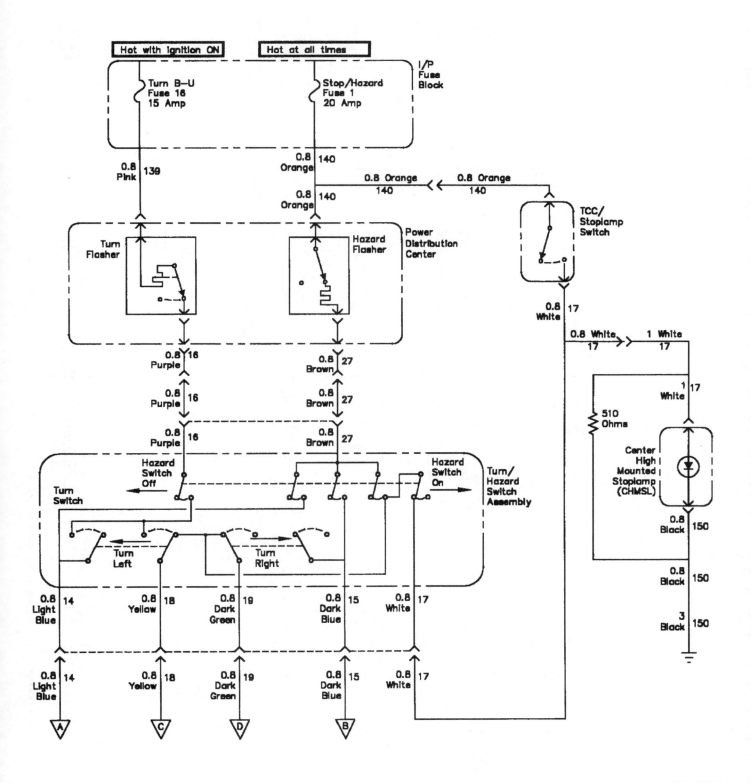

Typical exterior lighting system wiring diagram (1 of 3)

Chapter 12 Chassis electrical system

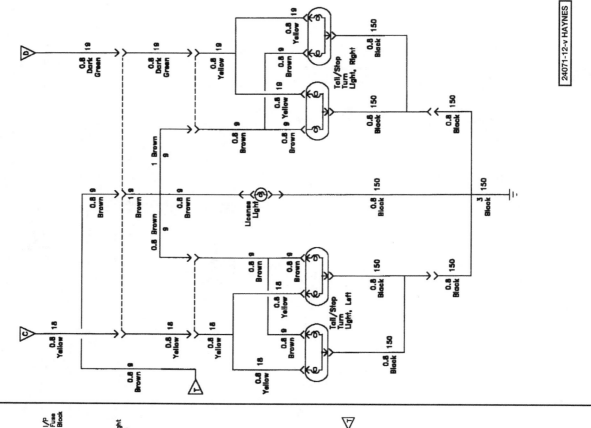

Typical exterior lighting system wiring diagram (3 of 3)

Typical exterior lighting system wiring diagram (2 of 3)

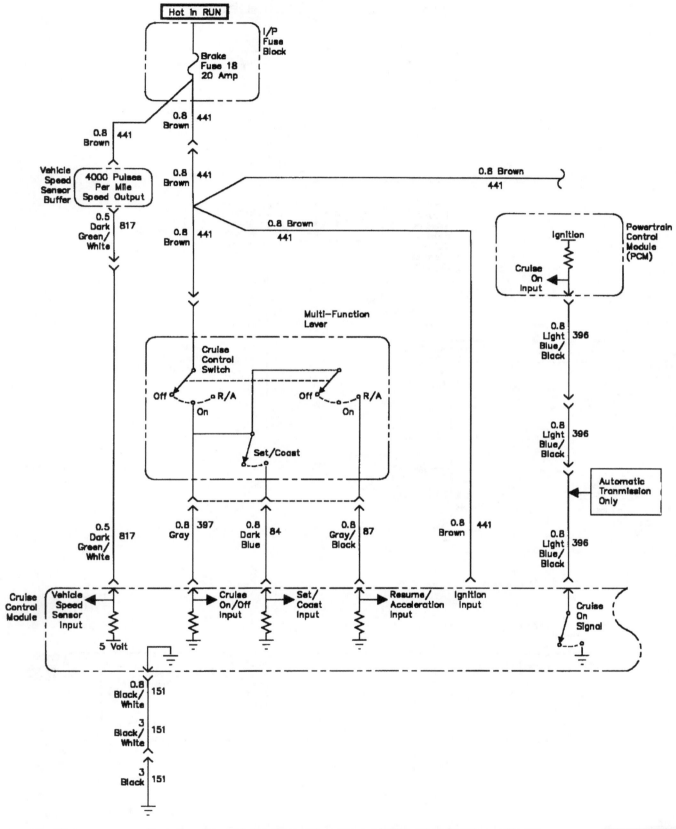

Typical cruise control system wiring diagram (1995 and earlier shown) (1 of 2)

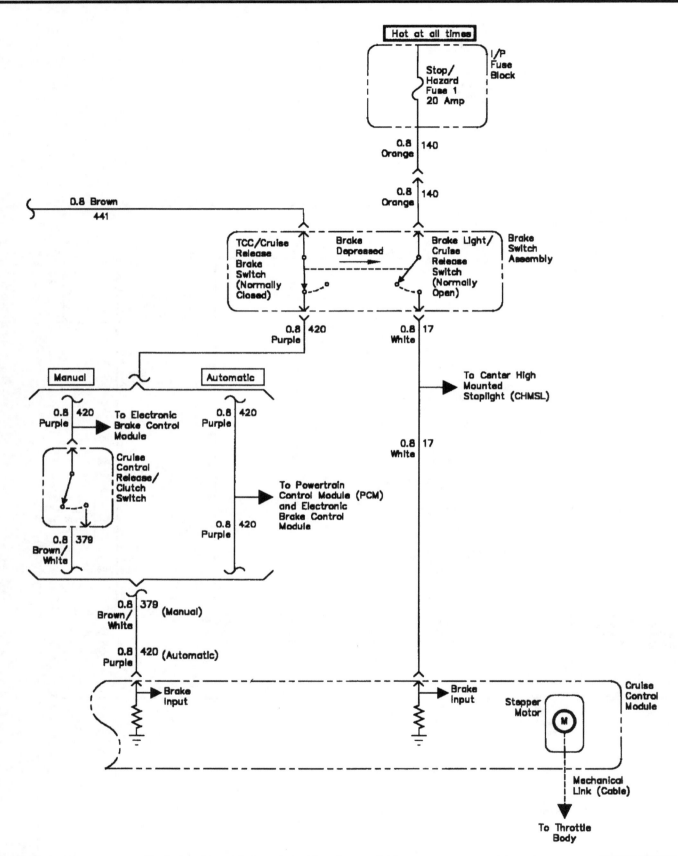

Typical cruise control system wiring diagram (1995 and earlier shown) (2 of 2)

Chapter 12 Chassis electrical system

12-39

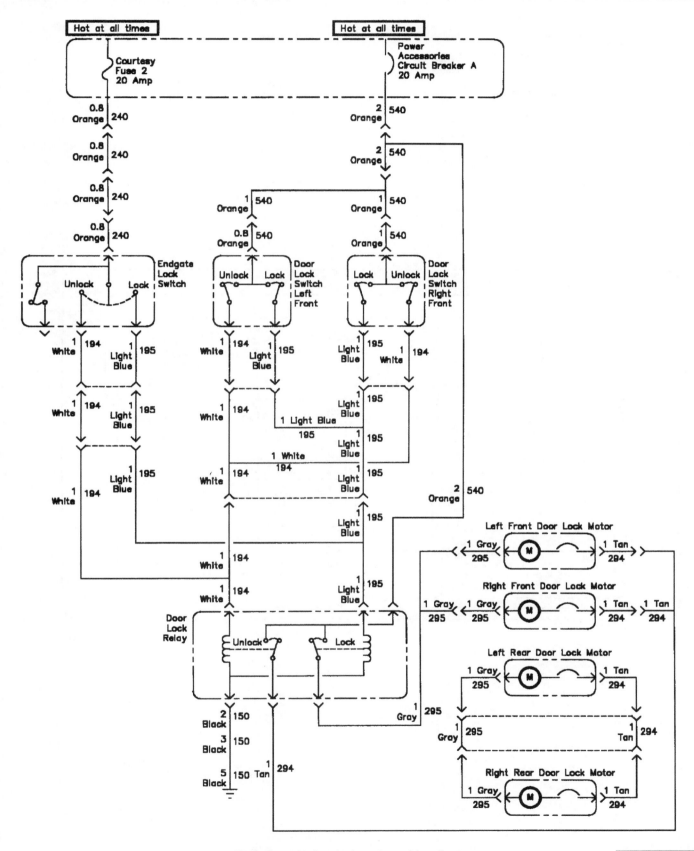

Typical power door lock system wiring diagram

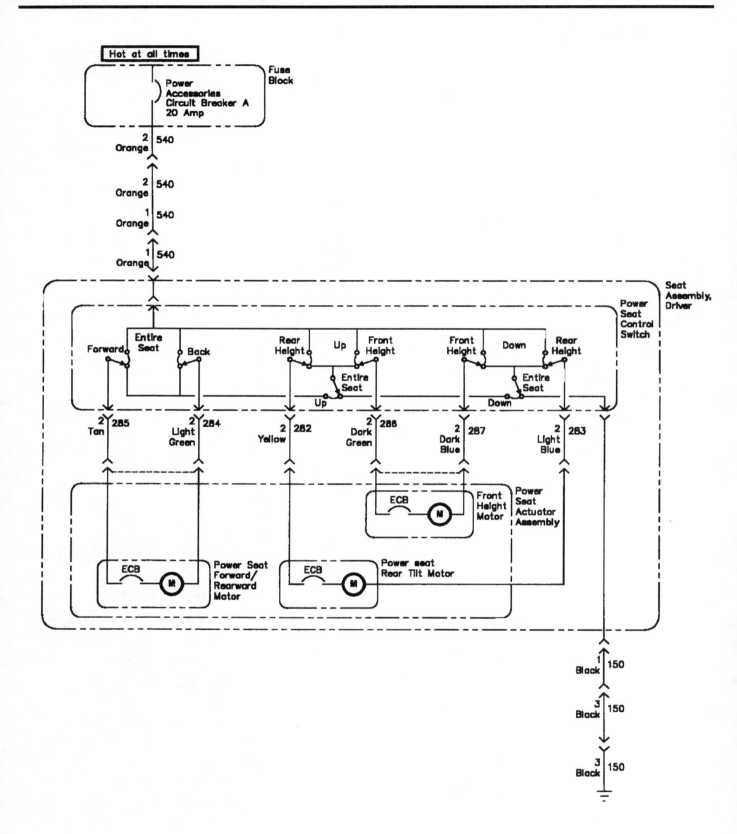

Typical power seat system wiring diagram

Chapter 12 Chassis electrical system

12-41

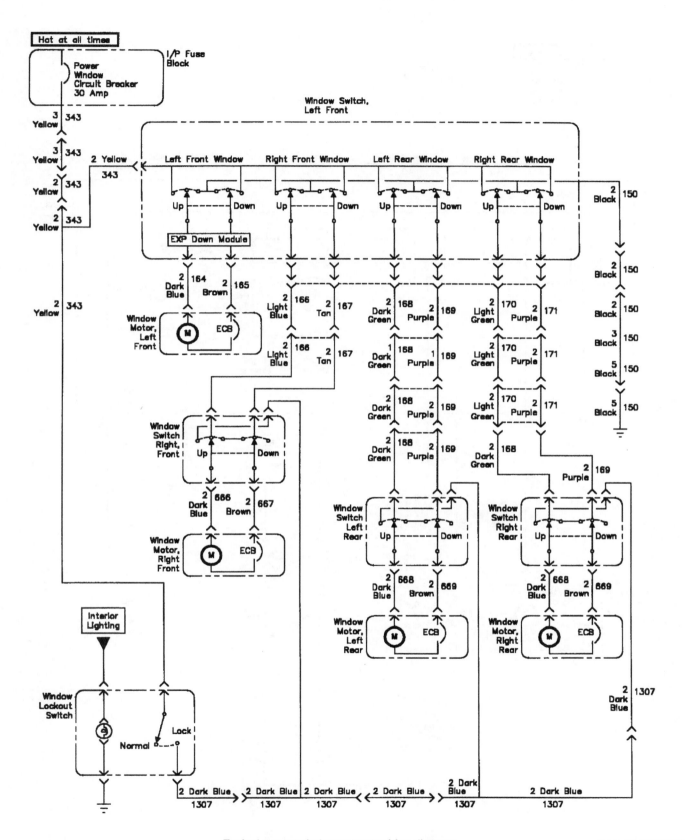

Typical power window system wiring diagram

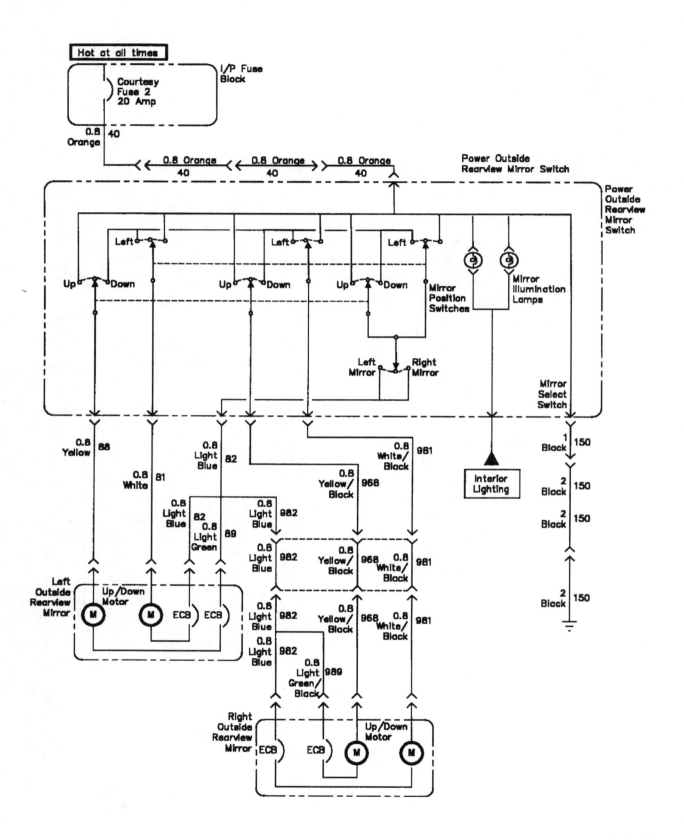

Typical power outside mirror system wiring diagram

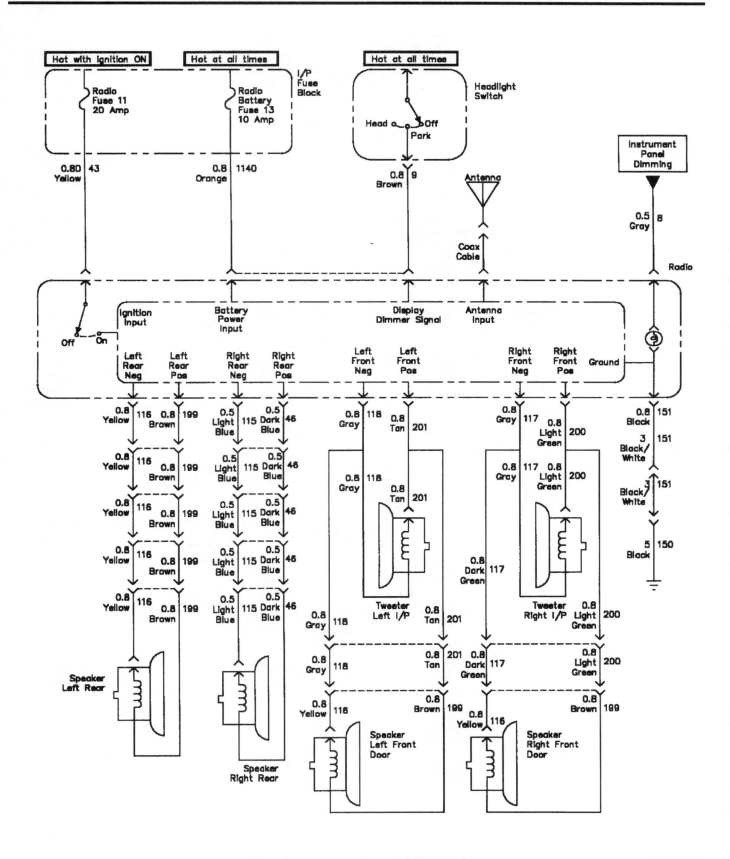

Typical radio and audio system wiring diagram

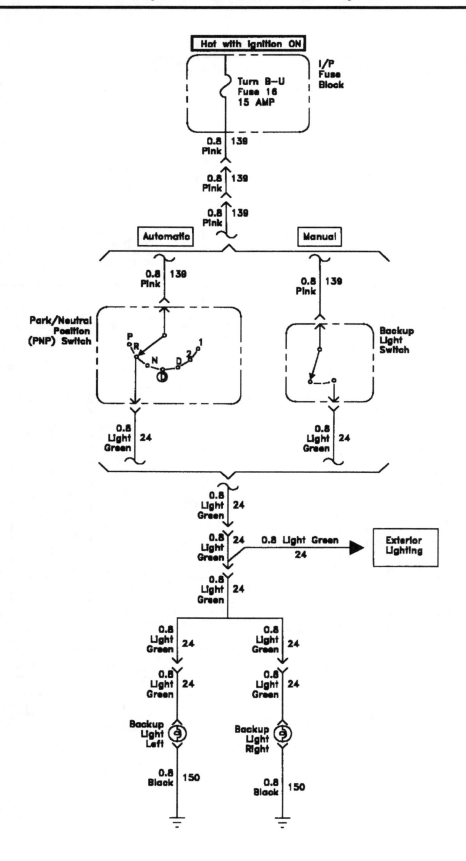

Typical backup light system wiring diagram

Index

A

About this manual, 0-2
Accelerator cable, removal and installation, 4-8
Accumulator, removal and installation, 3-13
Acknowledgements, 0-2
Air cleaner assembly, removal and installation, 4-8
Air conditioning and heating system, check and maintenance, 3-11
Air filter replacement, 1-19
Airbag, general information, 12-14
Alignment, general information, 10-16
Alternator, removal and installation, 5-9
Antenna, removal and installation, 12-10
Antifreeze/coolant, general information, 3-1
Anti-lock brake system, general information, 9-2
Anti-theft audio system, 0-13
Automatic transmission, 7B-1 through 7B-6
   diagnosis, general, 7B-1
   extension housing seal, replacement, 7B-4
   fluid and filter change, 1-22
   fluid level check, 1-10
   general information, 7B-1
   park/neutral switch, check, replacement and adjustment, 7B-4
   removal and installation, 7B-5
   shift cable (1995 and 1996 models), removal, installation and adjustment, 7B-3
   shift linkage (1994 models), removal, installation and adjustment, 7B-2
   transmission mount, check and replacement, 7B-5
Automotive chemicals and lubricants, 0-14
Axle (rear), removal and installation, 8-12
Axles, description and check, 8-9
Axleshaft
   bearing (rear), replacement, 8-11
   oil seal (rear), replacement, 8-11
   rear, removal and installation, 8-10

B

Balance shaft, installation (V6 engines only), 2C-19
Balljoints, check and replacement, 10-6
Battery
   cables, check and replacement, 5-2
   check, maintenance and charging, 1-12
   emergency jump starting, 5-1
   removal and installation, 5-1

Block
   cleaning, 2C-12
   inspection, 2C-13
Blower motor
   and circuit, check, 3-7
   removal and installation, 3-7
Body, 11-1 through 11-14
   bumpers, removal and installation, 11-6
   console, removal and installation, 11-11
   cowl vent grille, removal and installation, 11-13
   dashboard trim panels, removal and installation, 11-13
   door
      latch, lock cylinder and outside handle, removal and installation, 11-10
      removal and installation, 11-8
      trim panel, removal and installation, 11-6
      window glass, removal and installation, 11-10
      window regulator, removal and installation, 11-11
   front fender, removal and installation, 11-8
   general information, 11-1
   hinges and locks, maintenance, 11-1
   hood release latch and cable, replacement, 11-5
   hood, removal, installation and adjustment, 11-5
   instrument cluster bezel, removal and installation, 11-12
   maintenance, 11-1
   outside mirror, removal and installation, 11-8
   radiator grille, removal and installation, 11-6
   repair
      major damage, 11-5
      minor damage, 11-4
   seats, removal and installation, 11-14
   steering column cover, removal and installation, 11-11
   support struts, replacement, 11-9
   tailgate glass, removal and installation, 11-9
   tailgate, removal and installation, 11-9
   upholstery and carpets, maintenance, 11-1
   vinyl trim, maintenance, 11-1
   windshield and fixed glass, replacement, 11-5
Booster battery (jump) starting, 0-13
Brake and Transmission Shift Interlock (BTSI) system check (1996 and later models with automatic transmission), 1-17
Brake check, 1-18
   disc brakes (front), 1-18
   drum brakes (rear), 1-19
   parking brake, 1-19
Brake light switch, check and replacement, 9-15
Brakes, 9-1 through 9-16
   Anti-lock brake system, general information, 9-2
   brake disc, inspection, removal and installation, 9-7

brake light switch, check and replacement, 9-16
combination valve (non-ABS models), replacement, 9-13
disc brake
    caliper, removal, overhaul and installation, 9-6
    pads, replacement, 9-3
drum brake shoes, replacement, 9-7
general information, 9-1
hoses and lines, check and replacement, 9-12
hydraulic system, bleeding, 9-15
master cylinder, removal, overhaul and installation, 9-11
parking brake
    adjustment, 9-13
    cables, replacement, 9-14
    shoes, removal and installation, 9-16
power brake booster, removal and installation, 9-15
wheel cylinder, removal, overhaul and installation, 9-10

Bulb replacement, 12-8
Bumpers, removal and installation, 11-6
Buying parts, 0-6

C

Camshaft and lifters, removal, inspection and installation
    2.2L four-cylinder engine, 2A-9
    4.3L V6 engine, 2B-12
Camshaft, balance shaft and bearings, removal and inspection, 2C-10
Catalytic converter, 6-16
Charging system
    check, 5-9
    general information and precautions, 5-9
Chassis electrical system, 12-1 through 12-44
Chassis lubrication, 1-15
Circuit breakers, general information, 12-3
Clutch
    components, removal, inspection and installation, 8-3
    description and check, 8-2
    hydraulic system, bleeding, 8-6
    master cylinder, removal and installation, 8-2
    release bearing (1994 and 1995 models), removal, inspection and installation, 8-5
    release cylinder, removal and installation, 8-3
    start switch, check and replacement, 8-6
Clutch and driveline, 8-1 through 8-16
Coil spring (2WD models), removal and installation, 10-7
Composite headlight housing, removal and installation, 12-7
Compressor, removal and installation, 3-12
Condenser, removal and installation, 3-13
Console, removal and installation, 11-11
Conversion factors, 0-15
Coolant temperature sending unit, check and replacement, 3-6
Cooling system
    check, 1-14
    servicing (draining, flushing and refilling), 1-23
Cooling, heating and air conditioning systems, 3-1 through 3-16
    air conditioning
        accumulator, removal and installation, 3-13
        and heating system, check and maintenance, 3-11
        compressor, removal and installation, 3-12
        condenser, removal and installation, 3-14
        evaporator, removal and installation, 3-14
    antifreeze/coolant, general information, 3-1
    blower motor
        and circuit, check, 3-7
        removal and installation, 3-7
    coolant temperature sending unit, check and replacement, 3-6
    engine cooling fan and clutch, check, removal and installation, 3-3
    general information, 3-1
    heater and air conditioning control assembly, removal and installation, 3-10
    heater core, removal and installation, 3-8
    radiator and coolant reservoir tank, removal and installation, 3-4
    thermostat, check and replacement, 3-2
    water pump
        check, 3-5
        removal and installation, 3-5

Cowl vent grille, removal and installation, 11-13
Crankshaft front oil seal, replacement
    2.2L four-cylinder engine, 2A-6
    4.3L V6 engine, 2B-9
Crankshaft
    inspection, 2C-15
    installation and main bearing oil clearance check, 2C-18
    removal, 2C-12
Cruise control system, description and check, 12-13
Cylinder compression check, 2C-4
Cylinder head
    cleaning and inspection, 2C-8
    disassembly, 2C-8
    reassembly, 2C-10
    removal and installation
        2.2L four-cylinder engine, 2A-8
        4.3L V6 engine, 2B-13
Cylinder honing, 2C-14

D

Dashboard trim panels, removal and installation, 11-13
Differential
    carrier bushing, front (4WD models), replacement, 8-16
    carrier, front (4WD models), removal and installation, 8-16
    lubricant change, 1-22
    lubricant level check, 1-18
    output shaft pilot bearing (4WD models), replacement, 8-16
    pinion seal (rear), replacement, 8-11
Distributor (HEI and EDI systems), removal and installation, 5-6
Door
    latch, lock cylinder and outside handle, removal and installation, 11-10
    lock system, power, description and check, 12-14
    removal and installation, 11-8
    trim panel, removal and installation, 11-6
    window glass, removal and installation, 11-10
    window regulator, removal and installation, 11-11

Index

Driveaxle (4WD models), removal and installation, 8-12
Driveaxle boot
   check (4WD models), 1-19
   replacement and CV joint overhaul (4WD models), 8-12
Drivebelt check and replacement, 1-20
Driveline inspection, 8-7
Driveplate, removal and installation, 2C-20
Driveshaft(s) and universal joints, general information, 8-6
Driveshaft, removal and installation, 8-7
   front (4WD models), 8-8
   rear, 8-7

E

Electrical troubleshooting, general information, 12-2
Electronic shift motor (NP 233), replacement, 7C-3
Emissions and engine control systems, 6-1 through 6-16
   catalytic converter, 6-16
   Evaporative Emissions Control (EVAP) System, 6-14
   Exhaust Gas Recirculation (EGR) system, 6-13
   general information, 6-1
   information sensors, check and replacement, 6-6
      air conditioning "On" signal, 6-11
      air conditioning control, 6-11
      camshaft sensor (V6 CMFI and CSEFI engines), 6-12
      crankshaft sensor (four-cylinder MFI and V6 CMFI and CSEFI engines), 6-12
      Engine Coolant Temperature (ECT) sensor, 6-6
      Intake Air Temperature (IAT) sensor, 6-8
      knock sensors (V6 CMFI and CSEFI engines), 6-12
      Manifold Absolute Pressure (MAP) sensor, 6-7
      Neutral Start switch, 6-11
      oxygen sensor, 6-8
      Throttle Position Sensor (TPS), 6-10
      Vehicle Speed Sensor (VSS), 6-11
   Positive Crankcase Ventilation (PCV) system, 6-15
   Secondary Air Injection (AIR) system, 6-16
   Self-diagnosis system and trouble codes, 6-2
   Transmission Converter Clutch (TCC) system, 6-15
Engine cooling fan and clutch, check, removal and installation, 3-3
Engine electrical systems, 5-1 through 5-10
   alternator, removal and installation, 5-9
   battery
      cables, check and replacement, 5-2
      emergency jump starting, 5-1
      removal and installation, 5-1
   charging system
      check, 5-9
      general information and precautions, 5-9
   distributor (HEI and EDI systems), removal and installation, 5-6
   general information, 5-1
   ignition
      coil, check, removal and installation, 5-4
      module, check and replacement, 5-7
      pick-up coil (HEI systems), check and replacement, 5-8
   starter motor
      removal and installation, 5-10
      testing in vehicle, 5-10
   starter solenoid, removal and installation, 5-10
   starting system, general information, 5-9

Engine oil and filter change, 1-11
Engines
   2.2L four-cylinder engine, 2A-1 through 2A-12
      camshaft and lifters, removal, inspection and installation, 2A-9
      crankshaft front oil seal, replacement, 2A-6
      cylinder head, removal and installation, 2A-8
      exhaust manifold, removal and installation, 2A-5
      flywheel/driveplate, removal and installation, 2A-11
      general information, 2A-2
      intake manifold, removal and installation, 2A-4
      mounts, replacement, 2A-12
      oil pan, removal and installation, 2A-11
      oil pump, removal and installation, 2A-11
      rear main oil seal, replacement, 2A-12
      repair operations possible with the engine in the vehicle, 2A-2
      rocker arms and pushrods, removal, inspection and installation, 2A-3
      timing chain cover, chain and sprockets, removal, inspection and installation, 2A-7
      valve cover, removal and installation, 2A-2
      valve springs, retainers and seals, replacement, 2A-4
   4.3L V6 engine, 2B-1 through 2B-16
      camshaft and lifters, removal, inspection and installation, 2B-12
      crankshaft front oil seal, replacement, 2B-9
      cylinder heads, removal and installation, 2B-13
      exhaust manifolds, removal and installation, 2B-8
      flywheel/driveplate, removal and installation, 2B-16
      general information, 2B-3
      intake manifold, removal and installation, 2B-6
      mounts, replacement, 2B-16
      oil pan, removal and installation, 2B-15
      oil pump, removal and installation, 2B-15
      rear main oil seal, replacement, 2B-16
      repair operations possible with the engine in the vehicle, 2B-3
      rocker arms and pushrods, removal, inspection and installation, 2B-4
      timing chain cover, chain and sprockets, removal, 2B-11
      valve covers, removal and installation, 2B-3
      valve springs, retainers and seals, replacement, 2B-5
   general engine overhaul procedures, 2C-1 through 2C-20
      balance shaft, installation (V6 engines only), 2C-19
      block
         cleaning, 2C-12
         inspection, 2C-13
      camshaft, balance shaft and bearings, removal and inspection, 2C-10
      crankshaft
         inspection, 2C-15
         installation and main bearing oil clearance check, 2C-18
         removal, 2C-12
      cylinder compression check, 2C-4
      cylinder head
         cleaning and inspection, 2C-8
         disassembly, 2C-8
         reassembly, 2C-10
      cylinder honing, 2C-14
      driveplate, removal and installation, 2C-20

engine removal, methods and precautions, 2C-5
engine, removal and installation, 2C-6
general information, 2C-3
initial start-up and break-in after overhaul, 2C-20
main and connecting rod bearings, inspection, 2C-16
overhaul
    disassembly sequence, 2C-7
    engine reassembly sequence, 2C-17
    general information, 2C-3
piston rings, installation, 2C-17
piston/connecting rod assembly
    inspection, 2C-14
    installation and bearing oil clearance check, 2C-19
    removal, 2C-11
rear main oil seal, installation, 2C-19
rebuilding alternatives, 2C-7
vacuum gauge diagnostic, checks, 2C-5
valves, servicing, 2C-10

Evaporative emissions control system check, 1-24
Evaporative Emissions Control (EVAP) System, 6-14
Evaporator, removal and installation, 3-14
Exhaust Gas Recirculation (EGR) system, 6-13
Exhaust manifold, removal and installation
    2.2L four-cylinder engine, 2A-5
    4.3L V6 engine, 2B-8
Exhaust system
    check, 1-17
    servicing, general information, 4-20
Extension housing seal, replacement, 7B-4

F

Fender, front, removal and installation, 11-8
Filling and painting, 11-4
Fluid level checks, 1-7
    battery electrolyte, 1-8
    brake and clutch fluid, 1-8
    engine coolant, 1-8
    engine oil, 1-7
    windshield washer fluid, 1-8
Flywheel/driveplate, removal and installation
    2.2L four-cylinder engine, 2A-11
    4.3L V6 engine, 2B-16
Front wheel bearing check, repack and adjustment (2WD models), 1-21
Fuel and exhaust systems, 4-1 through 4-20
    accelerator cable, removal and installation, 4-8
    air cleaner assembly, removal and installation, 4-8
    Central Multiport Fuel Injection (CMFI) and Central Sequential Electronic Fuel Injection (CSEFI) systems, component check and replacement, 4-16
    exhaust system servicing, general information, 4-20
    fuel injection system
        check, 4-10
        general information, 4-8
    fuel filter replacement, 1-20
    fuel level sending unit, check and replacement, 4-7
    fuel pressure relief procedure, 4-2
    general information, 4-1
    lines and fittings, repair and replacement, 4-4
    Model 220 Throttle Body Injection (TBI), removal, overhaul and installation, 4-11
    Multiport Fuel Injection (MFI), component check and replacement, 4-14
    pump, removal and installation, 4-6
    pump/fuel system pressure, check, 4-3
    tank
        cleaning and repair, general information, 4-6
        removal and installation, 4-5
Fuel injection system
    Central Multiport Fuel Injection (CMFI) and Central Sequential Electronic Fuel Injection (CSEFI) systems, component check and replacement, 4-16
    check, 4-10
    general information, 4-8
    Model 220 Throttle Body Injection (TBI), removal, overhaul and installation, 4-11
    Multiport Fuel Injection (MFI), component check and replacement, 4-14
Fuel system check, 1-19
Fuses, general information, 12-2
Fusible links, general information, 12-3

G

General engine overhaul procedures, 2C-1 through 2C-20
    balance shaft, installation (V6 engines only), 2C-19
    block
        cleaning, 2C-12
        inspection, 2C-13
    camshaft, balance shaft and bearings, removal and inspection, 2C-10
    crankshaft
        inspection, 2C-15
        installation and main bearing oil clearance check, 2C-18
        removal, 2C-12
    cylinder compression check, 2C-4
    cylinder head
        cleaning and inspection, 2C-8
        disassembly, 2C-8
        reassembly, 2C-10
    cylinder honing, 2C-14
    driveplate, removal and installation, 2C-20
    engine removal, methods and precautions, 2C-5
    engine, removal and installation, 2C-6
    general information, 2C-3
    initial start-up and break-in after overhaul, 2C-20
    main and connecting rod bearings, inspection, 2C-16
    overhaul
        disassembly sequence, 2C-7
        engine reassembly sequence, 2C-17
        general information, 2C-3
    piston rings, installation, 2C-17
    piston/connecting rod assembly
        inspection, 2C-14
        installation and bearing oil clearance check, 2C-19
        removal, 2C-11
    rear main oil seal, installation, 2C-19
    rebuilding alternatives, 2C-7
    vacuum gauge diagnostic, checks, 2C-5
    valves, servicing, 2C-10

H

Headlight(s)
  adjustment, 12-7
  replacement, 12-6
  switch, removal and installation, 12-6
Heater and air conditioning control assembly, removal and installation, 3-10
Heater core, removal and installation, 3-8
Hinges and locks, maintenance, 11-4
Hood release latch and cable, replacement, 11-5
Hood, removal, installation and adjustment, 11-5
Horn, check and replacement, 12-13
Hub and bearing assembly (4WD models), replacement, 10-9
Hydraulic system, bleeding, 9-15

I

Ignition
  module, check and replacement, 5-7
  pick-up coil (HEI systems), check and replacement, 5-8
  coil, check, removal and installation, 5-4
    Distributorless Ignition System (DIS), 5-6
    Enhanced Distributor Ignition (EDI) systems, 5-5
    High Energy Ignition (HEI) systems, 5-4
  key lock cylinder, replacement, 12-5
  system
    check, 5-2
      Distributorless Ignition System (DIS), 5-4
      High Energy Ignition (HEI) systems, 5-3
    general information, 5-2
      Distributorless Ignition System (DIS), 5-2
      Enhanced Distributor Ignition (EDI) system, 5-2
      High Energy Ignition (HEI) system, 5-2
  timing check and adjustment (V6 VIN Z only), 1-27
Information sensors, check and replacement, 6-6
  air conditioning "On" signal, 6-11
  air conditioning control, 6-11
  camshaft sensor (V6 CMFI and CSEFI engines), 6-12
  crankshaft sensor (four-cylinder MFI and V6 CMFI and CSEFI engines), 6-12
  Engine Coolant Temperature (ECT) sensor, 6-6
  Intake Air Temperature (IAT) sensor, 6-8
  knock sensors (V6 CMFI and CSEFI engines), 6-12
  Manifold Absolute Pressure (MAP) sensor, 6-7
  Neutral Start switch, 6-11
  oxygen sensor, 6-8
  Throttle Position Sensor (TPS), 6-10
  Vehicle Speed Sensor (VSS), 6-11
Initial start-up and break-in after overhaul, 2C-20
Instrument cluster
  bezel, removal and installation, 11-12
  removal and installation, 12-13
Intake manifold, removal and installation
  2.2L four-cylinder engine, 2A-4
  4.3L V6 engine, 2B-6
Introduction to the Chevrolet S-10 and Blazer, GMC Sonoma and Jimmy, Oldsmobile Bravada and Isuzu Hombre, 0-4

J

Jacking and towing, 0-12

L

Leaf spring/shackle, removal and installation, 10-11
Lower control arm, removal and installation, 10-8

M

Main and connecting rod bearings, inspection, 2C-16
Maintenance schedule, 1-6
Maintenance techniques, tools and working facilities, 0-6
Manual transmission, 7A-1 through 7A-4
  general information, 7A-1
  lubricant change, 1-22
  lubricant level check, 1-18
  overhaul, general information, 7A-4
  removal and installation, 7A-2
  shift lever housing (NV1500/3500), removal and installation, 7A-2
  shift lever, removal and installation, 7A-2
Master cylinder, removal, overhaul and installation, 9-11
Mount, transmission, check and replacement, 7B-5
Mounts, engine, replacement
  2.2L four-cylinder engine, 2A-12
  4.3L V6 engine, 2B-16

O

Oil pan, removal and installation
  2.2L four-cylinder engine, 2A-11
  4.3L V6 engine, 2B-15
Oil pump, removal and installation
  2.2L four-cylinder engine, 2A-11
  4.3L V6 engine, 2B-15
On Board Diagnostic (OBD) system and trouble codes, 6-2
Output shaft, tube and tube seal (4WD models), removal and installation, 8-15
Outside mirror, removal and installation, 11-8
Overhaul, engine
  disassembly sequence, 2C-7
  reassembly sequence, 2C-17
  general information, 2C-3

P

Park/neutral switch, check, replacement and adjustment, 7B-4
Parking brake
  adjustment, 9-13
  cables, replacement, 9-14
  shoes, removal and installation, 9-16
Pilot bearing, inspection and replacement, 8-5
Piston rings, installation, 2C-17

P

Piston/connecting rod assembly
  inspection, 2C-14
  installation and bearing oil clearance check, 2C-19
  removal, 2C-11
Positive Crankcase Ventilation (PCV) system, 6-15
Positive Crankcase Ventilation (PCV) valve check and replacement, 1-24
Power brake booster, removal and installation, 9-15
Power mirror control system, description and check, 12-13
Power steering
  fluid level check, 1-11
  pump, removal and installation, 10-15
  system, bleeding, 10-16
Powertrain Control Module (PCM), check and replacement, 6-5

R

Radiator and coolant reservoir tank, removal and installation, 3-4
Radiator grille, removal and installation, 11-6
Radio and speakers, general information, removal and installation, 12-10
Rear main oil seal
  installation, 2C-19
  replacement
    2.2L four-cylinder engine, 2A-12
    4.3L V6 engine, 2B-16
Rear window defogger
  check and repair, 12-12
  switch, removal and installation, 12-12
Rear wiper switch, removal and installation, 12-6
Rebuilding alternatives, engine, 2C-7
Relays, general information, 12-3
Repair operations possible with the engine in the vehicle
  2.2L four-cylinder engine, 2A-2
  4.3L V6 engine, 2B-3
Rocker arms and pushrods, removal, inspection and installation
  2.2L four-cylinder engine, 2A-3
  4.3L V6 engine, 2B-4

S

Safety first!, 0-16
Seat belt and restraint system check, 1-17
Seats
  power, description and check, 12-14
  removal and installation, 11-14
Secondary Air Injection (AIR) system, general description and component replacement, 6-16
Selector switch (NP 231), check and replacement, 7C-2
Shift cable (1995 and later models), removal, installation and adjustment, 7B-3
Shift cable (NP 231), replacement and adjustment, 7C-2
Shift lever (NP 231), replacement, 7C-2
Shift lever housing (NV1500/3500), removal and installation, 7A-2
Shift lever, removal and installation, 7A-2
Shift linkage (1994 models), removal, installation and adjustment, 7B-2

Shock absorber, removal and installation
  front, 10-5
  rear, 10-11
Spark plug
  replacement, 1-25
  wire, distributor cap and rotor check and replacement, 1-24
Stabilizer bar, removal and installation
  front, 10-5
  rear, 10-12
Starter motor
  removal and installation, 5-10
  testing in vehicle, 5-10
Starter safety switch check, 1-17
Starter solenoid, removal and installation, 5-10
Starting system, general information, 5-9
Steering column cover, removal and installation, 11-11
Steering knuckle, removal and installation
  2WD models, 10-7
  4WD models, 10-9
Steering system
  alignment, general information, 10-16
  gear, removal and installation, 10-14
  general information, 10-3
  linkage, inspection, removal and installation, 10-13
  power steering
    pump, removal and installation, 10-15
    system, bleeding, 10-16
  steering wheel, removal and installation, 10-12
  wheels and tires, general information, 10-16
Support struts, replacement, 11-9
Suspension and steering check, 1-16
Suspension and steering systems, 10-1 through 10-16
Suspension system
  balljoints, check and replacement, 10-6
  coil spring (2WD models), removal and installation, 10-7
  general information, 10-3
  hub and bearing assembly (4WD models), replacement, 10-9
  leaf spring/shackle, removal and installation, 10-11
  lower control arm, removal and installation, 10-8
  shock absorber, removal and installation
    front, 10-5
    rear, 10-11
  stabilizer bar, removal and installation
    front, 10-5
    rear, 10-12
  steering knuckle, removal and installation
    2WD models, 10-7
    4WD models, 10-9
  Torsion bar (4WD models), removal and installation, 10-10
  track bar (ZR2 models), removal and installation, 10-12
  upper control arm, removal and installation, 10-7

T

Tailgate glass, removal and installation, 11-9
Tailgate, removal and installation, 11-9
Tank, fuel
  cleaning and repair, general information, 4-6
  removal and installation, 4-5

Index

Thermostat, check and replacement, 3-2
Timing chain cover, chain and sprockets, removal and installation
   2.2L four-cylinder engine, 2A-7
   4.3L V6 engine, 2B-11
Tire and tire pressure checks, 1-9
Tire rotation, 1-17
Tools, 0-8
Torsion bar (4WD models), removal and installation, 10-10
Track bar (ZR2 models), removal and installation, 10-12
Transfer case, 7C-1 through 7C-4
   control module (NP 233), replacement, 7C-3
   electronic shift motor (NP 233), replacement, 7C-3
   general information, 7C-1
   lubricant change (4WD models), 1-23
   lubricant level check, 1-18
   oil seals, replacement, 7C-1
   removal and installation, 7C-4
   selector switch (NP 231), check and replacement, 7C-2
   shift cable (NP 231), replacement and adjustment, 7C-2
   shift lever (NP 231), replacement, 7C-2
   switch (NP 233), replacement, 7C-4
   vacuum switch, check and replacement, 7C-2
Transmission
   automatic, 7B-1 through 7B-6
      diagnosis, general, 7B-1
      extension housing seal, replacement, 7B-4
      fluid level check, 1-10
      general information, 7B-1
      park/neutral switch, check, replacement and adjustment, 7B-4
      removal and installation, 7B-5
      shift cable (1995 and 1996 models), removal, installation and adjustment, 7B-3
      shift linkage (1994 models), removal, installation and adjustment, 7B-2
      transmission mount, check and replacement, 7B-5
   manual, 7A-1 through 7A-4
      general information, 7A-1
      lubricant level check, 1-18
      overhaul, general information, 7A-4
      removal and installation, 7A-2
      shift lever housing (NV1500/3500), removal and installation, 7A-2
      shift lever, removal and installation, 7A-2
Transmission Converter Clutch (TCC) system, 6-15
Trouble code identification, 6-3
Troubleshooting, 0-17
Tune-up and routine maintenance, 1-1 through 1-28
Tune-up general information, 1-7
Turn signal switch assembly, removal and installation, 12-4
Turn signal/hazard flasher, check and replacement, 12-3

U

Underhood hose check and replacement, 1-14
Universal joints, removal, overhaul and installation, 8-8
Upholstery and carpets, maintenance, 11-1
Upper control arm, removal and installation, 10-7

V

Vacuum actuator and shift cable (4WD models), replacement, 8-15
Vacuum gauge diagnostic, checks, 2C-5
Vacuum switch, check and replacement, 7C-2
Valve cover, removal and installation
   2.2L four-cylinder engine, 2A-2
   4.3L V6 engine, 2B-3
Valve springs, retainers and seals, replacement
   2.2L four-cylinder engine, 2A-4
   4.3L V6 engine, 2B-5
Valves, servicing, 2C-10
Vehicle identification numbers, 0-5
   automatic transmission number, 0-5
   engine identification numbers, 0-5
   manual transmission number, 0-5
   rear axle number, 0-5
   service parts identification label, 0-5
   transfer case number, 0-5
   vehicle certification label, 0-5
   Vehicle Emissions Control Information label, 0-5
   Vehicle Identification Number (VIN), 0-5
   VIN engine and model year codes, 0-5
Vinyl trim, maintenance, 11-1

W

Water pump
   check, 3-5
   removal and installation, 3-5
Wheel, steering, removal and installation, 10-12
Wheel cylinder, removal, overhaul and installation, 9-10
Wheels and tires, general information, 10-16
Window system, power, description and check, 12-14
Windshield and fixed glass, replacement, 11-5
Wiper blade inspection and replacement, 1-15
Wiper motor, removal and installation, 12-10
Wiring diagrams, general information, 12-15
Working facilities, 0-11

Haynes Automotive Manuals

NOTE: If you do not see a listing for your vehicle, please visit **haynes.com** *for the latest product information and check out our* **Online Manuals!**

ACURA
- 12020 Integra '86 thru '89 & Legend '86 thru '90
- 12021 Integra '90 thru '93 & Legend '91 thru '95
- Integra '94 thru '00 - see HONDA Civic (42025)
- MDX '01 thru '07 - see HONDA Pilot (42037)
- 12050 Acura TL all models '99 thru '08

AMC
- 14020 Concord/Hornet/Gremlin/Spirit '70 thru '83
- 14025 (Renault) Alliance & Encore all models '83 thru '87

AUDI
- 15020 4000 all models '80 thru '87
- 15025 5000 all models '77 thru '83
- 15026 5000 all models '84 thru '88
- Audi A4 '96 thru '01 - see VW Passat (96023)
- 15030 Audi A4 '02 thru '08

AUSTIN-HEALEY
- Sprite - see MG Midget (66015)

BMW
- 18020 3/5 Series '82 thru '92
- 18021 3-Series including Z3 '92 thru '98
- 18022 3-Series including Z4 models '99 thru '05
- 18023 3-Series '06 thru '14
- 18025 320i all 4-cylinder models '75 thru '83
- 18050 1500 thru 2002 except Turbo '59 thru '77

BUICK
- 19010 Buick Century '97 thru '05
- Century (front-wheel drive) - see GM (38005)
- 19020 Buick, Oldsmobile & Pontiac Full-size (Front wheel drive) '85 thru '05
- 19025 Buick, Oldsmobile & Pontiac Full-size (Rear wheel drive) '70 thru '90
- 19027 Buick LaCrosse '05 thru '13
- Regal - see GENERAL MOTORS (38010)
- Skyhawk - see GM (38015)
- Skylark - see GM (38020, 38025)
- Somerset - see GENERAL MOTORS (38025)

CADILLAC
- 21015 CTS & CTS-V '03 thru '14
- 21030 Cadillac Rear Wheel Drive '70 thru '93
- Cimarron, Eldorado & Seville - see GM (38015, 38030, 38031)

CHEVROLET
- 10305 Chevrolet Engine Overhaul Manual
- 24010 Astro & GMC Safari Mini-vans '85 thru '05
- 24015 Camaro V8 all models '70 thru '81
- 24016 Camaro all models '82 thru '92
- Cavalier - see GM (38016)
- Celebrity - see GM (38005)
- 24017 Camaro & Firebird '93 thru '02
- 24018 Camaro '10 thru '15
- 24020 Chevelle, Malibu, El Camino '69 thru '87
- 24024 Chevette & Pontiac T1000 '76 thru '87
- Citation - see GENERAL MOTORS (38020)
- 24027 Colorado & GMC Canyon '04 thru '12
- 24032 Corsica & Beretta all models '87 thru '96
- 24040 Corvette all V8 models '68 thru '82
- 24041 Corvette all models '84 thru '96
- 24042 Corvette all models '97 thru '13
- 24044 Cruze '11 thru '19
- 24045 Full-size Sedans Caprice, Impala, Biscayne, Bel Air & Wagons '69 thru '90
- 24046 Impala SS & Caprice and Buick Roadmaster '91 thru '96
- Impala '00 thru '05 - see LUMINA (24048)
- 24047 Impala & Monte Carlo all models '06 thru '11
- Lumina '90 thru '94 - see GM (38010)
- 24048 Lumina & Monte Carlo '95 thru '05
- Lumina APV - see GM (38035)
- 24050 Luv Pick-up all 2WD & 4WD '72 thru '82
- 24051 Malibu '13 thru '19
- 24055 Monte Carlo all models '70 thru '88
- Monte Carlo '95 thru '01 - see LUMINA (24048)
- 24059 Nova all V8 models '69 thru '79
- 24060 Nova/Geo Prizm '85 thru '92
- 24064 Pick-ups '67 thru '87 - Chevrolet & GMC
- 24065 Pick-ups '88 thru '98 - Chevrolet & GMC
- 24066 Pick-ups '99 thru '06 - Chevrolet & GMC
- 24067 Chevy Silverado & GMC Sierra '07 thru '14
- 24068 Chevy Silverado & GMC Sierra '14 thru '19
- 24070 S-10 & S-15 Pick-ups '82 thru '93
- 24071 S-10 & Sonoma Pick-ups '94 thru '04
- 24072 Chevrolet TrailBlazer, GMC Envoy & Oldsmobile Bravada '02 thru '09
- 24075 Sprint '85 thru '88, Geo Metro '89 thru '01
- 24080 Vans - Chevrolet & GMC '68 thru '96
- 24081 Chevrolet Express & GMC Savana Full-size Vans '96 thru '19

CHRYSLER
- 10310 Chrysler Engine Overhaul Manual
- 25015 Chrysler Cirrus, Dodge Stratus, Plymouth Breeze, '95 thru '00
- 25020 Full-size Front-Wheel Drive '88 thru '93
- K-Cars - see DODGE Aries (30008)
- 25025 Chrysler LHS, Concorde & New Yorker, Dodge Intrepid, Eagle Vision, '93 thru '97
- 25026 Chrysler LHS, Concorde, 300M, Dodge Intrepid '98 thru '04
- 25027 Chrysler 300 '05 thru '18, Dodge Charger '06 thru '18, Magnum '05 thru '08 & Challenger '08 thru '18
- 25030 Chrysler & Plymouth Mid-size '82 thru '95
- Rear-wheel Drive - see DODGE (30050)
- 25035 PT Cruiser all models '01 thru '10
- 25040 Chrysler Sebring '95 thru '06, Dodge Stratus '01 thru '06 & Dodge Avenger '95 thru '00
- 25041 Chrysler Sebring '07 thru '10, 200 '11 thru '17 Dodge Avenger '08 thru '14

DATSUN
- 28005 200SX all models '80 thru '83
- 28012 240Z, 260Z & 280Z Coupe '70 thru '78
- 28014 280ZX Coupe & 2+2 '79 thru '83
- 300ZX - see NISSAN (72010)
- 28018 510 & PL521 Pick-up '68 thru '73
- 28020 510 all models '78 thru '81
- 28022 620 Series Pick-up all models '73 thru '79
- 720 Series Pick-up - see NISSAN (72030)

DODGE
- 30008 Aries & Plymouth Reliant '81 thru '89
- 30010 Caravan & Plymouth Voyager '84 thru '95
- 30011 Caravan & Plymouth Voyager '96 thru '02
- 30012 Challenger & Plymouth Sapporro '78 thru '83
- 30013 Caravan, Chrysler Voyager & Town & Country '03 thru '07
- 30014 Grand Caravan & Chrysler Town & Country '08 thru '18
- 30020 Dakota Pick-ups all models '87 thru '96
- 30021 Durango '98 & '99 & Dakota '97 thru '99
- 30022 Durango '00 thru '03 & Dakota '00 thru '04
- 30023 Durango '04 thru '09 & Dakota '05 thru '11
- 30025 Dart, Challenger/Plymouth Barracuda & Valiant 6-cylinder models '67 thru '76
- 30030 Daytona & Chrysler Laser '84 thru '89
- Intrepid - see Chrysler (25025, 25026)
- 30034 Neon all models '95 thru '99
- 30035 Omni & Plymouth Horizon '78 thru '90
- 30036 Dodge & Plymouth Neon '00 thru '05
- 30040 Pick-ups full-size models '74 thru '93
- 30041 Pick-ups full-size models '94 thru '01
- 30042 Pick-ups full-size models '02 thru '08
- 30043 Pick-ups full-size models '09 thru '18
- 30045 Ram 50/D50 Pick-ups & Raider and Plymouth Arrow Pick-ups '79 thru '93
- 30050 Dodge/Plymouth/Chrysler RWD '71 thru '89
- 30055 Shadow & Plymouth Sundance '87 thru '94
- 30060 Spirit & Plymouth Acclaim '89 thru '95
- 30065 Vans - Dodge & Plymouth '71 thru '03

EAGLE
- Talon - see MITSUBISHI (68030, 68031)
- Vision - see CHRYSLER (25025)

FIAT
- 34010 124 Sport Coupe & Spider '68 thru '78
- 34025 X1/9 all models '74 thru '80

FORD
- 10320 Ford Engine Overhaul Manual
- 10355 Ford Automatic Transmission Overhaul
- 11500 Mustang '64-1/2 thru '70 Restoration Guide
- 36004 Aerostar Mini-vans '86 thru '97
- 36006 Contour & Mercury Mystique '95 thru '00
- 36008 Courier Pick-up all models '72 thru '82
- 36012 Crown Victoria & Mercury Grand Marquis '88 thru '11
- 36016 Escort & Mercury Lynx '81 thru '90
- 36020 Escort & Mercury Tracer '91 thru '02
- 36022 Escape '01 thru '17, Mazda Tribute '01 thru '11 & Mercury Mariner '05 thru '11
- 36024 Explorer & Mazda Navajo '91 thru '01
- 36025 Explorer & Mercury Mountaineer '02 thru '10
- 36026 Explorer '11 thru '17
- 36028 Fairmont & Mercury Zephyr '78 thru '83
- 36030 Festiva & Aspire '88 thru '97
- 36032 Fiesta all models '77 thru '80
- 36034 Focus all models '00 thru '11
- 36035 Focus '12 thru '14
- 36045 Ford Fusion '06 thru '14 & Mercury Milan '06 thru '10
- 36048 Mustang V8 all models '64-1/2 thru '73
- 36049 Mustang II 4-cylinder, V6 & V8 '74 thru '78
- 36050 Mustang & Mercury Capri '79 thru '93
- 36051 Mustang all models '94 thru '04
- 36052 Mustang '05 thru '14
- 36054 Pick-ups and Bronco '73 thru '79
- 36058 Pick-ups and Bronco '80 thru '96
- 36059 F-150 '97 thru '03, Expedition '97 thru '17, F-250 '97 thru '99, F-150 Heritage '04 & Lincoln Navigator '98 thru '17
- 36060 Super Duty Pick-up & Excursion '99 thru '10
- 36061 F-150 full-size '04 thru '10
- 36062 Pinto & Mercury Bobcat '75 thru '80
- 36063 F-150 '15 thru '17
- 36064 Super Duty Pick-ups '11 thru '16
- 36066 Probe all models '89 thru '92
- 36070 Ranger & Bronco II gas models '83 thru '92
- 36071 Ranger '93 thru '11 & Mazda Pick-ups '94 thru '09
- 36074 Taurus & Mercury Sable '86 thru '95
- 36075 Taurus & Mercury Sable '96 thru '05
- 36076 Taurus '08 thru '14, Five Hundred '05 thru '07, Mercury Montego '05 thru '07 & Sable '08 thru '09
- 36078 Tempo & Mercury Topaz '84 thru '94
- 36082 Thunderbird & Mercury Cougar '83 thru '88
- 36086 Thunderbird & Mercury Cougar '89 thru '97
- 36090 Vans all V8 Econoline models '69 thru '91
- 36094 Vans full size '92 thru '14
- 36097 Windstar '95 thru '03, Freestar & Mercury Monterey Mini-van '04 thru '07

GENERAL MOTORS
- 10360 GM Automatic Transmission Overhaul
- 38005 Buick Century, Chevrolet Celebrity, Olds Cutlass Ciera & Pontiac 6000 '82 thru '96
- 38010 Buick Regal, Chevrolet Lumina, Oldsmobile Cutlass Supreme & Pontiac Grand Prix front wheel drive '88 thru '07
- 38015 Buick Skyhawk, Cadillac Cimarron, Chevrolet Cavalier, Oldsmobile Firenza Pontiac J-2000 & Sunbird '82 thru '94
- 38016 Chevrolet Cavalier/Pontiac Sunfire '95 thru '05
- 38017 Chevrolet Cobalt & Pontiac G5 '05 thru '11
- 38020 Buick Skylark, Chevrolet Citation, Olds Omega, Pontiac Phoenix '80 thru '85
- 38025 Buick Skylark & Somerset, Olds Achieva, Calais & Pontiac Grand Am '85 thru '98
- 38026 Chevrolet Malibu, Olds Alero & Cutlass, Pontiac Grand Am '97 thru '03
- 38027 Chevrolet Malibu '04 thru '12
- 38030 Cadillac Eldorado, Seville, Oldsmobile Toronado & Buick Riviera '71 thru '85
- 38031 Cadillac Eldorado, Seville, DeVille, Fleetwood, Oldsmobile Toronado & Buick Riviera '86 thru '93
- 38032 Cadillac DeVille '94 thru '05, Seville '92 thru '04 & Cadillac DTS '06 thru '10
- 38035 Chevrolet Lumina APV, Olds Silhouette & Pontiac Trans Sport all models '90 thru '96
- 38036 Chevrolet Venture, Olds Silhouette, Pontiac Trans Sport & Montana '97 thru '05
- 38040 Chevrolet Equinox '05 thru '17, GMC Terrain '10 thru '17 & Pontiac Torrent '06 thru '09

GEO
- Metro - see CHEVROLET Sprint (24075)
- Prizm - '85 thru '92 see CHEVY (24060), '93 thru '02 see TOYOTA Corolla (92036)
- Storm all models '90 thru '93
- 40030 Tracker - see SUZUKI Samurai (90010)

GMC
- Vans & Pick-ups - see CHEVROLET

HONDA
- 42010 Accord CVCC all models '76 thru '83
- 42011 Accord all models '84 thru '89
- 42012 Accord all models '90 thru '93
- 42013 Accord all models '94 thru '97
- 42014 Accord all models '98 thru '02
- 42015 Accord '03 thru '12 & Crosstour '10 thru '14
- 42016 Accord '13 thru '17
- 42020 Civic 1200 all models '73 thru '79
- 42021 Civic 1300 & 1500 CVCC '80 thru '83
- 42022 Civic 1500 CVCC '75 thru '79
- 42023 Civic all models '84 thru '91
- 42024 Civic & del Sol '92 thru '95
- 42025 Civic '96 thru '00 & CR-V '97 thru '01 & Acura Integra '94 thru '00
- 42026 Civic '01 thru '11 & CR-V '02 thru '11
- 42027 Civic '12 thru '15 & CR-V '12 thru '16
- 42030 Fit '07 thru '13
- 42035 Odyssey models '99 thru '10
- Passport - see ISUZU Rodeo (47017)
- 42037 Honda Pilot '03 thru '08, Ridgeline '06 thru '14 & Acura MDX '01 thru '07
- 42040 Prelude CVCC all models '79 thru '89

HYUNDAI
- 43010 Elantra all models '96 thru '19
- 43015 Excel & Accent all models '86 thru '13
- 43050 Santa Fe all models '01 thru '12
- 43055 Sonata all models '99 thru '14

INFINITI
- G35 '03 thru '08 - see NISSAN 350Z (72011)

ISUZU
- Hombre - see CHEVROLET S-10 (24071)
- 47017 Rodeo '91 thru '02, Amigo '89 thru '94 & '98 thru '02 & Honda Passport '95 thru '02
- 47020 Trooper '84 thru '91 & Pick-up '81 thru '93

JAGUAR
- 49010 XJ6 all 6-cylinder models '68 thru '86
- 49011 XJ6 all models '88 thru '94
- 49015 XJ12 & XJS all 12-cylinder models '72 thru '85

JEEP
- 50010 Cherokee, Comanche & Wagoneer Limited all models '84 thru '01
- 50011 Cherokee '14 thru '19
- 50020 CJ all models '49 thru '86
- 50025 Grand Cherokee all models '93 thru '04
- 50026 Grand Cherokee '05 thru '19 & Dodge Durango '11 thru '19
- 50029 Grand Wagoneer & Pick-up '72 thru '91
- 50030 Wrangler all models '87 thru '17
- 50035 Liberty '02 thru '12 & Dodge Nitro '07 thru '11
- 50050 Patriot & Compass '07 thru '17

KIA
- 54050 Optima '01 thru '10
- 54060 Sedona '02 thru '14
- 54070 Sephia '94 thru '01, Spectra '00 thru '09, Sportage '05 thru '20
- 54077 Sorento '03 thru '13

LEXUS
- ES 300/330 - see TOYOTA Camry (92007, 92008)
- RX 300/330/350 - see TOYOTA Highlander (92095)

LINCOLN
- Navigator - see FORD Pick-up (36059)
- 59010 Rear-Wheel Drive Continental '70 thru '87, Mark Series '70 thru '92 & Town Car '81 thru '10

MAZDA
- 61010 GLC (rear wheel drive) '77 thru '83
- 61011 GLC (front wheel drive) '81 thru '85
- 61012 Mazda3 '04 thru '11
- 61015 323 & Protegé '90 thru '03
- 61016 MX-5 Miata '90 thru '14
- 61020 MPV all models '89 thru '98
- 61030 Pick-ups '72 thru '93
- Pick-ups '94 thru '09 - see Ford (36071)
- 61035 RX-7 all models '79 thru '85
- 61036 RX-7 all models '86 thru '91
- 61040 626 (rear-wheel drive) all models '79 thru '82
- 61041 626 & MX-6 (front-wheel drive) '83 thru '92
- 61042 626 '93 thru '01 & MX-6/Ford Probe '93 thru '02
- 61043 Mazda6 '03 thru '13

MERCEDES-BENZ
- 63012 123 Series Diesel '76 thru '85
- 63015 190 Series 4-cylinder gas models '84 thru '88
- 63020 230, 250 & 280 6-cylinder SOHC '68 thru '72
- 63025 280 123 Series gasoline models '77 thru '81
- 63030 350 & 450 all models '71 thru '80
- 63040 C-Class: C230/C240/C280/C320/C350 '01 thru '07

MERCURY
- 64200 Villager & Nissan Quest '93 thru '01
- All other titles, see FORD listing.

MG
- 66010 MGB Roadster & GT Coupe '62 thru '80
- 66015 MG Midget & Austin Healey Sprite Roadster '58 thru '80

MINI
- 67020 Mini '02 thru '13

MITSUBISHI
- 68020 Cordia, Tredia, Galant, Precis & Mirage '83 thru '93
- 68030 Eclipse, Eagle Talon & Plymouth Laser '90 thru '94
- 68031 Eclipse '95 thru '05 & Eagle Talon '95 thru '98
- 68035 Galant '94 thru '12
- 68040 Pick-up '83 thru '96 & Montero '83 thru '93

NISSAN
- 72010 300ZX all models incl. Turbo '84 thru '89
- 72011 350Z & Infiniti G35 all models '03 thru '08
- 72015 Altima all models '93 thru '06
- 72016 Altima '07 thru '12
- 72020 Maxima all models '85 thru '92
- 72021 Maxima all models '93 thru '08
- 72025 Murano '03 thru '14
- 72030 Pick-ups '80 thru '97 & Pathfinder '87 thru '95
- 72031 Frontier, Xterra & Pathfinder '96 thru '04
- 72032 Frontier & Xterra '05 thru '14
- 72037 Pathfinder '05 thru '14
- 72040 Pulsar all models '83 thru '86
- 72042 Roque all models '08 thru '20
- 72050 Sentra all models '82 thru '94
- 72051 Sentra & 200SX all models '95 thru '06
- 72060 Stanza all models '82 thru '90
- 72070 Titan pick-ups '04 thru '10 & Armada '05 thru '10
- 72080 Versa all models '07 thru '19

OLDSMOBILE
- 73015 Cutlass V6 & V8 gas models '74 thru '88
- For other OLDSMOBILE titles, see BUICK, CHEVROLET or GM listings.

PLYMOUTH
- For PLYMOUTH titles, see DODGE.

PONTIAC
- 79008 Fiero all models '84 thru '88
- 79018 Firebird V8 except Turbo '70 thru '81
- 79019 Firebird all models '82 thru '92
- 79025 G6 all models '05 thru '09
- 79040 Mid-size Rear-wheel Drive '70 thru '87
- Vibe '03 thru '10 - see TOYOTA Corolla (92037)
- For other PONTIAC titles, see BUICK, CHEVROLET or GM listings.

PORSCHE
- 80020 911 Coupe & Targa models '65 thru '89
- 80025 914 all 4-cylinder models '69 thru '76
- 80030 924 all models including Turbo '76 thru '82
- 80035 944 all models including Turbo '83 thru '89

RENAULT
- Alliance & Encore - see AMC (14025)

SAAB
- 84010 900 all models including Turbo '79 thru '88

SATURN
- 87010 Saturn all S-series models '91 thru '02
- Saturn Ion '03 thru '07 - see GM (38017)
- 87020 Saturn L-series all models '00 thru '04
- 87040 Saturn VUE '02 thru '09

SUBARU
- 89002 1100, 1300, 1400 & 1600 '71 thru '79
- 89003 1600 & 1800 2WD & 4WD '80 thru '94
- 89080 Impreza '02 thru '11, WRX '02 thru '14 & WRX STI '04 thru '14
- 89100 Legacy all models '90 thru '99
- 89101 Legacy & Forester '00 thru '09
- 89102 Legacy '10 thru '16 & Forester '12 thru '16

SUZUKI
- 90010 Samurai/Sidekick & Geo Tracker '86 thru '01

TOYOTA
- 92005 Camry all models '83 thru '91
- 92006 Camry '92 thru '96 & Avalon '95 thru '96
- 92007 Camry, Avalon, Solara, Lexus ES 300 '97 thru '01
- 92008 Camry, Avalon, Lexus ES 300/330 '02 thru '06 & Solara '02 thru '08
- 92009 Camry, Avalon & Lexus ES 350 '07 thru '17
- 92015 Celica Rear-wheel Drive '71 thru '85
- 92020 Celica Front-wheel Drive '86 thru '99
- 92025 Celica Supra all models '79 thru '92
- 92032 Corolla all models '75 thru '79
- 92032 Corolla all rear wheel drive models '80 thru '87
- 92035 Corolla front-wheel drive models '84 thru '92
- 92036 Corolla & Geo/Chevrolet Prizm '93 thru '02
- 92037 Corolla '03 thru '19, Matrix '03 thru '14, & Pontiac Vibe '03 thru '10
- 92040 Corolla Tercel all models '80 thru '82
- 92045 Corona all models '74 thru '82
- 92050 Cressida all models '78 thru '82
- 92055 Land Cruiser FJ40/43/45/55 '68 thru '82
- 92056 Land Cruiser FJ60/62/80/FZJ80 '80 thru '96
- 92065 Matrix '03 thru '11 & Pontiac Vibe '03 thru '10
- 92065 MR2 all models '85 thru '87
- 92070 Pick-up all models '69 thru '78
- 92075 Pick-up all models '79 thru '95
- 92076 Tacoma '95 thru '04, 4Runner '96 thru '02 & T100 '93 thru '08
- 92077 Tacoma '05 thru '18
- 92078 Tundra '00 thru '06, Sequoia '01 thru '07
- 92079 4Runner all models '03 thru '09
- 92080 Previa all models '91 thru '95
- 92081 Prius '01 thru '12
- 92082 RAV4 all models '96 thru '12
- 92085 Tercel all models '87 thru '94
- 92090 Sienna all models '98 thru '10
- 92095 Highlander '01 thru '19 & Lexus RX330/330/350 '99 thru '19
- 92179 Tundra '07 thru '19 & Sequoia '08 thru '19

TRIUMPH
- 94007 Spitfire all models '62 thru '81
- 94010 TR7 all models '75 thru '81

VW
- 96008 Beetle & Karmann Ghia '54 thru '79
- 96009 New Beetle '98 thru '10
- 96016 Rabbit, Jetta, Scirocco & Pick-up gas models '75 thru '92 & Convertible '80 thru '92
- 96017 Golf, GTI & Jetta '93 thru '98, Cabrio '95 thru '02
- 96018 Golf, GTI & Jetta '99 thru '05
- 96019 Jetta, Rabbit, GLI, GTI & Golf '05 thru '11
- 96020 Rabbit, Jetta, Pick-up diesel '77 thru '84
- 96021 Jetta '11 thru '18 & Golf '15 thru '19
- 96023 Passat '98 thru '05, Audi A4 '96 thru '01
- 96030 Transporter 1600 all models '68 thru '79
- 96035 Transporter 1700, 1800, 2000 '72 thru '79
- 96040 Type 3 1500 & 1600 '63 thru '73
- 96045 Vanagon Air-Cooled all models '80 thru '83

VOLVO
- 97010 120, 130 Series & 1800 Sports '61 thru '73
- 97015 140 Series all models '66 thru '74
- 97020 240 Series all models '76 thru '93
- 97040 740 & 760 Series all models '82 thru '88
- 97050 850 Series all models '93 thru '97

TECHBOOK MANUALS
- 10205 Automotive Computer Codes
- 10206 OBD-II & Electronic Engine Management
- 10210 Automotive Emissions Control Manual
- 10215 Fuel Injection Manual '78 thru '85
- 10225 Holley Carburetor Manual
- 10230 Rochester Carburetor Manual
- 10305 Chevrolet Engine Overhaul Manual
- 10320 Ford Engine Overhaul Manual
- 10330 GM and Ford Diesel Engine Repair Manual
- 10331 Duramax Diesel Engines '01 thru '19
- 10332 Cummins Diesel Engine Performance
- 10333 GM, Ford & Chrysler Engine Performance
- 10334 GM Engine Performance
- 10340 Small Engine Repair Manual, 5 HP & Less
- 10341 Small Engine Repair Manual, 5.5 thru 20 HP
- 10345 Suspension, Steering & Driveline Manual
- 10355 Ford Automatic Transmission Overhaul
- 10360 GM Automatic Transmission Overhaul
- 10405 Automotive Body Repair & Painting
- 10410 Automotive Brake Manual
- 10411 Automotive Anti-lock Brake (ABS) Systems
- 10425 Automotive Heating & Air Conditioning
- 10435 Automotive Tools Manual
- 10445 Welding Manual

Over a 100 Haynes motorcycle manuals also available

Haynes North America, Inc. • (805) 498-6703 • www.haynes.com